Motor Control Electronics Handbook

Other Reference Books of Interest by McGraw-Hill

Handbooks

BENSON • *Television Engineering Handbook, Revised Edition*
CHEN • *Fuzzy Logic and Neural Network Handbook*
COOMBS • *Printed Circuits Handbook, 4/e*
COOMBS • *Electronic Instrument Handbook, 2/e*
CHRISTIANSEN • *Electronics Engineers' Handbook, 4/e*
HARPER • *Electronic Packaging and Interconnection Handbook, 2/e*
JURAN AND GRYNA • *Juran's Quality Control Handbook*
JURGEN • *Automotive Electronics Handbook*
JURGEN • *Digital Consumer Electronics Handbook*
OSA • *Handbook of Optics, 2/e*
RORABAUGH • *Digital Filter Designer's Handbook, 2/e*
SERGENT AND HARPER • *Hybrid Microelectronic Handbook*
SHAMA • *Programmable Logic Handbook*
WAYNANT • *Electro-Optics Handbook*
WILLIAMS AND TAYLOR • *Electronic Filter Design Handbook*
ZOMAYA • *Parallel and Distributed Computing Handbook*

Other

ANTOGNETTI AND MASSOBRIO • *Semiconductor Device Modeling with SPICE*
BEST • *Phase-Locked Loops, 3/e*
GRAEME • *Optimizing OP Amp Performance*
GRAEME • *Photodiode Amplifiers Op Amp Solutions*
KIELKOWSKI • *Inside SPICE, 2/e*
KIELKOWSKI • *SPICE Practical Device Modeling*
PERRY • *VHDL, 3/e*
SEALS • *Programmable Logic*
SMITH • *Thin-Film Deposition*
SZE • *VLSI Technology*
TSIVIDIS • *Mixed Analogy-Digital VLSI Devices and Technology*
TSUI • *LSI/VLSI Testability Design*
VAN ZANT • *Microchip Fabrication, 3/e*
WOBSCHALL • *Circuit Design for Electronic Instrumentation*
WYATT • *Electro-Optical System Design*

Motor Control Electronics Handbook

Richard Valentine

McGraw-Hill

New York San Francisco Washington, D.C. Auckland Bogotá
Caracas Lisbon London Madrid Mexico City Milan
Montreal New Delhi San Juan Singapore
Sydney Tokyo Toronto

Library of Congress Cataloging-in-Publication Data

Motor control electronics handbook / [edited by] Richard Valentine.
p. cm.
Includes bibliographical references and index.
ISBN 0-07-066810-8
1. Electronic controllers. 2. Electric motors—Electronic control. I. Valentine, Richard.
TK7881.2.M68 1998
621.46—dc21 98-5074
CIP

1 2 3 4 5 6 7 8 9 0 DOC/DOC 9 0 3 2 1 0 9 8

ISBN 0-07-066810-8

The sponsoring editor for this book was Steve Chapman, the editing supervisor was Bernard Onken, and the production supervisor was Pamela Pelton. It was set in Century Schoolbook by Dina E. John of McGraw-Hill's Professional Book Group composition unit.

Printed and bound by R. R. Donnelley & Sons Company.

This book is lovingly dedicated to my wife, Linda Eymann Valentine, who worked diligently and patiently as my coeditor, and to my parents, Richard and Mary Bell.

Contents

Preface

The inventors of electric motors probably never imagined that their basic motor designs would evolve into complex motion control systems managed by computers. Controlling motors with electronics technology allows precise control of the motor in any application—from small computer disk drives to large factory production equipment. The recent introduction of high-performance and cost-effective microcontroller devices has displaced analog motor controls almost completely. As power devices, such as insulated gate bipolar transistors (IGBTs), become more cost-effective, the proliferation of electronic motor controls will continue to grow. Several chapters of this handbook focus on the application of microcontrollers and power devices for motor control designs.

Chapters 2 to 6 discuss basic motor speed control designs. Advanced motor speed controls for automotive electronic and appliance controls as well as electric vehicle motor systems are reviewed.

Digital motor speed controls are covered in depth in Chaps. 7 to 12 and include actual design examples and software code. Chapter 9, on deadtime correction, is especially significant for some motor drive designs. Motor control development tools are described in detail in Chap. 10.

The chapter that discusses electrical noise emphasizes the importance of careful printed-circuit board layout. Chapter 17, on motor control testing, shows how to implement a motor test lab and specifies the requirements for motor test instrumentation.

Nonmicrocontroller products, such as rectifiers, transistors, power modules, and sensors, are essential components of a motor control circuit and are discussed in some detail. Several chapters focus on how these devices are rated to work in power control circuits.

Chapter 18, on future motor controls, includes an ultra-high-speed motor design. Most motor controls are designed around low-speed technology. As this section emphasizes, electronic devices that can drive a motor at extreme speeds are now available.

A motor control electronics glossary is included to help you understand the meaning or general description of motor control terms. It should be noted that motor theory or descriptions were not included, since this handbook focuses on the electronics of a motor control system.

We also give you information on how to make things work—the editor and contributors are sharing their "real life" application experiences. The editor has determined, through many years of experience with technical and engineering related projects, that the sharing of knowledge can make a big difference in the success of any project. Understanding how to solve complex math equations is part of designing a complex circuit, but knowing where and when to apply the math equations to fix a problem is more important. Design examples in this handbook that involve math equations are usually presented in a spreadsheet format.

The editor would like to thank all the contributors, especially his wife, Linda Eymann Valentine, who helped make this book possible. The editor would also like to extend his gratitude to Neil Krohn and the management team at Motorola Semiconductor Products Sector for their support during the development of this handbook.

Disclaimer

Neither the editor nor the authors or contributors assumes any liability arising out of the designs or use of any product or circuit described herein; nor do they convey any license under any patent rights or the rights of others. The products and circuits described are not designed, intended, or authorized for use as components in systems intended for surgical implant into the body, or applications intended to support or sustain life, or for any other application in which the failure of design could create a situation in which personal injury or death may occur.

Richard Valentine

Contributors

Ken Berringer, Motorola Semiconductor Products (Chap. 14)

Gary Dashney, Motorola Semiconductor Products (Chap. 13)

Scott Deuty, Motorola Semiconductor Products (Chap. 13)

Randy Frank, Motorola Semiconductor Products (Chaps. 1, 16)

Jim Gray, Motorola Semiconductor Products (Chap. 10)

Thomas Huettl, Motorola Semiconductor Products (Chaps. 6, 11, 15)

Bill Lucas, Motorola Semiconductor Products (Chap. 10)

Peter Pinewski, Motorola Semiconductor Products (Chaps. 6, 8)

Chuck Powers, Motorola Semiconductor Products (Chap. 11)

Pablo Rodriguez, Motorola Semiconductor Products (Chap. 15)

Warren Schultz, Motorola Semiconductor Products (Chaps. 7, 10)

Richard J. Valentine, Motorola Semiconductor Products (Chaps. 2, 3, 4, 5, 6, 10, 12, 17, 18)

David Wilson, Motorola Semiconductor Products (Chap. 9)

Part

1

Introduction

Chapter

1

Motor Control: Driving Forces and Evolving Technologies

Randy Frank
Motorola Semiconductor Products

Introduction

Electric motors, widely used throughout our industrial society in factories and homes, have been evolving for several decades. In this chapter, we will discuss various ways in which electronics technology will further the evolution of motors.

1.1 Increasing Use of Motors

Recent advances in high-energy batteries, combined with the development of smaller and more powerful motors, have opened new markets for a wide range of products, including portable appliances, entertainment equipment, and electric vehicles. The convenience of devices such as portable drills, weed cutters, and other rotating tools has resulted in the rapid growth of motor controls. The newest laser-based audio equipment [compact disks (CDs)], digital audio tape (DAT) and cameras of all kinds, including still, video, and the newest digital versions require sophisticated, high-performance motion control systems. One emphasis for new designs has concentrated on low power consumption in order to extend battery life, providing longer hours of operation.

Designers' concerns, however, for improved performance and more efficient operation are not limited to portable applications. The next generation of motors in many stationary applications will take further advantage of changes that are occurring in both the motors and the drive electronics that control and protect them. Motor drive electronics is experiencing improvements in the packaging, control, and power, as well as in the interconnectivity and communication that will allow motors to run more efficiently, adapt to new applications more quickly, and operate with fewer down-time hours in home, office, and industrial applications. Motor control requirements for specific applications are addressed throughout this book.

The size of the motor can be as small as a fraction of a watt to several kilowatts, depending on the application. Control techniques are changing from none or analog to digital-based. Improved semiconductor technology and control schemes can be implemented by advanced integrated circuits and increasingly efficient power devices. The motors are also changing because of new magnetic materials, laminations, and winding insulation, and new designs and new approaches to old designs, such as switched reluctance made practical by electronic technology.

As shown in Fig. 1.1, a total of over 4 billion motors were produced worldwide in 1994. This is an equivalent manufacturing rate of nearly 11 million motors per day. With an expected 8.5 percent combined average growth rate, the number will increase to 16 million motors per day before the end of this century. These motors have a direct impact on the quality of life, providing essentials such as heating, cooling, and water, work-saving items such as vacuum cleaners, clothes washers, dishwashers, food storage, shop tools, and even luxury items, including various forms of entertainment. The number of motors that can be found in a home easily exceeds 50, while some

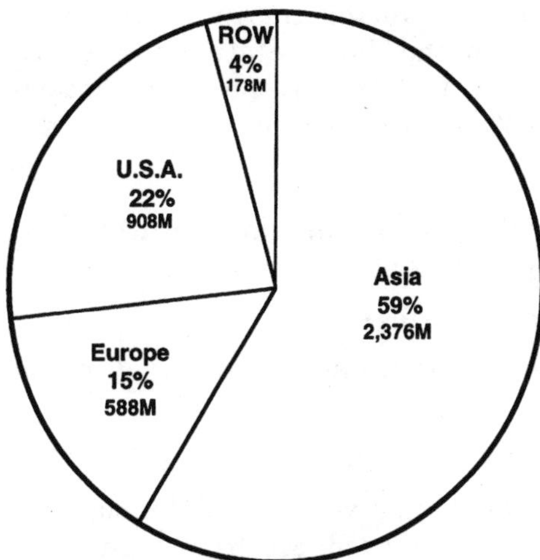

Figure 1.1 Worldwide motor production. (*Source: Motion Tech Trends*).

automobiles can have over 60 motors performing anything from routine tasks, such as windshield wipers and fan control, to more unusual functions, such as headlight wipers, remote-control mirrors, and memory seat adjustments.

Some form of electronic motor controls was employed in 15 to 20 percent of the motor units manufactured in 1994. However, electronic variable-speed drives (VSDs) can produce energy savings from an estimated 14 to 27 percent and, in some applications, over 50 percent. As a result, the number of units with electronic control is expected to increase at a growth rate approaching 20 percent. This increase, as well as the increasing number of motor control applications, makes motors and their control electronics an important area for investigation.

1.2 Need for Increasing Control

The timing could not be more ideal for new technology. Increased energy costs, public concern for unnecessary energy consumption and the resulting environmental impact, and legislation that requires manufacturers to design for improved efficiency are among the key forces that are driving the development and implementation of new motor controls. Motors consume approximately 50 percent of the electricity generated in the United States. An estimated 10 percent is consumed at idle, when no useful work is being accomplished. It should be no surprise that motors have been targeted for more efficient control.

Government regulations aimed at reducing energy consumption have been enacted for several areas in the United States and are having an impact on motor controls. New U.S. federal energy regulations that took effect on January 1, 1992 require heating, ventilation, and air conditioner (HVAC) manufacturers to have

- Furnaces with a minimum of 78 percent annual fuel utilization efficiency (AFUE) compared to prior 62 percent minimum
- Unitary air-conditioning minimum ratings
- Heat pump minimum rating seasonal energy efficiency ratios (SEERs)

Improved motor designs and motor controls are a significant portion of the solution for these legislated improvements. Motors operate most efficiently at higher speeds. Noise produced from high-speed fans, however, is detrimental to an employee's ability to concentrate. Excessive noise from fans or dampers can be reduced by lowering the speed of drive motors and by avoiding resonant operating points. Units operating at optimum speeds can produce less noise, providing an additional environmental improvement in the workplace. As a result, variable-speed fans are attractive for many new applications. Variable-frequency drives (VFDs) provide an unlimited number of operating points and the potential for noise reduction.

Sensors created using semiconductor technology, including the newest micromachining processes, can be used to measure undesirable noise levels. These sensors can also detect bearing failures so that maintenance can be performed during off-hours without disruption of normal work activity. Fan or blower cavitation is often audible and irritating. A closed-loop control with a VFD and a sensitive accelerometer can keep the fan operating outside of natural harmonic frequencies. In effect, the closed-loop control arrives at a compromise between the maximum effective air exchange and audible noise.

Indoor air quality (IAQ) is an increasing concern for buildings sealed to minimize heat loss. The newest building construction techniques have reduced energy-wasting paths through improved sealing. Unfortunately, this sealing process is so effective that a problem called the "sick building syndrome" (SBS) has resulted. As a consequence, the energy management system is also required to exchange air periodically to "freshen" the air even if heating or cooling is not required. IAQ legislation promises to provide additional incentive for improved sensing and control in HVAC systems.

The increasing usage of air conditioning in developing countries is providing additional market potential for motor controls. Consumption for air conditioners in China, for example, is expected to double from 5 million units to 10 million by the year 2000. Other regions of the world that have not been significant users of air conditioning and other motor-controlled equipment are also expected to provide increased consumption in the future.

The U.S. Department of Energy has also established standards (1994) for a number of motor-controlled appliances including dishwashers, washing machines, and dryers. In 1993, refrigerators were required to be 30 percent more efficient, and by 1998 another 30 percent increase must be obtained. This means that the average refrigerator now consumes no more energy than does a 75-W lightbulb. Today, 12 to 15 percent of the 46 million kitchen and laundry appliances currently used in the United States contain microcontroller units (MCUs). Refrigerators, however, are on the low end, with a digital electronics content of only 5 percent. This will change as higher efficiencies are required.

Airflow management is one of the key elements to increased efficiency in high-efficiency clothes dryers. Variations of 50 to 170 cubic feet per minute (cfm) in airflow and even drum speed variations from 2 to 60 revolutions per minute (rpm) are made possible by variable-speed motors. Energy savings also result from improved drying cycles and the implementation of sensors to measure temperature and humidity.

In the United States, the Energy Act of 1992, scheduled for implementation by October 1997, defines efficiencies of several electrical devices, including motors. A minimum efficiency of 80 to 95 percent [except for enclosed 1-horsepower (hp) motors] is required for motors with 1- to 500-hp ratings. The difference in cost for a high-efficiency (95.5 percent) motor over a standard efficiency (91 percent) unit is recovered in operating cost within 3 months for a 100-hp motor. The improvement can be accomplished by motor design alone. Increased savings can result with adjustable-speed drives (ASDs) in applications where dampers and vanes were previously used to control airflow. The improvement that results from replacing a valve with a VSD means an energy savings of more than 50 percent. With more efficient controls, motors can also be sized smaller in many applications, contributing to system cost savings.

In some industrial systems, the use of VSDs allows the elimination of flow control valves and reduces electrical and mechanical maintenance. Power consumption rises exponentially to drive a typical pump. Therefore, a small reduction in speed can result in significant reduction in required motor output. The combination of a variable-displacement hydraulic pump with a brushless motor drive allows the pump to operate continually near maximum displacement, where it is most efficient. This results in a significant efficiency improvement over fixed-displacement, constant-speed systems.

Motor-controlled power valves have reduced compressed air leaks in a factory environment. An energy savings of approximately 77,000 kWh/yr has been documented by the use of motors to perform a task previously accomplished manually. This extension of motor controls to provide increased energy savings and conservation is not limited to electricity. Motor controls have been implemented to conserve water in applications such as toilets. A 3.3-in motor and a 150-W pump in water-conserving toilets clean effectively with less water. The annual cost of electricity of about $4 was more than offset by the $53/yr savings on water bills. Furthermore, a city with a population of 100,000 can save treating over 2.65 billion gal of sewage each year. The savings in these and other new applications can justify the additional cost of the motor and its associated controls; it, too, provides additional growth potential for the motor control market.

1.2.1 Implications of more extensive motor controls

The benefits provided by motor controls do not occur without some controversial implications, including the need for electromagnetic compatibility (EMC) and to minimize electromagnetic interference (EMI), deteriorating power quality based on the increasing use of digital control, concern for electrostatic discharge as a source of potential reliability problems for semiconductor components, and a new aspect to safety. No matter what terminology is used to describe the switched, digital control of the motor—VSD, variable frequency drive (VFD), or ASD—the switching of the voltage can produce electromagnetic interference. The electronic control of motors and inductive loads, when used without regard to the effect on power lines, has caused damaging voltage transients, lowered the power factor, and subsequently resulted in legislation and standards from organizations such as the International Standards Organization (ISO).

Electrostatic discharge (ESD) can cause device failure in unprotected integrated circuits (ICs) and power semiconductors. Fortunately, electronic controls, including circuitry and additional semiconductor components, can address EMC, power quality, and ESD. These solutions add another element of complexity to the motor control system.

Safety is also affecting motor controls. Underwriters Laboratories' Safe Software standard requires that a code be readable for safety situations. This requirement is currently in effect in Europe and targeted for U.S. implementation in 1998. As a result, safety influences both the potentially dangerous power levels of motor control and the logic control portion as well. Because of their impact on motor controls, these issues will be discussed in various chapters throughout this book.

1.3 Availability of New Technologies

Motor controls are advancing with improved motor technology, new and revised motor control techniques, integrated circuits designed specifically for motor control applications, and improvements in semiconductor power devices. Many of these improvements allow the use of systems that were previously too expensive or technically too complex. As a result, there has been a shift in the types of motors employed in different applications.

Cost reductions are an important aspect of the increase in digital motor drives. For example, the cost for a 750-W industrial motor drive has decreased from $685 in 1981 to $465 in 1994. Similar cost reductions are occurring at all power levels.

In factory automation in the United States, electronically controlled motors produced over $3 billion in sales in 1995, as seen in Fig. 1.2. Types of motors in these applications included brushless DC, permanent magnet and switched reluctance, brush DC permanent magnet, stepper motors, and AC three-phase. The technology will shift in favor of AC three-phase and brushless DC motors at the expense of brush-equipped motors by the year 2000, when the market is expected to exceed $4.3 billion. In North America, 34

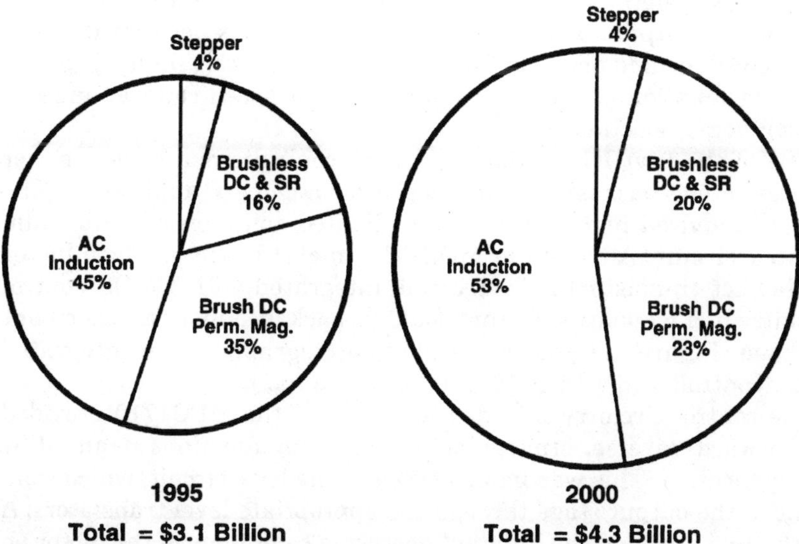

Figure 1.2 Growth in U.S. factory automation for motor controls.

percent of the AC motors are used for fans or pumps in HVAC or in flow applications.

Considerable advances have been made in control technology and control ICs within the past 10 years. The remaining portions of this section will put these improvements in perspective. Areas covered include motor control ICs, microcontrollers with integrated power and control, MCUs, and digital signal processors (DSPs) in motor control, power technologies, and hybrid power modules.

1.3.1 Motor control ICs

One of the initial efforts to simplify motor controls was the introduction of motor-control-specific ICs, first appearing in the mid-1980s. They exemplify some of the key requirements for switching and protecting power transistors.

A brushless DC motor control IC, such as Motorola's MC33034, provides a single-chip open-loop three- or four-phase motor control system. This particular controller includes undervoltage lockout, cycle-by-cycle current limiting with selectable time-delayed latched shutdown, internal thermal shutdown, and a no-fault output for interfacing to a microcontroller. The IC was designed to use the output from Hall effect phase sensors and drive external bipolar transistors and power field-effect transistors (FETs).

By combining motor control ICs with discrete power devices or a modular power stage, a complete brushless DC motor control can easily be designed. The MC33034P (or MC33035) control IC decodes Hall effect sensor signals that indicate motor position and turns on the appropriate output transistors. Additional functions performed by the MC33034P include braking, fault recognition, current limiting, pulse-width modulation of the lower transistors for speed control, and forward/reverse selection.

One of the first ICs to combine motor control circuitry and power devices in one package was developed for a camera application. The camera required high efficiency in a limited space on a flexible-film printed circuit. A complete MOSFET (metal oxide semiconductor field-effect transistor) H-bridge was integrated with CMOS control circuitry in a monolithic IC that could be packaged in a surface mount package. Figure 1.3 shows the photomicrograph of the integrated motor controller (a MPC1710 from Motorola Inc.).

The control circuitry included in the IC of the MPC1710 provided the forward, reverse, brake, stop, and standby functions required by the application. This was implemented in the logic circuit, which commanded the output stage through the appropriate level translators. A truth table provided the different operating conditions of the motor on the basis of the input states.

Figure 1.3 Photomicrograph of monolithic motor control combining control logic and a MOSFET H-bridge. (*Courtesy of Motorola Inc.*)

During turnoff, the spinning motor becomes a generator. To achieve dynamic braking, the two top devices of the H-bridge were turned off, while both lower devices were turned on. In this mode, the motor conducts the current through the lower devices and into ground, effectively shorting the generator. All pins in the MPC1710 were protected to avoid ESD problems. Since the package was a surface-mounted type, and automatic assembly was used for pick and place, the machine model for ESD testing was used.

Motor control ICs have also been designed with multiple motor controls for low-voltage applications. A multifunction IC controller for portable CD players incorporated four H-bridges and control circuitry as well as voltage regulation and a DC/DC converter. The ultimate extension of an integrated approach is combining motor interface circuitry, including the power devices, into a microcontroller. This is possible in areas of low power. For higher-power applications, package-level integration is required.

1.3.2 Microcontrollers and digital signal processors for motor controls

MCUs are being designed for a variety of motor control applications. A motor-control-specific MCU measures the speed and rotor position, calculates the resulting torque, accepts input, and calculates the timing and amplitude of the current pulses that energize the windings. Recently introduced MCUs for three-phase, brushless DC motor controls, such as Motorola's MC68HC705MC4, contain the necessary elements—six PWM output lines and timer inputs—for the Hall sensors. Higher-performance MCUs can also control open-loop three-phase induction motors.

Both the AC induction motor and brushless DC motor can be improved through vector-controlled motion control strategies. Open- and closed-loop techniques are used to implement vector control. The closed-loop technique uses feedback from sensors to provide an accurate speed input to the VSD. The open-loop approach uses a software algorithm, in the form of an estimator, to drive magnetic flux, torque, and shaft speed.

In addition to basic MCU control, more sophisticated digital control is being implemented with DSP and reduced instruction set computers (RISCs). These more complex systems involve control of several variables, including sensorless vector control. By eliminating shaft-position sensing, a cost savings can be achieved; even more important is the reduction in complexity and reduced potential for failure. A variety of sensorless vector drives has been introduced.

Since the mid-1980s DSP technology has been investigated and subsequently implemented in several more complex motor drive schemes. DSP techniques have been used for analog filtering, sampling and zero-crossing detectors, implementing fast Fourier transforms, spectral estimation methods, filtering, and manipulation of speed-related current harmonics. Sensorless speed detection algorithms combined with a DSP provide an accurate estimate of speed that is used to tune load-dependent motor parameters or to construct a rotor velocity observer.

Another sensorless drive uses a 32-bit RISC-based MCU to implement a flux transformer and a flux observer. The adaptive flux observer is actually a sophisticated motor model that calculates the vector components of the motor's primary current. The flux transformer distinguishes between the torque- and flux-producing currents flowing to the motor. The flux observer contains observer and voltage error compensation circuits that continuously compare the calculated current value with the actual motor current and compensates for differences in control parameters.

Direct torque control is yet another technique that provides flux vector control for induction motors and achieves accuracy comparable to that in units with encoder feedback. This approach relies on a DSP-type microcontroller making 40,000 calculations per second to determine optimum power switching. The method uses a very accurate adaptive motor model capable of calculating motor torque directly, providing an output that represents actual motor torque and flux.

The switched reluctance motor (SRM) control technique, which is over 100 years old, has found new life with advanced power electronics. The SRM is simple in construction, as well as rugged and economical, making it ideal for many general-purpose adjustable-speed applications. The primary disadvantage, however, is higher-torque ripple, which causes acoustic noise and vibration. The combination of an SRM and a digitally controlled converter allows the desired performance advantages to be achieved while minimizing the disadvantages. This is an area that is receiving considerable attention. Advancements in MCUs have enabled higher-complexity motor controls and allow the motor's dynamic performance range to be extended.

1.3.3 Power devices in motor controls

Power devices in industrial drives typically constitute about one-third of the cost. Note, however, that the cost of the power devices can rise substantially for motors larger than 5 kW. Motor topologies dictate the power device configuration, and the supply voltage affects the choice of power devices. For example, in low-voltage motors (<100 V), power MOSFETs, or simply FETs, are the current choice for power switching in digitally controlled applications. A single-direction motor driving a fan, for example, can be pulse-width modulation (PWM), controlled from a single power FET. If the motor is grounded, the FET must be a high-side switch, which means that a P-channel FET or an N-channel FET with a special gate biasing power source is required. If the motor is not grounded, an N-channel FET can be used. To reverse a brush DC motor, an H-bridge is required with four FETs. The top two legs of the bridge have the same requirements as a high-side switch—either P-channels or N-channels with a special gate bias power supply must be used. Inverters for three-phase AC or brushless DC motors utilize six FETs with the same high-side requirements. Table 1.1 shows a summary of a variety of motor applications, the supply voltage, and power technologies that are currently being employed for new designs.

The operating frequency (up to 50 kHz) and desired efficiency (as high as possible) for modern high-voltage motor control application eliminates bipolar power transistors as a possible technology. Power MOSFETs are high-frequency devices but lack the current density needed to make their use practical in these high-voltage applications. Power MOSFETs would require a very large area to reduce the ON resistance for the power module and create serious reliability concerns due to the number of potential opportunities for failure. To pro-

TABLE 1.1 Motor Applications and Critical Requirements

Motor usage	Supply voltage	Transistor voltage rating, V	Requirements and features
Camera focus	<10 V DC	<20	H-bridge
Remote-control toys	<10 V DC	<20	H-bridge
Disk-drive spindle motor	5–12 V DC	20–30	Integrated controls or SMT
Office automation (copiers, printers)	5–24 V DC	40	Integrated
Car mirror control	14 V DC	40–60	Integrated H-bridge
Car window or seat motor	14 V DC	40–60	Relay or H-bridge
Car fan	14 V DC	40–60	Highly efficient discrete FET
Portable tools	7–24 V DC	20–50	Discrete FET
Home appliances	120/240 V AC	300/600	FETs or IGBTs
Industrial motor control	240/440 V AC	600/1200	Three-phase IGBTs
Electric vehicles	24–350 V DC	60–600	FETs or IGBTs

vide the necessary power-handling capability, insulated gate bipolar transistors (IGBTs) are the technology of choice. IGBTs use a cell design similar to that of power MOSFETs but have P^+ instead of N^+ substrate as part of the fabrication process. This provides a power device that is easy to drive, such as a power MOSFET, and has the area efficiency at higher voltages, such as the bipolar power transistor. These devices have a high voltage rating, as well as high current capability.

The IGBT does not significantly increase its power losses with junction temperature increases, whereas a power MOSFET will have about twice as much power loss at a maximum operating temperature of 150°C. IGBTs also have lower input capacitance than do similarly rated power MOSFETs, which simplifies the drive circuitry. Furthermore, some IGBTs have greater short-circuit capability than do bipolar or MOSFETs, which provides greater fault tolerance. However, one of the drawbacks of the process change used to make IGBTs instead of power MOSFETs is the lack of an internal, antiparallel diode. IGBTs require an external diode to withstand the reverse current that occurs when the motor winding is switched off.

Semiconductor manufacturers' specific thrust for motor control in power semiconductors is occurring in the following areas:

- Higher voltage, higher power, and increased efficiency through improved power devices and improved control techniques, as well as cost reduction.
- Discrete-packaged FETs and IGBT products with emphasis on surface mount packages. Also, new, larger packages such as TO-264 (formerly TO-3PBL) and ISOTOP package provide a larger area for more efficient power switches.
- Hybrid power modules that incorporate multiple output devices and increasing levels of integration from catch diodes to ICs and even microcontrollers.

Bipolar power and thyristors continue to be used in DC and AC applications, respectively. New device structures, such as MOS-controlled thyristors (MCTs), promise increased efficiency in high-voltage applications. While vertical output devices [diffused MOS (DMOS)] are used in discrete power devices, lateral MOS outputs are easily isolated and efficient in monolithic smart power devices, especially with decreasing, submicrometer geometries.

1.3.4 Hybrid power modules for motor controls

Increasing integration at the module level has created smart power modules that combine control circuitry and power devices at the package level. The packaging provides mechanical durability, isolation, and improved thermal performance. An integrated motor drive system can be purchased for AC motors up to 2.2 kW. In a smart motor system the control electronics is an integral part of the motor. The choice of power devices still depends on the application.

The power electronic building blocks (PEBBs) effort sponsored by the U.S. Department of Defense has a target of increasing the current density of power switching to 10 MA/m^2 by the year 2000. This is about 10 times the level possible today. If this project is successful, the voltage will also increase to several kilovolts and the switching frequency to more than 60 kHz. Dual industrial and military use is being pursued. Eliminating or reducing the thermal and electromagnetic interference should have a drastic impact on the packaging for these devices.

Hybrid power modules typically use a ceramic as the isolation material. Recent advances in metal matrix silicon carbide have improved power switch performance, especially under temperature and power cycling.

1.4 Network Implications of Smart Control Systems

Motor controls are being integrated into the factory, building, home, and even automotive environments. The communication networks in these applications have several different protocols—frequently many competing for the same application. Some protocols, such as Controller Area Network (CAN), developed by Robert Bosch, GmbH, are finding acceptance in automotive systems.

The integrated factory or factory automation with distributed control will mean the interconnection of motor-controlled processes to remote monitoring stations. A large plant could have hundreds of low-speed and many high-speed communication buses. The low-speed buses will operate at frequencies of approximately 31.25 kbytes with minimum response time of 100 ms or less. The high-speed buses will operate at 1 to 2.5 Mbyte with a response time of 1 ms.

The power of a digital communication system is easily demonstrated in factory automation. It has been found that an installation that previously took a crew of electrical technicians several days to wire could be completed by one or two people in a day with a digital communication system. Just running a twisted pair of wires, as well as power and ground wires, greatly simplified the interconnection process. These installations also worked successfully the first time, resulting in considerable cost reduction. Once a system strategy is developed, nodes can be added or moved easily without reengineering the system. These nodes can include sensors, as well as valves, motors, and lamp loads. A reduction in point-to-point wiring, installation, and maintenance costs, as well as system noise, including EMI/RFI, has been demonstrated by a distributed motion control system in a factory environment. The key for communication is an open standard and plug-and-play capability.

Buildings in the United States account for 35 to 40 percent of the nation's energy consumption. Reduced energy consumption is achieved by using improved efficiency power devices and control techniques, such as PWM, that allow a VFD to be implemented in building automation. A number of factors, including temperature zones, humidity, airflow, outside/inside air mix, light level, and air quality, must be considered for achieving maximum comfort in buildings. The interaction of these factors provides a comfort level that is predictable and, therefore, controllable. Individual comfort levels, however, vary considerably requiring localized control to optimize comfort. Key factors for motor controls are humidity, air movement, outside-air ventilation rate, and outgas pollution. Interconnecting the motor control to the sensed inputs is simplified in an automated building environment.

One power-reducing strategy would be to have the variable-frequency-drive motors in the HVAC system run only when required at their lowest speed for minimum energy consumption. An alternate strategy would be to run the motor continuously at an optimum efficiency point and direct the airflow to specific zones. A variable-air-volume (VAV) system controls the volume of air that is heated or cooled, depending on the user's demand, reducing the amount of air that must be heated or cooled and consequently reducing energy consumption.

The smart house intelligently communicates control information. For example, the many motor applications, as well as lighting and solenoid controls, could be connected for automatic and remote programming for improved efficiency and flexibility for different lifestyles. Safety and security can be increased in homes by sensing problems, such as smoke or harmful chemicals, and activating motors to restrict or increase airflow. Homes with motor-activated sunscreens can be positioned to increase or minimize heating from the sun according to sensor input to a central control system or interaction between remote smart nodes. A number of protocols have been defined and are available to address these applications in the home.

Vehicle multiplex (another form of digital communication networking) or MUX (multiplexing) systems have been classified as low-, medium-, or high-speed buses. For vehicles with internal combustion engines, the low-speed bus is used for body controllers to interface to motor controls for motor-controlled mirrors, three-way seat adjustments, as well as power windows and door locks. Several motor-controlled components and the drive-activating switches are typically mounted in the door, requiring as many as 60 wires passing between the door and the chassis. The reduction of the number of wires connecting these components was one of the initial driving forces for a vehicle network. Other vehicle loads, such as lighting, have also been included in MUX systems. The traction motor system used in EVs and hybrid EVs require a medium-speed MUX bus for diagnostics and other communications.

1.5 Motor Control Topics of Special Interest

Other areas not previously addressed are critical to the successful application of motor controls: (1) modeling and tools to reduce design and manufacturing time, (2) faults and diagnostics in motor control systems, and (3) the futuristic possibilities of motor controls.

The increasing complexity of motor controls has required improved modeling techniques. Suppliers are developing both the digital control and power control demonstration boards as reference designs to simplify the user's implementation of new technologies. These tools typically provide the digital controller, the power switching devices, and the interfacing circuitry that simplifies the motor control design, allowing focus on the application.

Even the best-designed motor control system can encounter faults during its lifetime. Diagnostics that detect problems early and take action, or alert operators to take action, can mean the difference between interrupted operation and costly failures with extended downtime.

Finally, in the last chapter of this book, we will attempt to visualize the direction in which today's developments and those of the near future could take motor controls. Some of these advances are actually used in low quantities at this time or are in the experimental stage. Additional development efforts and technology breakthroughs may be required to bring about some of these future approaches. In other cases, increased competition and manufacturing attention to details will provide cost breakthroughs that promise to change motor controls of the future.

Summary

This chapter has provided an overview of the driving forces for new technologies in motor control. These driving forces include the legislated and competitive need for increased efficiency. New global markets are an integral part of increasing volumes for developed products, as well as a source of new technology and competitively priced products. The variety of advanced control techniques and power switching considerations for micro- to megawatt-range applications are propelling many designers to take technology to the next level—out of the R&D labs and into industrial and consumer products. The remaining chapters will provide additional details and guidelines for taking these next steps. Key points about this chapter are listed below.

- New markets for electronic motor controls are being created by the arrival of higher-power-density batteries and more powerful motors for portable tools and lawn equipment. Home electronics, such as digital videodisks, computer printers, disk drives, and automatic cameras, also require complex electronic motor systems.
- About 11 million motors are produced per day, and are expected to increase to about 16 million by the year 2000. Of these, 15 to 20 percent will use electronic controls.
- Motors consume over 50 percent of the electricity produced in the United States.

- The exact amount of savings will vary for each application, but an overall savings of 10 percent may be realized with electronic motor controls.
- Variable-frequency drives save electricity because they can be operated at optimal speeds; specifically, the motor is operated at just the right speed for a particular application.
- Electromagnetic compatibility (EMC), electromagnetic interference (EMI), deteriorating power quality, and electrostatic discharge (ESD) need to be considered when designing an electronic motor control.
- Three types of semiconductor devices are normally used in motor controls: analog ICs, microcontrollers, and power transistors.
- Normally, only the power transistors or power modules must be matched or sized to the motor's power rating. Sometimes the interface design must be upgraded to accommodate increased EMI from large motors.
- Besides varying the motor's speed or torque, electronic motor controls can also be used to measure and track motor temperature, bearing noise, and so on, and to communicate with other electronic systems.
- It is becoming important for motor-based systems to have the ability to communicate with other electronic equipment because motors, whether they are in the factory or in the home, consume a significant amount of electrical power. If several motors happen to switch on at the same time, the electrical power may sag or place high peak demands on the electrical infrastructure. A load manager can communicate with the larger electrical appliances, which helps optimize their operation, while providing information about the operation status of each unit.

Further Reading

Alberkrack, J., "A New Brushless Motor Controller," *Motor-Con Proceedings,* Oct. 1986, pp. 269–282.

Anderson, R., "The Open Door: Efficient Appliance Technologies," *Appliance Engineer,* Dec. 1991, p. 49.

Artusi, D., Jorvig, J., and Shaw, M., "Industry's First Monolithic Smart Power Microcontroller Handles Over Six Watts," *PCIM,* Nov. 1990, pp. 12–18.

Automation Research Corporation Report, 1994.

Bartos, F., "Sensorless Vector Drives Strive for Recognition," *Control Engineering,* Sept. 1996, pp. 99–105.

Berardinis, L., "SERCOS Lights the Way for Digital Drives," *Machine Design,* Aug. 22, 1994, pp. 52–63.

Cave, D., Frank, R., and Sukurai, T., "A Monolithic Motor Controller for Low Voltage Applications," *Proceedings of Motor-Con,* April 1–3, 1987, Hannover, Germany, pp. 159–164.

Cole, R., and Thome, T., "Getting the Most from High-Efficiency Motors," *Power Transmission Design,* May 1995, pp. 27–31.

Divan, D., "Low-Stress Switching for Efficiency," *IEEE Spectrum,* Dec. 1996, pp. 33–39.

Dzierwa, R., "The Responsible Approach," *Appliance,* Feb. 1997, pp. B-17–19.

Emerson's Inside Technology Making Your World A Better Place to Live, Emerson Motor Company, 1997.

Emerson's Motor Technology News, Sept. 1994, July 1996.

Frank, R., and Powers, C., "Microcontrollers with Integrated CAN Protocol," *Proceedings of Power Conversion & Intelligent Motion,* Sept. 17–19, 1994, pp. 1–10.

Frank, R., and Walters, D., "MEMS Applications in Energy Management," *Proceedings of Sensors Expo,* Boston, MA, May 16–18, 1995, pp. 11–18.

Hartman, T., "New Zone Controls Help Achieve Total Environment Quality," *Heating/Piping/Air Conditioning,* Nov. 1993, pp. 43–46, 92–93.

Humbert, D., Hooser, D., and Frank, R., "Hybrid Power Modules Advance Motor Controls," *Control Engineering,* Jan. 1996, pp. 68–72.

Hurst, D., and Habetler, T., "Sensorless Speed Measurement Using Current Harmonic Spectral Estimation in Induction Machine Drives," *IEEE Transactions on Power Electronics,* Jan. 1996, vol. 11, no. 1, pp. 66–73.

Husain, I., and Ehsani, M., "Torque Ripple Minimization in Switched Reluctance Motor Drives by PWM Current Control," *IEEE Transactions on Power Electronics,* Jan. 1996, vol. 11, no. 1, pp. 83–88.

Jones, D., "Motion Control in Global Dynamic Change," *Control Engineering,* Nov. 1995, pp. 67–72.

Jorgenson, B., "Smart Appliances Open a New Market for Semiconductors," *Electronic Business,* Jan. 1993, p. 17.

Kaplan, G., "Industrial Electronics," *IEEE Spectrum,* Jan. 1997, pp. 79–83.

Motion Tech Trends Report, 1995, Inglewood, CA.

Murray, C., "Motor Simplifies HVAC Zone Control," *Design News,* March 28, 1994, pp. 42–44.

Murray, C., "Energy-Efficient Design," *Design News,* March 11, 1991, pp. 62–65.

Remich, Jr., N., "Energy-Smart Laundry," *Appliance Manufacturer,* July 1994, pp. 30–31, 36.

Romero, G., Fusaro, J., and Martinez, Jr., J., "Metal Martix Composite Power Modules: Improvements in Reliability and Package Integration," *IEEE IAS Annual Proceedings,* Oct. 8–12, 1995, Orlando, FL, pp. 916–922.

Somheil, T., "No Shanghai Surprises," *Appliance,* June 1996, pp. 63–67.

Sraeel, H., "The Quest for Energy-Efficient Building," *Buildings,* Feb. 1992, pp. 58–60.

Stefanides, E. J., "New Pumps Get Power-Stingy," *Design News,* April 8, 1991, pp. 86–88.

Widell, B., "Saving Energy Through Pump Speed Control," *Design News,* Feb. 20, 1995, p. 80.

Part

2

Motor Speed Controls

Chapter

2

Motor Control Attributes

Richard J. Valentine
Motorola Semiconductor Products

Introduction

In the preceding chapter it should have become apparent that electric motors are employed in a wide variety of applications and that many of these applications are evolving from electromechanical to electronic designs. As will be seen in this chapter and the chapters that follow, electronic motor circuits interface easily with digital logic. An obstacle to wide-scale use of electronic motor controls has been the high cost of the control computer and power electronics. This obstacle is diminishing as the semiconductor industry improves the technology of microcomputers and power devices.

In this chapter we deal with many of the issues that affect the electronic design of motor controls. Reliable electronics for a motor control requires careful attention to the effects of factors such as inductance, excessive temperatures, short circuits, and locked rotor conditions. Products that are specifically rated or designed for motor control applications minimize potential risks. The design criteria shown in this chapter and following chapters can be applied to many different types of motor control designs.

Following are a few notes about the technical aspects of the materials presented here. Crucial specifications or values are noted and need to be heeded when contemplating an electronic motor design. *Items that affect the long-term reliability or stability of the electronic motor design are italicized.* The authors are writing primarily from experience and an applications vantage; they know how to make something work and which details, on the basis of actual lab testing, require special consideration. Many of the designs shown have been simulated but may not be 100 percent laboratory-tested. Conceptual designs mean just that—they are new design ideas that need to be proven.

2.1 Motor Control Design Principles

In this chapter we will discuss electronic motor control fundamentals and provide an essential review of electronic motor design characteristics. Since this book's focus is on the design and application of motor control electronic circuits, basic motor operation and theory will be covered only when the control electronic designs are directly affected by the motor type. (Several motor books that cover motor operation and types are listed in the reference section.)

Some math equations necessary in designing any electronic motor control are given. Throughout this book, whenever an equation is shown, a practical example of how to use it is usually given. Tips for utilizing standard desktop computer spreadsheet software to implement the math are also included. [Note that some math equations use an asterisk (*) to signify a multiply function and a slash (/) for division; this complies with most computer spreadsheet keystroke standards and makes it easier to type in the equations.]

Before discussion begins on electronic motor control design criteria, it is useful to understand some objectives of motor control. Motor control designs are intended to allow a user to manage the motor's speed, torque, or direction and should be user-friendly, avoiding complex steps and procedures. Some motor control systems are highly automated with the motor control preprogrammed to complete certain tasks. Whether the motor control is a simple open-loop speed control or a computerized closed-loop torque control, the foremost motor control design goal should be reliability. Motors, by their nature and application, are used to convert electrical energy into mechanical force. In some applications, the mechanical force involves driving things such as gears, fans, pumps, and belts, especially important since they may cause damage or even bodily harm if the motor control fails.

2.2 Motor Control Overview

The basic theory behind electronic motor controls is that the motor's speed, torque, and direction are managed by electronically switching or modulating the voltages to the motor. The current level to the motor can also be managed indirectly by modulating the motor's voltage. Pulse-width modulation (PWM) is the most commonly used method to vary the average voltage to the motor. Figure 2.1 illustrates both edge- and centered-aligned PWM waveforms. Note that the waveforms are either in one of two states—full OFF or full ON. This two-state, or binary, operation is ideal for controlling power transistors with digital controls. The motor's inductance, which is partially set by the number of turns used in the motor's windings, will integrate or smooth out the PWM voltages. For example, if 12 V is applied to the motor at 50 percent duty cycle, the average motor voltage will be 6 V. This is the basic principle used to vary the average voltage in most electronic motor control systems. It is important to

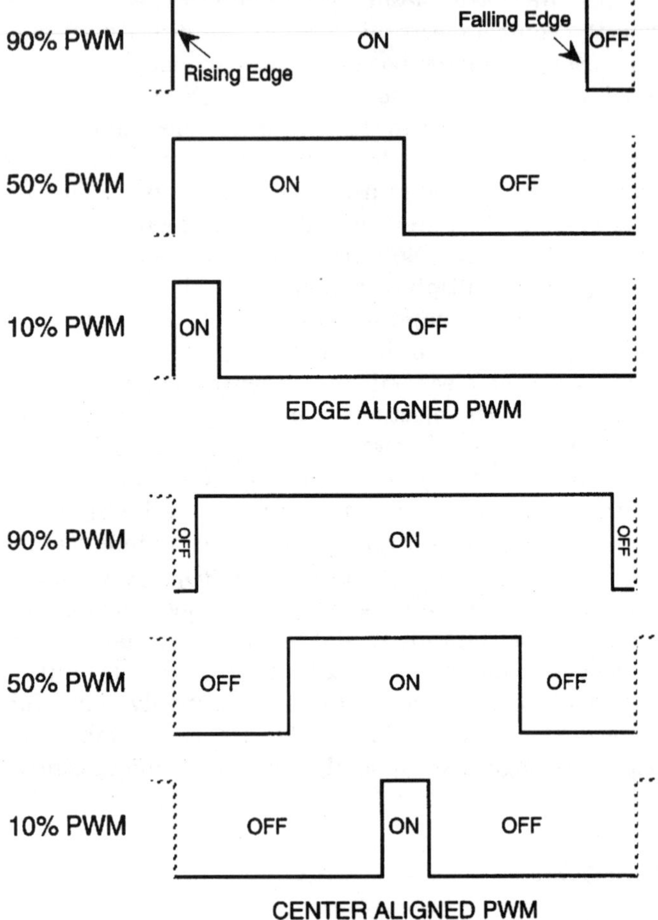

Figure 2.1 Pulse-width modulation is used to control the motor's voltage.

thoroughly understand PWM waveforms, not only how they are generated but also the basics of the actual PWM signal. The PWM's transition edges are especially crucial and have a direct effect on the motor control electronic design. Other methods to vary the voltage to the motor include linear amplifiers, which act like a variable resistance. The linear method is highly power-inefficient and is seldom recommended for new designs.

2.2.1 PWM considerations

There are many aspects to PWM for motor control. As previously noted, the duty cycle of the PWM directly affects the amount of energy applied to the motor. The frequency of the PWM waveform will also influence the motor's operation and the long-term reliability of the power electronics. In most motor controls, the PWM frequency remains constant while its duty cycle varies from 0 to 100 percent. Since the motor is a dynamic machine with the armature and mechanical loads acting as a flywheel, the PWM frequency can be fairly low—100 Hz or less, before the motor starts to pulsate noticeably in synchronization with the PWM frequency. The acoustic noise generated from the motor's windings interacting with the motor's rotor, bearings, shaft, and mechanical load can be a problem. This audible motor sound may not be tolerable for noise-sensitive environments, such as those surrounding home appliances. At higher PWM frequencies the audible noise is minimized because the average human hearing range diminishes rapidly above 15 kHz.

There are other concerns associated with high-voltage PWM signals that drive motors. Some of the energy is coupled to the rotor or shaft of the motor during the PWM's transitions. This energy flows or tries to flow to common or ground across the shaft's bearings, which may not behave as good conductors. There has been some research into bearing degradation caused by PWM signals. *The PWM signals also place more stress on the motor's winding insulation. Selecting a motor specifically designed or rated for PWM-type applications is recommended.* Many motor manufacturers offer *inverter-duty* models that have improved insulation and construction.

2.2.2 PWM frequency effects

The most important fact to remember about selecting the PWM's operating frequency is that a power loss occurs every time a transition occurs. A 10-kHz PWM frequency will have 10 times more transitions per second than a 1-kHz PWM frequency. This means that the motor's power stage incurs at least 10 times more switching power loss than will a power stage operating at 1 kHz. The PWM frequency should be as low as the application will allow.

2.2.3 PWM switching edges and RFI

Since the PWM waveforms are squarewaves, significant radio-frequency interference (RFI) can be generated. The RFI is proportional to the transitional speed of the PWM waveform edges and the PWM frequency. For example, if the PWM switching edges are 100 ns, radio frequencies will be generated that can usually be detected by a standard AM (amplitude-modulation) radio. Without delving into frequency domain theory and Fourier analysis, a simple rule of thumb is that RFI will occur whenever PWM switching edges are faster than 10 μs; this means that most PWM motor drives produce RFI to some extent. Methods to reduce RFI include snubbers and filters connected to the appropriate RFI sources. One important consideration to reduce RFI is to locate the power electronics as close as possible to the motor's windings. Mounting the power stage in a metal box, located on the motor, is a recommended practice, provided adequate cooling is available to the power transistors.

2.2.4 PWM switching edges and power loss

There is more to PWM transitions than the RFI aspects. Slowing down the switching edges helps reduce the RFI. The switching speed parameters also play a crucial role in the overall motor control system. Reasonably fast transitional speeds—0.5 μs, for example—are desirable to minimize switching losses and improve reliability, but very high switching speeds become impractical beyond a certain point because of the inherent inductance in the wire and component leads. A compromise must be made between the edge speeds and the power device heat loss. The drawing in Fig. 2.2 illustrates the switching speed and power transistor dissipation relationship at a fixed PWM frequency and duty cycle.

PWM switching speeds have an important effect on the heat loss of the motor's power transistor. Since this is a crucial issue involving reliability, a close examination is in order. As shown in Fig. 2.2, the voltage and current levels cross over during each transition in the PWM waveform. The voltage, current values, and time-domain values representing the turnon and turnoff of the power transistor are entered into a spreadsheet (see App. A). The spreadsheet allows a quick estimate of switching power loss for simple single-switch motor drives. The switching power loss of each transition can then be calculated and graphed from the spreadsheet, as shown in Fig. 2.3. Close examination of this graph shows that the transistor's collector-to-

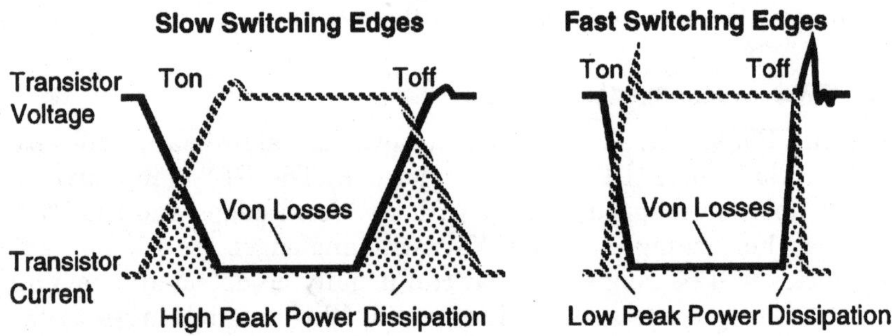

Figure 2.2 Comparison of fast and slow PWM switching transitions.

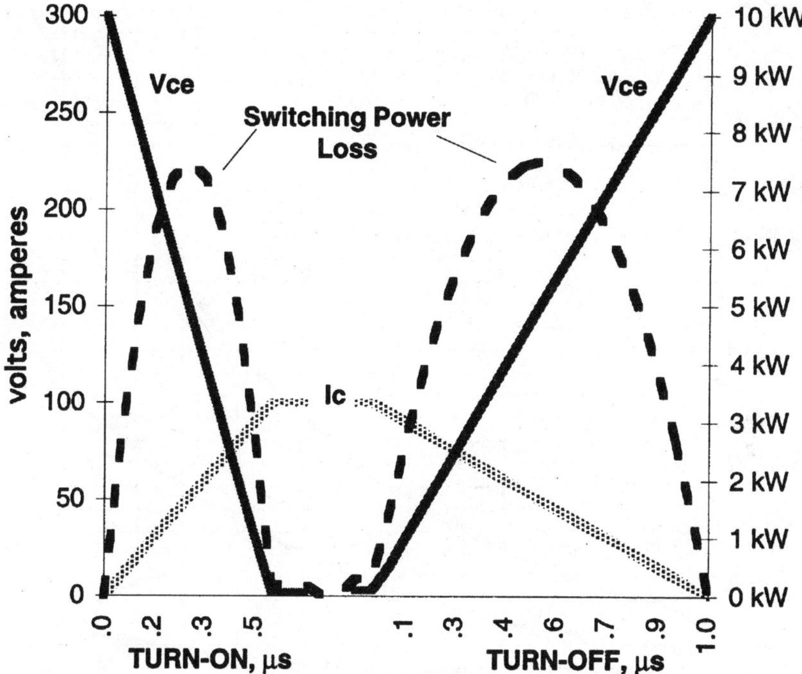

Figure 2.3 Peak power switching loss.

emitter voltages and collector current are triangle-shaped. By first dividing their V_{ce} and I_c values by 2 and then multiplying V_{ce} by I_c, one can calculate the peak power loss of either turnon or turn off transitions. It is worth noting that the peak power dissipation is a fairly high value and directly affects the reliability of the power transistor, as will be explained in other chapters.

The average switching power dissipation is much lower than the peak power loss, as shown in Fig. 2.4, which also illustrates the effect of changing the PWM frequency. In order to minimize audible noise, the 20 kHz frequency would be favored, but, unfortunately, the 0.1 μs switching speeds required to maintain reasonable switching loss cause the intrinsic inductance in the hookup wiring, components, and, most notably, the motor, to become critical. These faster edge speeds will generate significant voltage spikes just across the hookup wiring and may lead to reliability problems unless precautions are taken to minimize stray inductance. This leads us to a review of the PWM signal and the effect of inductance from the electronics and motor wiring.

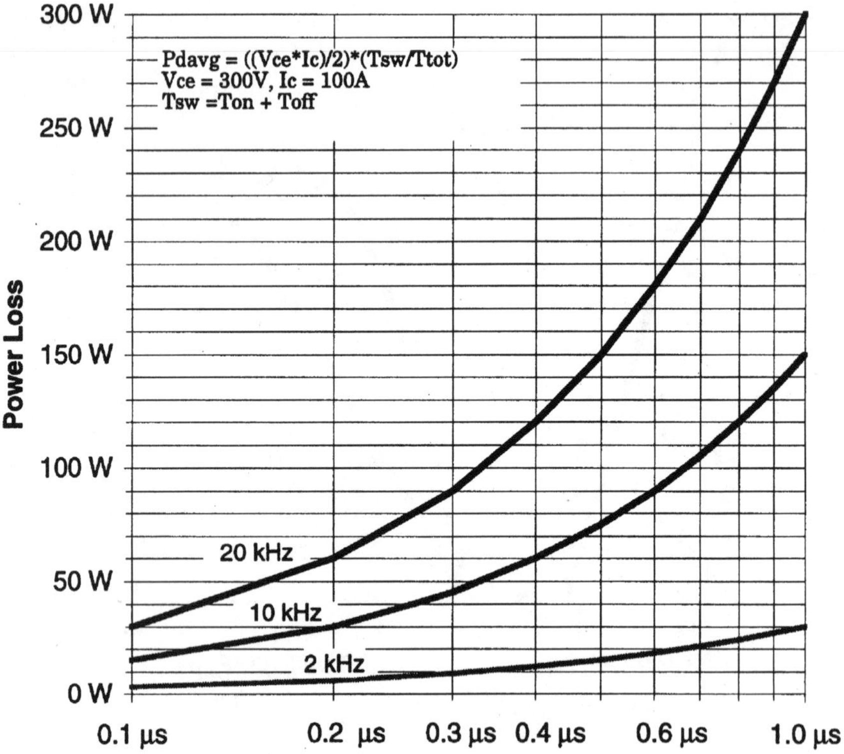

Figure 2.4 Average switching power loss.

2.2.5 Effects of inductance

Inductance is a fact of life with electronic designs and must be accounted for if a motor control design is to be successful. All electronic and electrical components exhibit inductance: resistors, capacitors, semiconductors, wires, printed-circuit traces, connectors, and usually anything that is used to control electricity. Transformers, chokes, and coils are highly inductive devices. The amount of inductance is the important factor. Many motor drive electronics failures can be traced back to an incorrect estimate of the effects of high currents flowing through a seemingly "low-inductance pathway," such as a short length of 1/0 wire. When high current levels are switched from ON to OFF in less than a microsecond, an inductive kickback voltage is generated. (*Note:* This is different from the counter electromotive voltage generated from a motor, which will be discussed later.) Keeping the 1/0 wire as an example, a significant kickback voltage can be generated, which will be explained below.

An inductance value of 0.12 μH can be obtained in about 152 mm (6 in) of wire length with a 1/0 wire size. Straight-line wire inductance is calculated with an equation [Eq. (2.1)] that should be part of an engineer's reference notebook. This equation can contribute a critical piece of data that is usually missing from most wire tables and that is, namely, the wire's inductance. (See App. B.)

$$L = (0.0002*l)*2.3026*\mathrm{LOG}((4*l/d)-0.75)) \qquad (2.1)$$

where L = straight-line wire inductance, μH
d = wire radius, mm
l = wire length, mm; note $l \gg d$

A graph in Fig. 2.5 based on the straight-wire equation shows two important factors with reference to its inductance—the wire's diame-

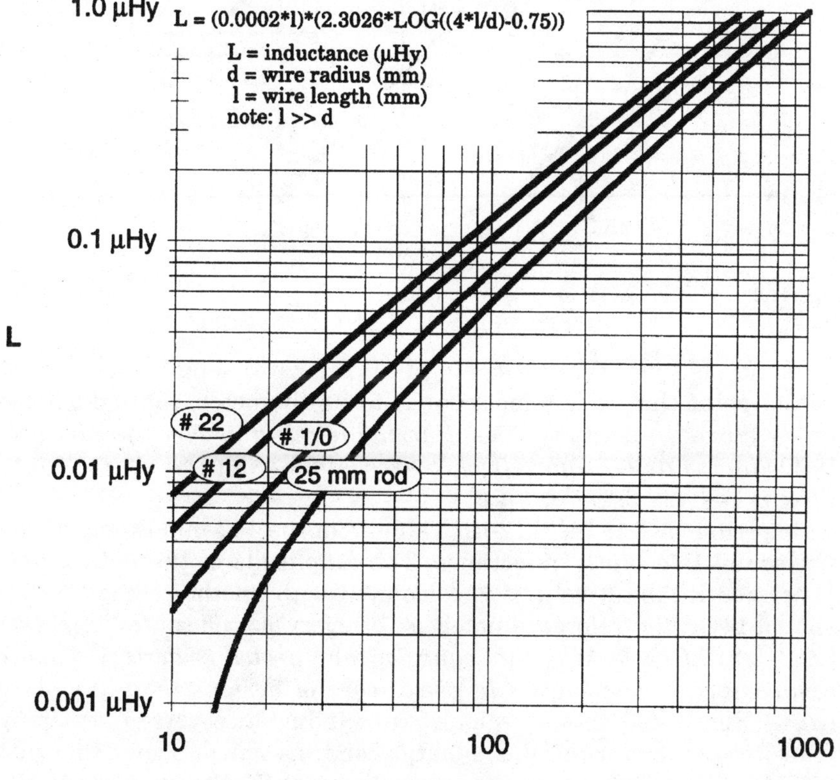

Figure 2.5 Straight-line wire inductance graph.

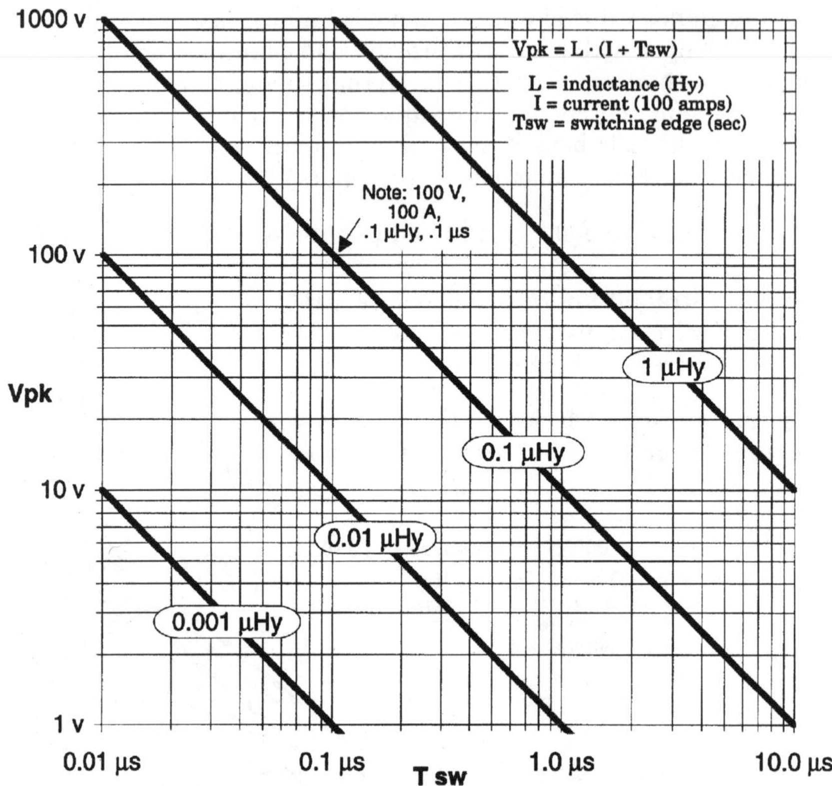

Figure 2.6 Inductive-generated kickback voltage.

ter is not as significant as its length. If the wire's length is shortened to the point that it is almost equal to its diameter, the inductance really starts to decrease. *Doubling the wire's diameter to reduce its inductance has a minor effect, whereas reducing the wire's length directly reduces its inductance.*

Referring back to the 0.12-μH value from the 152-mm example may not sound like much inductance. The amplitude of the inductance-generated kickback voltage is calculated with another equation that should be in the reference notebook of every power control engineer [Eq. (2.2)]. *Overlooking this equation has probably destroyed more semiconductor components in power control designs than any other cause.* Figure 2.6 shows a plot of wire inductance versus switching transition speeds using this equation and, as noted, shows that just 0.1 μH of lead inductance can generate a 100-V voltage spike (at 100 A motor current) when switching edges are 0.1 μs.

$$V_{\text{pk}} = L * \frac{Di}{Dt} \tag{2.2}$$

where V_{pk} = inductive kickback voltage
Di = current variation, A
Dt = time period of current variation, s

If the motor's leads are several meters away from the control electronics, severe and potentially damaging voltage spikes will occur. There has been some research on methods to counter this problem. The best method is to locate the control electronics on the motor and to use a remote-control user interface. By locating the power electronics right on the motor, the inductive kickback issue is minimized, with better containment of electromagnetic interference (EMI) as well.

There are several possible conclusions about PWM-type power control design and inductance:

- *If the motor's lead lengths exceed 25 mm (1 in), their inductance values must be taken into account.* The effect of the motor's hookup wire's inductance can be dramatic, especially in motor control systems where the motor and drive electronics are located several meters apart.
- PWM switching transition speeds affect the power transistor's heat loss. High speeds reduce heat loss, but very fast speeds will generate significant inductive kickback voltages. Very fast switching will also contribute to RFI and EMI problems. High transitional speeds are desirable to minimize switching losses and to improve reliability, but very high switching speeds become impractical beyond a certain point because of the inherent inductance in the wire and component leads. A compromise must be made between the edge speeds and power device heat loss.

2.3 Motor Inductance Clamping

As has been discussed, all electrical components can produce inductance-generated kickback voltages. Motors, by nature, are highly inductive and when driven from an electronic power control stage will generate significant kickback voltages. In addition, some motors will inherently act as generators when they coast down, producing a counterelectromotive (CEMF) voltage or regeneration. Both inductive kickback and CEMF voltages must be contained for reliable operation by the power control electronics.

The motor's inductance value will be dominant over the hookup wiring and other electronic components. When the motor's voltage is switched off by the electronic power stage, the resultant high-voltage spike generated primarily by the motor's inductance must be dealt with, or the power electronics will be destroyed. (Specific design methods to tackle CEMF and regeneration will be covered in later chapters.)

The energy content of the motor's inductive kickback voltage is determined principally by the motor's current, the motor's internal and external wire inductance value, and the rate at which the current is switched off. This effect resembles that of the motor's connecting wires, which was reviewed in the previous section. The main difference is that the motor's inductance value is much larger than that of wires or components, and the motor's inductance will vary according to the position of the motor's rotor. To determine the motor's inductive kickback voltage spike, an inductance measurement of the motor's windings is required. When the motor's winding's inductance is known, the maximum kickback spike can be delivered. In most cases, the kickback voltage will greatly exceed the power transistor's voltage ratings that are used in the power stage.

If a rectifier is placed across the motor, as shown in Fig. 2.7, the motor's kickback voltage will be clamped to a level set by the recti-

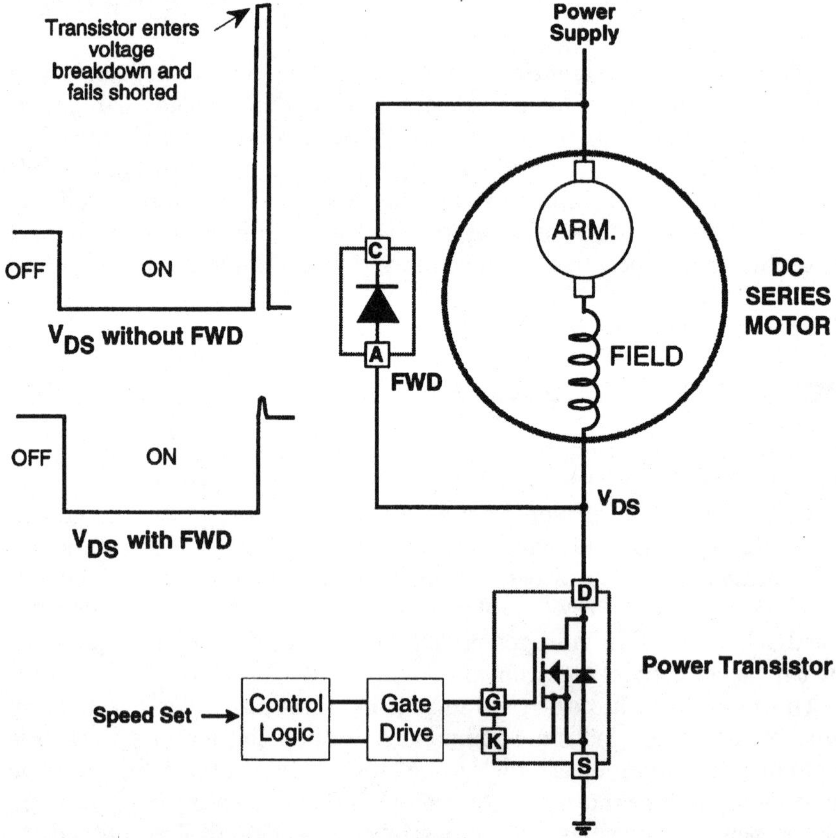

Figure 2.7 FWD clamps motor's kickback voltage.

fier's forward voltage drop, usually around one volt, and the power supply voltage bus. *It is important to know that the rectifier is usually called a free-wheeling-diode (FWD) and is used to clamp the motor's kickback voltage, as well as to steer the motor's current during normal PWM operation.*

Other protective devices are made specifically to handle voltage transients of a random nature. There are two main groups of transient-voltage suppressors (TVSs) available that can protect semiconductor devices: metal oxide varistors (MOVs) and zener diodes. The MOV devices offer low cost and high joule/size packaging. The zener devices offer fast transient response and fairly flat voltage clamping. Figure 2.8

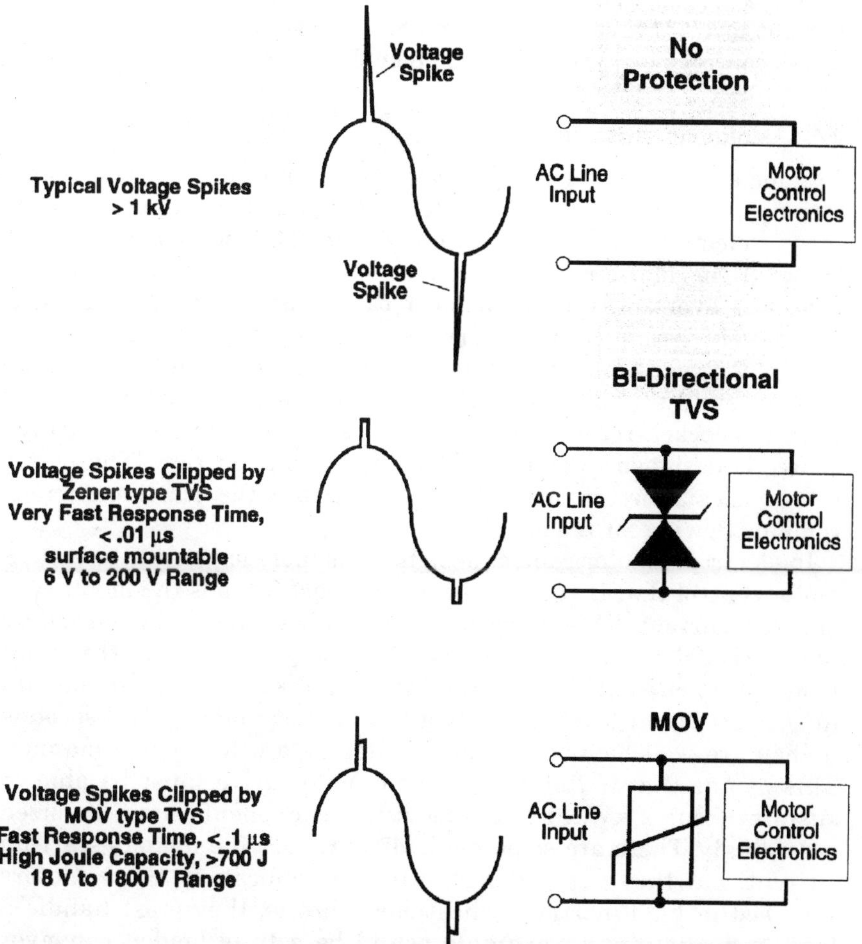

Figure 2.8 Transient suppressors clamp voltage spikes.

shows both types of devices in action. *For prototyping or experimental devices, it is a good idea to use TVSs across all power semiconductor devices and DC supply voltages, including the main power DC supply and the lower-voltage logic/MCU supplies.* It is also a good idea to employ TVS devices on 120-V AC power line supply connections, which can be a source of voltage transients of up to 6000 V, with 1000-V spikes being fairly common. Voltage spikes of these magnitudes will destroy electronic circuits.

2.4 Motor Current

Thus far, we have discussed excessive voltages caused by inductive kickback or external factors. In this section, we will consider motor current levels. Excessive motor current levels almost always lead to power control electronic failures and/or the destruction of the motor.

2.4.1 Motor current faults

In nonelectronic motor control systems, a motor overcurrent condition can be caused by many faults, such as a short-circuited winding, a defective bearing, or a jammed load. The motor may have an internal thermal circuit breaker or some form of circuit breaker that will eventually trip. In most cases, motors draw at least 3 times their normal running current during start-up, which may last for a few seconds until the motor's speed stabilizes. Motor manufacturers usually indicate the current draw that a motor will incur during a locked-rotor condition. In most cases, the motor's power source is designed to supply sufficient starting current. The power source uses a circuit breaker that will trip if the starting current remains flowing for too long.

In electronic motor control designs, it is more expensive to build a motor control that is rated to supply at least 3 times the motor normal run current. This is because electronic components, especially power transistors, must be sized to safely handle the threefold motor start current. Power transistors, unlike motors, can sustain huge overcurrent levels for only a few microseconds or milliseconds if they are well heatsunk before failing in a catastrophic manner. This means that a 750-W electronic motor drive must be able to sustain about 2 kW, and its electronic components must be sized accordingly. There are some methods to reduce the motor's start-up current, but they vary according to the application. Some motors are used in high-inertia applications—that is, they must handle a load immediately; an example would be a fully loaded conveyor belt. Other applications, such as fans, can start up with minimal power. If allowed to start at maximum speed, a large fan with a high-starting-torque motor may shear off the fan blades. In all cases, direct short circuits, as caused by a motor internal winding failure or incorrect wiring hookup, will immediately destroy the power electronics, unless some form of current detection and shutdown has been implemented.

There is an old adage among power semiconductor applications engineers: "The power transistor will always protect any fuse by blowing first." Recently, faster-responding fuses have been developed that may help, especially to keep the power transistor from exploding after it has failed. A fuse will generally not protect the power transistor from overcurrent levels. Fuses are intended primarily to prevent catastrophic electrical fires, which may occur if the wire's insulation ignites or other components overheat. A fuse should protect the power electronics and motor from catastrophic damage, but the electronics still require protection from momentary overcurrents, such as a stalled motor will produce, as well as protection from a short-circuited motor or wiring leads.

2.4.2 Calculating maximum motor current

Determining the maximum motor current levels is dependent on the motor's specifications, the motor's application, and the power supply. In the design of an electronic motor control, the unit's current rating should be valid for a wide environmental temperature range, −40 to + 85°C, for example, and at low or high power line voltages.

Motor current levels will vary with their ambient temperature. This is because the motor's copper windings exhibit a positive temperature coefficient of 0.00393 per degree Celsius (°C^{-1}). This means that the motor's internal resistance will decrease with a drop in temperature and increase with a temperature rise. For example, if a motor has a 5.0-Ω motor resistance at 25°C or normal room temperature, the motor's resistance will decrease to about 3.5 Ω at −50°C. The same motor's resistance will increase to 5.9 Ω at +75°C ambient temperature. In percentage terms, the motor's resistance decreased 30 percent at cold temperatures and increased 20 percent at hot temperatures. *The motor's resistance changes over temperature is significant when a locked-rotor condition occurs at −50°C. This means that the locked-rotor current draw can increase significantly, which will overload the electronic power stage.* Figure 2.9 shows a graph of a DC motor's resistance and locked-rotor power consumption over a wide ambient-temperature range.

Motors, in general, can sustain momentary locked-rotor or hard-starting conditions without failing for at least one second. Power elec-

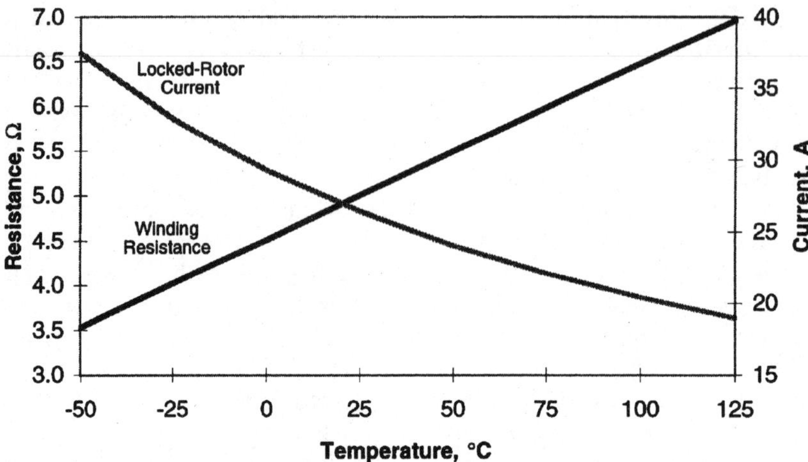

Figure 2.9 Ambient-temperature effects on motor operation.

tronics, on the other hand, cannot safely sustain 30 percent current levels over their maximum design values. When designing an electronic motor control, one must calculate the motor's maximum current. A good starting point for most motors is to measure the motor's terminal resistance and then apply the temperature coefficient for the windings for the coldest temperature in which the motor may be operated. (*Note:* Some motors use aluminum instead of copper windings; the temperature coefficient of aluminum, 0.0038, is slightly less than that of copper, 0.0039.)

A motor, for example, with a resistance of 5 Ω at 25°C that operates from 120 V will draw about 24 A on starting. The current is calculated using Ohm's law or $I = E/R$. The 24-A starting current value, however, is valid only during normal conditions. If the motor's ambient temperature is decreased to −50°C, and its power supply voltage rises to 132 V, then the motor's current will increase to 37 A, as calculated in Eq. (2.3). This 37-A value will serve as a reasonable design point for selecting the current ratings of the components used in the electronics power stage. The effect of the cold temperature on the motor's copper windings is significant and can be calculated [see Eq. (2.4)].

$$I_{\max} = \frac{E_{\max}}{R_{\min\ \text{mtr}}}$$

$$37 = \frac{132}{3.5} \tag{2.3}$$

where E_{max} is the maximum power supply voltage, 132 V, and $R_{min\ mtr}$ is the minimum motor resistance (−50°C). Then

$$R_{min\ mtr} = [(\Delta T * R_{coef}) * R_{mtr}] + R_{mtr}$$

$$3.5 = [(-75 * 0.00393) * 5] + 5 \tag{2.4}$$

2.4.3 Short-circuit protection

Selecting power transistor devices that can sustain the motor's maximum current levels will suffice for normal operating conditions in a perfect world. Unfortunately, haphazard events happen, such as the motor's leads short-circuiting (shorting) together, or the motor winding's insulation failing and shorting to the motor's frame. These events spell instant disaster for the power stage electronics unless some form of electronic current limiting is used; if the current is not limited, enormous current levels will transpire, quickly destroying the semiconductor power devices.

The short-circuit current levels are set mostly by the power supply resistance and the short-circuit resistance. For discussion purposes, short circuits will be considered to be less than 10 percent of the motor's nominal load resistance. A dead short circuit will be considered to be less than 1 percent of the nominal load resistance. In most motor controls, a dead short circuit means that hundreds to thousands of amperes may flow until the fuse or circuit breaker interrupts the flow. Few motor control power transistors are large enough to sustain these huge currents without failing.

The important point to remember about all this motor-current-level analysis is that the electronic power stage will require a fast current-sensing element and shutdown function. By measurement and tracking of the motor's current level, the motor control can be programmed to react to specific load conditions, in addition to protecting against catastrophic short-circuit failures. Two elements are required for electronic current protection: (1) a fast and accurate current sensing device and (2) a shutdown circuit.

2.4.4 Current sensing

A *current-sensing element* is any device used to indicate the amount of current flowing in a circuit. In a simple motor control, a series sense resistor can be employed. Higher-performance motor controls use more advanced current sensors such as current transducers. The main attributes of a current sensor for motor controls that you should consider are its response speed, accuracy, and cost of implementation.

Most power transistors can sustain short-circuit current levels for only about 10 μs. *A current sensor and shutdown function should respond in less than the 10 μs; one-half of that speed, or 5 μs, would be a good design point.* Several factors of a current sensor and shutdown function design are crucial:

- Adjustable or fixed-trip point-type current sensing.
- Adjustable or fixed speed delay from short-circuit event to actual shutdown.
- Latching or momentary-type shutdown—whether the motor stays off until a manual restart occurs, or automatically tries to restart. Some motor controls latch off under a shorted load condition but autorestart during an overcurrent condition.
- Nuisance tripping caused by RFI or EMI noise from the motor's power stage, external power supply, or external sources.
- Isolated or in-circuit current sensing.
- Resolution and type of output signal; analog or digital.

As can be seen, designing a current sensor is a serious effort, but it is a critical part of any electronic motor's control. Different types of current sensors and shutdown methods will be reviewed in the chapters on motor control designs.

Thus far, overvoltages and excessive currents in the electronic motor control have been reviewed. One other reliability issue involves excessive motor and power electronics temperatures.

2.5 Temperature Considerations

Electronic components are more sensitive to temperature extremes than are electrical parts such as switches or relays. Electronic motor controls, by nature, must manage the total energy used by the motor. Unfortunately, electronics motor controls are not 100 percent efficient and incur some power loss. The control's power loss is related to the PWM operating frequency, as discussed in Sec. 2.2.4. The power transistor and FWD specifications also affect power loss.

Semiconductor long-term reliability is directly affected by operating temperatures. *The maximum junction temperature is the critical design point for motor control power stages. Reliability research suggests that for each 10°C rise in junction temperature, the long-term reliability is decreased by about 50 percent.*

It has been the author's experience that standard silicon power transistors will almost always fail when their junction temperatures approach 230°C, even if the transistor's current and voltage levels are within the device's safe operating area (SOA). Heatsink and power semiconductor mounting or fastening are key components of reliability.

This issue is even more complex, because another failure mode in power semiconductors is thermal cycling, or when wide junction-to-ambient temperature swings occur at a periodic rate. An example of thermal cycling would include a motor that runs at full load for 5 min, is off for 10 min and then repeats this sequence every 15 min for 365 days, adding up to four power cycles per hour, 96 per day, or 35,040 per year. This can present significant reliability problems for power semiconductor devices.

Power semiconductor dissipation can be reduced by several methods, but usually compromises must be made, and certain parameters, such as the PWM operating frequency, are set by audible noise factors. The main compromise is between heatsink size or type and power semiconductor costs. If one power transistor costs twice as much but reduces the power loss by more than twice, heatsink costs must be considered. A condensed design review is presented below for those who want to grasp the basics of handling power dissipation in the motor electronics power stage.

2.5.1 Power dissipation basics

When the maximum motor current and ambient temperature extremes have been established, the power transistor device and heatsink specifications can be determined. A spreadsheet (Table 2.1) has been developed that can provide assistance. This list of equations can be directly entered into a spreadsheet on a desktop computer to perform the calculations. In this table, a 120-V-DC 5-Ω motor, a 55-A 200-V MOSFET, and a medium-sized heatsink are used to demonstrate how to calculate one of the critical design points, power transistor junction temperature, for a reliable electronic motor control. The junction temperature of the MOSFET device is calculated at both hot and cold ambient temperatures during a locked-rotor-mode operation. Note that the thermal resistance of each path (junction-to-case, case-to-heatsink, and heatsink-to-ambient) is used to calculate each temperature. In this example, the worst-case junction temperature would be about 117°C under a locked-rotor condition in high operating temperatures. Selecting a larger heatsink or power transistor can reduce junction temperature, which will increase long-term reliability.

TABLE 2.1 Motor Control Power Transistor Dissipation Calculations

Parameter	Symbol	Value	Notes
	Motor System Specifications		
Maximum supply voltage	V_{sup}	*120 V*	Maximum motor supply voltage
DC motor current, locked rotor	$I_{mtr,pk}$	*5.0 A*	Locked rotor motor current at 25°C
DC motor resistance	R_{mtr}	*5 Ω*	Resistance at 25°C
Maximum ambient temperature	$T_{a,hot}$	*85°C*	Maximum operating temperature
Minimum ambient temperature	$T_{a,cold}$	*−40°C*	Minimum operating temperature
Hot DC motor resistance	$R_{mtr,hot}$	6.2 Ω	$= \{[(T_{a,hot} - 25)*Cu_{\alpha}] * R_{mtr}\} + R_{mtr}$
Hot locked-rotor current	$I_{max,hot}$	**19 A**	$= V_{sup}/R_{mtr,hot}$
Cold DC motor resistance	$R_{mtr,cold}$	3.7 Ω	$= \{[(T_{a,cold} - 25)*Cu_{\alpha}]* R_{mtr}\} + R_{mtr}$
Cold locked-rotor current	$I_{max,cold}$	**32 A**	$= V_{sup}/R_{mtr,cold}$
	Power Stage Specifications		
Power transistor type	N-FET	*MTY55N20E*	Device part number
$R_{DS,on}$ at 25°C	$R_{DS,on}$	*0.028 Ω*	Per device data sheet
Junction-to-case resistance	$R_{øjc}$	*0.42°C/W*	Per device data sheet
	Heatsink Specifications		
Heatsink thermal resistance	$R_{øsa}$	*1.5°C/W*	Per heatsink data
Case-to-heatsink thermal resistance	$R_{øcs}$	*0.1°C/W*	Per mounting material data
	Hot Operation, Locked Rotor		
Estimated T_{jc}, hot	$\approx T_{jc}$	4°C	$= [(I_{max,hot}^2)*R_{DS,on}]*R_{øjc}$
Estimated $T_{j,max}$, hot	$\approx T_{j,max}$	105°C	$= [(I_{max,hot}^2)*R_{DS,on}]* R_{øsa} + (\approx T_{jc} + T_{a,hot})$
$R_{DS,on}$ temperature variation factor	$R_{DS,on,x}$	*1.52*	Per device data sheet
$R_{DS,on}$ at $T_{a,hot}$	$R_{DS,on,hot}$	0.043 Ω	$= R_{DS,on}*R_{DS,on,x}$
$V_{DS,on}$ voltage at $I_{max,hot}$	$V_{DS,on,hot}$	0.83 V	$= I_{max,hot}*R_{DS,on,hot}$
Power transistor dissipation	PD_{hot}	**16.1 W**	$= I_{max,hot}*V_{DS,on,hot}$

TABLE 2.1 (Continued) Motor Control Power Transistor Dissipation Calculations

Parameter	Symbol	Value	Notes
	Hot Operation, Locked Rotor		
Junction-to-case temperature rise	T_{jc}	7°C	$= R_{øjc} * P_{D,hot}$
Case-to-heatsink temperature rise	T_{cs}	1.6°C	$= R_{øcs} * P_{D,hot}$
Heatsink-to-ambient temperature rise	T_{ha}	24°C	$= R_{øsa} * P_{d,hot}$
Junction temperature, hot	$T_{j,max}$	**117°C**	$= (T_{jc} + T_{cs} + T_{ha}) + T_{a,hot}$
	Cold Operation, Locked Rotor		
Estimated T_{jc}, cold	$\approx T_{jc}$	12°C	$= [(I_{max,cold}{}^2) * R_{DS,on}] * R_{øjc}$
Estimated $T_{j,max}$, cold	$\approx T_{j,max}$	16°C	$= [(I_{max,cold}{}^2) * R_{DS,on}] * R_{øsa} + (\approx T_{jc} + T_{a,cold})$
$R_{DS,on}$ temperature variation factor	$R_{DS,on,x}$	*0.92*	Per device data sheet
$R_{DS,on}$ at $T_{a,cold}$	$R_{DS,on,cold}$	0.026 Ω	$= R_{DS,on} * R_{DS,on,x}$
$V_{DS,on}$ voltage at $I_{max,cold}$	$V_{DS,on,cold}$	0.83 V	$= I_{max,cold} * R_{DS,on,cold}$
Power transistor dissipation	$P_{D,cold}$	**26.8 W**	$= I_{max,cold} * V_{DS,on,cold}$
Junction-to-case temperature rise	T_{jc}	11°C	$= R_{øjc} * P_{D,cold}$
Case-to-heatsink temperature rise	T_{cs}	2.7°C	$= R_{øcs} * P_{D,cold}$
Heatsink-to-ambient temperature rise	T_{ha} °	40°C	$= R_{øsa} * P_{d,cold}$
Junction temperature, cold	$T_{j,max}$	**14°C**	$= (T_{jc} + T_{cs} + T_{ha}) + T_{a,cold}$

NOTES: *Italicized text* indicates design value entries; **boldface text** indicates critical results of calculations. Reference data: copper temperature coefficient, mean, $Cu_{\alpha} = 0.00393\ \alpha\ 1/°C$.

2.6 Power Semiconductor Selection

It should be emphasized that the maximum ratings given in the power semiconductor sheets are not to be used if you need a design with high reliability. The semiconductor maximum ratings are the levels at which the product is operating at 100 percent of its ability. Many power semiconductor application engineers suggest that the power semiconductor device should operate at only 50 percent of its maximum ratings in order to obtain reasonable long-term reliability. The ratings that are listed as typical in the products data sheet can also be used as a guide to establish the maximum values at which the motor design will operate.

Another way to select the most reliable power semiconductor is by determining the heat dissipation that a particular device will exhibit in the worst-case motor operation. Minimizing the power semiconductor's power dissipation lowers its junction temperature and also lowers thermal cycling temperature swings. The lowest power dissipation device will almost always be the most expensive, but it will reduce heatsink size and offer good long-term reliability.

Selecting a fast-switching power semiconductor also helps reduce switching loss, which, in turn, reduces junction temperature. This has a point of diminishing return on power semiconductor switching speed, since the stray inductance in the power stages connections will set an upper limit on switching speeds. At the time of this book's publication, power devices with switching speeds of 100 to 500 ns are readily available. Many electronic motor power stages employ IGBT or MOSFET devices that switch in the 300- to 500-ns range, with some approaching 1 μs. It is fairly easy to slow down the switching speeds of a IGBT or MOSFET, but it is seldom possible to switch the devices faster than their data sheet specifications indicate for typical switching speeds. Be aware that semiconductor manufacturers tend to be conservative in determining their product's electrical parameters, especially dynamic parameters such as switching speeds. Therefore, the slowest switching speeds may, in actuality, rarely be seen in the products; the typical value shown in the product data sheet is usually the average. A circuit designer should pay close attention to this practice, but also be aware that sometimes semiconductor manufacturers do have wafer lots that are just on the edge of meeting process tolerances, and slow parts may occur and may be shipped.

IGBT and MOSFET power devices are used for the power stage control in the design examples throughout this book. Older power semiconductor technologies, such as bipolar junction transistors (BJTs) and silicon controlled rectifiers (SCRs), have been used in lower-frequency motor controls.

2.7 Motor Control Topologies

Several types of motor electronic control topologies are possible, and new ones seem to appear every few months. Different motor types can often be driven from the same power electronic designs. The main differences occur in the microcomputer, software design, and feedback elements. There are three main elements or blocks in an electronic motor design: the motor and its power electronics; the PWM control element or microcomputer; and the sensors used to measure speed, current, temperature, user commands, and so forth. A general description of each of these elements is given below.

2.7.1 Basic motor power stages

The most prevalent control method for DC motors is a single low-side switch with a PWM drive signal. *Low-side switching* refers to the placement of the active power semiconductor switching device between the motor and common, as shown in Fig. 2.10. Low-side switching is preferred for ease of driving the power transistor. The IGBTs and MOSFETs most commonly used for motor controls require

Low-Side Switching (Low Voltage)

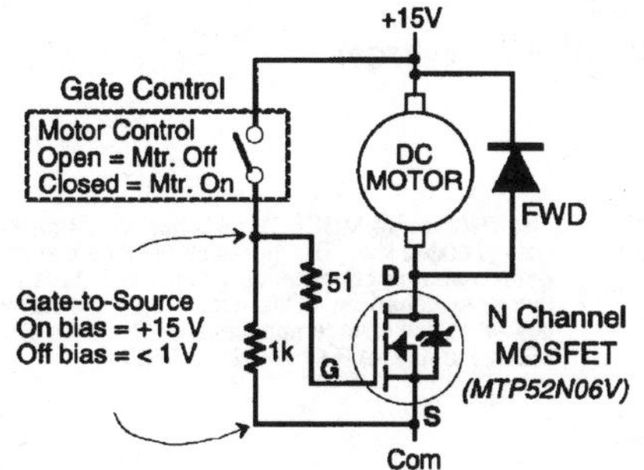

The N-Channel MOSFET switches on when its gate is pulled high. The 1k resistor discharges the gate-to-source capacitance when the gate is no longer switched on. A 51 ohm series gate resistor limits the MOSFET's switching speed. This circuit is simple but is limited to low voltage applications.

Low-Side Switching (Medium Voltage)

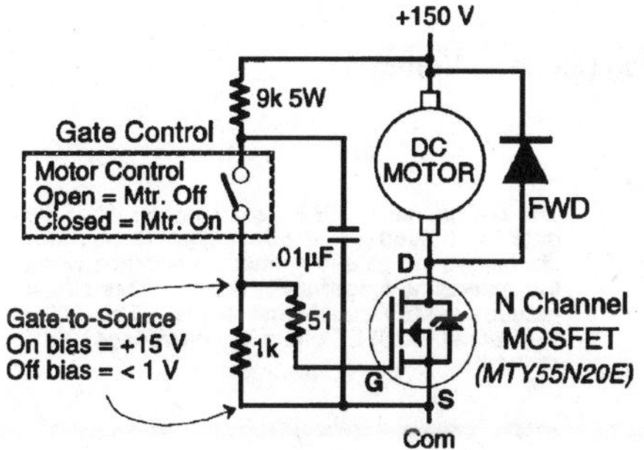

The N-Channel MOSFET switches on when its gate is biased at +15V. The 1k resistor discharges the gate-to-emitter capacitance when the gate is no longer switched on. The +15V supply is obtained with a resistor divider from the 150V bus. Note the high power loss in the 9k voltage dividing resistor. A capacitor is required to assist in charging the MOSFET's gate input capacitance. This circuit shows the disadvantages of using a simple resistor voltage divider to obtain the +15V supply.

Low-Side Switching (High Voltage)

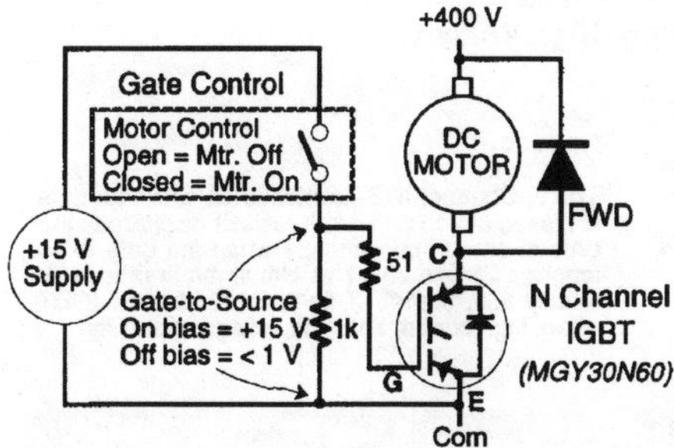

The N-Channel IGBT switches on when its gate is biased at +15V. The 1k resistor discharges the gate-to-emitter capacitance when the gate is no longer switched on. This circuit requires a separate +15V supply. This 600V N-Channel IGBT offers high current and high voltage capability.

Figure 2.10 Low-side gate drive for DC motor control.

a positive gate signal with respect to the IGBT or MOSFET emitter or source line. A low-side switch topology is easier to drive than if the device is connected between the power source's positive lead and the motor, which is called *high-side switching.*

High-side switching is appropriate for applications in which it is advantageous to remove the power supply's positive voltage from the load. Providing the correct gate voltage to a high-side switch transistor is usually difficult because the gate voltage must float above the main power supply voltage, as shown in Fig. 2.11.

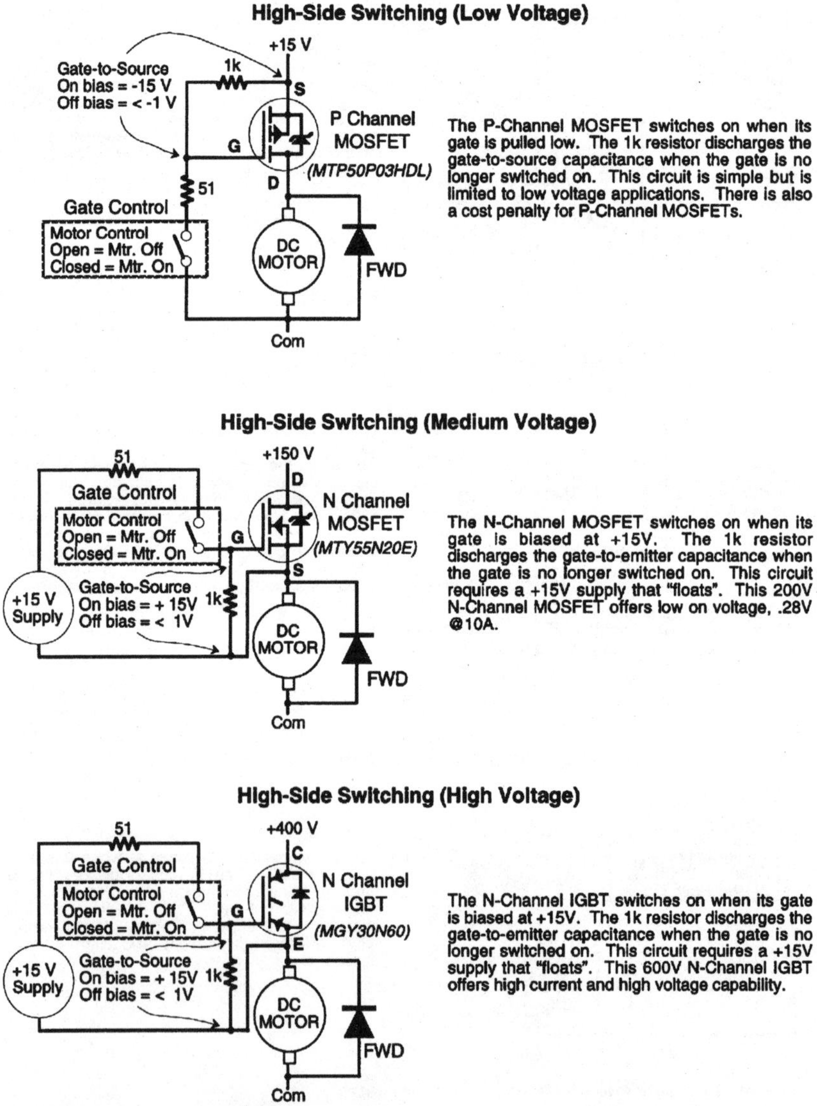
High-Side Switching (Low Voltage)
+15 V
Gate-to-Source
On bias = -15 V
Off bias = < -1 V
1k
S
G
D
P Channel
MOSFET
(MTP50P03HDL)
51
Gate Control
Motor Control
Open = Mtr. Off
Closed = Mtr. On
DC
MOTOR
FWD
Com
The P-Channel MOSFET switches on when its gate is pulled low. The 1k resistor discharges the gate-to-source capacitance when the gate is no longer switched on. This circuit is simple but is limited to low voltage applications. There is also a cost penalty for P-Channel MOSFETs.
High-Side Switching (Medium Voltage)
+150 V
51
D
Gate Control
N Channel
MOSFET
(MTY55N20E)
Motor Control
Open = Mtr. Off
Closed = Mtr. On
G
S
+15 V
Supply
Gate-to-Source
On bias = + 15V
Off bias = < 1V
1k
DC
MOTOR
FWD
Com
The N-Channel MOSFET switches on when its gate is biased at +15V. The 1k resistor discharges the gate-to-emitter capacitance when the gate is no longer switched on. This circuit requires a +15V supply that "floats". This 200V N-Channel MOSFET offers low on voltage, .28V @10A.
High-Side Switching (High Voltage)
+400 V
51
C
Gate Control
N Channel
IGBT
(MGY30N60)
Motor Control
Open = Mtr. Off
Closed = Mtr. On
G
E
+15 V
Supply
Gate-to-Source
On bias = + 15V
Off bias = < 1V
1k
DC
MOTOR
FWD
Com
The N-Channel IGBT switches on when its gate is biased at +15V. The 1k resistor discharges the gate-to-emitter capacitance when the gate is no longer switched on. This circuit requires a +15V supply that "floats". This 600V N-Channel IGBT offers high current and high voltage capability.

Figure 2.11 High-side gate drive for DC motor control.

The gate drive circuits thus far have been rudimentary and are meant to show their basic elements. More efficient gate driver designs are discussed in the chapters on motor control applications.

2.7.2 H-bridge basics

If the DC motor application requires bidirectional or fast braking control, a more complex power stage is needed. The H-bridge topology resembles the letter H and allows the power electronics a high degree of control over the motor. Each of the four power switches in the H-bridge is driven to accomplish a specific motor response: forward, reverse, brake, regeneration, and coastoff. The H-bridge also allows full four-quadrant motor control. Each quadrant refers to the motor's state of operation. The four possible states are forward motor, reverse motor, forward generation, and reverse generation. The motor's direction is controlled by the logic of the control signals, as shown in Fig. 2.12.

2.7.3 Half H-bridge basics

A variation of the H-bridge, a half H-bridge, is the standard power stage topology for AC and brushless motors. Three half H-bridge stages are used for three-phase (3ø) motor controls as shown in Fig. 2.13. As in the previous DC motor H-bridge design, the logic that drives the half H-bridges allows full control of the motor. Note that the logic is more complex and does not show the effects of microstepping. There are some motors, such as switched reluctance (SR) motors, that require a single power stage for each winding end and a separate FWD.

2.7.4 H-bridge problems

One fundamental danger of the half or full H-bridge topology is that a direct short circuit can occur if the top and bottom switches are turned on at the same time. As observed in the previous logic tables, the top and bottom stages of one-half of the H-bridge are never on at the same time unless activated by an out-of-control MCU program. Another failure mode occurs when one power switch has failed shorted and its other half is turned on. The problem of a short-circuited power switch is more difficult to solve because the current-limiting function relies on the power switches to turn off the motor. If just one of the power switches has failed, shutting off the remaining switches will work. If, however, both top and bottom switches have failed shorted, the current shutdown function has no effect, and a catastrophic event will occur whose magnitude will be set by the fuse or circuit breaker reaction times.

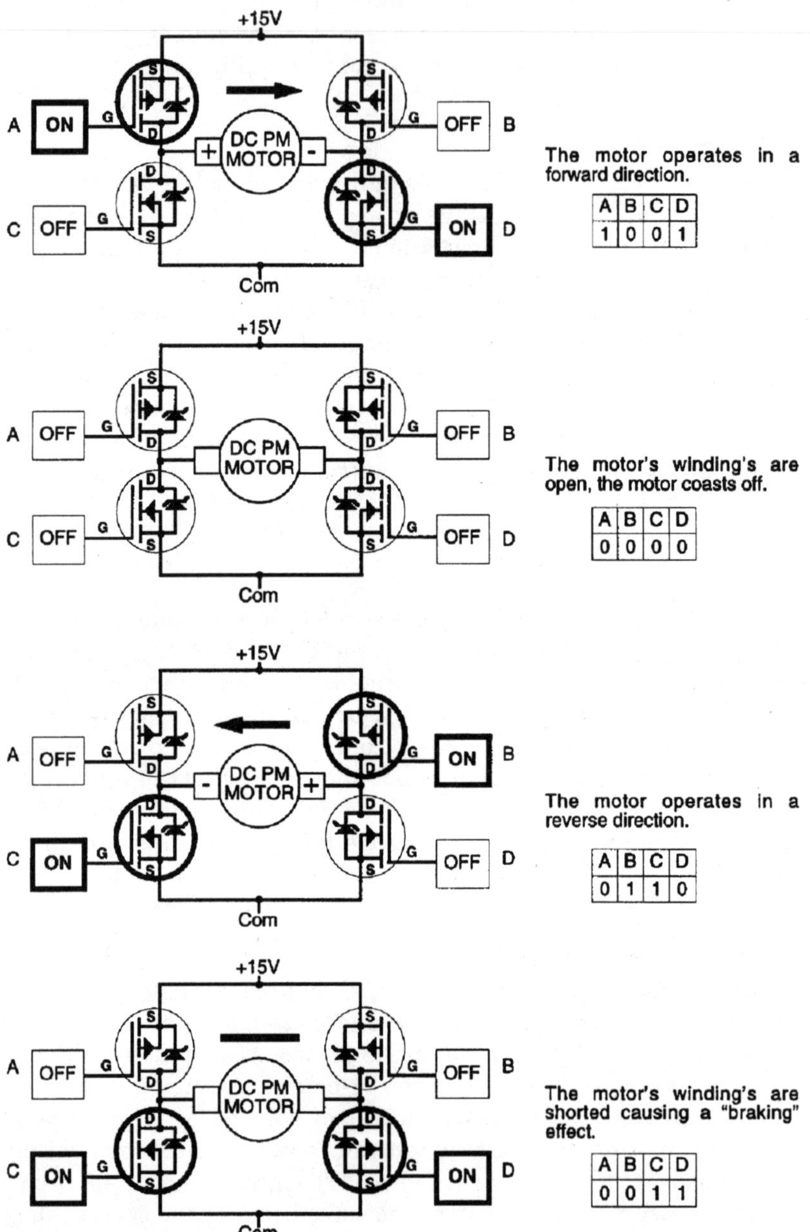

Figure 2.12 H-bridge allows four-quadrant control.

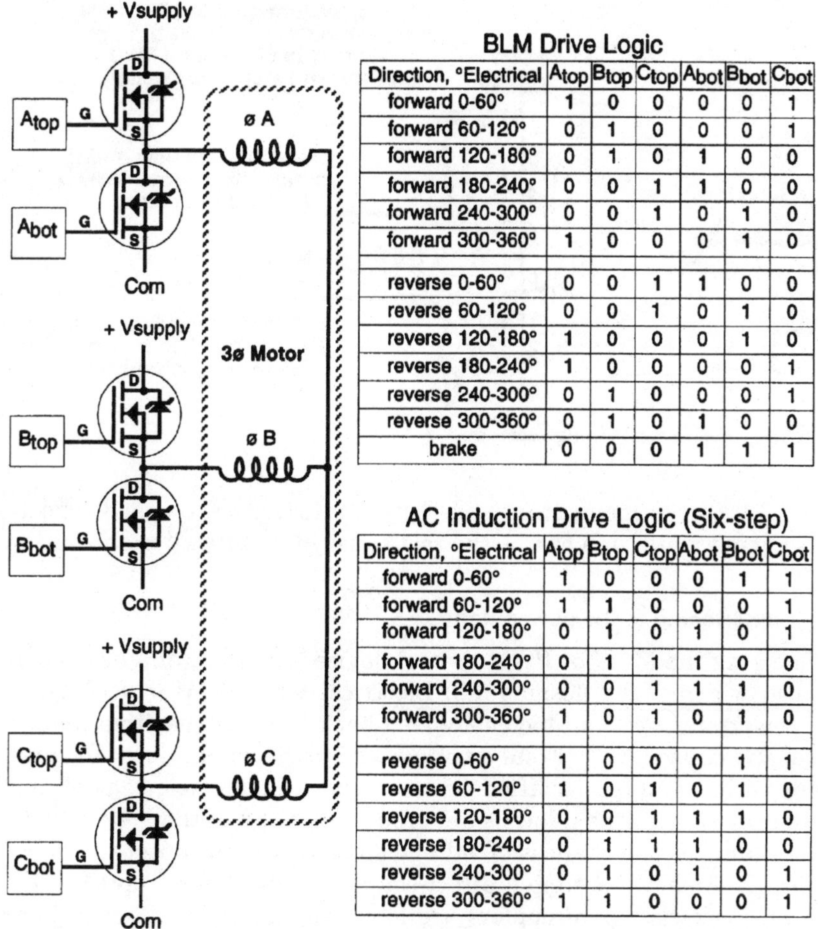

BLM Drive Logic

Direction, °Electrical	A_{top}	B_{top}	C_{top}	A_{bot}	B_{bot}	C_{bot}
forward 0-60°	1	0	0	0	0	1
forward 60-120°	0	1	0	0	0	1
forward 120-180°	0	1	0	1	0	0
forward 180-240°	0	0	1	1	0	0
forward 240-300°	0	0	1	0	1	0
forward 300-360°	1	0	0	0	1	0
reverse 0-60°	0	0	1	1	0	0
reverse 60-120°	0	0	1	0	1	0
reverse 120-180°	1	0	0	0	1	0
reverse 180-240°	1	0	0	0	0	1
reverse 240-300°	0	1	0	0	0	1
reverse 300-360°	0	1	0	1	0	0
brake	0	0	0	1	1	1

AC Induction Drive Logic (Six-step)

Direction, °Electrical	A_{top}	B_{top}	C_{top}	A_{bot}	B_{bot}	C_{bot}
forward 0-60°	1	0	0	0	1	1
forward 60-120°	1	1	0	0	0	1
forward 120-180°	0	1	0	1	0	1
forward 180-240°	0	1	1	1	0	0
forward 240-300°	0	0	1	1	1	0
forward 300-360°	1	0	1	0	1	0
reverse 0-60°	1	0	0	0	1	1
reverse 60-120°	1	0	1	0	1	0
reverse 120-180°	0	0	1	1	1	0
reverse 180-240°	0	1	1	1	0	0
reverse 240-300°	0	1	0	1	0	1
reverse 300-360°	1	1	0	0	0	1

Figure 2.13 Half H-bridge motor power stage.

2.7.5 H-bridge and FWD problems

Besides the problem of top/bottom switches turned fully on, another difficulty occurs with free-wheeling diodes. Diodes or rectifiers are devices that block DC voltage in one direction and conduct in the opposite direction. When the FWD is in the forward conduction mode (anode positive with respect to cathode), and the voltage is reversed, the FWD does not instantaneously stop conduction and revert to its blocking state. This is called *reverse recovery time* in semiconductor terminology. The FWD's reverse recovery time is affected by the semiconductor technology and manufacturing processes. During the FWD's recovery time, a momentary short circuit occurs, as seen in

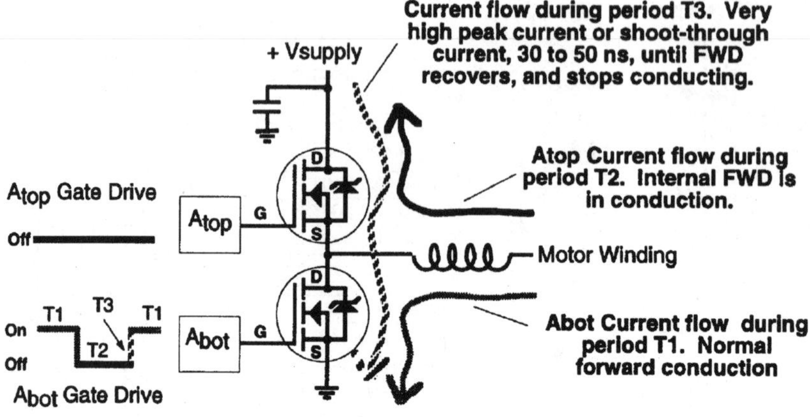

Figure 2.14 FWD shoot-through currents.

Fig. 2.14. These momentary short-circuit currents are called "shoot-through currents" and can be the root cause of additional problems.

2.7.6 Interfacing to the power stage

Connecting the MCU or PWM control device to the power stage is one area in which many designs fail. Electrical noise (EMI or RFI) from the power stage can all too easily feed back into the sensitive control stage and cause havoc. From a voltage safety viewpoint, it is advantageous to use voltage isolation between the user or logic control and the power stage. This allows safer troubleshooting of the logic control and minimizes the chances of voltage spikes that will upset the logic. Figure 2.15 shows a simple direct-coupled design and a optocoupler design. Pricewise, optocouplers are low-cost insurance. Even in low-voltage battery-powered motor equipment, an optocoupled interface will protect the MCU or PWM control element from catastrophic power stage failures.

Just 10 pF of capacitance across the interface element's input-to-output can cause problems in high-voltage motor control systems. This is because the top switch driver element is switching between common and the full supply voltage at a fast transitional rate, usually between 0.2 and 1 μs. The feedback current spike's magnitude can be calculated [Eq. (2.5)]. (*Note:* This is another one of those equations that should be part of a motor control engineer's library.) In this example, the motor's supply voltage is set at 400 V and the PWM transition speeds are 0.2 μs. The capacitance across the optocoupler or other type of isolater device (pulse transformer, for example) is only 10 pF. *A peak current of 0.02 A is more than enough to disturb many low-level PWM logic devices and can affect the optocoupler's*

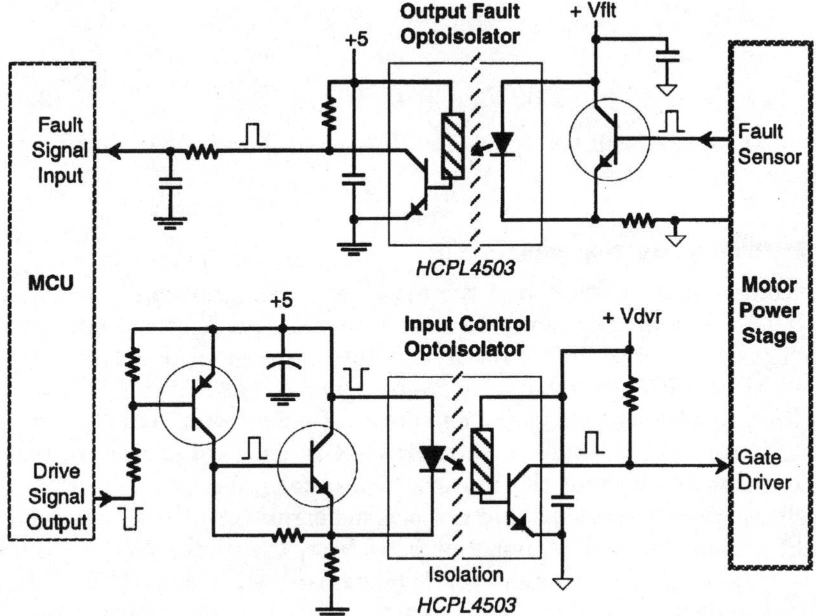

Figure 2.15 Interfacing to the power stage.

operation. Selecting an interface element requires careful consideration to this problem.

$$I_{pk} = C * \frac{E}{T}$$

$$0.02 = 0.00001 * \frac{400}{0.2} \tag{2.5}$$

where I_{pk} = capacitor peak current through interface element, A
C = capacitance, μF
E = motor's power supply voltage
T = transition time, μs

Many optocoupler devices are rated to withstand high voltage and switching speed. This is called a *Dv/Dt* rating and is used in electronic power control terminology. The *Dv* refers to the voltage delta or value of voltage change, while *Dt* refers to the delta of time period in which the voltage changes occur. *Dv/Dt* is commonly expressed in volts or kilovolts per microsecond. In the previous example, where 400 V was switched in 0.2 μs, this would be equal to 2000 V/μs *Dv/Dt* rating per Eq. (2.6). Therefore, if an interface device is rated at 15,000 V/μs, it should more than suffice.

$$\frac{Dv}{Dt} = \frac{1}{Dt} \times Dv$$

$$2000\ \text{V}/\mu\text{s} = 1/0.2 * 400 \tag{2.6}$$

where Dv is the peak voltage swing, V, and Dt is the time duration of voltage swing, μs.

2.7.7 PWM control elements

The control and generation of the motor's PWM signal can be accomplished with many types of products. A dual CMOS flip-flop logic gate can be configured as an oscillator and duty cycle controller. There are several PWM ICs available that are designed to generate a PWM signal and, in some cases, are made specifically for motor control applications. Any PWM device or circuit that is used for motor control must be stable during the motor's worst-case operation. In other words, the PWM signal should not become erratic or jittery because a slight amount of ground noise or EMI from the motor power stage causes the PWM device to stutter. When choosing a stable PWM part, study the device's product sheet to see if it lists motor control as the part's intended application. CMOS devices draw little power but tend to be sensitive to EMI, whereas bipolar linear integrated circuit (ICs) consume more energy but usually offer EMI tolerance.

Numerous motor control PWM IC devices (see App. C) are available from several semiconductor manufacturers. Some motor control IC devices contain internal power transistor elements that can drive small motors directly. Large motors, as will be shown in later chapters, require the use of external power transistor stages. There are also a few microcontroller products that offer powerful or dedicated timers for PWM motor control.

2.7.8 User interface elements

Most electronic motor controls require some method for the user or operator to issue commands to the PWM control unit. This can be a simple forward/off/reverse switch and potentiometer for speed adjustments. Digital PWM controls work more efficiently with on/off-type signals, so a keypad-type user interface is preferred rather than potentiometers (pot.) that require an analog-to-digital (A/D) converter. However, many MCUs have built-in A/D converters that can be utilized for pot.-type controls. Potentiometer controls are generally less than one-turn-type units whose repeatability and resolution may not be adequate. For example, if an exact speed of 525 rpm is required in a speed control that has a range of 10 to 2500 rpm, adjusting a pot. to this value would be difficult. Think of it this way: Divide

300°, which is a typical potentiometer's turning radius into 2490 points, which is the required speed resolution. The best resolution would be in the 5 percent range, which is about 125 rpm; adjusting the pot. to 525 rpm will be difficult, especially if it must be reset from another speed back to 525 rpm. Entering the numbers 525 with a keypad solves this problem. The issues of precision, resolution, and MCU bit size are interrelated.

2.7.9 MCU precision and resolution

Since it will directly affect the motor's performance, it is useful to understand how to figure the precision of an MCU. There are two main factors to think about when determining an MCU device performance: (1) how precisely it can compute a number and (2) how fast it can perform calculations with that number. *The MCU's precision is set by its bit number, and the speed of computations is governed mostly by its internal-clock bus rate.* External oscillator frequencies do not always indicate the actual processing speed of the MCU. There are other criteria that also are important, such as how fast the MCU can read an input signal, make its computations, and output the correct response. In closed-loop motor control applications, this is a critical parameter. A powerful timer unit architecture inside the MCU is required for motor control PWM generation.

2.8 Motor Feedback Control

There are several methods to maintain a constant motor speed under varying load conditions. Most of the methods require a sensor that measures the motor's speed; the motor's actual speed is then compared to the desired speed, and a correction signal is generated, if necessary. A microcontroller can be programmed to measure the motor's speed and decide the direction in which it is drifting. A proportional-integral (PI) routine produces one output signal—"p"—that is proportional to the difference between the desired speed and the actual motor speed (this is the error value) by a proportional gain block labeled "Kp." Another signal—"i"—is generated that ramps up or down at a rate set by the error signal magnitude. The gains of both "Ki" and "Kp" are chosen to provide a quick response with little instability. The PI system adds up the error rate over time, and, if an underspeed condition accumulates, the error signal will begin to increase in order to compensate. Under normal, steady-load conditions the integral control block "ki" will tend toward zero because there is less error over time. The motor's load, inertia, and torque performance are the factors that determine the PI gain constants. In

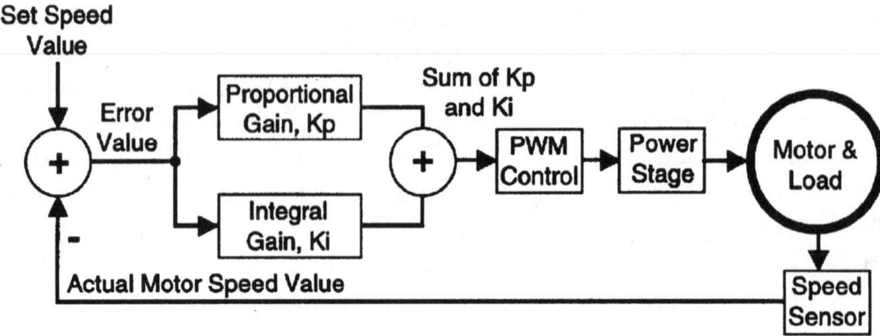

Figure 2.16 Speed control using PI method.

summary, the PI method allows fast response to abrupt load changes and stable operation under light loads. Figure 2.16 shows a traditional PI speed control diagram.

Fuzzy logic is another method that can be used for constant speed control. The design of a fuzzy program is somewhat easier to implement than normal programming. The fuzzy speed error calculation uses simple linguistics, such as "IF speed difference negative and small THEN increase PWM slightly."

The motor's voltage is then raised by slightly expanding the pulse width modulation duty cycle, which increases the motor's speed. The motor's speed update rate, motor torque performance, and other user input parameters can also be added to the fuzzy program. Figure 2.17 shows one part of a fuzzy logic design used for computing motor speed. Another program design requirement includes the need to verify that the input signals fall within expected boundaries. A broken or intermittent speed sensor, for example, could be detected. Another program can test the motor's speed to the user's speed setting to prevent unsafe acceleration under certain conditions. For example, if a high-torque motor were initially running at a low speed and the user changed its speed setting to 10 times faster, the motor might accelerate too rapidly, damaging the load. A fuzzy design can limit the acceleration, using simple rules, such as "IF new speed setting and big speed error THEN increase PWM slightly."

Adaptive programming can be used to change the motor control system to match unique needs. This is done by varying the response time and/or gain of the motor control design. Some applications, for example, may require the motor to slow down its response time when handling an increased load and then gradually return to its set-speed value; other applications may require a constant speed at all times. In some cases, it may be preferable to bring the motor to a stop when an

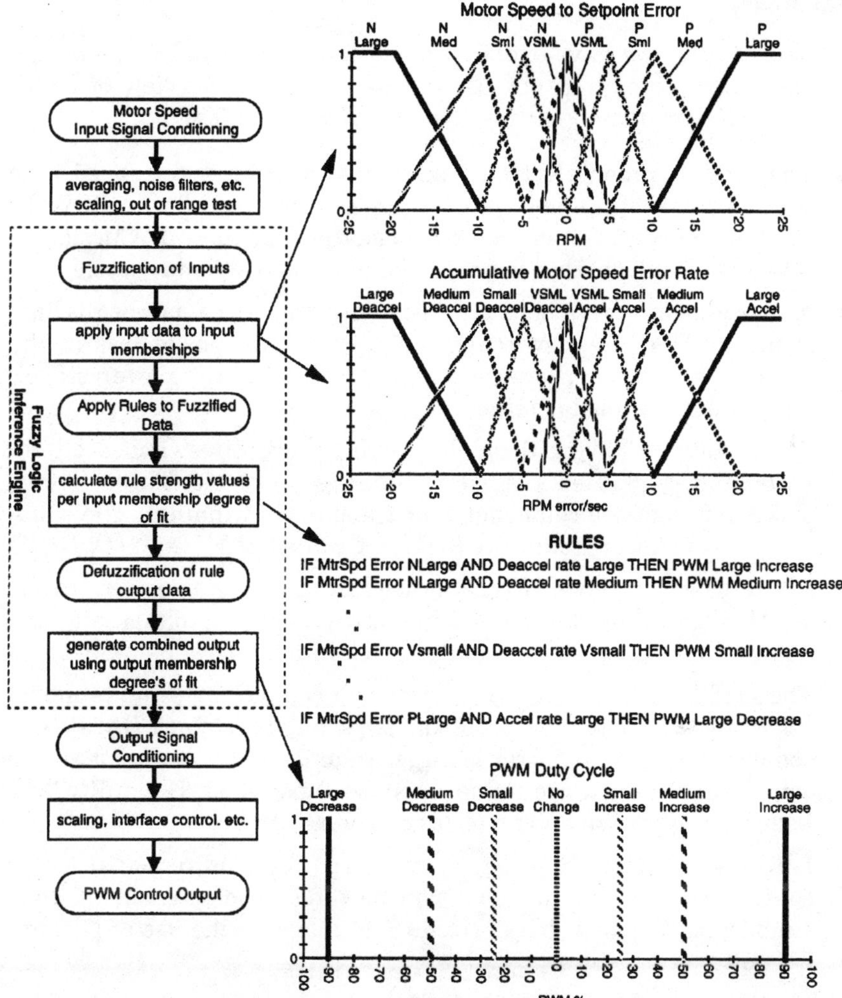

Figure 2.17 Speed control using fuzzy method.

abnormal speed correction is detected, which might happen if the motor's load were jammed or excessively loaded down. The motor drive system can be designed to adapt to the user's needs by adding a response selection switch (slow, medium, fast) or by analyzing the motor's acceleration/deceleration rates during normal operation. When these rates are analyzed, they can be stored as reference points.

Summary

In this chapter we have explored many factors that affect the design of electronic motor controls, as well as a general overview of motor controls. Key points about this chapter are listed below.

- The most common electronic method to vary a motor speed is to change the motor's voltages by pulse-width modulation (PWM). The motor's internal inductance smoothes out these PWM signals into a voltage level that is proportional to the PWM duty cycle.
- A relationship exists between fast-switching edges, power dissipation, and EMI. Fast-switching edges minimize power loss in the power stage but can create significant EMI if the power stage's stray inductance is too large.
- When current flowing through a wire is abruptly switched off, an inductance-generated kickback voltage is generated. These voltage spikes can damage components or the motor's windings. The equation for calculating inductive kickback voltage is $V_{pk} = L*(Di/Dt)$.
- A free-wheeling diode (FWD) is used in most motor power stages. The FWD clamps the motor's inductive kickback voltage and can allow regeneration current to flow back into the power supply bus.
- The lead length of a wire or an internal component exhibits inductance and can become a problem, especially at high *Di/Dt* rates. If the length of the conductor is larger than its width or diameter, the conductor's inductance value must be considered. Generally, any length greater than 25 mm (1 in) can create problems.
- Low-side switching topology is preferred for unidirectional motor control because the gate bias can be referenced to common, thus simplifying the gate driver circuit and allowing the use of N-channel power transistors.
- If both top and bottom power transistors in a full or half H-bridge are activated at the same time, a direct short circuit occurs, and the power transistors will be destroyed.
- The four most important design criteria for obtaining the highest reliability in a motor's power electronic stage are (1) the reduction of the power semiconductor's junction temperature will always pay dividends for long-term reliability; (2) reduced junction temperatures will minimize temperature variations, which will help lower the mechanical stress levels in the power semiconductor materials; (3) transient-voltage suppressors placed in critical points in the motor's electronic circuits ensure that the motor electronics will survive when voltage spikes occur; and (4) current limiting should

always be used in electronic motor controls to protect against stalled or shorted motor conditions.

- Half or full H-bridge power stages exhibit noise spikes in the power supply bus and common leads because of two events that occur during the normal operation of the bridge: (1) when a power transistor is turning on while the FWD is in conduction, the reverse recovery time of the free-wheeling diode allows a momentary shoot-through current; and (2) the motor current transitions and stray inductance in the power transistor modules, leads, and motor also cause electrical noise spikes.
- A 40-MHz 8-bit microcontroller will not be more precise than a 20-MHz 16-bit microcontroller. The 8-bit MCU has a bit resolution of only 1 in 256, whereas the 16-bit MCU has a bit resolution of 1 in 65,536. A higher-speed MCU of the same bit size offers faster computational time and throughput. The timer resolution is also affected by clock frequency. A faster clock will usually increase the number of counts the timer generates in a second, thereby giving it a finer resolution.

Further Reading

Fitzgerald, A. E., Kingsley, C., and Umans, S., *Electric Machinery,* 5th ed., McGraw-Hill, 1990, p. 457.

Jouanne, A., "Application Issues for PWM Adjustable Speed in AC Motor Drives," *IEEE Industry Applications Magazine,* Sept./Oct. 1996, vol. 2, no. 5, pp. 10–18.

Kaufman, M., and Seidman, A., *Handbook of Electronics Calculations,* 2d ed., McGraw-Hill, 1988, pp. 4–13.

Motor books (partial listing)

Anderson, Edwin P., and Miller, Rex, *Electric Motors,* 5th ed., Macmillan, 1991.

Baker, Robert C., *Linear Electronics in Control Systems,* Barks Publications, 1988.

Fitzgerald, A. E., Kingsley, Charles, Jr., and Umans, Stephen D., *Electric Machinery,* 5th ed., McGraw-Hill, 1990.

Gottlieb, Irving, *Electric Motors and Control Techniques,* 2d ed., Tab Books, 1994.

Kenjo, Tak, *Electric Motors and Their Controls,* Oxford Science Publications, 1994.

Nailen, Richard L., *Managing Motors,* Barks Publications, 1991.

Rosenburg, Robert, and Hand, August, *Electric Motor Repair,* 3d ed., Harcourt Brace Jovanovich College Publishers, 1988.

Chapter

3

DC Motor Control Designs

Richard J. Valentine

Motorola Semiconductor Products

Introduction

Direct-current permanent magnet (PM) motors provide high performance while maintaining low cost and are produced by the millions in automotive equipment, consumer electronics, power tools, toys, and appliances. Electronic DC motor control designs described in this chapter are divided into two main groups. The first group consists of standard PM DC motors with brush/commutator assemblies that can operate from a direct current power supply and do not necessarily utilize control electronics. The second group consists of stepper motors, servo motors, brushless DC motors (BLM), and other motor types that are powered from a DC supply and require electronics to function.

The control circuitry for the first group of DC brushed motors can vary from simple electrical switches to electronic speed controls. Examples include an on/off switch (vacuum cleaner), a reversal switch for direction control (automobile power windows), and speed or torque control electronics (tool motors). The availability of low-cost power electronic devices and associated integrated circuits allows many types of motor control designs for DC brushed motors. Most DC motors (<1 kW) use a permanent magnet structure instead of a wound field for higher efficiency. PM motors exhibit high-starting torque, and the motor's speed yields an inverse relationship to the load's torque requirements; loading down the PM motor causes the speed to drop but the torque to increase. The speed of a PM motor is directly related to the applied voltage, and, even at zero load, the motor's speed is usually limited to a safe value. Two drawbacks of the PM motor for speed control implementation include the requirement that the entire motor current be modulated or switched and the possibility that permanent magnets may be degraded under certain abnormal operating conditions. (*For example, if the armature power level is allowed to create excessive magnetic flux fields, the magnets may be irreversibly weakened, especially when the motor's operating temperature range is also exceeded.*)

The second group consists of motors that do not use brush/commutator assemblies and require power transistors to commutate their windings, such as brushless DC motors and steppers. BLM are commonly found in fan applications where the fan's speed can be easily controlled by pulse-width modulation. Stepper motors are used when sequential mechanical control is required, as in computer printers, disk drives, or copy machines. Stepper motor systems tend to be digitally based.

Servo motor systems may employ either brushed PM or brushless DC motors and are typically used in positioning systems with the motor's shaft directly or indirectly connected to a potentiometer. This provides a voltage level proportional to the motor's shaft position. Servo control systems, historically, have used analog-type control electronics.

Choosing the best or most optimum motor control semiconductor devices can be a challenge because of the numerous products available. There are two fundamental choices—use off-the-shelf motor control analog integrated circuits or use microcontrollers. This chapter reviews the employment of off-the-shelf motor control analog ICs and notes some of their advantages and inherent limitations. (Motor-control-specific MCUs for DC motor applications are covered in later chapters.)

3.1 Brushed DC Motor Control Design

Operating from low voltages and using standard parts, the first DC motor speed control design described below would be a good circuit to build in order to understand basic DC motor controls. We will examine each component of the design. More complex DC motor circuits will also be discussed in this chapter.

3.1.1 DC motor speed control

The DC motor control circuit shown in Fig. 3.1 incorporates the three main elements found in most electronic motor speed controls. These elements—user inputs and/or motor feedback signals, logic control (which includes a PWM generator), and a power stage—will vary in complexity and size but form the basic motor control system architecture. More complex motor control systems may use additional elements to sense several of the motor's attributes, such as average current, speed, shaft position, winding temperatures, phase current, phase voltage, and vibration. The DC motor circuit uses a PWM motor control IC, an MC33033, chosen to allow easy transition to a brushless motor test design that is covered later in this chapter. A MC33035 could also be used if a brake input and fault output function are required. (Note that the MC33035 is a 24-pin device, while the MC33033 is a 20-pin package, both available from Motorola, Inc. There are also similar BLM control ICs available from other companies—see App. C.)

3.1.1.1 User input. The test circuit uses three user input controls: speed adjust potentiometer, FORWARD/REVERSE switch, and an ENABLE switch. The speed adjust pot. forms a voltage divider from the MC33033's 6.2-V reference supply. This ensures that the pot.'s refer-

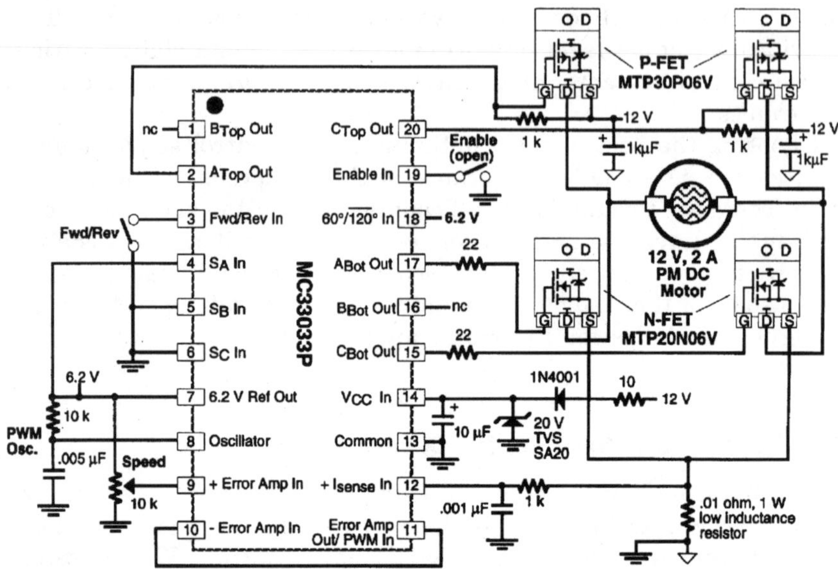

Figure 3.1 DC brush motor speed control test circuit.

ence voltage is stable. If, for some reason, the 6.2-V reference supply or the V_{cc} sags, the MC33033's built-in undervoltage detector kicks into action, shutting down the output drivers, which then shuts off the power stage and motor. The undervoltage lockout prevents unstable operation of the IC during power supply fluctuations and normal power ON/OFF states. If the supply voltage to the IC drops below 8.9 V or the reference drops to less than 4.5 V, the drive outputs are switched off—that is, the top open-collector drivers go open, allowing the P-FET top transistors to switch off, and the bottom drive outputs go into a low stage, which switches off the N-FET bottom transistors.

The pot.'s adjustable output voltage in this design can vary from 6.2 V to zero, thereby changing the PWM duty cycle. Limiting the speed range is possible by adding series resistors between the pot.'s connections to the 6.2-V reference and common, as shown in Fig. 3.2. (The MC33033's data sheet shows that a 1.5-V to 4-V PWM input changes the duty cycle from 0 to 100 percent.)

An interesting test is to vary the PWM frequency from a very slow 100 Hz to a very fast 100 kHz. Note how the slow PWM frequency becomes audible, while the high-frequency PWM increases the power loss of the power stage. (See Chap. 2, Sec. 2.2.4.)

Another worthwhile test is to induce noise into the speed control input voltage line. This can be done by capacitor-coupling a signal

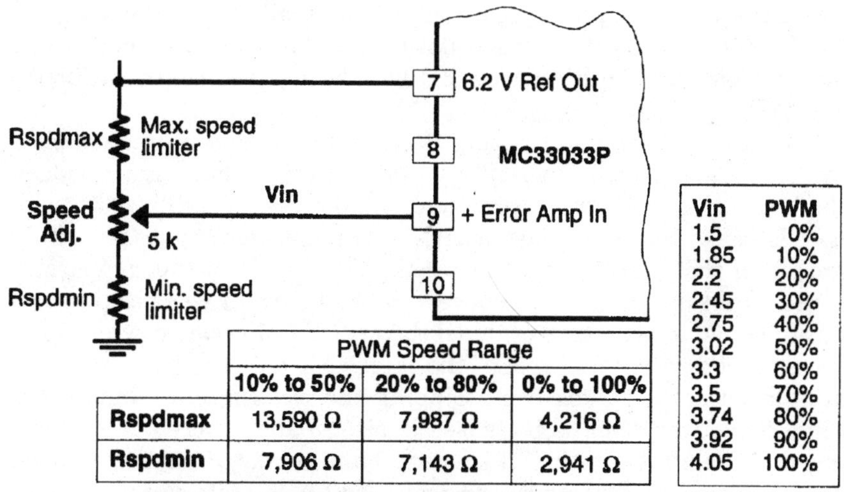

	PWM Speed Range		
	10% to 50%	20% to 80%	0% to 100%
Rspdmax	13,590 Ω	7,987 Ω	4,216 Ω
Rspdmin	7,906 Ω	7,143 Ω	2,941 Ω

Vin	PWM
1.5	0%
1.85	10%
2.2	20%
2.45	30%
2.75	40%
3.02	50%
3.3	60%
3.5	70%
3.74	80%
3.92	90%
4.05	100%

Rspdmax = (Vref - Vin@max.PWM) / Ipot
Rspdmin = Vin@min.PWM) / Ipot
Ipot = (Vin@max.PWM - Vin@min.PWM) / Rpot

Figure 3.2 Speed range limiting circuit.

from an external generator into pin 9. The motor's speed will become unstable and will produce audible sounds related to the external noise generator's frequency and amplitude.

A FORWARD/REVERSE switch changes the logic output to the H-bridge power stage, which reverses the voltage to the motor. An internal pull-up resistor in the IC raises pin 3 to a high stage when the FORWARD/REVERSE switch is open. All the MC33033's logic inputs (pins 3 to 6, 18, 19) are TTL (transistor-transistor-logic)-compatible. The inputs have typical thresholds of 2.2 V for a high state and 1.7 V for a low state. Pin 19 controls the output drivers. Pulling this pin low disables the output drivers and turns off all the transistors in the H-bridge. This action would allow the motor to coast down.

3.1.1.2 PWM generator and logic control. The MC33033 IC, like other motor control ICs of this type, has an internal PWM oscillator whose output is connected to a comparator. The other input to this comparator is set by an external voltage (pin 11), which determines the comparator's pulse width. The PWM oscillator frequency (approximately 25 kHz) is determined by an *RC* (resistance-capacitance) network connected to pin 8. An amplifier buffers the external PWM control voltage. Motor speed is controlled by adjusting the input voltage to the noninverting input of this amplifier (pin 9), whose output establishes the PWM's slice or reference level.

The three inputs (pins 4, 5, 6) that are normally used for decoding a BLM rotor position, have been hard-wired for a bidirectional brush motor. Essentially, a logic code of 100 is hard-wired to the MC33033 inputs.

Cycle-by-cycle current limiting of the motor current is accomplished by using a small-value current sensing resistor in the common side of the H-bridge so that the entire motor winding's current flows through it. The sense resistor drives a 100-mV comparator input (pin 12) to ground of the H-bridge motor current. The output of the comparator turns off all the outputs whenever the load current reaches a predetermined value and stays off until the next PWM cycle, at which time the sawtooth oscillator resets the latch. A small *RC* filter network minimizes noise spikes from tripping the overcurrent comparator. This *RC* network can be increased in size to allow more current during the motor's start-up, which may be necessary in some applications; without the increase, the motor will accelerate very slowly or may not start at all.

It is possible to switch the motor's direction on the fly, using the normal FORWARD/REVERSE switch. *Changing the motor's direction by instantly reversing the motor's voltage* (*this is also called "plugging the motor"*) *creates significant current levels in the power stage.* The cycle-by-cycle current limiter reduces the plugging current but requires more time for the reversal to take place than would be true if no current limiting were used. If fast motor reversal is required, the power stage must be sized to accommodate the increased current levels that occur during the motor's directional transition. The motor's peak current may easily double during the time period that the motor terminal voltages are reversed. This higher current level can be verified by connecting a voltage probe across the current-sensing resistor and adding a temporary jumper from pin 12 to common to disable the current limiter. The voltage drop across the sense resistor will increase significantly when the motor goes into reverse, and the current limit is disabled.

The three top-side outputs (pins 1, 2, 20) can sink an output current of up to 50 mA to drive P-FET power devices. The other three bottom side outputs (pins 15, 16, 17) can sink and source up to 100 mA to drive N-FET power devices. Both the rise and fall switching times of the bottom side outputs are less than 200 ns when driving a 1000-pF load. If very large power MOS devices are to be controlled from the MC33033, you will need a gate amplifier that can sink or source current peaks in the ampere range (MC33051, 2, 4, etc.).

A thermal shutdown circuit is included in the MC33033, which will turn off all the output drivers if the maximum high-temperature rating of the IC is exceeded. It can also protect the motor and other components, such as the power output stage, if the IC is assembled in close proximity with the motor and the power components.

3.1.1.3 Power stage. An H-bridge topology is employed to allow bidirectional motor control. N-FETs are used in the bottom legs, while P-FETs are used in the top legs. (See Chap. 2, Sec. 2.7.2 for H-bridge basics.) The MC33033 output drivers consist of three open-collector outputs for switching on the top P-FET power devices and three totem-pole outputs for switching on and off the bottom N-FET power devices. Note that the open-collector output allows a fast turnon time for the P-FET but leaves the P-FET gate open-circuited during the turnoff mode. This circuit relies on a 1-kΩ resistor across the P-FET's gate-to-source to discharge the gate capacitance, which decreases its turnoff time. As the PWM frequency increases, the P-FET's turnoff time will become more important and may actually be the limiting factor for the PWM's maximum frequency.

The power FET transistors used in the test circuit exhibit a forward ON resistance of 0.08 Ω. This means that they will dissipate about 2.4 W each when conducting 5 A and will require a medium-sized heatsink. Since the power stage should not cause burns when touched, the power FET's heatsink temperature needs to be less than 60°C. Assuming that the test circuit will operate in a location that has a maximum ambient temperature of 40°C, the heatsink size can be calculated, as shown in Table 3.1.

This table contains several equations that calculate the heatsink's thermal requirements, plus several other important parameters. The italicized values are the data entries. One section of the table that may be confusing is the $R_{DS,on}$ variation factor. To obtain this value, it is necessary to first calculate the transistor's power dissipation using $R_{DS,on}$ at 25°C and the junction-to-case thermal resistance value, providing an estimate of the actual junction temperature. This estimated junction temperature value is then used to find the $R_{DS,on}$ variation factor per the transistor's data sheet specifications (usually shown as a graph of $R_{DS,on}$, normalized against the junction temperature).

Note that the motor's DC resistance, which directly affects the motor's locked-rotor current value, is a critical design point, in addition to the FET transistor specifications. The heatsink's thermal resistance value is for one device, but both the upper P-FET and lower N-FET that constitute one-half of the H-bridge can be connected to the same heatsink since only one FET is powered on at a time. The FET's drain terminals are electrically connected to the heatsink.

TABLE 3.1 Heatsink Calculations for Test Circuit

Parameter	Symbol	Value	Notes
	Motor Specifications		
Maximum supply voltage	V_{sup}	*12 V*	Maximum power supply voltage
DC motor current, locked rotor	$I_{mtr,pk}$	*5.0 A*	Locked rotor motor current at 25°C
DC motor resistance	R_{mtr}	2.4 Ω	$= V_{su}/I_{mtr,pk}$
Allowable heatsink temperature	$T_{hs,max}$	*60°C*	Design point
Maximum ambient temperature	$T_{a,hot}$	*40°C*	Environmental temperature
Minimum ambient temperature	$T_{a,cold}$	*−20°C*	Environmental temperature
Hot DC motor resistance	$R_{mtr,hot}$	2.5 Ω	$= \{[(T_{a,hot} - 25)*Cu_{\Delta}]* R_{mtr}\} - 1 + R_{mtr}$
Hot locked-rotor current	$I_{max,hot}$	**4.7 A**	$= V_{sup}/R_{mtr,hot}$
	Power Stage Specifications		
Power transistor type	MOSFET	*MTP20N06V*	Device part number
$R_{DS,on}$ at 25°C	$R_{DS,on}$	*0.08 Ω*	Per device data sheet
Junction-to-case resistance	$R_{øjc}$	*2.50°C/W*	Per device data sheet
	Heatsink Specifications		
Heatsink thermal resistance calculations	$R_{øsa}$	**8.4°C/W**	$= (T_{hs,max} - T_{a,hot})/P_{D,hot}$
Case-to-heatsink thermal resistance	$R_{øcs}$	*0.1°C/W*	Per mounting material data
	Hot Operation, Locked Rotor		
Estimated T_j rise	$\approx T_{jc}$	4°C	$= [(I_{max,hot}^2)*R_{DS,on}]* R_{øjc}$
Estimated $T_{j,max}$	$\approx T_{j,max}$	64°C	$= [(I_{max,hot}^2)*(R_{DS,on}]* R_{øsa} + (\approx T_{jc} + T_{hs,max})$
$R_{DS,on}$ temperature variation factor	$R_{DS,on,x}$	*1.33*	Per device data sheet for $\approx T_{j,max}$

TABLE 3.1 Heatsink Calculations for Test Circuit (Continued)

Parameter	Symbol	Value	Notes
$R_{DS,on}$ at $T_{a,hot}$	$R_{DS,on,hot}$	0.106 Ω	$= R_{DS,on} * R_{DS,on,x}$
$V_{DS,on}$ voltage at $I_{max,hot}$	$V_{DS,on,hot}$	0.50 V	$= I_{max,hot} * R_{DS,on,hot}$
Power transistor dissipation	PD_{hot}	**2.4 W**	$= I_{max,hot} * V_{DS,on,hot}$
Junction-to-case temperature rise	T_{jc}	5.93°C	$= R_{øjc} * P_{D,hot}$
Case-to-heatsink temperature rise	T_{cs}	0.24°C	$= R_{øcs} * P_{D,hot}$
Heatsink-to-ambient temperature rise	T_{ha}	20°C	$= R_{øsa} * P_{D,hot}$
Maximum junction temperature	$T_{j,max}$	**66°C**	$= (T_{jc} + T_{cs}) + T_{hs,max}$

NOTES: *Italicized text* indicates design value entries; **boldface text** indicates critical results of calculations. Reference data: copper coefficient, mean $Cu_{\alpha} = 0.00393\ \alpha\ 1$ °C.

3.1.1.4 Motor. A small 12-V 2-A brush PM motor is a good choice for this test circuit. If a larger-amperage motor is used, lower $R_{DS,on}$ power stage transistors will be required. It is important not to raise the supply voltage above 20 V in order to protect the P-FET devices from exceeding their gate-to-source ratings. Once the test circuit is operational, an interesting motor test would include locking the motor's shaft. This should induce an overcurrent condition that the MC33033 detects and then causes the PWM duty cycle to decrease. If locked-rotor testing is not conducted carefully, the motor and/or power stage may be destroyed.

One sneaky failure mode that can occur with a locked-rotor condition is a power supply that goes into a current-limiting mode and lowers the supply voltage to the point that the MC33033 undervoltage circuit activates. This, in turn, causes the supply voltage to jump back to normal, reactivating the MC33033 and the motor, which starts the whole sequence all over again. Using a high current supply helps solve this problem, but the use of fuses in the power supply are mandatory to prevent catastrophic electrical fires. If a 12-V automotive battery is used as a power source, an in-line fuse should be placed near the battery's terminal. (*Remember that a fully charged auto battery can supply over 1000 A of current, which will burn most hookup wires or motors if a short circuit occurs.*)

3.1.2 Other DC motor speed controls

There are numerous circuit designs possible for driving DC PM motors. A simple nonreversing speed control can be achieved by using

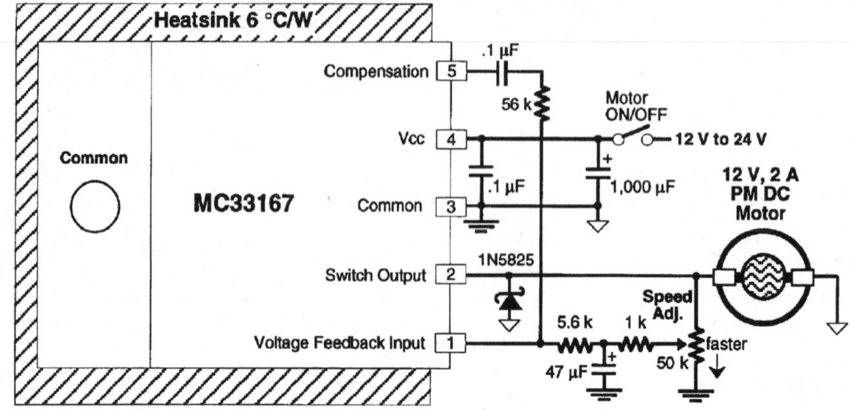

Figure 3.3 DC motor control using power switching regulator.

a power switching regulator IC, such as the MC33167. This device contains an oscillator and PWM control, as well as a high-side power stage, that can switch up to 5 A of peak current. Figure 3.3 shows one experimental motor control application. Speed regulation is determined by the motor's back EMF.

3.2 DC Motor Control Problems

DC speed control designs are usually more straightforward than AC induction motor drives, but problems still arise. The nature of these problems are not unique to motor electronics but apply to many types of power control electronic systems.

3.2.1 Noise in the user input control

A potentiometer is commonly used to vary the speed of DC motor controls. It is connected to a regulated power supply with an adjustable line feeding the IC's PWM control, which sets the motor's speed. The stability of the speed adjustment is dependent on the power supply voltage regulation and the potentiometer's wiper-contact quality. Some potentiometers do not make smooth contact when adjusted, which causes their output voltage to be noisy. Adding a capacitor to the potentiometer's output line can help, but a large capacitor value does delay the motor's response time when the pot. is adjusted.

A common design error would be to connect the potentiometer to the motor's supply, eliminating the expense of a voltage regulation device. But a problem can develop because the motor's power supply will vary somewhat with the motor's current draw. As the current increases, the PWM control voltage input decreases, causing the speed to change, which may alter the motor's current, which, in turn, changes the PWM voltage again. *This condition can cause the motor's speed to oscillate or vary, and this is why it is important to ensure that the input control signal is well isolated from the motor's effects on the supply voltage that powers the control electronics.*

The analog speed control input is also subject to electrical noise from the motor's wiring and external noise generated by nearby radio transmitters or other EMI sources. Adding a low-pass filter to the PWM control voltage input line minimizes the chances for EMI to affect the PWM control circuitry and helps reduce the potentiometer's internal wiper-contact noise.

Another area of design that is often overlooked is the manner in which the analog input and PWM control IC are connected to power common or ground. If the common lines of the motor's power stage are mixed improperly into the control stage, a large "ground noise" problem will occur, affecting the stability of the motor control. If the motor and control electronics operate from the same power source, as in most small motor systems, the interconnection of the control and power stages common lines becomes critical. In the DC motor test circuit, the power stage common and the control stage common are connected only at the power transistor's source, which is also where a power supply filter is connected. A closer examination of the test circuit (Fig. 3.4), reveals the rationale behind this hookup method. Total isolation of the

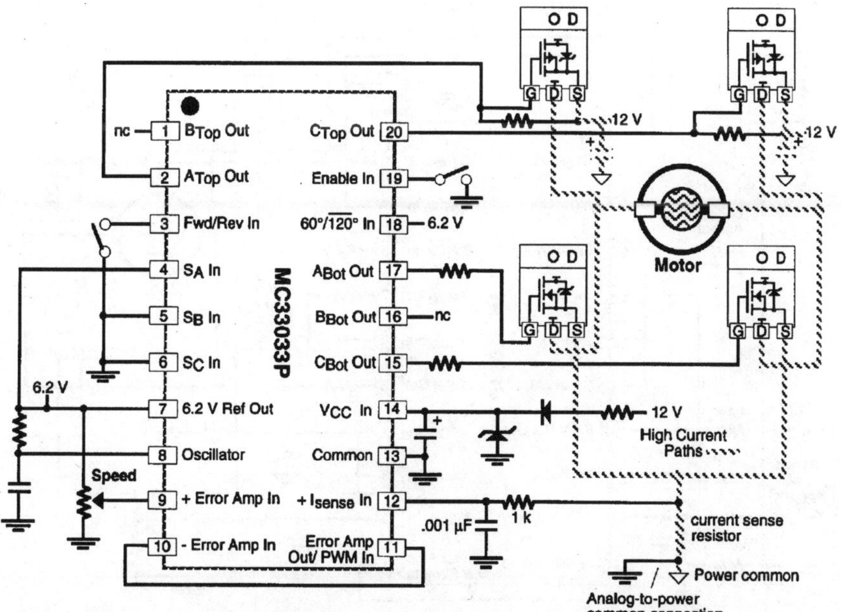

Figure 3.4 Power supply common connection minimizes ground noise.

control stage and the power stage will be necessary in large motor drives because of ground noise, as will be seen in later chapters.

One common wiring harness error is to combine the motor's power lines with the control input signal lines. Placing the speed control pot. wires with the motor's power leads or mixing a motor speed sensor's wires with the motor's power lines will invite stability problems. Remember that the input control lines are high-impedance and therefore act as antennas. The motor's power lines are low-impedance and will transmit switching spikes to nearby lines by magnetic induction and/or capacitive coupling.

3.2.2 Open connections

If the speed potentiometer connection to common used in the previous test circuit should open, the V_{in} rises to a maximum value and turns on the motor at full speed. This can be a very serious problem in certain motor control systems, and protective elements need to be incorporated to prevent a runaway motor condition. Open wire connections occur because of poor contacts in connectors, broken wires caused by sustained flexing, and so forth. Two elements can help prevent the "floating-potentiometer common" problem, as shown in Fig. 3.5. One is to add a comparator that monitors the current through the common lead of the pot., disabling the PWM control device if no pot. current is sensed. The second element is to add a large value pull-down resistor

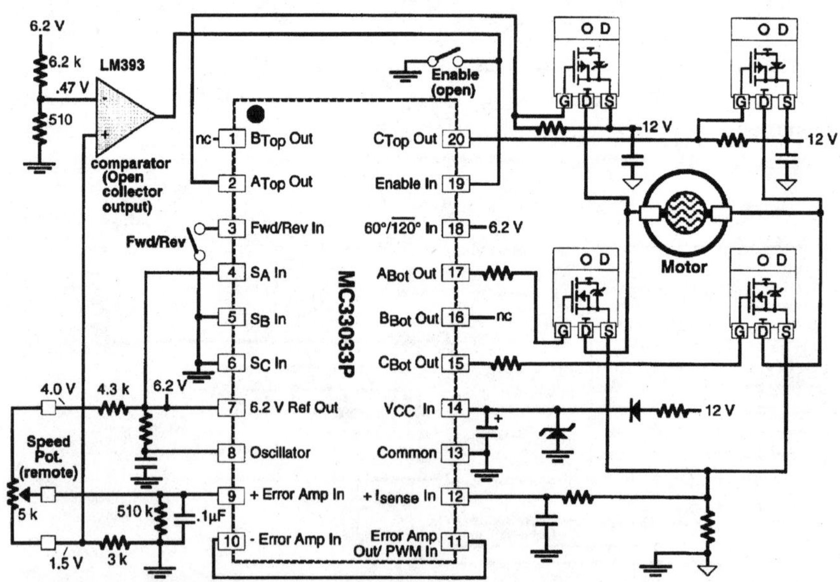

Figure 3.5 Open speed potentiometer protection circuit.

near the V_{in} pin that controls the PWM duty cycle. This will ensure that the V_{in} will go to zero if all the leads to the speed pot. are opened, as might happen if the pot. were placed in a remote location.

3.2.3 Unstable power supply

As mentioned in Sec. 3.1.1, excessive motor currents can initiate power supply problems, causing the PWM logic control to shut off the motor, allowing the power supply to recover, which, in turn, produces a sequence of events that is initiated over and over again. There are no easy design methods to correct this motor system oscillation when both the PWM logic and motor are powered from the same voltage supply. It can help to add a series rectifier to the PWM logic as well as a large capacitor value that can "hold up" the PWM logic control supply voltage for a few seconds after the motor's power supply voltage has collapsed. One method that does work is to use a separate power supply for the PWM logic control. Besides solving the control system oscillation problem, the independent supply also allows the PWM logic control to run diagnostics on the power stage before and after the power stage is energized.

Since fast-switching edges are used in the motor's power stage, a capacitor filter network is mandatory in the power stage because of the power supply wiring inductance. This filter network smoothes out both the PWM frequencies, as well as the much higher frequencies associated with the 200- to 1000-ns PWM switching edges. The filter consists of capacitors that are specified to operate in electronic motor control applications. Metalized-film capacitors are recommended for filtering high-frequency currents, in addition to electrolytic capacitors that are rated for use in PWM motor drives. The bus interconnections or printed-circuit lines in the power stage should be constructed to minimize inductance. Figure 3.6 shows a good-versus-poor layout for

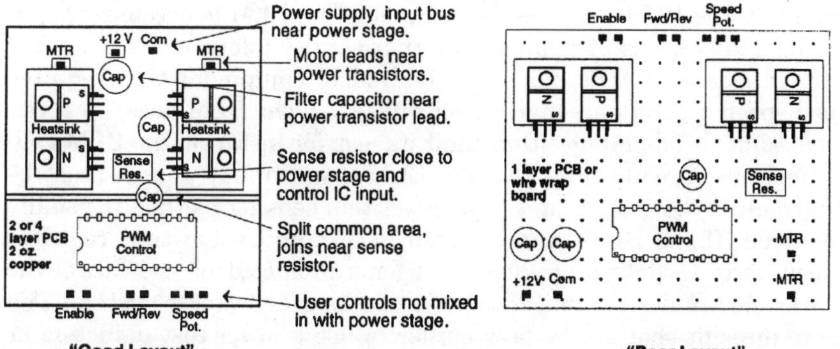

Figure 3.6 Power stage component layout.

the capacitors and other associated devices in the motor's power stage. Note that the good layout is also the most compact, which means that the lead or bus inductances are minimized.

3.2.4 Power stage oscillations

Power MOS devices such as FETs and IGBTs will oscillate at radio frequencies if improperly driven or used in a poor PCB (printed-circuit-board) layout. The most common design error occurs in the gate driver connections to the MOS transistor's gate and source. The gate driver circuit should be located as close as possible to the power MOS transistor, and the gate driver's source connection should be the line closest to the power transistor. Very large power MOS transistors implement a special source line called a *Kelvin contact* for the gate driver's source connection.

3.3 Brushless DC Motor Control Design

A disadvantage in brush-type motors is their inherent brush and commutator requirement. The brushes tend to arc, which generates radio-frequency interference (RFI). The mechanical action of the brush sliding along the commutator can also create audible noise that may be objectionable in some applications. The brushes in some low-cost DC motors may also need to be replaced in as little as 1000 h of operation. The commutator-and-brush assembly can fail in a catastrophic manner under worst-case conditions, such as during a stalled or locked rotor. (The brushes-to-commutator contact generates enough heat to cause permanent damage in a locked-rotor condition.) When electronic controls are added to a brush motor, it is sometimes worth upgrading to a brushless motor for increased performance.

A BLM eliminates the brush/commutator assembly and operates indirectly from DC. Unlike the brush motor, whose armature resides on the rotor, the BLM rotor carries magnets. The BLM is becoming popular for many types of applications that cannot tolerate brushes (low-voltage computer fans, for example) and in equipment that requires good speed control plus high power efficiency. The BLM does, however, have some disadvantages that limit its use; for instance, the BLM cannot operate directly from DC, the BLM's two to four phases must be electronically switched, and a rotor position sensing method is usually required. If the BLM is to be used in an application that requires speed, torque, control, braking, or regeneration control, the cost of the power electronics is similar to that of a brushed motor system. The more difficult choice may be whether to use a lower-cost induction or switched reluctance motor instead of the more power-efficient and more costly BLM. The control electronic design, as will be shown in the later chapters, does not vary significantly for BLM, 3Ø (three-phase) induction, or switched reluctance motors.

3.3.1 BLM speed control

The BLM control circuit shown in Fig. 3.7 is similar to the test circuit used for the brushed DC motor covered in Sec. 3.1.1. The bipolar analog MC33033 IC is retained as the PWM control logic device. All the other input controls remain the same. One main difference is that the rotor position inputs are connected to the BLM's internal Hall-effect sensors and the power stage is changed to three half H-bridge topologies. (Again, a MC33035 could be used if a brake input and fault output function are required.)

A somewhat more sophisticated three-phase BLM motor controller IC is the Si9979CS (TEMIC Semiconductors, Santa Clara, CA). This device is packaged in a 48-pin quad flat pak. The Si9979CS offers several features: internal bootstrap/charge pump supplies, which allow an all N-FET power stage; quadrature selection for choosing whether both top and bottom MOSFETs or just the bottom MOSFETs will respond to PWM; and a tachometer output that may be used for BLM closed-loop control. This device is specified to operate over a supply

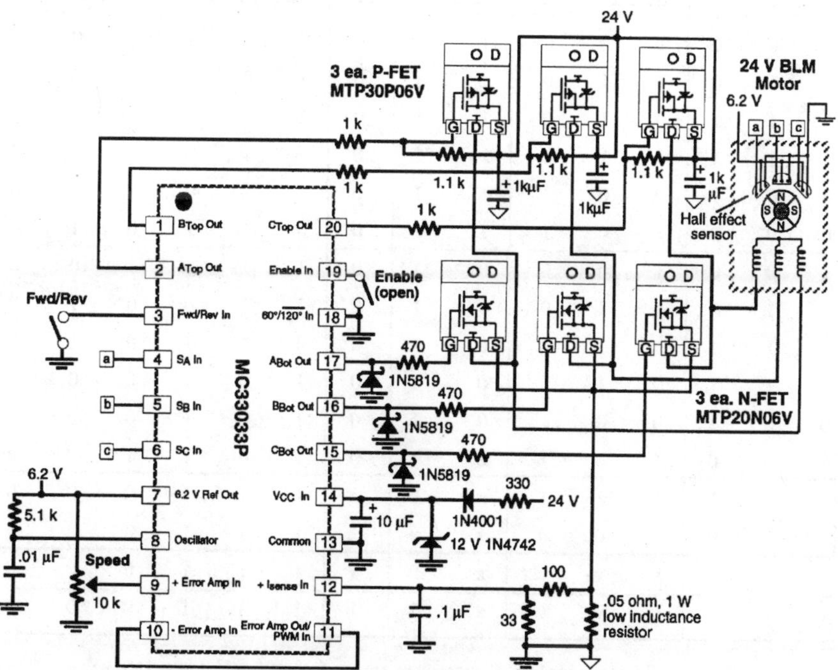

Figure 3.7 DC brush motor speed control test circuit.

voltage range of 20 to 40 V and, therefore, cannot be directly used in 12-V systems.

3.3.1.1 Rotor encoding. The MC33033 uses a built-in logic decoder for determining the correct sequence required to drive the top and bottom power transistors in the three half H-bridges. The rotor position sensor inputs (pins 4 to 6) are designed to interface directly with open-collector-type Hall-effect detectors or slotted optocouplers. The position sensor inputs are also TTL-compatible and use internal pull-up resistors. The three sensor's output signals indicate six possible positions of the rotor. Of the eight possible codes generated by the three sensors, two are invalid codes, and, if one of these two codes is received, all the output drives are turned off.

The MC33033 can be configured to handle either 60° or 120° electrical sensor phasing per the logic state on pin 18. Table 3.2 shows the

TABLE 3.2 MC33033 Logic Truth Table

Inputs									Outputs					
Sensor electrical phasing						User		Load	Top drives*			Bottom drives		
60°			120°											
S_A	S_B	S_C	S_A	S_B	S_C	F/R	En.	I_L	A_T	B_T	C_T	A_B	B_B	C_B
1	0	0	1	0	0	1	1	0	0	1	1	0	0	1
1	1	0	1	1	0	1	1	0	1	0	1	0	0	1
1	1	1	0	1	0	1	1	0	1	0	1	1	0	0
0	1	1	0	1	1	1	1	0	1	1	0	1	0	0
0	0	1	0	0	1	1	1	0	1	1	0	0	1	0
0	0	0	1	0	1	1	1	0	0	1	1	0	1	0
1	0	0	1	0	0	0	1	0	1	1	0	1	0	0
1	1	0	1	1	0	0	1	0	1	1	0	0	1	0
1	1	1	0	1	0	0	1	0	0	1	1	0	1	0
0	1	1	0	1	1	0	1	0	1	0	1	0	0	1
0	0	1	0	0	1	0	1	0	1	0	1	0	0	1
0	0	0	1	0	1	0	1	0	1	0	1	1	0	0
1	0	1	1	1	1	x	x	x	1	1	1	0	0	0
0	1	0	0	0	0	x	x	x	1	1	1	0	0	0
v	v	v	v	v	v	x	0	x	1	1	1	0	0	0
v	v	v	v	v	v	x	1	1	1	1	1	0	0	0

NOTES: v = any valid sensor input; I_L = current sense, 0 = <85 mV, 1 = >115 mV; x = don't care.

*Top drive outputs are open collector; 0 = on, 1 = open.

logic truth table for the sensor inputs and the resultant outputs. Note that the 120° sensor phasing illustrates that all ones or all zeros are invalid Hall effect sensor states. These conditions, all ones or zeros, are easier to diagnose than the 60° 101 or 010 invalid Hall sensor states.

3.3.1.2 BLM power stage. The BLM power stage is nearly the same as in the previous brushed motor test circuit. Instead of two half H-bridges configured to form a full H-bridge, three half H-bridges are used to connect the motor's winding to either the power supply positive bus or to the power supply common bus. *It is important to note that each leg of the half H-bridge must be rated to sustain the motor's locked-rotor current.* A common misconception is that the three half H-bridges need to be rated to carry only one-third of the motor's current. In reality, each half H-bridge can carry the full locked-rotor current.

3.3.2 BLM closed-loop speed control

Since the BLM already contains rotor position sensors, the motor's actual shaft speed and angular position (to within 60°) can be determined. Required for basic BLM control, this information can also be used to maintain a constant speed under varying loads by adding a special 8-pin IC, the MC33039 closed-loop BLM adapter, as shown in Fig. 3.8. This device can allow precise speed regulation and uses digital detection of each input signal transition. The MC33039's output (pin 5) consists of a train of pulses that are integrated by the error amplifier in the MC33033. An open/closed loop switch is connected across the error amplifier to disable closed-loop operation, which may be desired when the motor is first turned on. One precaution about the closed-loop system is that the motor will run at its maximum speed until the feedback loop has stabilized. (Refer to the MC33033 data sheet's application section for design methods to set the motor's acceleration rate.)

3.3.3 BLM speed control with integrated power stage

To minimize PCB space, it is sometimes advantageous to combine different semiconductor technologies. The L6234 (SGS-Thomson) and HIP4011 (Harris Semiconductor) are two examples of how mixed-technology ICs combine the BLM PWM logic control with three half H-bridge power stages in one power IC. The L6234 is available in a 20-pin power dip or power SO20 package and operates from 7 to 52 V with up to 5 A peak currents. The HIP4011 operates from 10 to 13 V, switches up to 5 A peak current, and is packaged in a 15-pin SIP.

One important aspect of using either of these power BLM IC devices is their heatsinking requirements. This becomes especially

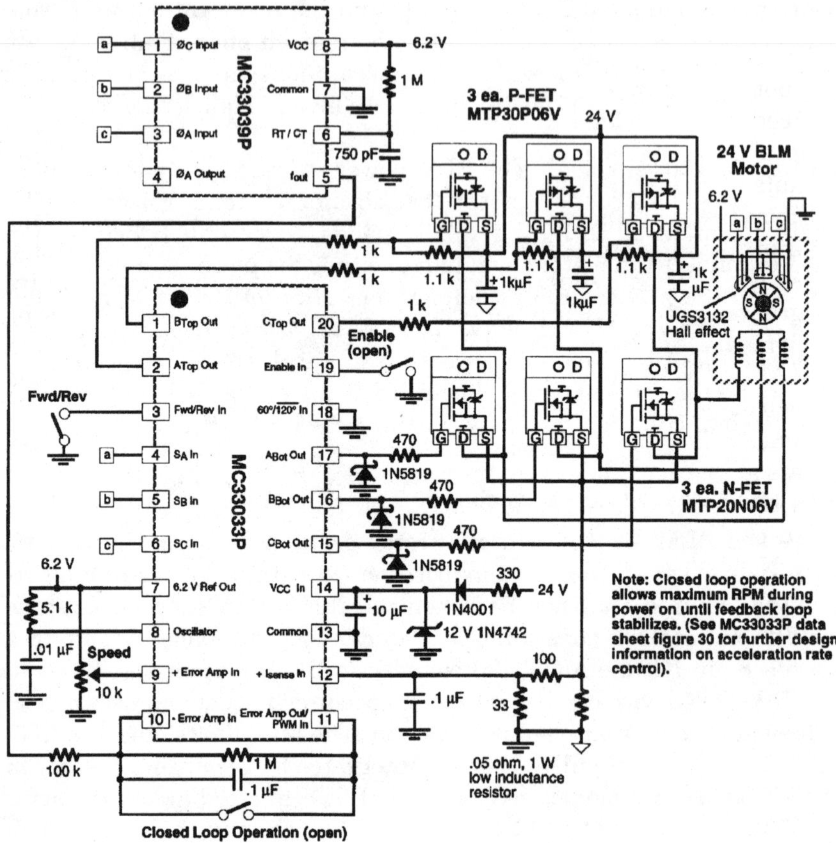

Figure 3.8 Closed-loop speed control using MC33039.

critical if the power IC is mounted inside a totally enclosed BLM. The BLM frame may easily reach temperatures of over 60°C, requiring a good thermal interface to an external heatsink and incorporating a fan driven by the motor's unused shaft end, if the BLM application can allow it. Otherwise, a large external heatsink will be needed. In a practical sense, the power BLM IC products work well for a low-wattage BLM, but it may be more efficient to use discrete low $R_{DS,on}$ MOSFETs for larger BLMs, thereby saving on heatsink requirements. At the time of this book's preparation, MOSFETs were available in TO-220 and D-PAK surface mount packages with 0.006-Ω $R_{DS,on}$ 75-A current rating, and voltage rating of 50 V. (SUP75N05 from TEMIC Semiconductors is one example.)

3.3.4 BLM system considerations

Thus far, the electronic motor control designs for DC motors, including both brush and BLM types, have incorporated analog IC-type devices. Discrete analog circuits have been avoided because of the wide availability of motor-specific analog IC products from several manufacturers. (See App. C.)

Advantages of analog IC products include low cost and inherent ease of use. Designed for noncomplex motor control applications, these products fall short when applied in applications that require programmability. A closed-loop constant-speed BLM controller can be implemented with the analog devices previously described. However, a BLM control that requires different acceleration/deceleration rates, as well as a specific sequence of different constant speed points that is dependent on certain input conditions, becomes difficult to utilize unless a microcontroller is employed.

In the past, a separate MCU was used to drive a BLM analog IC when complex speed patterns or user programmability were necessary. New motor control MCUs now allow PWM control logic and user programmability or complex speed and/or torque patterns to be implemented in one device. The main drawback in using MCUs is the need for an MCU development station and some background with MCU programming. *There is, however, a positive side to the programming: You can change the motor control's attributes and system behavior from a keyboard rather than with massive circuit and component changes.*

A microcontroller not only allows motion control systems to perform basic speed control but also, and more importantly, permits easy implementation of more complex functions. Examples include measuring several different inputs, such as pressure, temperature, speed, load current, and optosensors, and operating the motor in an optimal manner based on these data values.

BLM and brush motors do not normally require the complex calculations that an AC induction or SWR motor require and therefore can use an 8-bit MCU architecture that is optimized for BLM-type applications.

3.4 PM Stepper Motors

Stepper motors perform crucial mechanical tasks in the vast majority of electronic equipment used in offices and homes. Various stepper motor types include variable reluctance, hybrid, claw-tooth, and permanent magnet. PM stepper motors are commonly used in digital motion control systems, since they interface easily with digital wave-

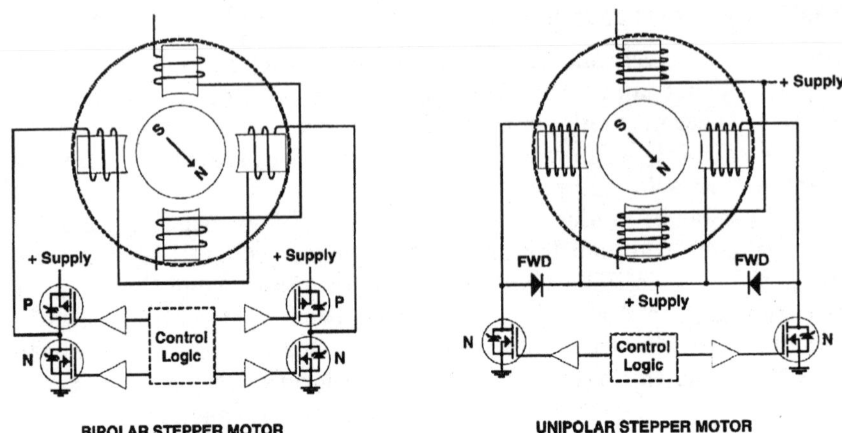

Figure 3.9 Stepper motor power stage topologies.

forms. Stepper motors do not require position sensing for commutation but may require some type of position indicator to calibrate or reset the stepper control logic.

Stepper motors are typically designed to have step angles of 0.9°, 1.8°, 2°, 7.5°, and 15°. Obviously, the smaller the step angle, the higher its position accuracy will be, as will be the number of step control signals required per second.

There are two main stepper types: unipolar and bipolar. Figure 3.9 illustrates both types and their electronic power stage topology requirements. Bipolar stepper motors use one field coil per H-bridge, while a unipolar stepper motor requires bifilar wound field coils and only one half H-bridge. Bipolar stepper motors offer higher power density than do unipolar types. However, with the integration of two or more H-bridges in a power IC, the advantage of the unipolar motor's reduced power stage complexity is less important, and the bipolar motor's torque versus size is a significant advantage. Power dissipation in the stepper motor can produce a major reliability issue but can be somewhat controlled by drive electronics.

3.4.1 Stepper motor design

A two-phase bipolar stepper motor test circuit (Fig. 3.10) utilizes a single IC device, a MC3479 stepper motor power IC. (Please note that there are several other stepper motor ICs available; see App. C.) The MC3479 consists of two main sections: decoding/sequencing logic and a dual H-bridge power stage. An inverted phase A logic state output is useful in some digital control systems for initializing the motor to a phase A state.

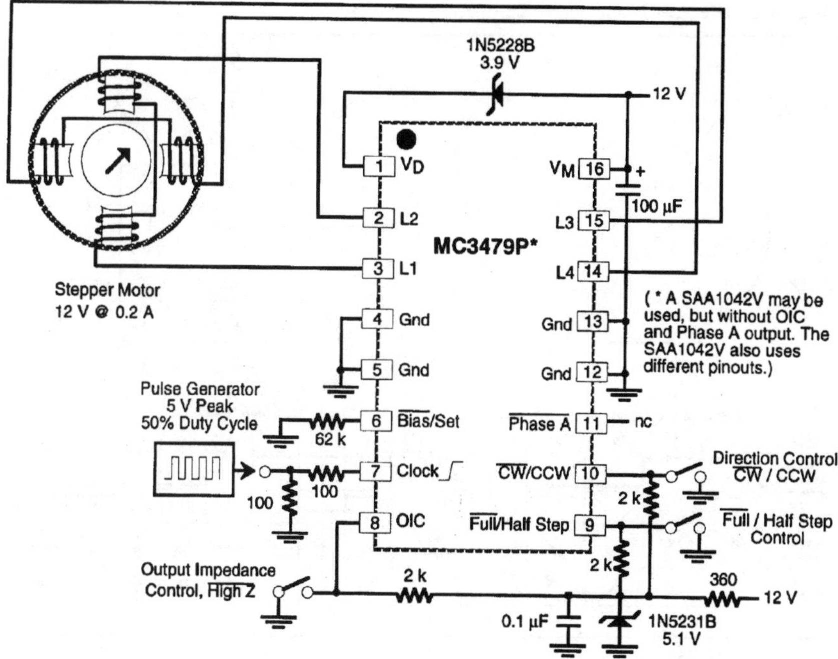

Figure 3.10 Stepper motor test circuit using MC3479.

The test circuit uses an external square-wave generator to supply the clock signal whose frequency relates to the number of steps per second. User input switches control the stepper motor's direction, step size, and power stage output impedance. The bias pin (pin 6) adjusts the bias current for the power stage and is dependent on the motor's current requirements. (Refer to the MC3479 data sheet for bias resistor value.)

A logic timing diagram (Table 3.3) shows the output sequence for both full-step and half-step modes. Note how the full-step operation lines up the rotor between the magnetic poles of the field windings; the half-step operation does the same and also lines up right on the field's magnetic pole for the half-step positions. The design of the stepper motor determines to a large degree how accurate its position will be as it steps along.

3.4.2 Stepper motor design variations

There are some "tricks" that can be used to increase the shaft's velocity of stepper motors. One is to add series resistors to the field windings (anywhere from 1*R* to 4*R* of the winding's resistance) and then

TABLE 3.3 Stepper Motor Drive Operation

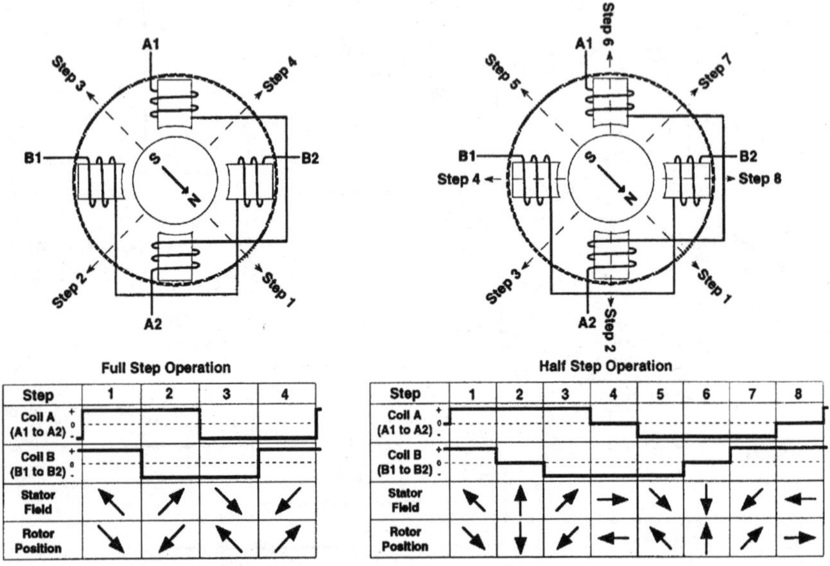

SOURCE: MC33192 data sheet.

use a higher voltage supply. This drives the stepper motor from a current source, which produces faster response because of the increased supply voltage. However, it also increases the cost of the stepper motor design and may affect the stability of the stepper as well. Another way to drive the stepper motor from a current source is to pulse the coil's voltage at a high frequency rate. Figure 3.11 shows an application example of stepper motor ICs incorporating constant current with switch mode regulation. (For more information, refer to SGS-Thomson Application Note AN235, *Stepper Motor Driving* and TEA3717, TEA3718 *Stepper Motor Driver* product specifications. National Semiconductors also has a LMD18200 3-A 55-V device that can drive stepper motors with switch mode regulation; refer to National Application Note AN428, *Increasing the High Speed Torque of Bipolar Stepper Motors.*)

3.5 Servo Motors

Contemporary servo motors depend on control electronics in order to operate. A servo motor system is useful for controlling mechanical movements such as satellite dish position, flow valves, or any motor driven mechanism that is designed to operate in a servo-type fashion. There are two main factors that control the accuracy of a servo motor

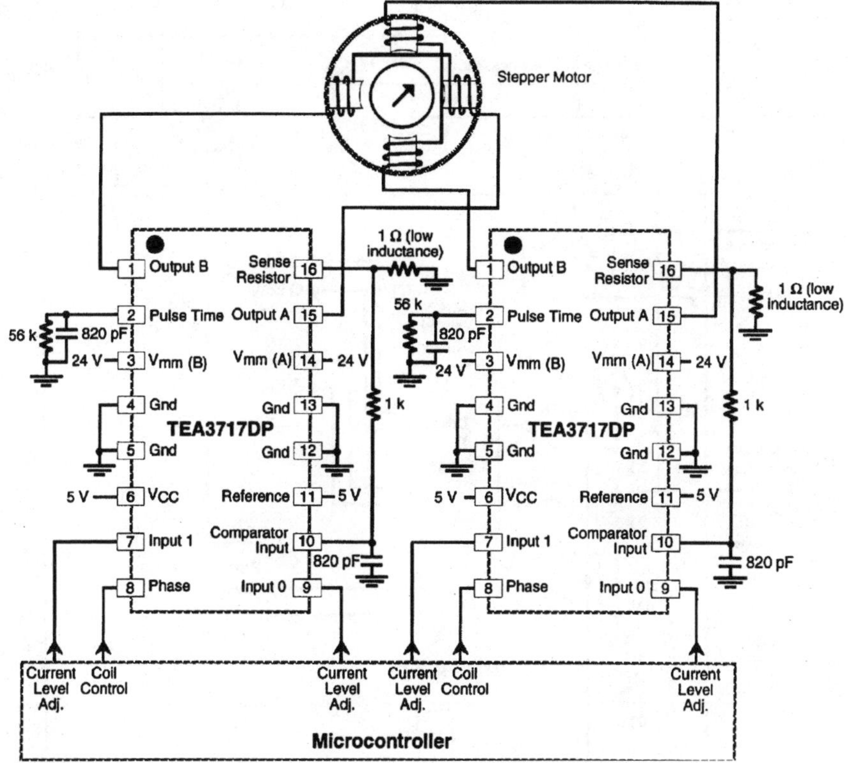

Figure 3.11 Stepper motor system using switch mode ICs.

system: (1) the resolution of the feedback signal, which relates to the type of analog sensor incorporated; and (2) the sensitivity of the servo controller device. One low-cost feedback method is to use a potentiometer that changes its resistance (voltage) according to the position of the motor's shaft or load position. A basic servo system could use travel-limiting switches that trip when the motor's load, such as a garage door, has reached a certain position (full-up or full-down).

3.5.1 Servo motor design

The design of a servo motor system is shown in Fig. 3.12. (Note that this design was derived from the MC33030 data sheet applications materials and should be considered experimental.) The feedback potentiometer is connected to either the motor's shaft (not recommended for large motors) or indirectly to the gearbox shaft, if the test motor has one. In most servo systems, the feedback device will probably be mounted on the load mechanism.

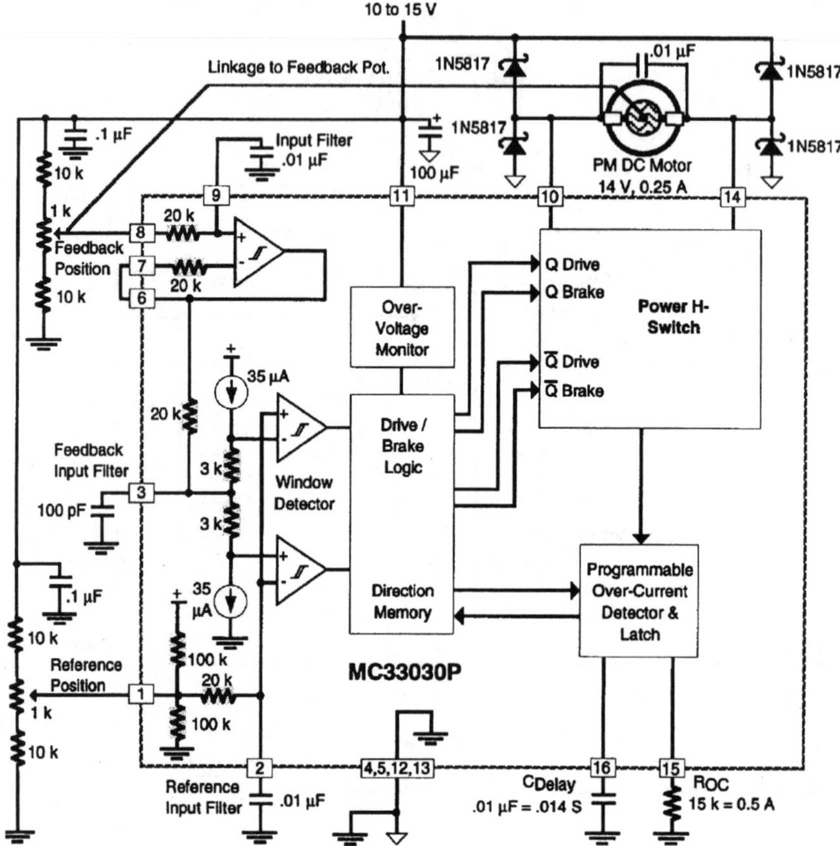

Figure 3.12 DC servo motor control test circuit using MC33030.

A power IC, the MC33030 DC servo controller/driver, serves three main functions. The first is to compare the reference position to the feedback position with an on-board op amp (operational amplifier) and window detector (two comparators). In the second function, the window comparators outputs are connected to a control logic section that determines what action is necessary to control the motor (run forward, brake forward, run reverse, brake reverse, etc.). The third function is a full H-bridge power stage that drives the servo motor. Other functions include adjustable overcurrent detection and shutdown delay. An overvoltage monitor is also incorporated to protect the power stage if the supply voltage rises above 18 V. It is especially important to use a good PCB layout with this device to ensure stable operation.

3.5.1.1 Operating description. On powerup, the motor will operate per the difference of the feedback voltage and the reference or user set-point position. Assuming that the feedback voltage (pin 8) is less than the reference input (pin 1), the motor will run in a forward direction until the feedback voltage is within approximately 35 mV of the reference input. When this occurs, the motor is braked, and, if it stops without overshooting the reference input, the power stage shuts off as long as the feedback voltage and reference voltage stay within the input dead zone. The *dead zone* is defined as the area where the feedback voltage and the reference position are within about 200 mV of each other. If the motor did not brake in time (because of high rotor inertia) and overshot the reference position, the motor runs in reverse direction until the feedback voltage is within about 35 mV of the reference input, at which time it is braked again and, hopefully, falls within the window detector's dead zone. This is where a problem can occur: If the motor continuously overshoots the dead zone in both directions, it will continuously "hunt" or oscillate back and forth.

Each servo application is unique in that the feedback loop will need to be adjusted for optimal operation.

Summary

The design of electronics for direct-current permanent magnet motors can vary from simple speed controls to complex closed-loop systems. Key items that should have been learned from this chapter are listed below.

- The main difference between brush-type PM DC motors and brushless DC motors is that the former can operate directly from a DC supply, while the latter require electronic switching for commutation and, therefore, cannot operate directly from a DC supply.
- PM motors, because of their internal magnets, act as generators and will continue to generate a DC voltage as they coast off. This can cause significant peak currents in the power stage when reversing the motor, especially if the motor's load has a high inertia factor or is connected to a large flywheel.
- The power stage design in a PM motor must be rated to handle instant changes of shaft rotation. The initial current through the power stage will be significantly higher than a normal start-up current as the motor changes its direction. This high-level current will flow for several milliseconds and may damage the power transistors if they are not large enough. One method to reduce the power stage's current during reversal is to first brake the motor, wait for it to stop, and then operate it in a reverse direction.

- When choosing power transistors for motor control, the single most important parameter for reliability, besides voltage breakdown, is the power transistor's heat dissipation. Choosing a power transistor that has low $R_{DS,on}$ or forward voltage drop may cost more initially, but it saves on heatsink size and minimizes thermal cycling. A small power transistor with a large heatsink is a poor design practice. Motors inherently draw high starting currents. These high currents will thermally stress a small power transistor, even when it has a large heatsink, since the transistor's smaller die area has less die contact area for heat conduction.
- There are advantages for using a separate power supply for the motor and the control electronics. If both the motor and control electronics operate from the same power supply, a locked-rotor condition can cause the control electronics to fail because the power supply voltage may sag. A separate power supply with its own fusing to power the motor will help ensure that the control electronics can maintain stable operation of the motor, even when motor faults occur.
- There are some basic differences between the control electronics for servo motors and stepper motors. Servo motor controls require the use of position feedback, whereas stepper motors may not. Stepper motors work well with digital electronics, whereas servo motors traditionally use analog control electronics.
- It is important to ensure that the motor control signals, such as speed or position feedback, are free from noise spikes. Electrical noise or EMI on the control signals will cause the motor to chatter and could initiate an unstable mode of operation. Filters and careful layout help minimize EMI. In certain situations, or in large motor controls, optoisolators are required to isolate the power stage from the control logic.
- The resolution in a stepper motor can be increased by electronics. This is accomplished by dividing full steps into smaller increments or fractions. The stepper motor must also be designed to accommodate fractional steps. Microstepping is possible with an MCU control system and allows high resolution in a stepper motor.
- There are a few problems associated with stepper motor drive electronics: The stepper motor drive electronics can continuously operate the motor, which may cause overheating. The use of smaller step sizes increases the angular resolution but reduces torque. The stepper motor will also create some noise that is related to the step size, motor design, and control method.

- The angular step response time for a stepper motor can be increased by raising the stepper motor's supply voltage. This will increase the stepper's current *Di/Dt,* and, unless some form of current limiting is used, the stepper will overheat. Reducing the stepper's coil inductance will also speed up the coils *Di/Dt,* but again, the current must be limited to avoid overheating.

Acknowledgments

The author would like to thank Jade Alberkrack for his assistance with material on linear integrated circuit applications.

Further Reading

Welch, Richard H., "Demagnetization Can Cause Reversible or Irreversible Changes in Permanent Magnet Motor Performance," *PCIM,* July 1995.

Chapter

4

Automotive Motor Controls

Richard J. Valentine
Motorola Semiconductor Products

Introduction

It is surprising to see how many motors can be found in cars and trucks. As shown in Table 4.1, over 50 motors may be used in a large,

TABLE 4.1 Automotive Motor Applications

Motor usage	Quantity used	Average power, W	Variable speed	Direction *F/R*	Torque limit	Position feedback	Motor type	Design notes
Body and Chassis								
Wipers (s)	2–5	100	Yes	No	No	Yes	PM	Complex mechanical linkage
Washer pump (s)	1–2	40	No	No	No	No	PM	—
Wiper arm pressure	1–2	<25	No	Yes	No	Yes	PM	Complex mechanical linkage
Door locks	2–4	30	No	Yes	No	No	PM	Locked-rotor operation
Window lifts	2–5	50	No	Yes	Yes	Yes	PM	Locked-rotor operation
Sunroof	1	100	No	Yes	Yes	Yes	PM	Locked-rotor operation
Trunk lock	1	30	No	Yes	No	No	PM	Locked-rotor operation
Antenna retractor	1	25	No	Yes	No	Yes	PM	Locked-rotor operation
Seat height adjustment	2	50	No	Yes	No	Yes	PM	Locked-rotor operation
Seat back adjustment	2	50	No	Yes	No	Yes	PM	Locked-rotor operation
Seat forward adjustment	2	50	No	Yes	No	Yes	PM	Locked-rotor operation
Lumbar air pump	2	50	No	No	No	No	PM	—
Pop-up headlamp	2	50	No	Yes	No	No	PM	Locked-rotor operation
Mirror adjustment, *X, Y*	4–6	<5	No	Yes	No	No	PM	Dual H-bridge per mirror

TABLE 4.1 Automotive Motor Applications (Continued)

Motor usage	Quantity used	Average power, W	Variable speed	Direction *F/R*	Torque limit	Position feedback	Motor type	Design notes
				Engine				
Fuel pump (s)	1	75	No	No	No	No	PM or BLM	Could vary pressure by speed
Throttle control	1	50	Yes	No	No	No	PM or BLM	Used in some engine controls
Idle speed adjust	1	<10	No	Yes	No	Yes	Stepper	Used in some engine controls
EGR* valve	1	<10	No	Yes	No	Yes	Stepper	Used in some engine controls
Radiator fan	1–2	180	Yes	No	No	No	PM or BLM	Water and dust environment
Starter (s)	1	2400	No	No	No	No	Wound DC	Switched by solenoid
Starter/alternator (a)	1	1800	No	No	No	No	AC or BLM	Combines starter with alternator
Water pump (a)	1	500	Yes	No	No	No	PM or BLM	Speed could vary temperature
Oil pump (a)	1	1500	Yes	No	No	No	PM or BLM	Critical to engine reliability
				HVAC				
Heater fan (s)	1	240	Yes	No	No	No	PM or BLM	Brush noise concern
Air-conditioning fan	1–2	300	Yes	No	No	No	PM or BLM	Brush noise concern
Duct airflow vanes	2–4	<10	No	Yes	No	No	Stepper	Outside-air mix, etc.
Compressor (a)	1	2000	Yes	No	No	No	AC or BLM	Electric vehicle

TABLE 4.1 Automotive Motor Applications (Continued)

Motor usage	Quantity used	Average power, W	Variable speed	Direction *F/R*	Torque limit	Position feedback	Motor type	Design notes
				Safety				
Seatbelt retractors	2–4	50	No	Yes	No	No	PM	Locked-rotor operation
				Suspension				
Antilock brakes	1	100	No	Yes	No	No	PM	Used in many ABS units
Ride control air pump	1	100	Yes	No	No	No	PM	—
Electric steering (a)	1	1000	Yes	Yes	No	Yes	AC or BLM	May require complex control
Electric braking (a)	2	100	Yes	Yes	—	—	PM	Under development
				Miscellaneous				
Winch	1	1000	Yes	Yes	No	No	Wound DC	Used by trucks
Air pump for tires	1	120	No	No	No	No	PM	Handheld or built-in

NOTES: The data shown here is estimated and will vary among automotive manufacturers and vehicle types.
KEY: a = advanced design; s = standard equipment.
*Exhaust gas recirculation.

fully equipped vehicle. Even a small economy vehicle may use more than five motors. Several additional applications, such as electric steering, are currently under development, which will incorporate complex motor control designs.

Automotive motors range in size from less than a few watts for remote mirrors to more than 2 kW for the starter motor. Electric-powered vehicles use motors that are quite large, up to 150 kW, and are discussed in Chap. 6. Most automobiles employ a 12-V electrical system, with some heavy-duty vehicles using 24-V systems. To design a reliable electronic motor control, one must understand various aspects of vehicular electrical systems.

4.1 Vehicular Electrical System

Most vehicles employ a medium to large 12-V battery whose size is determined mainly for its ability to start or crank the engine. A starter motor, usually a series field wound motor, is used to crank the engine. The starter motor draws fairly large current levels, somewhere between 100 and 400 A, depending primarily on the size of the engine and other factors, such as the engine's temperature, oil viscosity, and the battery's state of charge.

Because the battery can supply current levels in the 500- to 1000-A range, it is important to consider the effects of a shorted (short-circuited) motor or failures in the motor control electronics. Some type of fusing is normally used to protect the vehicle's electrical system. However, the vehicle's main energy source, the battery, is seldom fused directly at its terminals. The starter motor is ordinarily connected directly to the battery and is controlled by a solenoid-operated high-current switch. Therefore, any electrical mishap in the starter motor or its connections can lead to serious consequences. Most other loads, such as lights or motors, are fused or protected with circuit breakers.

The other energy-generating source, the alternator, is typically protected with a device called a *fuse link* or a special high-current fuse. The fuse link is a particular type of wire, usually a short piece of a smaller-gauge wire, enclosed in a flameproof material, that melts open if a direct short (circuit) occurs, thereby preventing a catastrophic wiring harness failure. A reverse battery hookup (which might occur during a jump-start event) can often blow the alternator's fuse link because the alternator's internal rectifiers become forward-biased when a negative voltage is applied to the alternator's normally positive output lead. (It is not uncommon for the alternator rectifiers to fail shorted, and then the fuse link to blow open.) *One important note about the fuse ratings in vehicular motor applications: The fuse is usually sized to prevent the motor's connecting wires from melting and not to protect the motor or its power electronics from a locked-rotor condition.*

4.1.1 Automotive electrical design considerations

Before examining the motor control designs, there are some electrical and environmental conditions in 12-V vehicles that must be considered. These can be divided into five groups: (1) 12-V bus voltage regulation, (2) voltage transients, (3) reverse battery hookup, (4) alternator "load dump" event, and (5) operating environment.

4.1.1.1 12-V bus voltage regulation. The typical 12-V bus DC voltage may range from 4 to 24 V. The 4-V condition occurs with a low battery charge during engine cranking at subzero temperatures, and the 24-V condition occurs with certain types of jump starts from an external power source. During normal engine operation, the battery voltage may vary from 10 to 16 V, depending on the battery's state of charge, temperature, and alternator loading. Most motors will function in some capacity, even as the voltage drops to near zero. When electronic control circuits are used, however, there is usually a voltage threshold at which the motor abruptly stops functioning. This threshold is normally between 1 and 8 V and is dependent on the control circuit's biasing networks and various internal transistor technologies. The electronic motor control must be designed to behave in a stable fashion when the power supply voltage varies between nominal levels and zero. A smooth transition during power supply ON/OFF conditions in the electronic designs can be implemented with some form of hysteresis element for the analog portions and a low-voltage interrupt or reset circuit for MCU circuits.

4.1.1.2 Voltage transients. Voltage transients in the 12-V electrical system tend to be of a high- or low-energy nature. The high-energy spikes, 40 to 150V, are especially harmful to electronics. High-energy transients are generated by the fast turnoff of high current inductive loads, such as air-conditioning compressor clutches or motors. The polarity of high-voltage inductive spikes depends on the configuration of the control switch. The low-energy spikes can be 2000 V or higher and are generated by static electricity, which is normally encountered in the manufacturing cycle.

4.1.1.3 Reverse battery hookup. Most auto manufacturers specify that all electronic modules, such as motor controls, must be able to withstand reverse battery connections. The exact magnitude of the

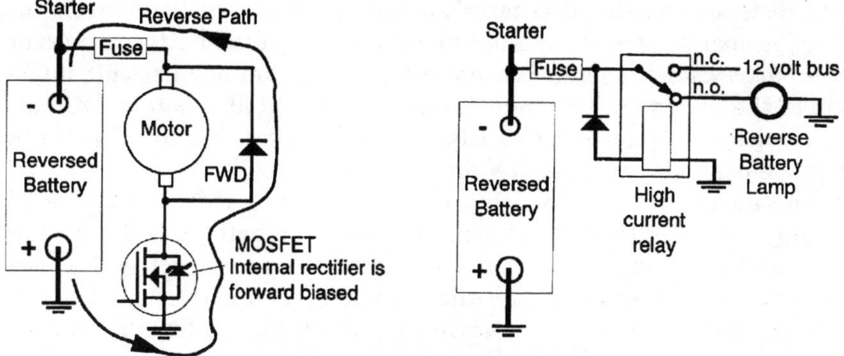

Figure 4.1 Reverse battery problems and protection method.

reverse voltage which can be tolerated varies with each auto manufacturer, but the extreme case appears to be −24 V for 10 min. The requirement to withstand reverse battery connections is especially troublesome with motor control power designs, since they usually present a high-current path when driven from a reversed power supply voltage. The MOSFETs intrinsic rectifier and the FWD become forward-biased, as shown in Fig. 4.1. The result is that the FWD fails or shorts first (it usually has the highest junction temperature in this situation), the transistor shorts second, and then the fuse blows, unless the rectifier or power transistor blows open first. A simple reverse battery isolator relay can be used to guard against incorrect battery polarity connections.

4.1.1.4 Alternator "load dump" event. The alternator's speed is governed by the engine speed (rpm), while the alternator's voltage output is controlled by varying its field current. The traditional alternator (also known as the *Lundell alternator*) is a three-phase claw-pole synchronous machine that uses a three-phase rectifier bridge to convert its AC voltage into DC. When the engine is running, the belt-driven alternator is used to charge the 12-V battery and to supply the normal electrical energy requirements of the vehicle's electrical loads. Under nominal conditions, the alternator charging system works well and usually supplies sufficient energy to keep the battery charged and to power the numerous electrical loads in use when the engine is operational. The alternator employs a rotating field and slip ring assemblies that conduct less than 5 A and therefore require minimal service.

The "load dump" problem can be more significant when the alternator is operating near its maximum ratings, which can happen, for instance, during an engine warmup at subzero temperatures: The battery requires heavy charging, and cold temperatures have reduced the resistance of the alternator's armature windings. The alternator may be operating at its maximum output, and, if the battery cables or the battery itself should become intermittent, an open circuit interrupts the alternator's current flow, allowing the alternator's DC output voltage to rise significantly. In some cases, this unsuppressed voltage peak may reach 125 V for a few milliseconds.

The automotive engineering community calls this event a "load dump," meaning that the alternator's load, the battery, is dumped or disconnected, and the alternator's field winding takes a few milliseconds to shut off, even though the voltage regulator shut off the field current the instant the open circuit occurred. Any motor control electronics connected to the 12-V bus is suddenly subjected to a voltage spike about 3 to 10 times the normal 12-V level. Because the alternator is a low-impedance voltage source, this voltage peak can supply about 1 kW of peak energy. This is more than enough to destroy electronic components that were only rated to function at 12 V, with perhaps a 24-V maximum rating. This is a significant reliability issue and has been addressed in some vehicle designs by using special avalanche-rated alternator rectifiers that clamp the alternator's DC output voltage to 40 V.

When designing automotive motor control electronics, it is important to allow for the load-dump condition. The motor designs shown in this chapter can be protected against load dump by using a high-power transient-voltage suppressor, which also protects against other voltage spikes.

4.1.1.5 Environmental considerations. An electronic motor control must be rugged in order to withstand the stress levels encountered in the automotive environment, including temperature, humidity, vibration, jammed rotor, and long-term use. For engine compartment locations, such as the radiator fan control, the motor electronics must operate reliably over a −40 to 125°C ambient-temperature range and endure thermal cycling as well. The control module will be subjected to high humidity, water splashes, and particulate or chemical contamination, such as salt, dirt, oil, brake fluids, battery acids, wax, and soap. Motor control electronics that are located inside the vehicle's passenger area are usually required to operate over a −40 to 85°C ambient-temperature range and are exposed to some water, soap, and dirt, but seldom chemicals. One aspect of designing automotive equipment is that the vehicle's warranty may last for 10 years, which means that the motor control electronics must be engineered for long-term reliability. The biggest challenge in automotive engineering is to minimize cost while maintaining a sturdy and long-lived design.

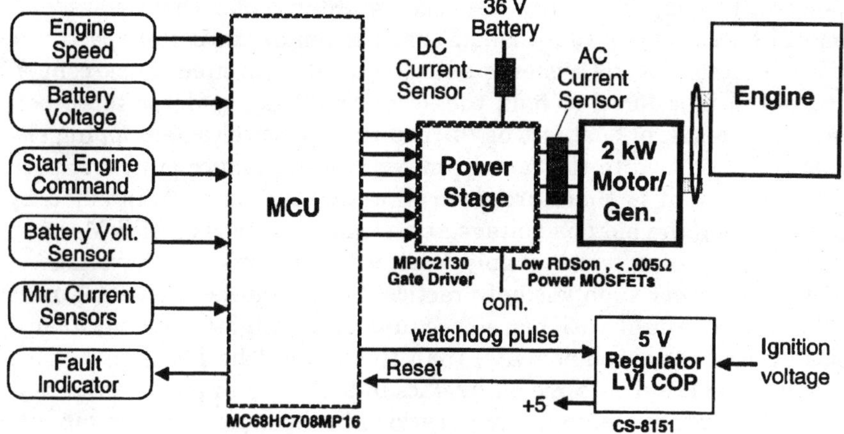

Figure 4.2 Starter motor and generator combination concept.

4.1.2 Advanced starter motor and charging systems

Electric steering, electric brakes, and other advanced automotive electronics require more electrical energy than most alternators can now provide. The present alternator can be somewhat improved by replacing the three-phase rectifier bridge with a transistor power stage. This design, called *synchronous rectification,* lowers the forward voltage drop and can increase the current output. Other types of generators which are under consideration combine both the starter motor and generator in one system. A combination starter motor and generator system concept is shown in Fig. 4.2. This conceptual design uses an AC induction machine that operates as a motor or generator, depending on the control strategy. Note that a microcontroller regulates the system. (Electronic valve control of the engine and higher battery voltages can reduce the engine cranking requirements, making this concept cost-effective.)

4.2 Variable-Speed Automotive Fan Motors

The auto industry uses DC permanent magnet motors for fans and other applications, since these types of motors are economical to produce and provide good performance. The PM motor's speed can be varied by either a simple voltage-dropping passive resistor or an active linear voltage regulator. The simplistic series resistor method is cost-effective and is still widely used for vehicular motors requiring crude speed control. It does have a serious drawback in terms of power efficiency: To control the speed of a fan motor that draws 20 A at full speed, about 10 A at half speed is required. At full speed, the overall motor control system's efficiency will be around 80 percent. If the speed is reduced to half, the system's efficiency drops to 40 percent as a result of heat loss of 70 W in the series voltage-dropping element, in addition to the normal motor losses. A more efficient speed control system is therefore desirable and can be accomplished by interrupting the motor's voltage at a variable duty cycle (PWM) or by using a switching power supply to alter the motor's supply voltage. A switching power supply is not practical because it has to control up to 300 W and would cost more (because of its high-frequency transformer and other components) than the motor. The PWM speed control method requires some electronics but offers high power efficiency.

There are three main fan applications in the typical car: an air-conditioning evaporator blower, a heater core blower, and a radiator cooling fan. Some of these applications may use more than one fan—a dual air conditioner (A/C), for example, has two evaporator blowers. In many vehicles the A/C blower and heater blower are combined. Add-on A/C units use their own blower fan. The design of an electronic variable speed fan control is fairly straightforward but will become more complex when used with a closed-loop temperature control system.

4.2.1 Variable fan motor control design

Designing electronic speed controls for automotive 12-V systems is less complex than for fan speed controls that operate from 120 or 240 V AC power. The low voltage allows direct interface of the control logic to the power stage. We will examine two designs here: a simple open-loop control circuit and a complete automatic temperature control system.

4.2.1.1 PM DC motor fan control. The circuit shown in Fig. 4.3 uses the MC33033 BLM control IC and a single power transistor stage to drive a PM motor. This circuit would be appropriate for HVAC or blower fans that do not require automatic control. The MC33033 device was chosen because it can run its PWM output from 0 to 100 percent, unlike most PWM control ICs, which limit the PWM to less than 100 percent. With this design the user is able to vary the speed in four steps per the multiposition switch. The voltage divider resistor string gives 25, 50, 75, and 100 percent speed reference points. In this open-loop type of circuit, the user must select the fan's speed to vary its effect on the air temperature.

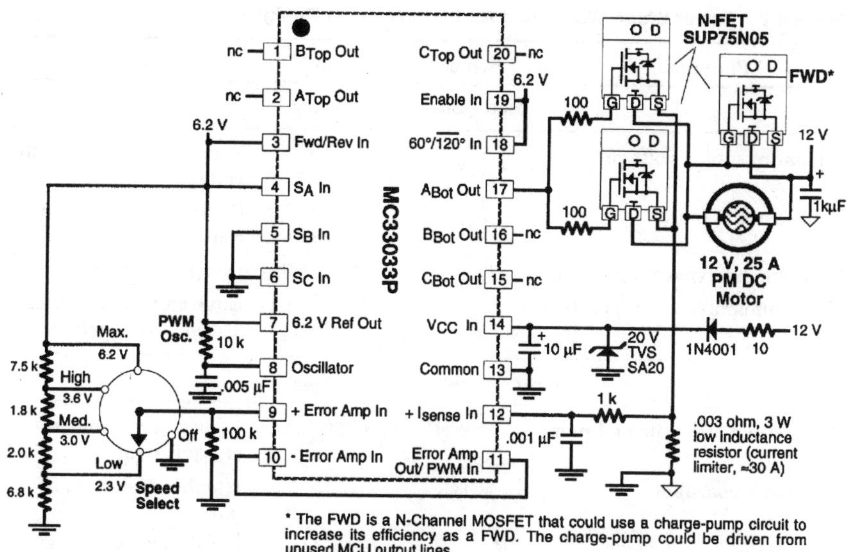

Figure 4.3 Open-loop variable-speed fan design.

The circuit includes a current limiter, which will shut off the PWM cycle when the motor's current levels exceeds about 30 A. The cycle-by-cycle current limit is set by the 0.003-Ω sense resistor and the 100-mV comparator inside the MC33033. The EMI generated by the power stage may be a problem, but it can be minimized by slowing the gate drive transitions with an *RC* network on the N-channel MOSFET's gate. Slowing the gate drive edges will increase the FET's power loss, and, therefore, must be approached with caution. A *RC* snubber across the FWD will also help reduce EMI. The FWD uses the same part type as the motor's power switch. This was done to minimize parts inventory, but other FWD devices could also be used, such as ultrafast or Schottky barrier rectifiers. The N-channel MOSFET power switching devices shown in the test circuit are rated at 50 V, 75 A, 0.006 Ω, and are in a TO-220 package. These were the lowest $R_{DS,on}$ devices available in the TO-220 package at the time of this book's publication. Two of these N-channel MOSFETs connected in parallel will dissipate less than 4 W at a motor current of 30 A.

A worksheet (see Table 4.2) is useful for determining the power stage requirements in the fan test circuit. Note that the maximum ambient temperature is set to 85°C, and the maximum allowable heatsink temperature is set to 110°C. This demonstrates one of the dilemmas of designing a power stage that must work reliably in a hot environment while maintaining a reasonable cost and size for the

TABLE 4.2 Power Stage Worksheet for the HVAC Fan Design

Parameter	Symbol	Value	Notes
	Motor Specifications		
Maximum supply voltage	V_{sup}	*12 V*	Maximum power supply voltage
Motor current limit	I_{max}	*30 A*	Current limit value per control circuit
Allowable heatsink temperature	$T_{hs,max}$	*110°C*	Design point ($>T_{a,hot}$)
Maximum ambient temperature	$T_{a,hot}$	*85°C*	Environmental temperature, lower dash area
	Power Switch Specifications		
Power transistor type and name	MOSFET	*SUP75N05**	Device part number, * or equivalent
Quantity in parallel	Qty	*2 each*	Number of devices in parallel
$R_{DS,on}$ at 25°C, each device	$R_{DS,on,each}$	*0.005 Ω*	Per device data sheet
$R_{DS,on}$ at 25°C, total	$R_{DS,on}$	**0.0025 Ω**	$= R_{DS,on,each}/\text{Qty}$
Junction-to-case resistance	$R_{øjc}$	0.80°C/W	Per device data sheet
	Heatsink Specifications		
Heatsink thermal resistance calculation	$R_{øsa}$	**7.4°C/W**	$= (T_{hs,max} - T_{a,hot}/P_{D,hot})$
Case-to-heatsink thermal resistance	$R_{øcs}$	*0.1°C/W*	Per mounting material data
	Hot Operation (Nonswitching, Locked-Rotor Condition)		
Estimated $T_{j,rise}$	$\approx T_{jc}$	1.80°C	$= \{[(I_{max}{}^2)/\text{Qty}]* R_{DS,on,each}\}*R_{øjc}$
Estimated $T_{j,max}$	$\approx T_{j,max}$	114°C	$= \{[(I_{max}{}^2)/\text{Qty}]*R_{DS,on,each}\}*R_{øjcsa}) + (\approx T_{jc} + T_{hs,max})$
$R_{DS,on}$ temperature variation factor	$R_{DS,on,x}$	*1.5*	Per device data sheet for $\approx T_{j,max}$
$R_{DS,on}$ at $T_{a,hot}$	$R_{DS,on,hot}$	0.0038 Ω	$= R_{DS,on}*R_{DS,on,x}$
$V_{DS,on}$ voltage at $I_{max,hot}$	$V_{DS,on,hot}$	0.11 V	$= I_{max}*R_{DS,on,hot}$
Power transistor dissipation	PD_{hot}	**3.4 W**	$= I_{max}*V_{DS,on,hot}$
Junction-to-case temperature rise	T_{jc}	1.4°C	$= R_{øjc}*(PD_{hot})/\text{Qty}$
Case-to-heatsink temperature rise	T_{cs}	0.17°C	$= R_{øcs}*(PD_{hot})/\text{Qty}$
Heatsink-to-ambient temperature rise	T_{ha}	25°C	$= R_{øsa}*P_{d,hot}$
Maximum junction temperature	$T_{j,max}$	**112°C**	$= (T_{jc} + T_{cs}) + T_{hs,max}$

NOTES: *Italicized text* indicates design value entries; **boldface text** indicates critical results of calculations.

heatsink. If only one N-channel MOSFET power switch were used, the heatsink's thermal requirements would increase. You must then decide whether it is more cost-effective to use a larger heatsink, a larger power transistor, or to use parallel transistors. Since power MOSFETs tend to operate properly in parallel because of their positive $R_{DS,on}$ temperature coefficient, it makes sense to look at the last option. *When paralleling MSOFETs, you must be careful to insert a series gate resistor to each MOSFET to minimize the chance of high-frequency oscillations.*

The 25-A fan motor could draw more than 60 A in a locked-rotor condition, but the control circuit is designed to limit the motor's current to about 30 A. Since the MC33033 limits the fan's current, this will increase the fan's start-up time, until some back EMF is generated, and the current drops below 30 A. (When a fan starts up with 100 percent speed input to the MC33033, and the motor current reaches 30 A, the MC33033 current limiter shuts off that PWM period.) The exact time it takes for the fan motor current to drop below 30 A depends on the motor's characteristics and the fan's inertia. One problem that may be observed is audible noise from the PWM duty cycle variations during the 30-A current-limiting event. Adjusting the PWM frequency and the current limit setpoint will reduce the problem.

4.2.1.2 BLM fan control. One drawback of using a PM motor for fan applications is its inherent brush-to-commutator audible noise and RFI. (The RFI is created by the brushes arcing and can interfere with the radio reception of weak stations.) The audible noise at low fan speeds can be significant in luxury cars, since their insulation from road noise allows sounds inside the vehicle to be easily discerned. Another drawback is the susceptibility of the brushes and commutator to water and dust. This can be a problem for the radiator fan because of its location. Obviously, choosing a motor that doesn't use brushes eliminates these specific problems but increases the cost of the fan system. (The BLM designs shown in Chap. 3 can be adapted for use with a 12-V brushless motor.)

4.2.1.3 Automatic fan control. Thus far, our motor control designs have focused only on open-loop motor speed controls. A complete climate control system is shown in Fig. 4.4, which manages temperature with the fan's speed and the other components in a HVAC system. A microcontroller is used instead of a linear IC. The MCU allows programmability, networking, and diagnostics. It is possible to use a linear IC, such as the MC33033, with a temperature sensor feedback line in its speed control (refer to the MC33033 data sheet). The 8-bit microcontroller device in this temperature control system was designed specifically for low-performance motor control systems and

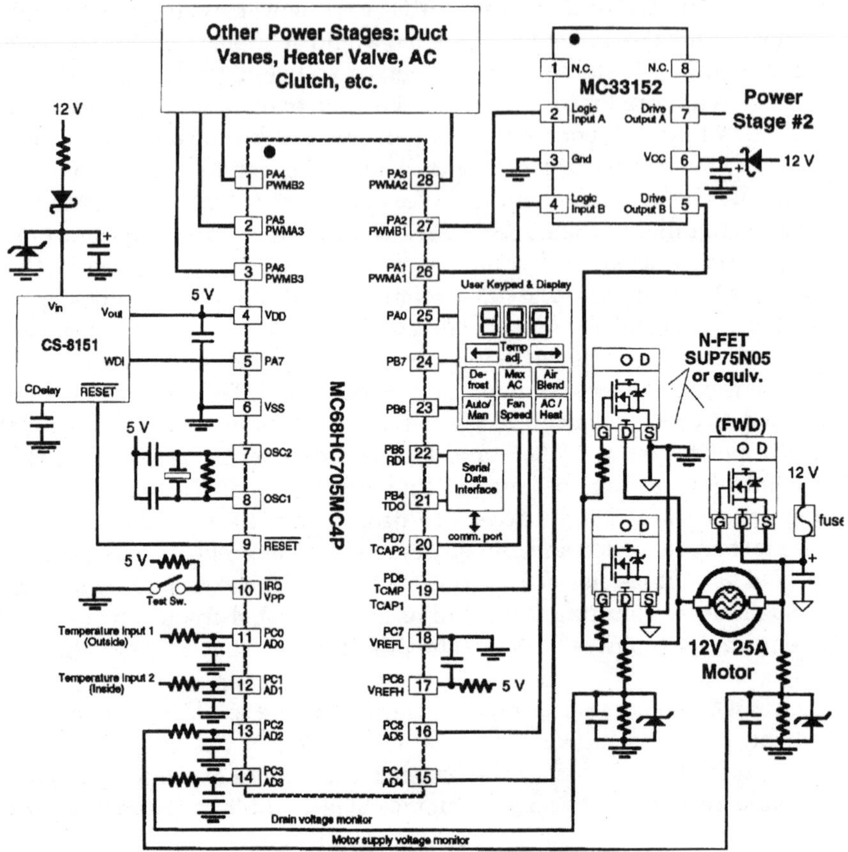

Figure 4.4 Automatic temperature control system.

is covered in more detail in a later chapter. The MCU motor drive outputs are buffered with a MOSFET gate driver, an MC33152.

The MC33152 logic inputs have a 30-kΩ internal pull-down resistor and are designed to toggle at 1.67 V with about 0.017 V of hysteresis. The 30-kΩ resistor helps ensure that the drive output stays low when the MCU is in a RESET state. The IC's output stage is a totem pole with 1.5-A peak ratings. The IC has an undervoltage lockout circuit that shuts off the output when the V_{cc} is below 5.8 V. The undervoltage lockout has about 500 mV of hysteresis.

The motor's power stage uses two low $R_{DS,on}$ N-channel MOSFETs connected in parallel. Two of the MCU's analog inputs monitor the power stage for fault conditions. The FET's drain-to-source voltage is monitored to indicate possible fault conditions: (1) if the supply voltage, 8 to 16 V, is not present when the N-channel MOSFET is switched off, the FET is shorted or the motor has an open winding or the power supply fuse is blown; or (2) when the FET is switched on, the drain-to-source voltage should be less than 0.12 V. If it is (0.18 to 0.3 V), the motor is in a locked rotor condition; if it is higher than 0.5 V, a shorted motor may have occurred or the FET has failed. This illustrates how the MCU can help detect a fault in the motor control power stage. Fault codes could be generated that can help the service technician trace the elements that have failed. Note that this conceptual design uses no current limiting and relies on the fuse to protect the power stage in the event of a short circuit. The two paralleled N-channel MOSFET devices can easily conduct over 400 A, which should blow out a fuse before the FETs fail from excessive heat dissipation.

In some automobiles, a more complex overall system control may be required that involves communication between the various electronic controls. For example, the engine management system may operate in a different manner if the air-conditioning compressor is switched on. The climatic control system would operate the heater control in a more comfortable manner if recognized that the engine coolant temperature is still cold and "understands" not to run the heater fan until the engine is approaching its normal operating temperature range. In automotive systems, module-to-module communication, referred to as *multiplexing* (MUX), has been implemented in many new vehicle designs. (Chapter 10 reviews the networking of different control systems.)

In future internal combustion engine designs, the water pump and other mechanically driven components may be electronically controlled that require communications with the engine management system.

4.3 Wiper Motors

Windshield wiper systems usually allow for multispeed speed operation, variable intervals, and blade parking and customarily work in conjunction with a washer pump. The wiper blade mechanism is connected by a worm-gear design in the motor, giving torque multiplication while allowing the motor to run over a practical speed range. An important design specification is that the motor must sustain a "stalled" wiper condition without damage. The motor must also provide sufficient torque to run the wiper mechanism if the blades are frozen to the windshield. Some form of electronic control is used for variable-interval operation. This can be a simple timer unit that drives a relay or a transistor power stage. The driver adjusts a time interval potentiometer or a multistep switch that is connected to a resistor divider network to set a wiper rate from one to several seconds. The electronic circuit activates the wiper motor with the appropriate time interval. The wiper's arm and linkage design vary from rather simple mechanical designs to fairly complex designs that allow more of the windshield to be wiped.

Most wiper systems use a travel-limit switch to ensure that the wipe cycle is completed and that the blades are returned to their nominal position when turned off. With today's technology, automatic windshield wiper designs are possible that utilize a moisture detector to determine the degree of wiper action required.

4.3.1 Wiper motor controls

The advanced wiper motor control system, shown in Fig. 4.5, uses a low-cost MCU to control the wiper system. The MCU allows precise control of the intermittent wipe mode and a one-touch wash-and-wipe operation. The MCU could also control other functions such as flashers and headlamp retractors. Power MOSFETs are used to drive the washer motor and wiper motor. A variable wiper speed rather than a fixed slow or fast speed is possible with an MCU PWM output. A limit switch detects when the wiper motor has reached its normal resting or OFF position. An additional motor can be added to change the wiper

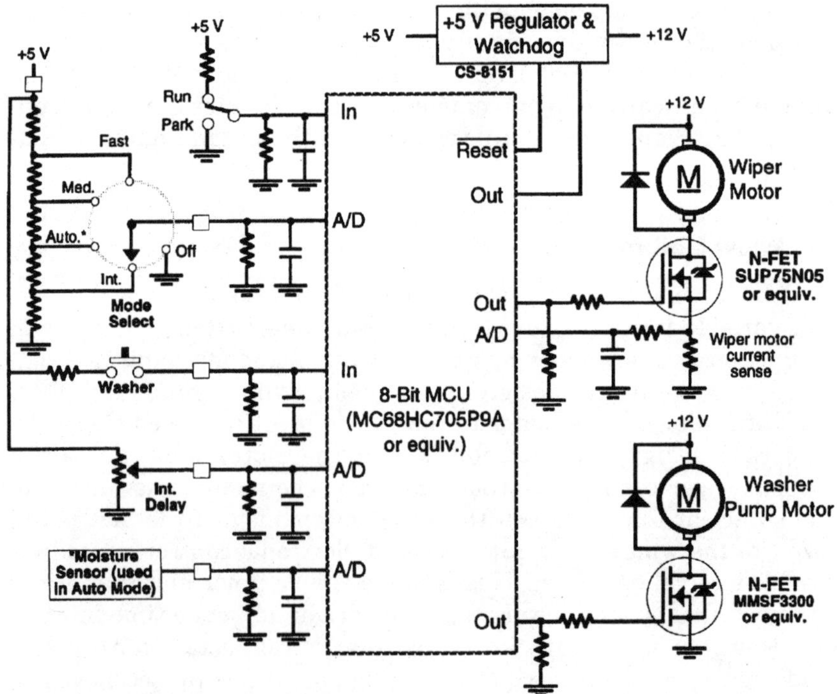

Figure 4.5 Advanced wiper control system.

blade pressure per the vehicle speed. This allows the blade pressure to be optimized and can reduce wiper edge deformation at rest.

4.4 Power Seat Motors and Miscellaneous Motors

Power seat motor controls are common in many luxury vehicles. With some seat control systems, the seat position is stored in the MCU's memory, which resets the seat's position to the driver's preference. The seat motors are permanent magnet brushed motors that run bidirectionally. Some sort of position feedback is necessary for the memory seat to find its exact position. One positioning feedback method is to mechanically link a potentiometer to the seat whose output is proportional to the seat's position. (This is similar to a servo-type motor system, discussed in Chap. 3.) The top of the potentiometer is connected to a fixed voltage (5 V, for example), the bottom connected to common, and the slider gives an output voltage according to the seat position.

The steering wheel column tilt, height, and distance can also be controlled by motors with a memory-type electronic control. It is possible to construct a smart module that uses just one motor to drive more than one mechanical load.

4.5 Electric Steering Motor Systems

Electric motors can be used to either drive the hydraulic pump or the steering mechanism. The design challenge for an electric-powered steering-assist system is to offer long-term safety and reliability and also remain cost-competitive with the existing hydraulic pump/actuator system. The safety dilemma of a direct-assist electric steering design is that the driver cannot always overpower the electric motor's running torque (or braking torque) in order to seize and retain control of steering in the event of a fault condition. Designing protective elements to prevent this condition can be expensive but is necessary. Both the electrohydraulic or direct-assist power steering system require highly efficient and cost-effective power devices. A 1.5-kW 12-V motor control in a 125°C environment requires extremely low $R_{DS,on}$ power MOSFETs.

Figure 4.6 shows block diagrams of two methods under implementation or consideration for electric-powered steering systems: the electrohydraulic and the direct-assist electric motor steering. Note that each method has similar functional requirements, which include the MCU and power stage. The MCU and power stage design specifications are directly affected by the motor type and its control strategy.

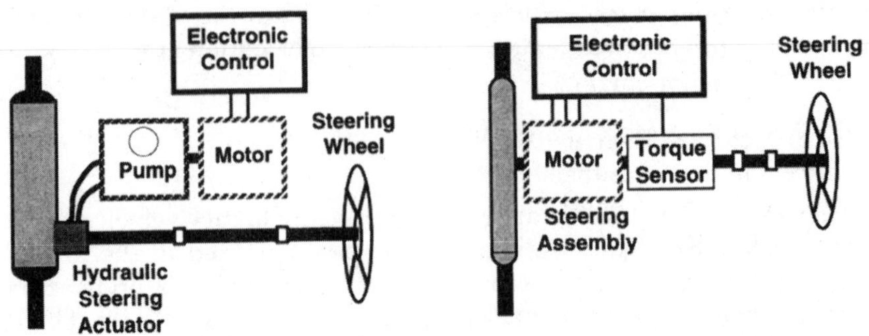

Figure 4.6 Electric motor power steering assist systems.

Low PWM frequency (<15 kHz) can create audible noise and decrease the motor's speed resolution but lower the power stage switching losses. A high PWM frequency (>15 kHz) is usually inaudible and increases the speed resolution (which requires a more powerful MCU) as well as switching losses. In an electric-assisted steering application, some PWM audible noise may be tolerable, since the engine noise will probably mask it; the motor's speed or slew-rate requirements, however, may demand higher PWM frequencies.

4.5.1 Electric steering power stage design considerations

Due to the safety aspects of the steering system, the design of a reliable motor power stage requires power transistor devices that can operate in a harsh environment. The power stage will be exposed to extreme temperatures, thermal cycling, and abnormal power supply voltages (i.e., reverse battery, load dump) while switching currents in the 100-A range. A failure in a seat motor control is an inconvenience, but an erratic steering system due to a motor control electronics failure is a serious safety issue.

Many factors affect the semiconductor costs and specifications for higher-voltage vehicular battery voltages, but, usually, the system cost of large motor drives can be reduced by increasing the supply voltage. Assuming that a 1.5-kW motor that is powered by a 12-V battery system, the motor current level is 125 A. Power dissipation increases by the square of the motor current and is calculated as

$$P_d = I^2R = (125 \cdot 125) \cdot 0.005 = 78.1 \tag{4.1}$$

where P = total power stage dissipation, W
R = total power stage on resistance, Ω
I = motor current, A (maximum load condition)

At a battery voltage of 12 V and a motor current of 125 A (1.5-kW motor), the power transistor stage's dissipation is about 78 W. The power transistor's heat dissipation is also directly affected by the PWM frequency and the transistor's switching times, but, for simple calculation purposes, these factors will be ignored. Raising the battery voltage to 36 V and using the same 1.5-kW motor would reduce the current levels to 42 A, which would lower the power transistors to less than 9 W if the same power transistors are used. In the case of a 36-V system, the 50-V-rated power transistors would be operated closer to their maximum ratings. *It is important to remember that the system cost of large motor drives used in automotive applications can usually be reduced by increasing the supply voltage. This assumes that the load dump and other voltage transients will be suppressed to less than 50 V, and that 50-V-rated power devices can be used, rather than 200-V-rated devices in the 12- to 36-V example.*

4.5.2 Electric motor hydraulic system

Removing the power steering pump from the engine offers several advantages. There is one less belt-driven load, there are no hydraulic hoses to the engine, an engine stall does not result in loss of power steering assist or boost, and there is integration of the hydraulic pump into the steering rack assembly. Cost is the major disadvantage for the electrohydraulic system, but development of a totally integrated steering system (the hydraulic pump, control module, and actuators are contained in one unit) may help decrease the cost factor. Essentially, the hydraulic pump is removed from the engine and driven from an electric motor.

4.5.2.1 Hydraulic motor design. The electric motor type chosen for driving the hydraulic pump determines the complexity of the motor's drive electronics. Table 4.3 compares the MCU requirements of four types of motors. The reliability of a brush motor (brush and commutator wear) may be a further consideration. While the PM motor is less expensive, it is subject to magnet degradation at high temperatures with excessive motor current levels.

The PM (permanent magnet) brush motor can be driven from a single power switch and external free-wheeling diode (FWD), as shown in Fig. 4.7. The 8-bit MCU (a MC68HC705MP4 or equivalent) in this

TABLE 4.3 Motor Electronics Comparison

Motor type	Power efficiency*, %	System cost	MCU type†	Power stage‡
PM brush	85	Lowest	8-bit low	1 sw + 1 FWD
Three-phase PM brushless	95	Highest	8-bit high	6 sw
Three-phase AC induction	90	Medium	16-bit high	6 sw
Three-phase sw reluctance	90	Medium	16-bit high	6 sw + 6 FWD

*Power efficiency is estimated and assumes optimum drive strategy.
†MCU performance relates to 8-bit or 16-bit architectures with PWM timers and A/D conversion.
‡The number of power MOSFET switches and external free-wheeling diodes.

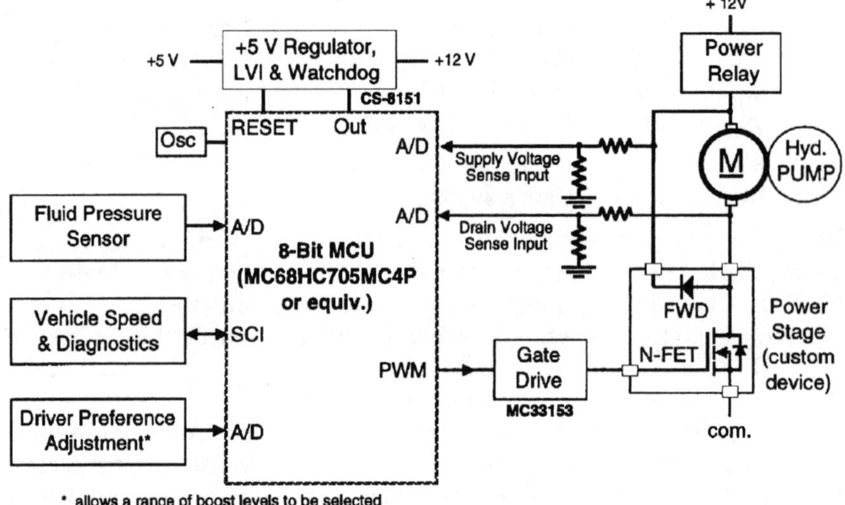

Figure 4.7 Hydraulic pump uses PM brush motor.

system drives the motor with a PWM signal to vary the motor's speed. This maintains a certain pump pressure level and allows the motor to operate only when required. The PWM signal is generated from an internal timer. The vehicle's speed value can also be used to adjust the boost pressure; at high road speeds, the boost is lowered, and, at low speed or parking, the boost is at maximum. The speed sensor could be eliminated by obtaining the speed data from the engine control system via a MUX interface. (See Chap. 10 for MUX interfaces.)

If the electric-motor-driven hydraulic pump system is operated open-loop, that is, without the pressure feedback signal, it is important to monitor the motor's supply voltage to compensate for varying voltage conditions. The speed of the motor will be directly affected by its supply voltage and may need to be adjusted by PWM control to maintain a nominal boost pressure. The ON voltage of the N-channel MOSFET is sensed to determine if it is operating in a normal range. The ON voltage of the FET will also indicate motor current but with a high error at −40 and +125° temperature points due to the FET's ON resistance temperature coefficient. The motor loading is highly intermittent in this system because steering events constitute a small percentage of normal driving. During ordinary driving, the pump only needs to maintain boost pressure and requires a small amount of power from the electric motor. The motor could operate at either a fixed speed or an adjustable speed. Some electrohydraulic systems implement methods to allow the motor to come to a complete stop during nonsteering time by using a hydraulic accumulator. This allows the boost pressure to be immediately available and gives time for the motor to start up and maintain the boost pressure.

The motor will need to work at maximum load to maintain the pump outlet pressure during extended vehicle maneuvering. The motor's power rating is chosen for this maximum case condition, which means that the motor and drive electronics must exhibit high-power efficiency during light loads (important for fuel savings) and supply ample power during steering maneuvers, especially during parking or at low road speeds.

4.5.3 Direct-assist electric motor steering system

The steering rack can be driven from an electric motor, eliminating all the hydraulic components. Obviously, reliability and safety are the top design considerations for the direct-assist steering system. The challenge is to keep the cost under control so that this system is still viable for standard vehicles. The motor affects the system's cost and performance and also plays a role in the safety aspects of the direct-assist design. PM brush motors, while inexpensive, present an inherent safety hazard if one-half of the bidirectional power stage shorts or is left on by an MCU glitch; the PM motor simply runs at maximum speed in whatever direction the short circuit or MCU glitch allows. This means that the vehicle is steering itself, and the driver will need to exert considerable physical strength to overcome the electric motor. A method to circumvent this particular type of failure is mandatory for the direct-assist electric steering system. Other motor types (BLM, induction, and SWR) require electronic commutation of the motor's windings for the motor to operate, and, if a power stage short circuits or a logic glitch occurs, the motor shaft stops rotating and locks up.

DC current flowing through the motor windings causes the lockup of the motor shaft, and, again, the driver must strain to overcome it with the steering wheel. The lock-up condition, however, may not immediately steer the vehicle off course, as will happen with a PM brush motor, so more time is available for a backup mode of operation to be invoked.

4.5.3.1 Motor control reliability design methods. Methods to detect power stage failures, microcontroller program crashes, erroneous sensor levels, and other electronic component faults require smart diagnostics and are necessary in order for the direct-assist electric motor system to succeed. In simple terms, the motor must be disabled by removing its power supply voltage when a fault condition occurs. Evidence of fault conditions are detected by using a second MCU and other gates, as shown in Fig. 4.8.

A second MCU is used to monitor the sensor inputs and main MCU's outputs for abnormalities. The main MCU also needs to send

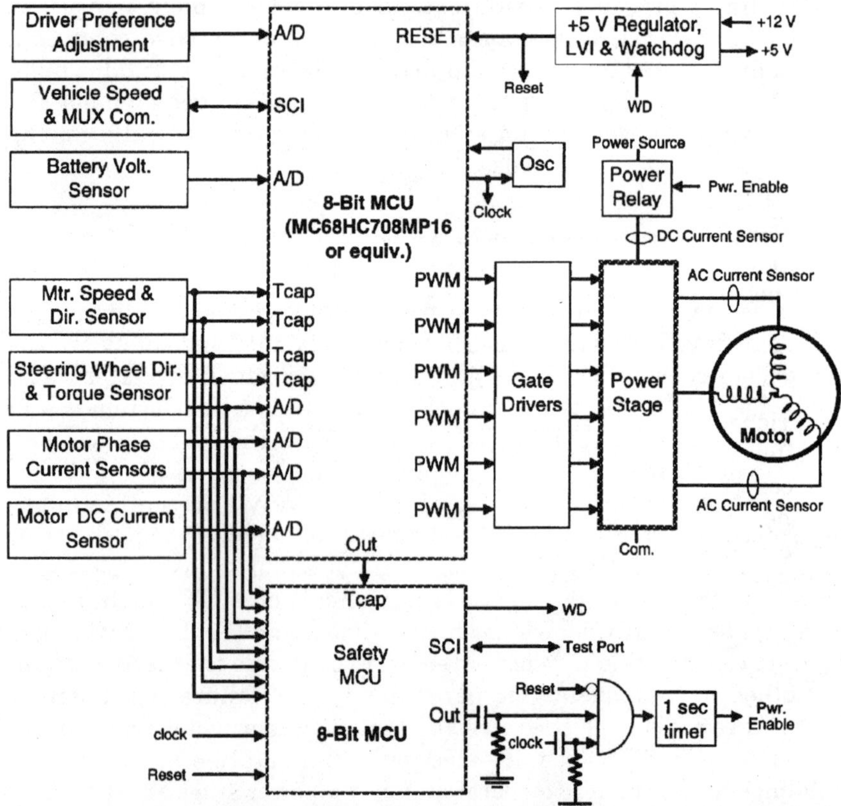

Figure 4.8 Direct-assist power steering system safety elements.

an appropriate signal to the second MCU which indicates that the main MCU is operating in a normal fashion. If the second MCU finds no errors, it sends a signal to an AND gate. The AND gate is also connected to the reset line and clock signal. If any one of these signals is absent for more than 1 s, a power relay is turned off, thus disconnecting the power from the motor. Note that the design of this power relay or contactor would need to ensure that the contacts would open without sticking or welding shut under worst-case conditions. The second MCU monitors all critical input/output signals; if any critical fault occurs, the motor's power supply is disabled and a diagnostic message is sent to the test port. The second MCU uses nonvolatile memory for recording the normal key parameters and any fault conditions, promoting faster diagnostics.

4.5.3.2 System performance requirements. When the hydraulic pump and actuator are completely replaced by an electronic motor, the direct-assist electric steering system needs to provide a smooth and rapid steering boost. This requires a high torque and fast-responding motor drive that can also be controlled to correlate with the vehicle's speed. The precision and smoothness of the direct-assist electric motor drive relate to the throughput (computing time) and bit size of the MCU, as well as to the steering wheel torque sensor specifications. DSP-based motor drives provide quick and precise computing power. Motor control drive software strategies directly affect the torque response of the motor. Vector control improves the motor's torque-speed range more than a *V/F* control strategy would, but the vector control program requires a higher performance MCU and peripheral elements, as discussed in Chap. 8.

4.5.3.3 Motor power stage design. Conventional AC induction and BLM designs need a half H-bridge for each phase of the direct-assist electric motor power stage. A SR (switched reluctance) motor requires a different power stage topology, as shown in Fig. 4.9. The FWD devices of the SR power stage need to be connected in a different manner, and the motor winding terminals need to be brought out separately. SR motors are reported to offer low-cost and high-torque characteristics.

Power MOSFET technology is preferred for use in 12-V power stages because of its low forward ON voltage and diminished bias requirements. Figure 4.10 compares different voltage-rated MOSFETs and a IGBT with similar die areas. Note that, as MOSFET technology improves, the die size for the lower-voltage devices will decrease. Some of the design factors that affect a MOSFET's $R_{DS,on}$ are planar or trench, bulk resistance, cell density, lead resistance,

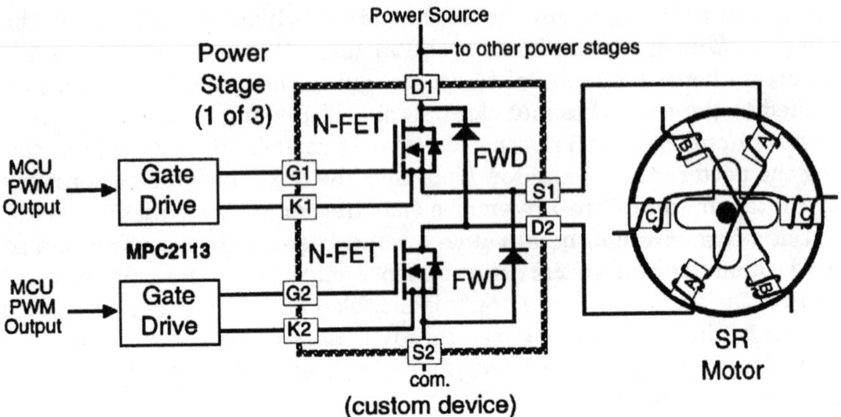

Figure 4.9 SWR motor power stage.

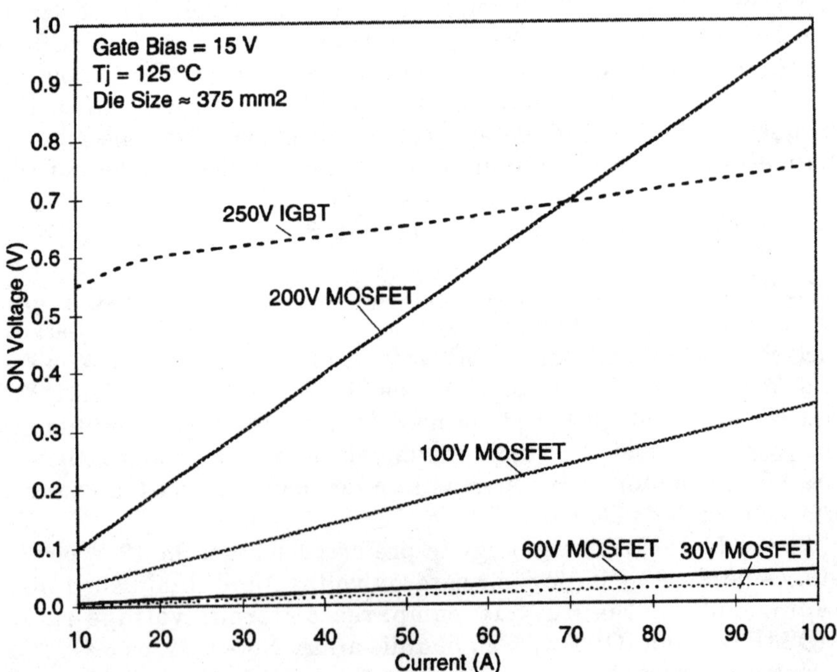

Figure 4.10 Comparison of MOSFET devices.

wire bond resistance, and gate design. Generally, the $R_{DS,on}$ for high-voltage MOSFETs is affected mostly by the bulk resistance and can be lowered only by using larger die areas. At 200 V, the IGBT technology offers good performance at high current levels.

The calculation of the power MOSFET's $R_{DS,on}$ requirement involves several factors, but a quick estimate can be made by just using the basic parameters. The first step is to determine how much heatsink temperature rise can be allowed. The power stage heat management aspects can be roughly calculated, as shown here:

$$T_{hs,rise} = T_{j,max} - T_{a,max}$$

$$50°C = 175 - 125 \qquad (4.2)$$

where $T_{h,rise}$ = maximum heatsink temperature rise, °C
$T_{j,max}$ = maximum allowable junction temperature, set by data sheet, °C
$T_{a,max}$ = ambient-temperature maximum, set by application, °C

When you know that the maximum allowable temperature rise of the heatsink is 50°C under worst-case conditions (i.e., with full load and maximum ambient temperature), the maximum heatsink dissipation can be roughly calculated. First, a heatsink must be selected. There are several methods for choosing a heatsink size. One method is to use the largest heatsink for the application that will fit in the area allowed. The maximum allowable dissipation will be directly affected by the heatsink size. Selecting the heatsink size and power transistor $R_{DS,on}$ is a compromise between heatsink cost and power MOSFET cost. Choosing the biggest heatsink that will fit into the size allotment is a good starting point since it will give an indication of the $R_{DS,on}$ requirements for the MOSFET power transistors. For this example, a large heatsink is chosen that has a 0.5°C/W thermal rating. This means that, for each watt of power dissipation, the heatsink's temperature will rise by 0.5°C. (*Note:* This is a rough estimate and does not take into consideration additional thermal resistance from the transistor's junction-to-case or case-to-heatsink interface, or switching losses.) Using the previous 50°C maximum heatsink temperature rise, the heatsink's maximum power dissipation is calculated as follows:

$$PD_{hs} = T_{h,rise}/HS_{spec}$$

$$100 = \frac{50}{0.5} \qquad (4.3)$$

where PD_{hs} = maximum heatsink power dissipation, W
$T_{h,rise}$ = maximum heatsink temperature rise, °C
HS_{spec} = heatsink thermal rating, °C/W

The current drain for a direct-assist electric motor steering system from a 12-V battery system will be in the 125-A range. The total $R_{DS,on}$ of the MOSFET power stage can be calculated using the 100-W maximum power dissipation rating as shown:

$$R_{DS,on} = PD_{hs} \div I_{max}^2$$

$$0.0064 = 100 \div (125 \cdot 125) \tag{4.4}$$

where $R_{DS,on}$ = nominal power stage total ON resistance, Ω
PD_{ns} = maximum heatsink power dissipation, W
I_{max} = maximum power stage current, amperes

A 100-W dissipation maximum with a 0.5°C/W heatsink operating in a 125°C environment means that the nominal $R_{DS,on}$ is 0.0064 Ω for the power stage. This value would need to be reduced by about 33 percent, or to 0.004 Ω to find the nominal 25°C data sheet $R_{DS,on}$ value. (Note that the exact $R_{DS,on}$ temperature coefficient will vary per each MOSFET type.) When a bridge circuit topology is used, as in a BLM or AC motor design, the $R_{DS,on}$ will again need to be reduced by 50% or to about 0.002 Ω because there are, in effect, two MOSFETs operating in series. This is a very low $R_{DS,on}$ value and will probably require the paralleling of MOSFETs. Going to a larger heatsink—for example, 0.25°C/W—would change the $R_{DS,on}$ requirement to 0.004 Ω. Conversely, using a really low $R_{DS,on}$ MOSFET, such as 0.0005 Ω, would allow the heatsink size to be reduced. This illustrates what we mean when we say that a compromise must be made between the cost of the heatsink and the power MOSFET device.

The preceding calculation of the MOSFET's $R_{DS,on}$ has revealed that a low $R_{DS,on}$ specification is required. Since the $R_{DS,on}$ is dependent on the MOSFET's breakdown voltage, it is wise to select a device with as low a breakdown voltage as the application will permit. When working with a supply voltage that can have voltage spikes, it is sometimes more cost-effective to use an external transient voltage suppresser rather than have a power device that is rated for higher voltages.

To ensure reliability, you need to reduce the MOSFET's power dissipation by using low-$R_{DS,on}$ devices. The power stage will be subject to serious thermal cycling, since the power steering system operates in a random pulse fashion. Thermal cycling is known to induce failures in power devices and is affected by the semiconductor die attachment methods, package design, and materials. The range of the thermal cycling temperature can be reduced with lower power stage dissipation.

4.5.3.4 MCU types for direct-assist steering system. This system requires a high-performance MCU and may actually require two—one high-performance unit (MC68HC708MP16, MC68HC16Y1, MC68332G, or equivalent) for real-time control and a second MCU that performs continuous fault detection. Other MCUs can be considered that utilize a reduced instruction set computer (RISC), or digital signal processor (DSP) architecture. (Chapter 9 reviews MCUs for motor controls.) The choice depends on the system application in which the motor control MCU is targeted. Less demanding systems, such as economy subcompact vehicles, may use a less expensive complex instruction set computer (CISC) architecture; the demands of the direct-assisted power steering, as in a high-speed sports car, may require a DSP or RISC-type MCU.

A high-performance MCU for the direct-assist power steering requires a powerful PWM generator. PWM generators are like the A/D converters in that there are high-precision PWM units and low-precision PWM units. The bit size of the PWM unit and its clock frequency determine its precision. Lab tests have shown that the PWM timer requires a faster clock period (25 ns) when using more bits of resolution (12 bits, for example) in high-performance motor drives. Ideally, the PWM unit would consist of at least six PWM channels to support the half H-bridge drive of a three-phase motor. These PWMs should have the ability to be paired with the inclusion of deadtime logic for the half H-bridge drives.

The direct-assist power steering (DAPS) system uses two kinds of A/D converters: the high-precision, high-speed (12-bits, 2-μs) A/Ds used in the control loops of the MCU and the lower-speed, lower-resolution (<10-bits, 10-μs) A/D converters used to process less critical signals, such as temperatures or pump pressure inputs. Unfortunately, high-speed, high-precision A/Ds are very costly to implement on an MCU. Therefore, the motor phase current sensing is done with external high-precision A/Ds, and other less demanding A/D elements are integrated into the MCU.

A conceptual DAPS system that offers more processing power, hardware support, an event timer, quadrature encoding, and either a DSP or RISC core is shown in Fig. 4.11. The A/D conversion unit has been increased to 10 bits. With this type of processor the entire motor control functions can be achieved in a single chip. The PWM generator supplies the gate signals, the A/D converter processes the feedback signals, and the high-speed RISC or DSP core executes the control algorithm.

Another option for advanced motor control electronics is in partitioning. If you need performance similar to that of a high-end processor but would like the benefits of a low-cost processor, another option

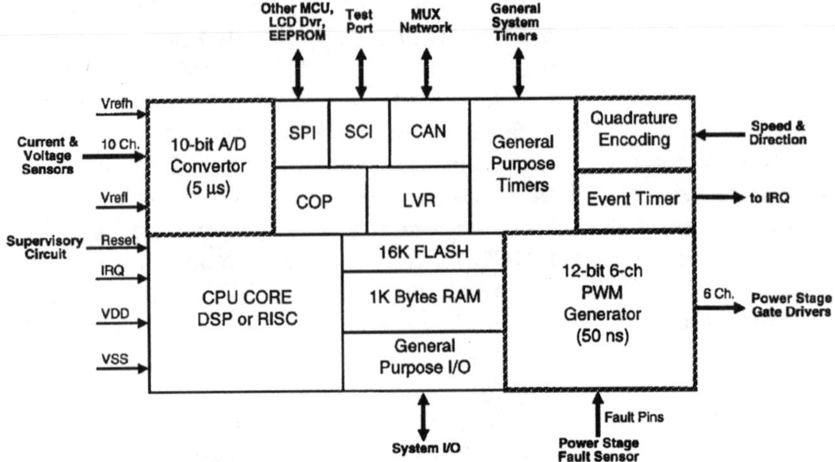

Figure 4.11 Conceptual direct-assist power steering MCU.

might be to utilize the low-cost motor control MCU to do the motor control functions and employ another low-cost general-purpose MCU to assist in the control functions. Whatever the desired system, integrating some of the motor control functions in an MCU can reduce system cost and complexity.

4.6 Electric Motor Brake Systems

An electric motor is used in hydraulic braking systems to drive a hydraulic fluid pump, which maintains high pressure in some advanced antiskid braking (ABS) systems. (This type of system is called electrohydraulic brake management.) Many conventional ABS systems also use a pump to recirculate the brake fluid during ABS operation. Electric motors can replace the hydraulic brake cylinders as well and have been used in the production of electric vehicles. (General Motors 1997 EV1 uses a rear electric brake system.) This type of system, which employs electric motors rather than hydraulic components to achieve braking, is a true electric brake-by-wire design.

4.6.1 Electrohydraulic brake system

The electrohydraulic brake (EHB) system relies on an electric motor to supply hydraulic pressure to the wheel brake cylinders. Figure 4.12 shows a system diagram of one EHB implementation. The high-pressure pump is controlled by the MCU. The motor's speed is used to adjust the pump power and is controlled to match the needs of the

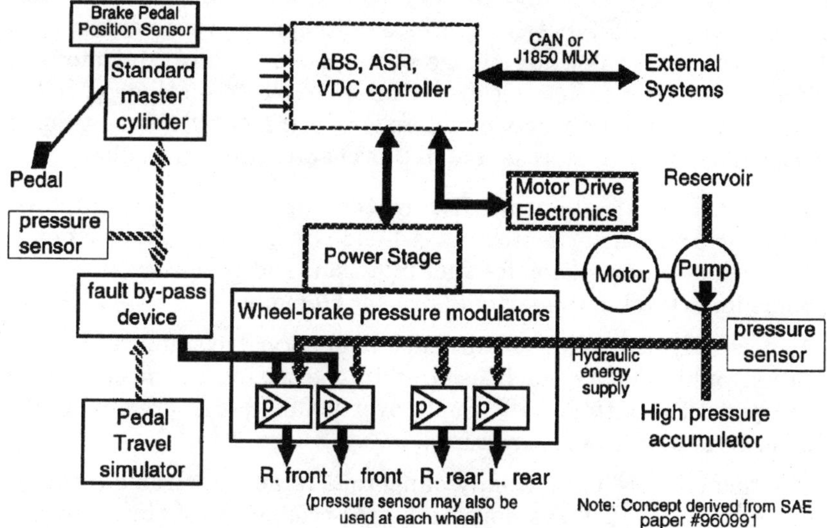

Figure 4.12 Conceptual electrohydraulic brake system.

brake operation. The advantage of this system is that the brake pedal remains stable during ABS operation, and the brake pedal response time can be improved.

4.7 Future Automotive Motor Applications

Any automotive application that requires mechanical force to function can be considered a candidate for an electric motor, including even the internal-combustion engine (ICE), as discussed in Chap. 6 on electric-powered vehicles. Many mechanical operations require a linear force or straight-line force, such as brake cylinders or intake/exhaust valves, in the ICE. Motors inherently produce rotary motion and thus are more difficult to employ in these types of applications. In less demanding linear force designs, such as power door lock mechanisms, a motor can be successfully used. A hybrid approach is possible when the motor drives a hydraulic pump that supplies hydraulic force to electronically operated valves or ABS solenoids. Because of the need to improve engine performance, electronic ICE valve controls are currently under development.

Summary

Designing automotive motor controls is a challenge because of the need to achieve high reliability at minimal cost in a hostile environment. As we approach the twenty-first century, motors and electronics will continue to evolve in all types of vehicles. Additional improvement in various systems will take advantage of more electronics and motors, including electric steering, electronic braking, electronically controlled valves for engines, and traction motors for electric or hybrid cars. Some of the significant points presented in this chapter are listed below.

- Most automobiles contain five motor applications as standard equipment: a starter motor, windshield wiper motor, washer pump motor, fuel pump motor for fuel injection, and a heater/fan motor. Some luxury vehicles may employ over 50 motors.
- A short circuit in the 12-V automotive electrical system can lead to catastrophic events, such as fires in the wiring harness. A fully charged 12-V battery can supply over 6000 W of energy, which is more than enough to start a fire.
- The "load dump" is especially dangerous to any electronic modules connected to a 12-V system. If a battery connection is intermittent when the alternator is charging the battery at a high current level, the alternator's output voltage will momentarily rise to over 60 V for several tens of milliseconds.
- A reversed battery connection, which might occur during a jump start, battery replacement, or reversed wiring hookup, can damage the motor drive electronics. The power stage transistor's internal rectifier and free-wheeling diodes (if used) will usually become forward-biased, resulting in excessive current flow, thus overheating the devices.
- The motor control electronics may need to sense the motor's power supply voltage to ensure correct operation under varying voltage conditions. The 12-V bus may vary over a wide voltage range, depending on the alternator's operation and the state of charge in the battery.
- The motor electronics in an automotive system must operate over a wide temperature range, typically −40 to 125°C, and withstand vibration.
- PM brush motors are very cost-effective for automotive applications. Their main disadvantage lies with their brushes and commutators, which can cause reliability problems in cases where the motor is operated for extended periods. Another disadvantage is the inherent arcing of the brush assembly, which generates RFI that may interrupt the normal operation of nearby electronic systems. If electronic speed and direction control are required, the cost advantage of a PM brush motor becomes less significant.

- The BLM design needs about one-third more electronics than a PM brush design. The advantage of the BLM is the total elimination of the brush/commutator assembly, which reduces audible noise and RFI problems.
- Memory seats or mirrors positions are sensed with a potentiometer that is connected to the motor's shaft mechanism. The potentiometer's voltage is sensed and stored in the motor's electronics. The motor is operated until the potentiometer voltage matches the stored value. Different stored values can be selected for multiple drivers.
- The electrohydraulic steering system is easy to implement because it uses the same basic hydraulic components as the original engine-driven power steering system. The major change is that the hydraulic fluid pump is now driven by an electric motor. This minimizes reliability problems, since the actual steering components remain basically the same. The direct-assist steering system requires all new steering assist components, which raises reliability questions since it is a new design.
- The direct-assist steering system requires a steering wheel torque and direction sensor to determine the driver's steering inputs. A microcontroller is needed to compute the best response and to manage an electric motor. The electric motor, operating in a closed-loop fashion, uses the steering wheel torque and direction information to drive the steering shaft. In order to protect against a "runaway" or locked-motor condition, it is necessary to verify the integrity of the direct-steering system.
- MOSFET power transistors are preferred for automotive power control electronics because they offer several advantages over bipolar transistors for low-voltage applications—they require low average gate energy, their switching times are stable over a varying temperature range, and the MOSFET's high current density allow smaller package sizes. (For example, 0.005-Ω 60-V MOSFETs are available in a TO-220 package.)
- Raising the automotive 12-V bus to higher voltages offers significant advantages for high-power applications. If the battery voltage is doubled, the current flow is halved and the power loss is only one-fourth of the original 12-V level. Tripling the battery voltage to 36 V reduces the power loss to one-ninth of the original level. Quadrupling the battery voltage to 48 V reduces the power loss to one-sixteenth of the original drive. (This assumes that the ON resistance of the power stage is constant.)
- When replacing automotive hydraulic systems, such as brake cylinders, with electric motors, the weight and size density of a

hydraulic actuator is usually better than an electric motor that is rated to produce the same amount of lineal force in the same time period. It is possible to use gears to increase the electric motor's torque, but at a diminished response time. In the case of brake actuators, drum brakes can be operated from an electric motor, since their force requirement is significantly less than that of disk-type brakes.

- A design issue with an all-electric motor braking system involves reliability. If the electrical power supply to the electric brakes fails, there is no backup, except, perhaps, a parking brake, which is rarely satisfactory for stopping a vehicle at high speed. Most hydraulic brake systems use two circuits, providing some backup if one circuit fails. It should also be noted that the hydraulic brake lines are constructed of metal tubing and reinforced hoses, which are probably more rugged than a wire harness.

Acknowledgments

The author would like to thank Randy Frank, Kim Gauen, Neil Krohn, Tom McDonald, and Pete Pinewski for their assistance with this chapter on automotive motor controls.

Further Reading

Adler, U., *Automotive Handbook,* 3d ed., Robert Bosch Gmbh, 1993.

Davis, B., Frank, R., and Valentine, R., *The Impact of Higher System Voltage on Automotive Semiconductors,* SAE 911659.

Denton, T., *Automobile Electrical & Electronic Systems,* SAE International, R-164, 1995.

Jonner, W. D., *Electrohydraulic Brake System—the First Approach to Brake-By-Wire Technology,* SAE 960991.

Jurgen, R., *Automotive Electronics Handbook,* McGraw-Hill, 1994.

Mack, J., ABS-TCS-VDC, *Where Will the Technology Lead Us?* SAE International, PT-57, 1996.

Mizutani, S., *Car Electronics,* Sankaidi Co. Ltd., 1992.

Phillips Semiconductors, *Semiconductors for In-Car Electronics,* data handbook IC18, 1996.

Ribbens, W. B., *Understanding Automotive Electronics,* 4th ed., SAMS, 1992.

Suzuki, K., *Integrated Electro-Hydraulic Power Steering System with Low Electric Energy Consumption,* SAE 950580.

Valentine, R., and Pinewski, P., *Advanced Motor Control Electronics,* SAE 961693.

Chapter

5

Appliance Motor Controls

Richard J. Valentine
Motorola Semiconductor Products

Introduction

Electric motors energize machines that clean our clothes, wash our dishes, and heat and cool our homes. Table 5.1 lists a number of household appliances that can use some type of electronic controls. Designing electronics for appliances is a challenging endeavor because of the nature of these products. An electronic appliance control, by itself, is seldom perceived as a key selling feature, unless it can make the appliance easier to use and more reliable. This is an area in which design challenges constantly occur, and important discoveries or inventions are created. It is not ordinarily an easy task to reduce or even maintain the cost of a design, while, at the same time, improving its performance and longevity.

Electronics can manage most of the appliance control functions and also improve the power efficiency of the motor. Since this book focuses on motor control electronics, appliance electronics that is not related to motors and motor theory itself will not be reviewed in detail. The designer should be aware, however, that if electronic systems are employed for nonmotor functions, especially an MCU, consideration should be given to expanding the role of the MCU to include some form of motor control or motor diagnostics.

New appliances will tend to contain more electronics, especially if higher energy efficiency or AC power line quality standards are mandated. Improving motor power consumption will help energy efficiency, but a high percentage of power is consumed for heating functions in many appliances. High-wattage resistance-type heater elements are used in dryers, dishwashers, and some clothes washers, which draw more energy than the motor. Other appliances, such as refrigerators and air-conditioning compressors, can gain higher power efficiency with electronic motor control systems.

Many contemporary appliances use open-loop type controls, which means that the user has to guess which to choose when setting the controls. In some instances, this means that the appliance may operate longer than necessary, which wastes energy, or not long enough, which may require another complete cycle. Adding some form of smart sensors to determine various factors can help optimize the appliance's operation. Examples of this type of design would include a clothes weight sensor and humidity sensor in a dryer to determine the best heat and speed settings. If the user overfilled the dryer, not

TABLE 5.1 Electronic Motor Controls for Appliances

Category / Motor usage	Average power,* W	Daily usage factor, %	Var. speed req.	Fwd. rev req.	Stall det. req.	Shaft pos. sen.	Motor type†	Notes
Dishwasher								
Drain/wash	281	3	Yes	No	Yes	No	AC, UNIV	Drain, wash different speed
Dry fan	40	2	No	No	No	No	AC, UNIV	
Clothes washer								
Front-loading drum	425	4	Yes	No	Yes	Yes	AC, UNIV, ECM	Wash, spin different speed
Top-loading agitator	425	4	Yes	Yes	Yes	Yes	AC, ECM	complex motion control
Dryer								
Drum/fan	281	8	Yes	No	Yes	No	AC, UNIV, ECM	Speed varies per need
Refrigerator								
Compressor	213	63	Yes	No	Yes	No	AC, ECM	Hard-starting, low speed limit
Fan	15	63	Yes	No	No	No	AC, ECM	−20°C environment
Icemaker dispenser	35	1	No	No	No	No	AC, ECM	−20°C environment
Floor cleaner								
Vacuum motor	960	2	Yes	No	Yes	No	UNIV, ECM	Noise and dust issues
HVAC								
Ceiling fan	120	100	Yes	No	Yes	No	AC, ECM	Speed varies per need
Compressor	7200	25	Yes	No	Yes	No	AC, ECM	Hard-starting, low speed limit
Condenser fan	281	25	Yes	No	Yes	No	AC, ECM	Speed varies per need
Evaporator fan	425	25	Yes	No	Yes	No	AC, ECM	Ramp up/down control
Garden-lawn equipment								
Electric lawn mowers	1400	4	Yes	No	Yes	No	PM, UNIV	Speed varies per need

TABLE 5.1 Electronic Motor Controls for Appliances (Continued)

Category / Motor usage	Average power,* W	Daily usage factor, %	Var. speed req.	Fwd. rev req.	Stall det. req.	Shaft pos. sen.	Motor type†	Notes
Shop tools								
Hand drill	425	4	Yes	Yes	Yes	No	UNIV	Speed varies per need
Hand drill, cordless	48	4	Yes	Yes	Yes	No	PM	Speed varies per need
Exercise equipment								
Treadmill	1275	4	Yes	No	Yes	Yes	AC, UNIV, ECM	Speed varies per need
Miscellaneous equipment								
Attic fan	255	50	Yes	No	Yes	No	AC, ECM	Speed varies per need
Garage door opener	425	1	Yes	Yes	Yes	Yes	AC, UNIV, ECM	Rampup/down speed control
Sewing machine	75	8	Yes	Yes	Yes	Yes	UNIV, ECM	May use stepper motors
Sump pump	213	4	No	No	Yes	No	AC, UNIV	Water-level sensor required

*Motor power only
†Motor type: AC = motors that can run only on AC; ECM = electronic commutated motors.
NOTE: The data shown here is estimated and will vary among manufacturers, model types, and individual households.

only would the unit fail to dry the clothes but the unit itself could fail. In this case, the extra cost of sensors and control electronics which protect against improper appliance operation is offset by reduced warranty costs. Another factor that affects clothes washers and dish washers is the relationship between water hardness and detergent requirements. A smart detergent dispenser that would probably be powered with a small motor could minimize guesswork on the part of the user. A water-hardness sensor and load-size sensor could be incorporated into the automatic dispenser, or it could just be a part of the user's preset selection to dispense the detergent.

Another aspect of appliance electronics is the need for easy-to-repair designs. This means incorporating self-test modes and even a diagnostic port for interfacing to a portable-type personal computer (PC). The PC would contain diagnostic software, as well as upgrades for the appliance's program if it is capable of being reprogrammed.

5.1 Appliance Systems

Applying electronics only to increase the power efficiency of an appliance motor is difficult to justify, unless mandated by law or instigated by savings in electricity that are high enough to favorably offset the increased design costs. For example, a 10-kW heat pump compressor motor that operates 1000 h per year, consuming 10,000 kW, will cost $1500 per year to operate (at $.15/kW). Increasing the compressor motor's power or system efficiency by just 10 percent will save $150 per year, or $3000 over a 20-year lifetime.

The employment of electronics to operate the appliance in a fashion that simplifies its use and protects against improper operation, which may allow longer warranty periods, is worth researching. As mentioned in previous chapters, motor control electronic devices are becoming more available, as well as more cost-effective. Several appliance motor control systems will be examined in this chapter to discover where and how electronics can improve the appliance's performance.

5.1.1 Appliance motor control types

Essentially, when discussing motor controls, appliances can be divided into four groups. The first group may vary the speed of the motor but does not use any form of motor shaft/load/position/pressure feedback; that is, the motor controls operate in an open-loop mode. Examples include ceiling fans, shop tools, and standard vacuum cleaners. The second group of motors, which require sensors to control them, are used in appliances such as refrigerators and HVAC systems. These sensors can turn the motor on or off and can adjust the motor's speed (when motor speed electronics are used). The third group of motor controls is similar to the second group but requires directional control of the motor or its load. A garage door opener is a good example. The fourth group of appliance motors includes high-performance sewing machines and home computer systems. (These may not always be considered to be appliances but fall in the appliance classification, since an "appliance" is generally defined as an electrical device that performs a specific task.) This fourth group uses electronically controlled precision motors and usually employs a MCU or similar devices. Most large appliances are in the second or third group; they employ sensors to control the motor, may use speed controls, and may require bidirectional motor controls.

Figure 5.1 shows a block diagram of these four groups. Note that the motor control is rather simple for the first group but increases in complexity with the remaining groups.

To determine if and how to employ complex electronics in an appliance, it is important to understand several aspects of the appliance's application. Possible marketing features must also be considered. One feature that is key to retaining market share is the long-term performance of an appliance. Many customers will not want to buy again from a manufacturer whose previous products have failed just beyond a 1- or 2-year warranty period. Consumers generally believe that appliances have been designed to operate for a fairly long time. (Refrigerators, for example, are expected to last for over 20 years.) This is one area in which electronics can have an impact by operating the appliance with computer controls and alerting the user to potential problems. The following design features interact with each other:

- *Long-term warranty,* which the consumer will perceive as positive
- *Ease of use,* which is dependent on the electronics
- *Low noise,* made possible by varying the appliance's motor speed
- *Energy efficiency,* improved by varying the motor's control strategy
- *Appearance,* a top consideration that is affected by flat-panel controls
- *Cost-to-performance,* affected by the electronic designs
- *Ease of service and installation,* improved with diagnostics

5.1.2 Appliance motor control design attributes

There are several factors that must be considered when designing motor control electronics for appliances. One significant factor is that

Motor control for simple non-feedback type application.

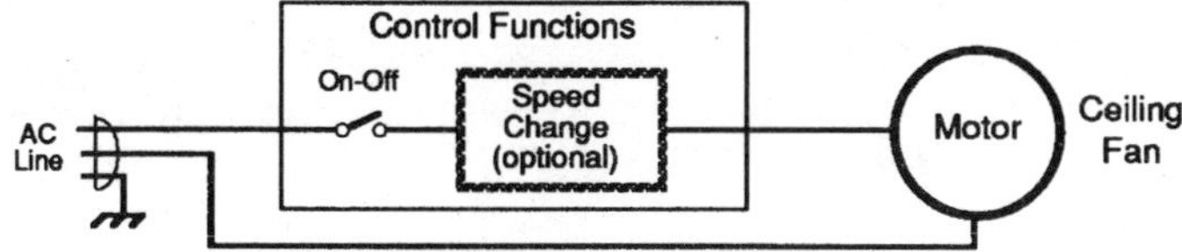

Motor control for application with some sensor feedback control.

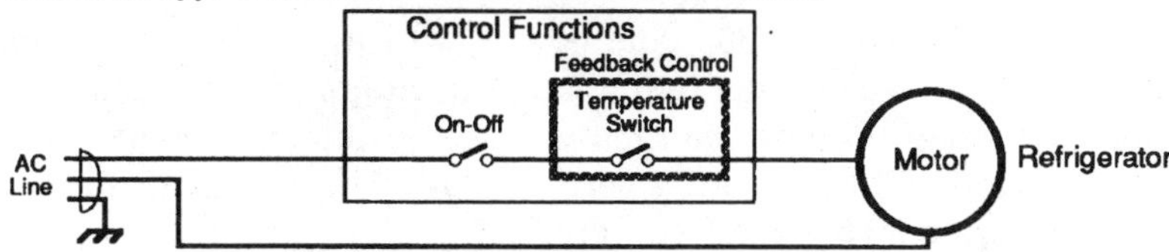

Motor control for application with sensor feedback and direction control.

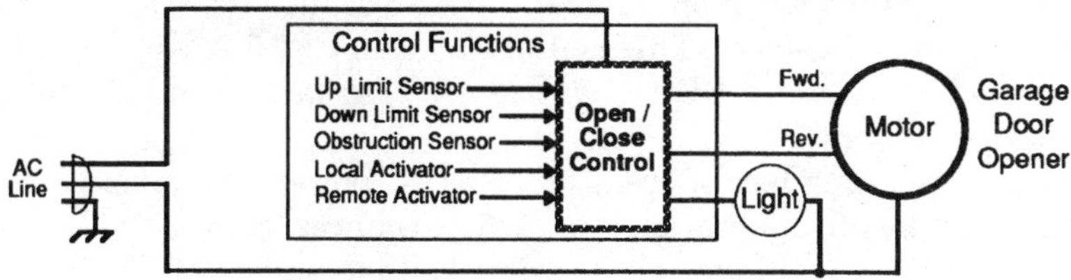

Motor control for complex applications.

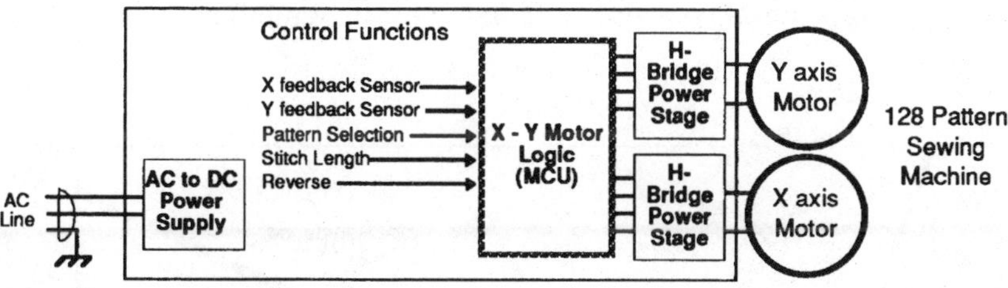

Figure 5.1 Appliance motor applications.

appliances are powered from AC voltages whose value will be either 120 or 240 V and whose frequencies are 50 or 60 Hz. This range of voltage levels usually requires two separate designs, one for the 108- to 132-V operation and another for 216 to 264 V. The difference in the line frequency, 50 or 60 Hz, can also affect an AC induction motor's performance. Going from 60 to 50 Hz will lower the motor's speed by about 16 percent, as can be demonstrated from the following equation:

$$\text{Sync rpm} = \frac{\text{line frequency}*120}{\text{poles}}$$

$$\text{For 60-Hz operation, } 1800 = \frac{60*120}{4}$$

$$\text{For 50-Hz operation, } 1500 = \frac{50*120}{4} \quad (5.1)$$

A speed difference of 300 rpm occurs, which is a 16 percent reduction. This is also the same reduction percentage of the line frequency. Note that most AC induction motors used in appliances are not synchronous and will run about 4 percent slower than the calculated synchronous speed. (This is called "slip.")

The motor's power consumption is also affected by the power line frequency. An AC induction motor designed for 60-Hz operation will draw more current when operated at 50 Hz because its AC resistance or reactance will be lower. This is illustrated in Fig. 5.2. Note that the motor's inductance and DC resistance are also factors in calculating the motor's current.

Besides designing for different types of AC power, one must be concerned with the effects that electrical disturbances from the AC power line can have on appliance electronics. Voltage spikes of 1000 V are common in AC power lines, with some spikes reaching 10,000 V from lightning strikes. If long-term reliability is to be achieved, electronics powered from the AC power line must contain some form of transient-

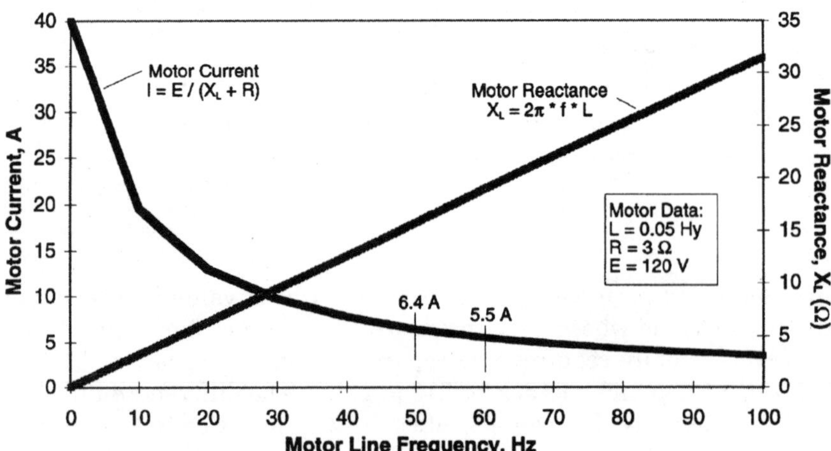

Figure 5.2 Line frequency effects on AC induction motor.

voltage protection. This protection must cover both low-power MCU circuits and motor power stages. Most semiconductor devices will fail in a catastrophic manner if their voltage ratings are exceeded. Even those devices that are "avalanche"-rated will fail because of heat dissipation if the voltage spike's duration lasts too long. Some devices will function incorrectly when operated beyond their voltage ratings. Thyristor-type devices, for example, may self-trigger on when their voltage rating is exceeded.

Other design concerns include safety requirements. There are numerous standards from several countries and international organizations that have established specific design practices and rules for household electrical machinery (App. D lists some of these organizations). These standards mainly affect the safety aspects of appliance wiring, flammability, fusing, terminal spacing, and other factors but also include EMI/RFI for electronics. There are testing laboratories that validate the appliance's compliance to a particular set of standards. EMI compliance can be a difficult design problem for motor power electronics, requiring careful layouts.

One final observation about designing electronics for appliances—the AC power line current capacity is limited by its source impedance. A faulty motor will trip a circuit breaker in the electrical panel in a few seconds, but not in time to prevent very high peak currents from flowing through the shorted motor and power electronics. The exact shorted-load peak current levels will be set by the outlet's wire resistance, the fuse or circuit breaker, the motor's short-circuit resistance, and the electrical supply source impedance. An estimate of the maximum peak current can be made [Eq. (5.2)] by adding the total resistance between the shorted load and the electrical supply source. Note that the source resistance is the prime factor in determining the short current peak level. The wire resistance from the appliance to the power source also limits the peak current but to a lesser degree.

$$I_{peak} = \frac{E}{R_{source} + R_{wire} + R_{short}}$$

$$218 = \frac{120}{0.4 + 0.1 + 0.05} \qquad (5.2)$$

where I_{peak} = short-circuit peak current (may last for several seconds)
E = supply voltage (120 to 240 V)
R_{source} = 240 V/600 A = 0.4 Ω (this assumes three houses per utility transformer)
R_{wire} = 0.1 Ω (20 m or 64 ft of 12-gauge wire)

$R_{short} = 0.05\ \Omega$ (depends on exact nature of short circuit)

It is important to remember that the electronic power stage in a typical appliance may be capable of safely conducting only the normal start-up or locked-rotor current. A major short circuit will usually cause a power stage failure. It is not practical to use very large power transistors in the motor's power stage to safely withstand a shorted motor. To protect against a failed motor and to ensure the survival of the power stage, some form of electronic current limiting besides the normal house fusing circuit and/or circuit breaker is necessary. If the power stage design is intentionally designed without protection from a shorted circuit, it must be capable of failing in a nonhazardous manner. In such a case, the motor and power stage electronics would be replaced by the repair technician.

5.1.3 AC power line and environment problems

To gain an appreciation of what the electronics of an appliance must contend with, we must examine various factors of its power source and environment. These factors can be divided into five categories: (1) power outages; (2) minimum and maximum operational voltages; (3) voltage transients, swells, and sags; (4) frequency variations and cycle dropouts; and (5) operating environment.

1. *Power outages:* One fundamental difference between electronic appliance controls and traditional electromechanical controls centers on the effect of a power failure. Most non-electronic-type appliances will resume their operation when AC power is restored, while the electronics-based appliances will not. This is because user settings are stored in the logic (or memory elements), which is reset whenever its power is removed. The electromechanical timer and snap switches, on the other hand, remain in their previous position, thereby allowing the appliance to continue its operation. To compensate for this disadvantage of electronic controls, memory elements can be added to them which retain their data during power outages. These nonvolatile memory elements are called *electrically erasable programmable read-only memory* (EEPROM) and can be purchased as separate devices or integrated into the MCU. There are other methods to retain the memory data in MCUs. For example, when a power outage is detected, the MCU is put in a standby or sleep state, and an ultracapacitor (such as a 1-F low-voltage unit) or long-life battery supplies a few microwatts to power the MCU. When the main power is restored, the MCU program can continue from its previous operation, or a smart

restart can occur. A smart restart routine looks at the effects that a power outage might have had on the appliance's operation and takes an appropriate action. For example, if the power has been out for several hours, the entire wash cycle of a washing machine might be restarted. One other feature that could be added with the electronics would be a delayed restart, which would minimize the current demand of the AC power line when first restored.

2. *Voltage regulation:* The AC line voltage may vary by ± 10 percent; in some cases, such as during a brownout, the voltage may sag by -20 percent, which can have serious effects on the appliance's motor operation. Obviously, a momentary voltage fluctuation that lasts for less than 1 s every few hours is not a concern, unless the appliance's electronics has a power supply design that is marginal. To ensure stable operation, large capacitors are normally used in the power supply that feeds the logic or MCU control elements. A power stage voltage sensor could be implemented to optimize the motor's control strategy over a wide range of power line voltages, depending on the motor's basic design specifications. For example, if a motor were designed to operate at 90 V, the electronics would pulse-width-modulate the motor's voltage to allow operation at much higher voltages. There is a practical limit to this idea, but it is not uncommon to find equipment that can operate, through the use of electronics, over a wide voltage range—between, for instance, 90 and 250 V. It is important in electronic motor control design to ensure that the motor's operation remains stable when the AC power supply voltage varies between nominal levels and zero. A smooth shutdown during abnormal power supply conditions in the electronics can be implemented with some form of hysteresis element. Certain devices can detect a sagging power supply voltage and cause a low-voltage interrupt or reset for the MCU.

3. *Voltage transients, swells, and sags:* Because of their high-voltage and fast-switching (Dv/Dt) characteristics, voltage transients in residential AC power lines, which can be generated from other nearby electrical loads or by lightning, are short in duration ($<$8 ms) but can be especially dangerous to appliance electronics. A voltage sag (brownout) can be followed by a voltage surge in an AC power line. Voltage surges or swells may last for a few seconds and can stress the appliance electronics. Voltage sags, which occur when several large loads are switched on simultaneously, may last for up to a minute. Electrostatic discharge (ESD) is another form of voltage transient that the electronics in an appliance must be able to safely sustain. Most semiconductor devices, MCUs, CMOS logic, and even power IGBTs are rated to withstand a limited amount of ESD but will be

destroyed by excessive levels. ESD damage can occur during manufacturing of the electronics or in the home. One subtle aspect to ESD damage is that the semiconductor devices may be only partially impaired by ESD during manufacturing and then, rather than failing immediately, will fail at a later date.

4. *Frequency variations and cycle dropouts:* As we noted before, operating a 60-Hz AC induction motor at 50 Hz can have significant effects on the motor's speed and power consumption. An AC induction motor design that incorporates an AC-to-DC power supply or converter makes the motor's operation independent of the AC line frequency. The motor's speed is managed by the control electronics. Universal motor speed controls may use a triac that varies the AC power line's sine wave on a cycle-by-cycle basis. If a cycle should drop out because of a power line disturbance, the triac will stop conducting, and the motor will start to decelerate. A 10-min burst of random dropouts will cause the motor speed to pulsate, producing some audible sound that is related to the cycle dropout's repetition rate.

5. *Operating environment:* Appliances normally operate indoors but are still subject to a wide variety of ambient temperatures and humidity. An upper limit of 50°C (122°F) ambient will normally cover the hottest climates (Phoenix, AZ, for example). The lower limit is usually set to 0°C (32°F), or freezing, for appliances that use water. The appliance itself will dissipate varying amounts of heat, which will contribute to heat buildup in the electronics. Most non-power-type electronic components are rated at −40 to +85°C, allowing 35°C headroom for their heat dissipation. Power electronic components, like IGBTs, are rated at 125°C or higher maximum junction temperatures. In a 50°C ambient, this allows a 75°C rise, which sounds significant, but it is quickly diminished by many factors. (See Chap. 15 on heat management.) Although most standard MCUs or similar devices are rated to operate from 0 to 70°C, many are rated to operate over a range of −40 to +105°C or higher.

The appliance's electronic controls and circuit boards can be affected by various conditions, including vibration, water splashes, detergent spills, and dust. Vibration can loosen electrical contacts and cause intermittent operation. Electronic controls that use an MCU or CMOS logic devices can be easily affected by water. The water usually contains enough minerals, soap, or dirt to cause leakage currents between PCB traces or device pins, and will cause erratic operation. High humidity can also have the same effect as water. Dust particles, which can come from clothes (lint) or dirty air, are seldom highly conductive unless wet. An accumulation of dust in the motor's power control electronics can be ignited if an electrical spark occurs. Some type of sealant is recommended for the appliance's motor control electronics. This sealant compound can also minimize the effects of vibration on the various electronic components. Although semiconductor components are somewhat hermetically sealed, they can be affected by prolonged exposure to moisture or other liquids. The semiconductor's package usually affects its moisture tolerance.

5.1.4 Small-appliance motor controls

Universal motors are widely used in small appliances and are also employed in some large appliances. The universal motor is a series-wound or compensated series-wound motor that can utilize either 50 to 60 Hz AC or DC power. The universal motor's speed is highest at zero load. The universal motor's torque is greatest at low speeds and decreases with rising speed. Because of this torque/speed relationship, the universal motor's speed is directly affected by load changes. Most universal motors are designed to operate at high rpm levels; 3000 to 10,000 rpm is common, while some shop tools, such as routers, run at over 20,000 rpm. The universal motor's speed can be adjusted by simply varying its supply voltage. If a constant speed is required under various load conditions, the universal motor is a poor choice, unless electronics are incorporated to maintain a stable speed.

Phase-angle control is commonly used to vary the universal motor's speed when operated from AC power sources. Phase-angle control shares a similarity with pulse-width modulation, since it, too, has an AC line frequency that is fixed, but the AC line frequency of the phase-angle control is many times lower than a PWM-based speed control and creates more audible noise in the motor. Phase-angle control also simply varies the on time of the AC sine wave in reference to its zero crossing, as shown in Fig. 5.3, while PWM changes the duty cycle.

The main advantage of phase-angle motor speed control is in its cost/performance ratio. A silicon controlled rectifier (SCR) and a simple gate triggering circuit can be used in small appliances. The most significant disadvantages of the phase-angle method include audible motor noise and high harmonic content. The harmonics affect both the motor's performance and the AC power line purity. Converting the AC power into pulsating DC with a bridge rectifier and using a PWM-type motor speed control can help minimize power line harmonics. If this approach is taken, other motor types that offer higher power efficiency when controlled with electronics can also be considered. (For example, when controlled with electronics, AC induction or switched reluctance motors are inexpensive to manufacture and offer high performance.)

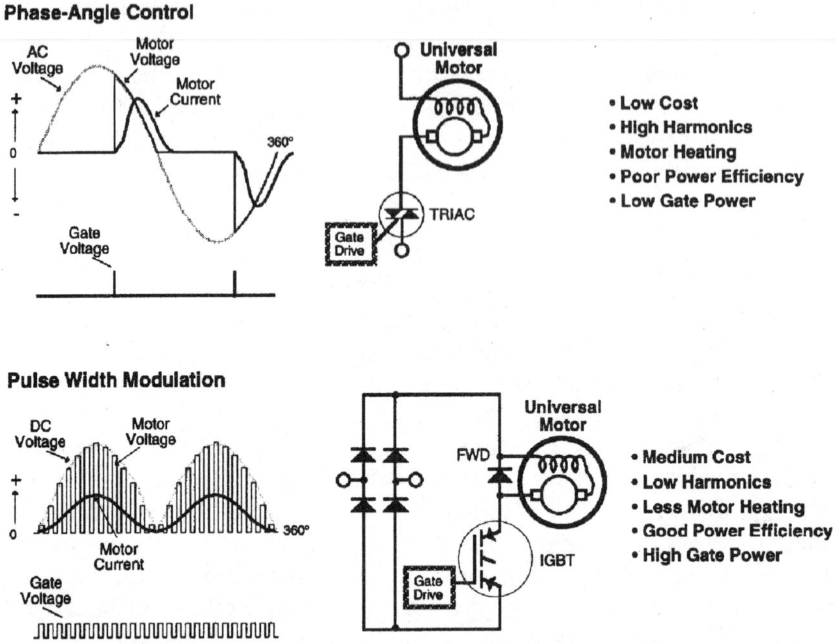

Figure 5.3 Phase-angle control versus pulse width modulation.

Small appliances such as food mixers can use a simple SCR motor control to vary the speed of the mixer's motor. (See Chap. 13 for SCR and triac control theory.) Figure 5.4 shows a block diagram of a universal motor speed control that is applicable to many small appliances such as mixers that feature some type of speed control. (Please note that this design was derived from several SCR application notes and is not lab-tested; it shows, however, the basic universal motor speed control concept.) This circuit is simple, but it does offer some torque feedback.

The circuit powers the universal motor on the positive half of the AC cycle. This changes the AC cycle into a 25- or 30-Hz (depending on the AC line frequency) pulsating DC voltage whose average voltage is set by the phase-angle control of the SCR. The motor will tend to smooth out the DC pulses, but, at low speeds, the narrow phase-angle pulses may produce audible motor noise. The operation of this simple SCR speed control circuit is straightforward. On start-up the SCR will be off until its gate-to-cathode voltage reaches about 0.7 V. The SCR shown is a sensitive gate type, requiring less than 200 μA of gate current. This gate trigger voltage can vary from 0.45 V at high junction temperatures to 0.95 V at subfreezing. The gate trigger voltage is controlled by the series resistor string, the 0.5-μF capacitor, and the

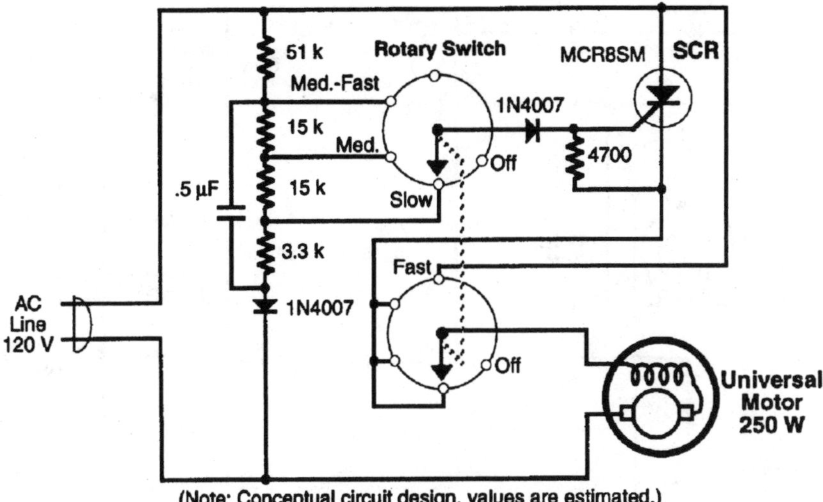

Figure 5.4 Simple universal motor speed control.

motor's back EMF. At start-up, there will be no back EMF. As the motor's speed increases, its back EMF will also rise, which will decrease the SCR's conduction time by changing the SCR's trigger firing phase angle. When the motor is loaded down, its back EMF decreases, causing the SCR's conduction time to increase and thereby increasing the motor's torque. The exact component values of the SCR's gate phase-angle control are dependent on the motor and its application.

If full-wave control is desired, a slightly more complex design can be used. Figure 5.5 shows a block diagram for an open-loop full-wave triac phase-angle control. The bipolar IC features automatic retriggering, 125-mA gate pulse, voltage and current synchronization, and internal low voltage reset, and draws about 2.5 mA average.

5.2 Clothes Cleaning Systems

The clothes washer was the first labor-saving household appliance to gain wide acceptance. It continues to evolve as savings in water and power become more important. Electric and gas clothes dryers are also being designed to accommodate a wider range of fabrics and to save power. There are two basic types of clothes washer designs: the front-loading machine, which uses a horizontal drum; and the top-loading type, which uses a vertical drum. Vertical drum units usually employ agitators. The front-loading types, which are widely used in Europe, typically employ universal motors that can be controlled with electronics.

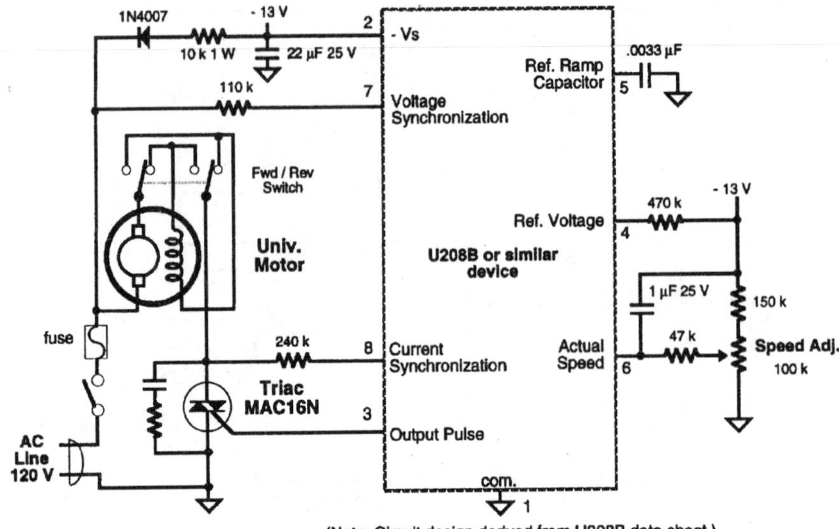

Figure 5.5 Triac universal motor speed control.

5.2.1 Clothes washer motor electronics

The motor in most clothes washers operate in two modes: washing and spinning. The washing mode requires low drum speed (about 50 rpm), high torque (10 N · m), and, in some cases, change of rotation. The spinning mode operates about 10 to 20 times faster than the wash mode and requires maximum motor power. The spinning speed directly affects the residual water in the clothes; higher spin speed removes more water, reducing drying times. Different speed and load requirements depend on some type of mechanical or electronic motor speed control.

Three approaches to electronic motor control design will be reviewed for clothes washers. The first uses a complex linear integrated circuit, and the second and third are MCU-based. It is interesting to note that the analog IC design is easy to implement, while the MCU control allows programmability to accommodate changes. The analog IC design uses fixed components to set various parameters, while the MCU relies on software to compute its operating parameters; instead of changing component values, the software program can be revised by changing a few lines of text.

5.2.1.1 Linear IC washing machine motor control. As shown in Fig. 5.6, the TDA1085C is a phase-angle triac analog controller for a clothes washer. This linear IC is designed to trigger a triac according to the

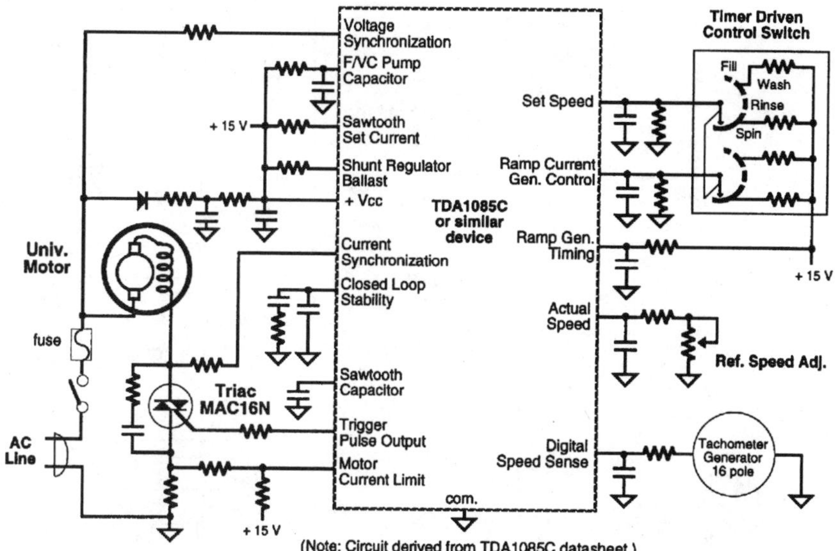

Figure 5.6 Analog IC clothes washer design.

speed regulation requirements. The speed set is externally set and is connected to an internal speed regulation circuit that also allows the motor's two different acceleration ramps to be set. An internal frequency-to-voltage converter for monitoring an external tachometer generator is used for the closed-loop speed control. Current sensing minimizes motor overloading by limiting the peak current. It should be noted that the current limiter method shown cannot instantaneously shut off the triac; it can only reduce the phase angle of the next cycle, thereby reducing the triac's peak currents. However, a short circuit in the motor may lead to triac failure. (For more detailed applications data, see the TDA1085C data sheet.)

5.2.1.2 MCU and triac washing machine motor control. The previous motor drive allows straightforward implementation, but it does not allow for easy changes to accommodate different models, nor does it control the entire clothes washer system. Figure 5.7 illustrates an MCU-based clothes washer conceptual design that controls a universal motor. The drum motor triac in the power stage will introduce line current harmonics, which may not pass some power line regulations. This form of phase-angle motor control is shown here for reference and discussion purposes. Any MCU-based appliance motor control raises several fundamental design issues, which are discussed below.

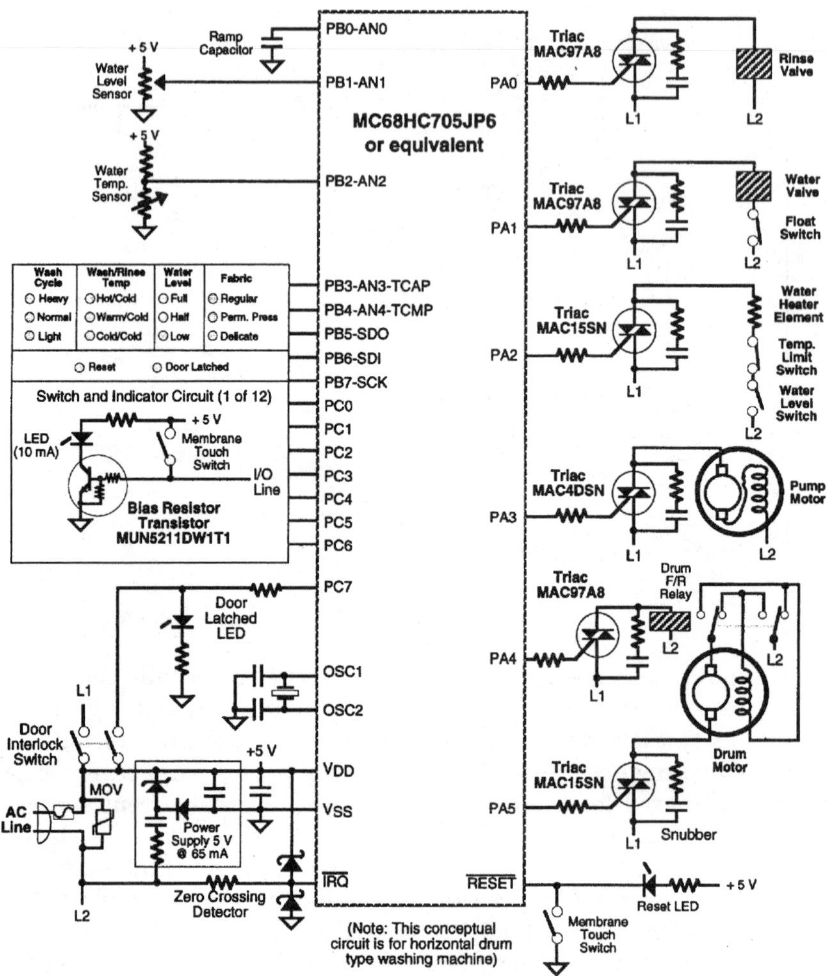

Figure 5.7 MCU-based clothes washer design using triacs.

5.2.1.3 MCU selection for washing machine motor control. Choosing the correct MCU for washing machines can be difficult, since there is usually more than one type of MCU that can accomplish the design's requirements. One way to simplify the MCU selection process is to first identify the system requirements, as shown in Table 5.2. A good place to start is with the switch functions. Notice that the clothes washer system has a significant number of switch functions which interact with several output devices and are normally controlled with certain time patterns. Switch functions will determine the number of input lines. There are many ways in which to encode the switch contacts, thereby reducing the number of MCU I/O lines. The method shown in Fig. 5.7 uses an I/O

TABLE 5.2 Clothes Washer MCU Considerations

Function requirements	Number required	Analog/ digital	Speed factor		Current requirements		Timer-based	Design notes
			Loop	Width	Peak	Average		
User input switch functions								
Fabric selection	3	Digital	0.1 s	50 μs	—	≈250 μA		Total switch scan <.2 S
Water level	3	Digital	0.1 s	50 μs	—	≈250 μA		—
Water temperature	3	Digital	0.1 s	50 μs	—	≈250 μA		—
Wash cycle	3	Digital	0.1 s	50 μs	—	≈250 μA		—
Reset all functions	1	Digital	Reset	—	—	≈250 μA		MCU reset line
System input functions								
Door-latched sensor	1	Digital	0.1 s	50 μs	—	≈250 μA		Safety element
Water temperature sensor	1	Analog	10.0 s	50 μs	—	≈250 μA		Safety element
Water-level sensor	1	Analog	1.0 s	50 μs	—	≈250 μA		Safety element
AC line zero detector	1	Digital	Interrupt	50 μs	—	≈250 μA		Sync to AC line, ±1%
Input line total: 14 digital, 2 analog, 1 external interrupt, 1 reset								
System output functions								
Rinse valve	1	Digital	0.1 s	5 min	10 mA	≈100 μA	Yes	Triac output stage
Pump motor	1	Digital	0.1 s	5 min	10 mA	≈100 μA	Yes	Triac output stage
Water inlet valve	1	Digital	0.1 s	5 min	10 mA	≈100 μA	Yes	Triac output stage
Water heater element	1	Digital	0.1 s	30 min	10 mA	≈100 μA	Yes	Triac output stage
Drum forward/reverse relay	1	Digital	0.1 s	5 min	10 mA	≈100 μA	Yes	Triac output stage
Drum motor	1	Digital	0.005 s	Adj.	10 mA	≈100 μA	Yes	Triac output stage
User output functions								
Fabric selection indicator	3	Digital	0.1 s	95%	10 mA	9.5 mA	—	LED indicator
Water-level indicator	3	Digital	0.1 s	95%	10 mA	9.5 mA	—	LED indicator
Water temperature indicator	3	Digital	0.1 s	95%	10 mA	9.5 mA	—	LED indicator
Wash cycle indicator	3	Digital	0.1 s	95%	10 mA	9.5 mA	—	LED indicator
Output line total: 6 digital high peak current, 12 digital high average current								
Overall MCU requirements: 14 inputs, 2 A/D inputs, 18 high current outputs, internal timer, external interrupt								

NOTE: The data shown here is estimated and will vary per the exact application requirements.

line for both a switch input and an indicator output. This places a burden on the software design but reduces the MCU's I/O pin count. Other input line considerations, besides switch contact sensing, concern analog signals. The important factors for analog processing include conversion speed, accuracy, and number of channels.

5.2.1.4 MCU interfacing considerations. The microcontroller's output port specifications are important design considerations for motor control applications. They are especially significant since the output lines of a typical MCU are low-current (<2 mA) and can rarely drive a large power device such as a 15-A thyristor in a direct manner. By employing various methods, certain MCUs which have output lines with high currents (≤20 mA) can drive sensitive-gate thyristors. It should be noted that a compromise must be made in the design of a sensitive-gate thyristor device: The increased gate sensitivity reduces the thyristor's *Dv/Dt* rating. The sensitive-gate thyristor will be more likely to false-trigger when operated in an electrically noisy environment. Adding a resistor-capacitor snubber across the sensitive-gate thyristor usually solves this problem. Thyristors that have better noise immunity generally require a 50-mA gate current level, which is beyond what most MCU output lines can safely handle. Some improvements are being made to sensitive-gate thyristors, but at the time of this book's publication, there still appears to be a 20:1 difference in noise immunity between standard and sensitive-gate thyristors.

A brief review of triac gate drive characteristics can help us understand why it is advantageous to float the MCU. Figure 5.8 shows three ways an MCU can be connected to drive a triac. One approach is to use an auxiliary isolated power supply for the MCU, allowing complete isolation of the MCU from the AC power. In many cases, the MCU and power stage must be isolated anyway to meet certain regulations or to minimize EMI problems. This approach of isolating the MCU from the power stage is usually followed for industrial motor drives and can supply ample current for indicators or interface devices. Its main disadvantage is cost, since a stepdown power supply is required, in addition to high-voltage isolated interface devices.

To save cost and to simplify the control of triacs, the MCU can be energized directly from the AC power line. This second method drives the triac as a low-side switch. Its main disadvantage is that the triac will be operated in the first and fourth quadrants, specifically, the gate will always be triggered with a positive voltage, while the triac's MT2 terminal will alternate from positive to negative per the AC line. Many triacs require the most gate current when triggered in the fourth quadrant. This can become a problem with an MCU output line that can usually handle only less than a few milliamperes. If the

MCU isolated interface to thyristor using step-down power supply.

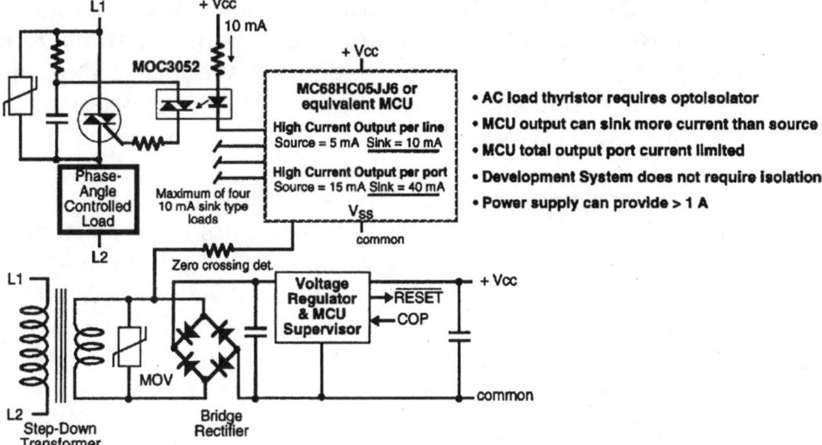

MCU non-isolated interface to low-side thyristor.

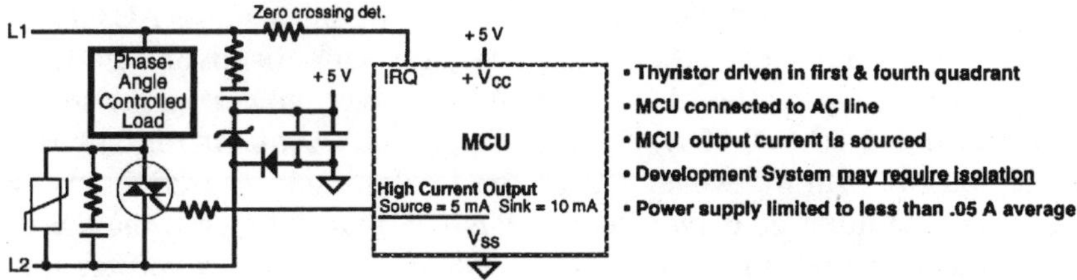

MCU non-isolated interface to high-side thyristor.

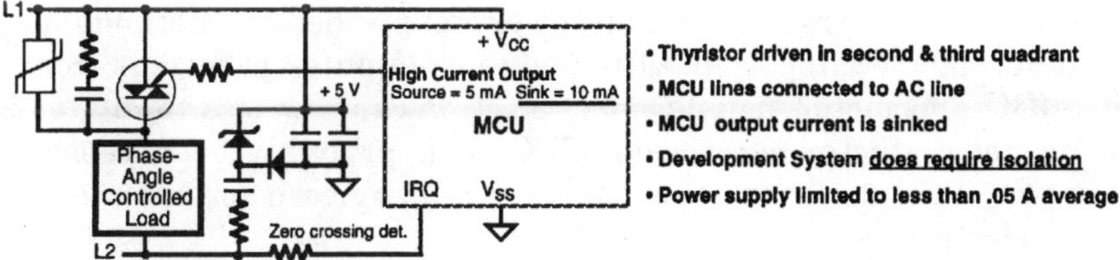

Figure 5.8 MCU-triac interface methods.

triac is connected as a high-side switch, a high-voltage interface is required, which can be a costly item, especially if several high-side triacs are needed. If the AC neutral line is connected to the MCU's common, the development tools will not be floating at full AC line voltage but still may operate erratically because of AC power ground loops and EMI.

The third method connects the MCU's V_{cc} directly to the AC power, and the triac is connected as a high-side switch. The triac's gate operates in the second and third quadrants reducing the current requirements of the MCU output line. It should also be noted that a standard MCU output port can usually sink more current than it can source. This method mandates that the MCU's development tools be fully isolated, and extreme care must be taken when troubleshooting the electronics.

5.2.1.5 MCU power supply methods. One design challenge is to decide how to power the appliance motor control electronics when more than a few milliamperes of current are needed, as in a design that employs IGBTs rather than triacs. The IGBT gate driver for a 500-W motor must control high peak currents in order to charge and discharge the IGBT's gate capacitance. This is one fundamental difference between a thyristor motor power stage and an IGBT version. The thyristor motor drive requires high current pulses at a 100-Hz rate while the IGBT motor drive requires high current pulses at up to a 20-kHz rate. The classic approach for a thyristor motor speed control has been to use a zener rectifier and a voltage-dropping network powered directly from the AC power line. This simple design can use a series AC capacitor rather than a large power series resistor. One disadvantage with this power supply method is that the output average current is somewhat limited (0.02 to 0.05 A) but can supply high peak currents. A difficulty occurs when several LED-type loads, such as indicators or optoisolators, must be driven continuously. Advanced electronic motor controls will require more current and two or more supply voltage levels. A standard AC power transformer can be used, but a more efficient approach might be a switching-type power supply. Switching-type power supplies have been successfully employed in millions of personal computers. Switching power supply design methods and simulations are supported with desktop computer programs. Since the MCU and motor signal-control electronics draw less than 1 or 2 W, no power factor correction (PFC) will probably be necessary. However, PFC and harmonic control are a concern for the motor's power electronics.

5.2.1.6 Advanced washing machine motor control. Electronic commutated motors, such as switched reluctance or brushless permanent magnet (BPM), offer high performance for some appliance applications including washing machines. The motor must operate in two different modes, wash and spin, with a reverse direction in the wash mode. Figure 5.9 shows a conceptual clothes washer motor control that employs a three-phase switched reluctance (SR) motor. The SR motor offers high starting torque and high-speed operation. It should be noted that other motors could also be used,

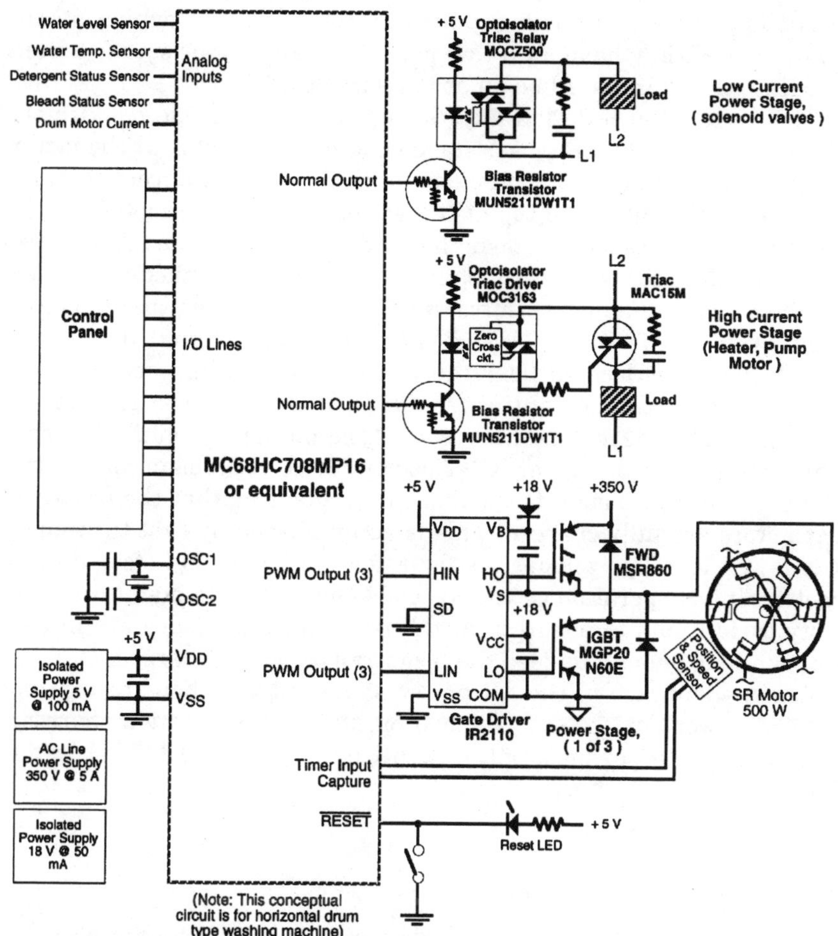

Figure 5.9 Advanced clothes washer design.

such as a BPM or AC induction motor. A high-performance 8-bit MCU, as shown, may suffice to control the SR motor but may fall short if complex control strategies or modulation methods are required. Note the manner in which the MCU is interfaced to the IGBT power stage and how the MCU and IGBT driver stage are powered. A transistor device with internal bias resistors buffers the MCU's output lines. Employing this type of external buffer device is much more cost-effective than enlarging the MCU's internal output stage. The external buffer transistors also reduce the MCU's internal power losses and keep high peak current levels away from the MCU's internal digital logic elements. Most MCUs are limited to less than 50 mA on their supply and common pins, which means that, even if they have an 8-line port that can sink or source 20 mA per line, they cannot all be on at the same time.

There are several design factors that must be considered with the SR drive stage. The SR power stage should be located near the motor. Doing this may present a problem, since the MCU and control panel may be located near the top of the appliance with the motor and its control electronics located near the bottom. It may make sense to use a second MCU for the control panel function that is linked with a serial data bus to the motor control MCU. One other area that may present problems is the MCU's PWM motor drive signal lines; they should never be located near the motor's position and speed sensor lines. Still another potential trouble spot involves the gate driver device. As shown, it uses a bootstrap-type method to obtain the gate power supply for the upper IGBT power device. A problem can occur if the power stage is operated without first ensuring that the bootstrap capacitors are fully charged. This is accomplished by first turning on the lower IGBT power devices for a few milliseconds before normal motor control operation occurs. It is also important to maintain PWM duty cycle limits to ensure that the free-wheeling diodes are in conduction long enough to keep the bootstrap capacitor charged. One significant advantage of the SR power stage is that when the top and bottom power devices are switched on, it does not create a short circuit across the power supply, as will happen with a half H-bridge power stage.

5.2.2 Clothes dryer motor electronics

A typical clothes dryer uses a single motor to rotate the drum and to blow heated air. This simple approach offers low manufacturing costs, but it may not offer the best power efficiency. There are optimum speeds for drum rotation and airflow, depending on fabric types and load sizes. To obtain maximum drying efficiency, separate drum and fan motors may be necessary. Figure 5.10 shows a conceptual advanced clothes dryer that uses one high-performance 8-bit MCU to control two brushless PM motors. This MCU (MC68HC08MP16) is normally used to only drive one motor but can drive two, if the rpm range of the motors is not excessive. The MCU has 6 timer input lines and 6 PWM output lines, which are divided between each motor. Standard output lines are used for driving the bottom part of the half H-bridge in the BLM power stage, while the MCU's PWM lines drive the upper part. These upper PWM lines provide speed control. A moisture sensor, a hot-air recirculation valve, a temperature sensor, a lint sensor, and a

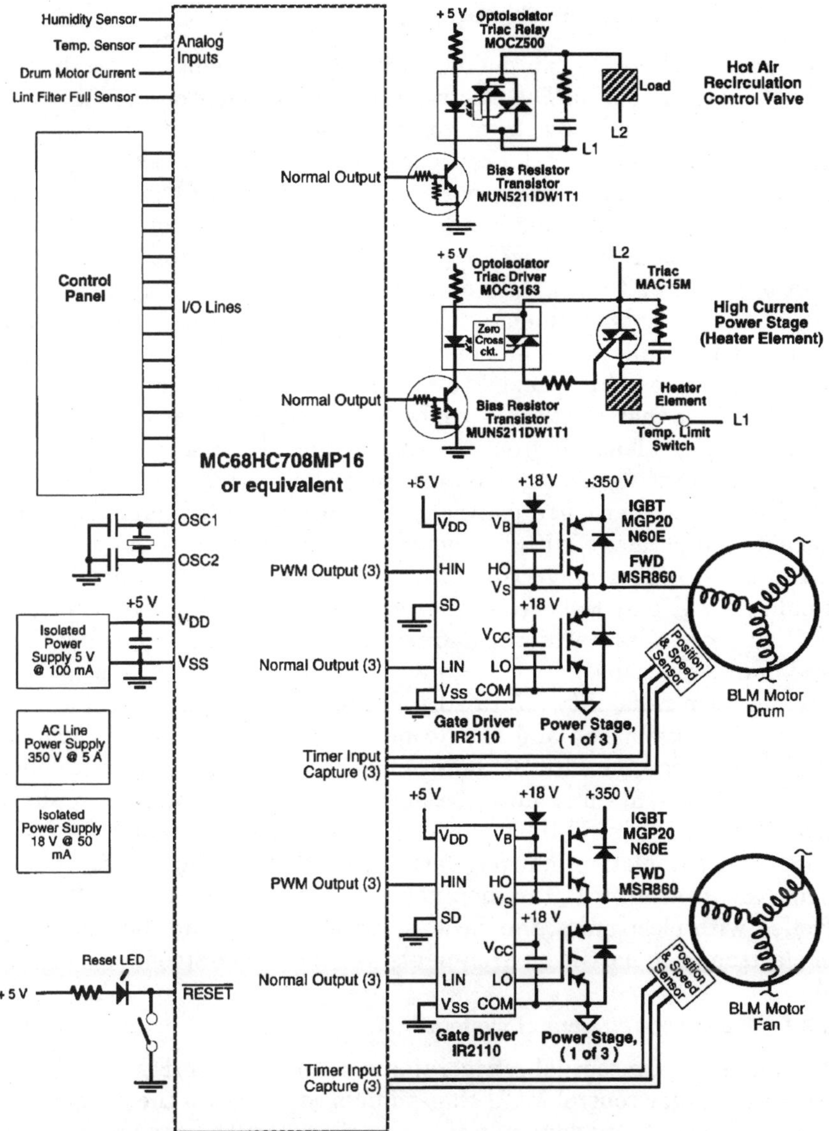

Figure 5.10 Advanced clothes dryer design.

drum current sensor are also employed to improve the dryer's ease of use. The drum current sensor gives a good indication of the clothes' weight, which can be used to determine the best drum and fan speeds. Also, as the clothes dry, their weights will decrease, which should be detectable and can help compute the optimum drum and fan motor speeds, as well as the best heater element temperature.

5.3 Refrigerator Motor Electronics

Electronic technology can make significant improvements in refrigerators in several areas. The electrical power efficiency of a refrigerator is controlled mostly by its motor and the refrigerant components. Since the basic purpose of a refrigerator is to preserve food, the motor control system needs to be very reliable and should provide some form of fault indication. A smart automatic defrost can also improve power efficiency since the freezer compartment will defrost only when required rather than at a fixed interval. The defrost cycle applies power to the heater elements, which are located in the freezer area in many (but not all) automatic-defrost models and then consumes more power to lower the freezer temperature after the defrost cycle. Another way that electronics could improve refrigerator design is to automatically blend air from the freezer and the fresh-food compartment, eliminating user guesswork when adjusting the freezer and fresh-food temperature controls. The user would just select the desired temperature with a keypad control panel for each compartment, and the control electronics would do the rest. The control panel would also display the actual freezer and fresh-food temperatures, as well as other data, including defrost mode, daily power consumption, and fault conditions.

Long-term reliability (15 to 20 years or longer) of refrigerator electronics is a key issue and should be addressed. A designer must be thoroughly familiar with the failure mechanisms associated with electronic components and must be able to select component suppliers who will stand behind their parts over a long period of time. (*Editor's note:* The automotive industry faced a similar challenge a few years ago. Now that over 100 million vehicle engines are successfully controlled with electronics, the various factors influencing the reliability of electronics in hostile environments are well understood.)

5.3.1 Advanced refrigerator system

The advanced conceptual refrigerator system shown in Fig. 5.11 incorporates a motor control MCU that adjusts the speed of the compressor to match the cooling demands of the refrigerator. The compressor may need to be designed to operate at lower speeds to ensure adequate low-speed internal lubrication. Several sensors are used to measure critical parameters. The temperature of the condenser and the temperature of the outlet refrigerant line of the evaporator can be sensed to give an indication of how well the refrigerant system is working. For example, a sensor could detect and then take the necessary steps to deal with a condenser coil that is clogged or an evaporator coil that is heavily iced. Another sensor could detect an improperly closed door.

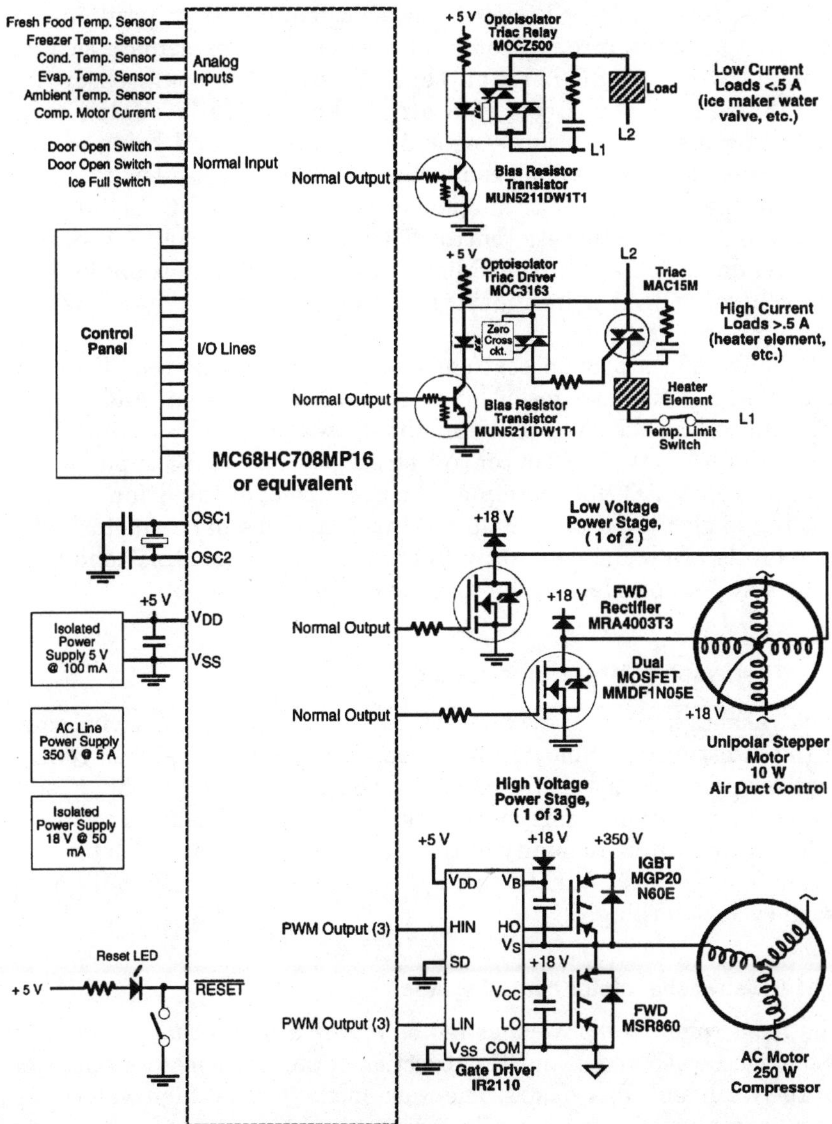

Figure 5.11 Advanced refrigerator design.

Although a permanent magnet brushless motor could be chosen for its high power efficiency, an SR motor may offer improved long-term reliability because of its power stage topology. If there is a noise glitch in the gate driver stage, the power stage of the SR motor cannot short-circuit the power supply. Such an electrical noise glitch can quickly destroy the half H-bridge power stage used in BLM and AC induction motors. The SR motor, however, requires shaft position sensors when used in a refrigerator compressor. This presents a location problem with the SR position sensors, since the compressor and motor are manufactured as a single hermetically sealed unit. Installing a shaft position sensor inside the unit and bringing five extra lines through a sealed connector may not be cost-effective.

If you use a permanent magnet brushless motor for this application, you need a sensorless control. There are specialized BLM control ICs available (ML4428 from Micro Linear Corp., for example) that can control a BLM if its inductance does not vary by more than 30 percent during rotation.

Because of its proven cost/performance ratio, a three-phase AC induction motor was chosen for this conceptual design. Adding current and voltage sensing to the motor's power stage allows the implementation of better motor control strategies. If the AC input power sags to 105 V AC, for example, it can be compensated for. Voltage sensing is also important to determine if changes in the motor's current levels are being caused by fluctuating AC power line conditions or by a system problem in the compressor/refrigerant.

5.4 Dishwasher Motor Electronics

Dishwashers have been evolving toward electronic controls with some of the latest units employing MCU-based controls. Flat-panel control switches improve the dishwasher's appearance, while the MCU simplifies its operation. Some models even sense the amount of food particles in the water (turbidity of the water), the presence of detergent, water temperature, washer arm speed, and automatically adjust the wash cycle accordingly.

5.4.1 Dishwasher motor control system

The most common dishwasher motor drives a pump that operates in two modes: wash and drain. The direction of the pump motor is changed for the wash and rinse modes. The wash mode requires high water pressure with a low flow rate, while the rinse mode requires low pressure with a high flow rate. A separate pump and motor for the wash and rinse mode can be used, as shown in Fig. 5.12, which is a conceptual design. AC induction motors (or universal motors) are used and controlled with triacs. Other motor types, such as a three-phase AC induction motor, may produce less audible noise and might be considered. A three-phase motor would need a more complex power stage and MCU.

The MCU in this conceptual design requires 6 A/D inputs, 8 high-current output lines (capable of sinking 5 mA), a synchronous serial

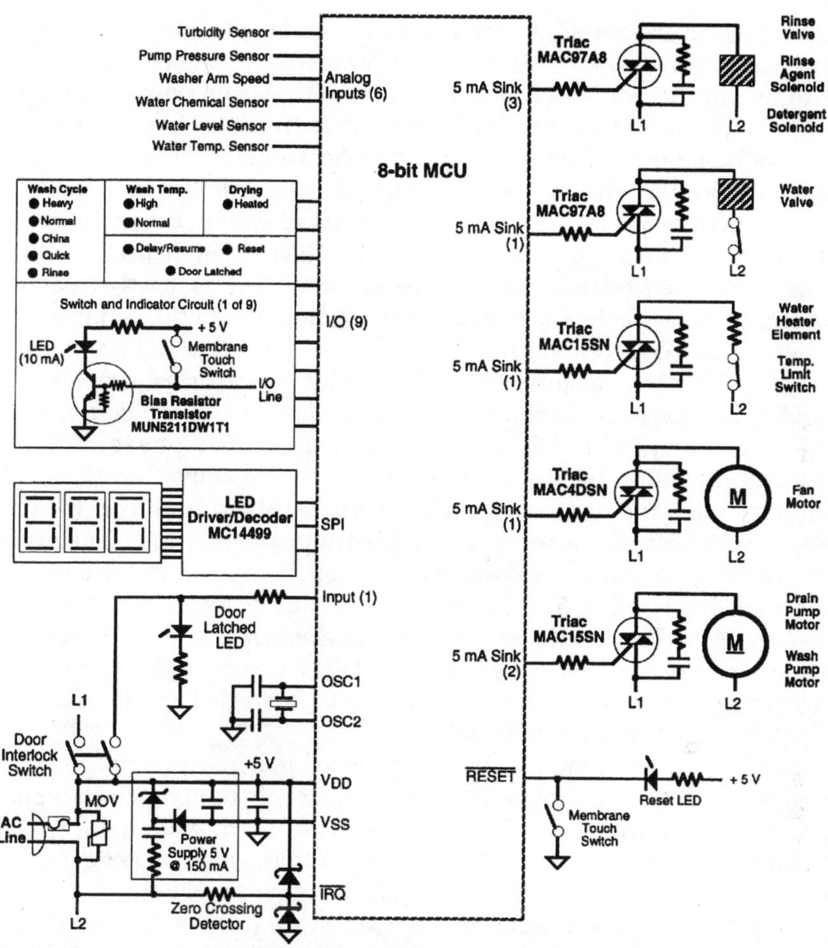

Figure 5.12 Electronic dishwasher conceptual design.

port (SPI), and about 10 standard I/O lines. An internal or external EEPROM can also be implemented to store user settings and the status of the dishwasher if a power failure occurs, allowing the dishwasher to restart from its previous operation when the power is restored.

5.5 Heating, Ventilation, and Air-Conditioning Motor Controls

Electronics can significantly improve heating, ventilation, and air-conditioning (HVAC) systems. The traditional bimetalic thermostat, for instance, can be replaced with a digital control that can also allow varied temperature setpoints for particular times of the day or night. Electronics can also manage the HVAC system in a manner that allows it to obtain the best power efficiency. HVAC systems typically employ large compressor motors (≤8 kW for residential units) and are used in air-conditioning and heating units or in heat pumps. Heat pump systems are popular and may employ some form of supplemental heating in more severe winter localities. The electrical energy consumption of HVAC systems can be considerable during severe hot or cold weather conditions. The electrical power efficiency of a HVAC is mostly centered on the compressor motor and the air-handler fan motor. Many large commercial HVAC systems use three-phase power, which allows the use of higher-efficiency motors. Since three-phase power is seldom used in residential areas, single-phase motors are necessary, which do not offer the inherent power efficiency of similarly rated phase motors. Residential 240-V-AC single-phase power can be converted to three-phase through the use of power electronics, as shown in Fig. 5.13. The 240-V-AC single-phase power is rectified into DC and then converted into three-phase power for driving the motor. The major obstacle to this approach is cost. The 5 to 15 percent boost in motor power efficiency may not be sufficient to justify the additional expense of the three-phase motor electronics. However, additional HVAC system performance may be obtained by adding more functionality to the logic portion (MCU) of the three-phase motor electronics.

The speed of the compressor can be varied to better match the heating/cooling load requirements. For example, if the outside-to-inside temperature differential is small, the HVAC can operate in a low-speed mode. The compressor motor's electronics would also be able to operate over a wide AC power input voltage, thereby minimizing problems during brownouts and erratic AC power conditions. Other conditions could be sensed, such as the differential temperature drops across both the evaporator and condenser coils. This would indicate how well the system is exchanging heat and whether a serious problem is occurring, such as an iced-up coil or an inoperative fan.

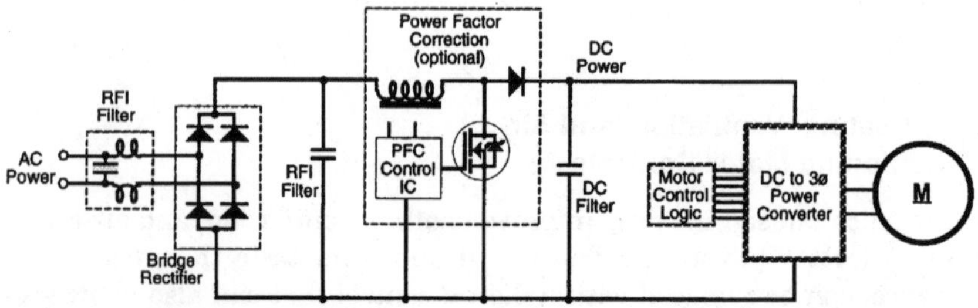

Figure 5.13 HVAC power conversion.

Pressure sensors on the compressor could test for extreme pressure conditions that may damage the system.

5.5.1 HVAC motor control system

A high-performance conceptual HVAC system is shown in Fig. 5.14. Several sensors are used in this centralized whole-house HVAC system to measure the parameters that directly affect the operation of

Control Panel & Thermostat

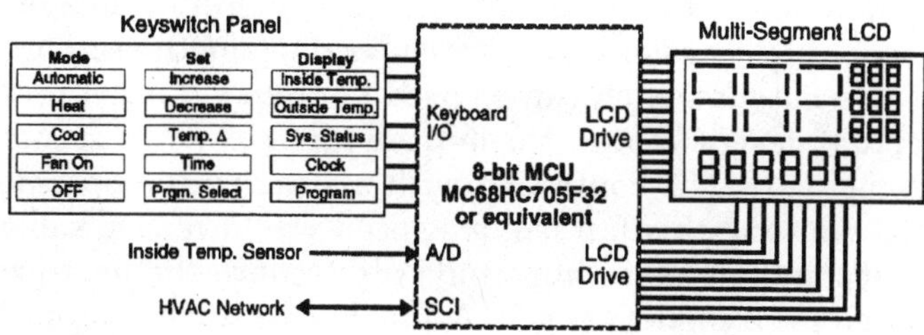

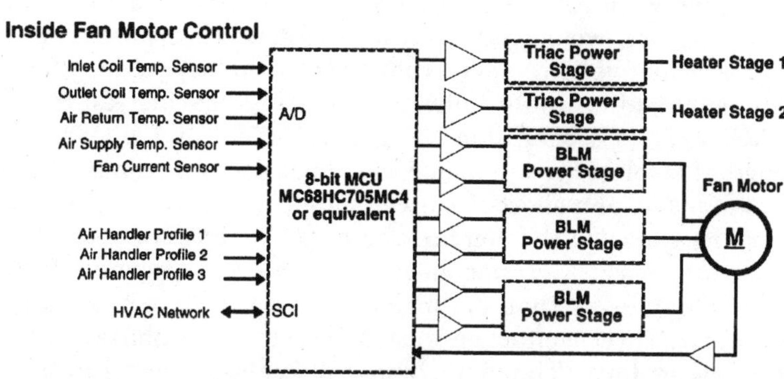

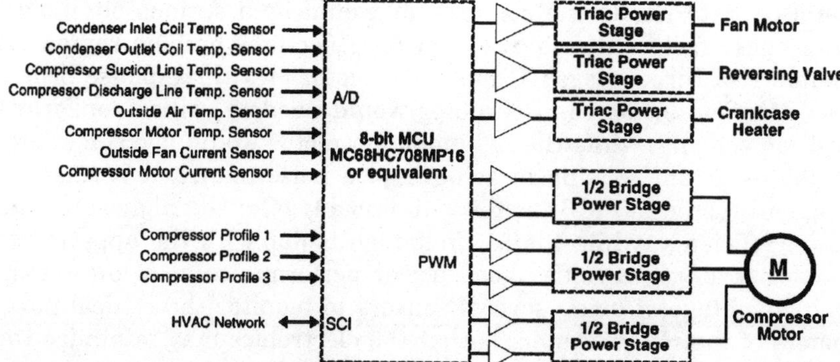

Figure 5.14 HVAC conceptual system.

HVAC. It should be noted that the software design should take into account the fact that a sensor may fail and that the HVAC should still be operable until that sensor is repaired. The HVAC should still rely on direct-acting sensors to protect against catastrophic failures. The HVAC electronics is split into three modules: the outside-fan and compressor motor control, the inside air-blower motor control, and the user control panel.

5.5.1.1 Outside-fan and compressor motor control. The outside-fan motor is normally configured to draw air across the condenser coil. If this airflow is blocked because of ice, an inoperative fan motor, or other mishap, high pressure may seriously damage the compressor. A high-pressure cutout switch can be used to protect the compressor. In the conceptual design shown, the outside fan motor is a standard AC induction motor that is controlled with a triac and is operated at a fixed speed. Varying this fan motor's speed would reduce some of the outside unit's audible noise but would also reduce the heat-exchange performance of the condenser coil. A simple current sensor on the fan motor could give valuable information on its operation. A zero or low current, for instance, would indicate that the motor is open-circuited or that the fan has been removed (no load). A slightly higher-than-normal current reading could indicate a pending bearing failure or that the fan blade is hitting something. A significantly higher-than-normal current level in the fan motor would suggest a frozen bearing, a jammed fan blade, or a short-circuited motor.

One consideration for the compressor motor selection is that the motor and compressor are normally combined in a hermetically sealed unit. This means that the motor's wires need a hermetically sealed connector. Three contacts are used for both three-phase motors or single-phase motors. (The single-phase motor has start, run, and common lines.) Adding extra wires for a shaft position/rpm sensor is expensive, and designing a cost-effective position/rpm sensor that could operate inside the compressor would be a serious challenge. With these factors in mind, the motor selections become somewhat limited to a three-phase AC induction motor or a three-phase brushless PM. The brushless DC motor would need to utilize sensorless position sensing, while the AC induction motor would use some form of phase-current sensing to achieve good performance. A permanent magnet brushless (PMB) motor will probably offer the highest overall power efficiency, while the AC induction motor with the appropriate control strategy will offer good power performance at a lower cost. One advantage of using several sensors to monitor the critical parameters of the HVAC system is that the electronics may minimize the chances that the compressor will pump liquid refrigerant or operate in a high-stress mode. (This condition is called *liquid floodback,* or "slugging," and can damage the compressor and motor.)

5.5.1.2 Inside (evaporator) fan motor control. The evaporator coil fan motor performs two important functions. It blows air through the evaporator coil, which promotes the liquefied refrigerant to change into a gaseous or vapor state. This process absorbs heat from the air of the evaporator coil, which, in effect, is then transferred to the outside condenser's coil. The other important function of the inside fan motor is to circulate or mix the air throughout the area to be cooled or heated. The performance of the evaporator fan motor is crucial to the overall HVAC system. The airflow rate can be varied by adjusting the fan motor's speed. The air-pressure peaks in the ductwork can also be minimized by incorporating a slow acceleration and deceleration for the fan motor. When the evaporator fan of a standard HVAC system, for example, is first turned on, the ductwork will be stressed with a fast air-pressure wave, producing a loud noise. If the fan motor's speed is slowly increased over a few seconds, the air will flow smoothly, reducing the initial air pressure peak in the ductwork. A slow rampup and rampdown fan speed minimizes ductwork flexing and its associated acoustical noise.

A permanent magnet brushless motor allows easy speed control for the fan, high power efficiency (which is important since this motor may operate 24 h a day), and offers low audible noise (assuming its PWM frequency is high). The running speed of the BLM fan can also be varied to match the user's requirements. For example, if the temperature differential between the user's setpoint temperature and the actual inside temperature is small, the fan can be operated at a lower speed. This saves energy, reduces audible noise, and minimizes the "cold-chill effect" when operating in a cooling mode. [The "cold-chill effect" is caused by the cold (4 to 10°C) air emitting from the ductwork.]

5.5.1.3 User control panel. The HVAC control panel allows the user to set desired temperatures and to select different operating modes. The control panel can also contain the master control MCU that is networked with the evaporator fan motor control and outside compressor unit. As shown in Fig. 5.14, an MCU in the user control box can interface to an LCD display and a keyboard and has serial communications capability. The MCU is fed information from the inside fan and outside compressor unit. This information can be processed with a fuzzy logic program to determine the best operation for the HVAC system. The user control panel needs some form of protection from a power failure, since a power interruption should not cause the

HVAC system to reset and stay in an OFF condition when the power is restored. The control unit can use an MCU with EEPROM or employ a battery (an ultra-high-value power supply filter capacitor may also work) to ensure that user settings are not lost during power outages.

5.6 Floor Cleaning Systems

An ideal vacuum cleaner would probably be a unit that emits minimal noise or dust and automatically adjusts to the surface that it is cleaning. It would have built-in protection against stalled motors and might even contain an active filter to contain unpleasant smells. This advanced vacuum cleaner could also alert the user to several useful factors, such as (1) the bag is full, (2) no more dirt is being picked up, or (3) the motor is overheating. Most of these features can or have already been implemented with electronics. The amount of audible noise is an exception but may yet be addressed with variable-speed motors, dual suction motors, or a different type of dirt removal method.

5.6.1 Vacuum motor control system

An advanced upright vacuum cleaner may employ three motors—one for the beater or agitator brush, one for the vacuum unit, and one for the drive wheels. Each of these motors has a variable-speed control, which accommodates a wide range of floor surfaces and user preferences. Figure 5.15 shows a conceptual vacuum cleaner system that uses three motors that are all electronically controlled. Current sensing of each motor is incorporated to protect the motor against overheating and can be used to monitor each motor's operation. The drive motor would be driven with a motor control strategy that would allow

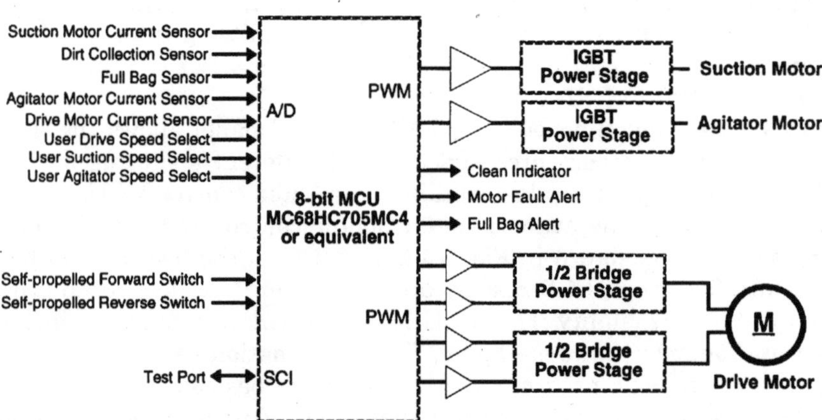

Figure 5.15 Advanced upright vacuum cleaner.

smooth acceleration. Note that the electronics of the drive motor needs to be rugged since the drive motor will experience rapid, repetitive directional changes. This will cause the motor to be "plugged" and its current levels to be significantly higher than normal. The drive, agitator, and suction motors are all speed controlled to allow the user to select preferences. A bridge rectifier on the AC power line supplies pulsating DC to the motor control power stages. This allows universal motors to be used while minimizing harmonics.

5.6.2 Water floor cleaner motor control system

Carpet and floor vacuum cleaners that utilize water require a water dispersal control and a suction system. This type of unit varies from the dry vacuum system discussed previously, since water, as well as dirt and debris, will be collected and stored in a container. A turbidity sensor may be employed to give some indication of how well the floor is being cleaned, minimizing unnecessary cleaning. Figure 5.16 shows a conceptual wet floor cleaner. A pump motor sprays the water soap solution, another motor drives an agitator or brush mechanism, and a suction motor is used to collect the dirt and water solution. The agitator motor is speed-controlled to allow the user to choose the best speed for varying types of floor coverings.

5.7 Garden-Lawn Equipment and Shop Tools

The move in lawn care equipment from gasoline-powered engines to electric-powered motors has resulted in some advantages and disad-

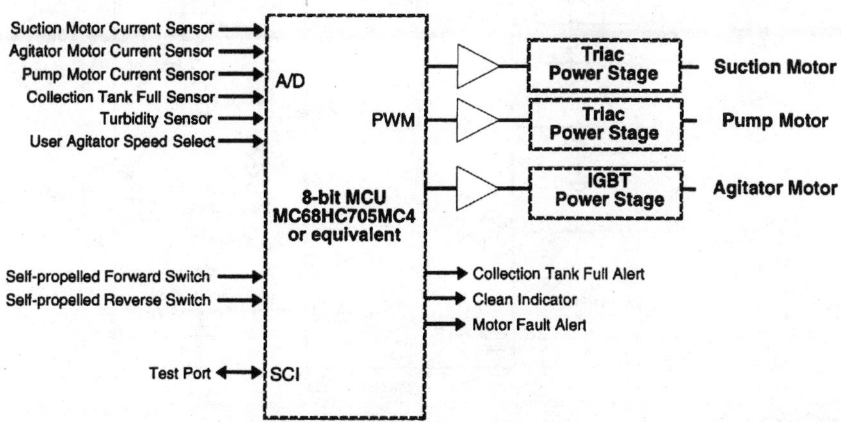

Figure 5.16 Advanced wet vacuum cleaner.

vantages. On the plus side, the storage and handling of flammable fuel is eliminated, and the electric motor instantly starts without emitting toxic emissions. The electric motor, however, does need a power source, which can be a rechargeable battery or an extension cord to the nearest power outlet. Some form of speed control of the electric motor may also be necessary in certain applications.

Electric-powered lawn equipment can be divided into two groups: those that operate directly from AC power and cordless units that use batteries as their operating power source. The AC-powered lawn equipment includes lawnmowers, string trimmers, hedgers, and so forth. These units usually employ a universal motor without speed control. Cordless lawn equipment typically uses low-voltage permanent magnet motors and rechargeable batteries. The size and weight of these batteries limit the power rating of handheld lawn equipment. This, however, may change as higher-power-density batteries, such as metal hydride or lithium types, are developed specifically for lawn equipment.

5.7.1 AC-powered lawn and shop tools

The motors and controls used in lawn equipment are similar to those employed in home shop tools. Universal motors are used for many AC-powered tools that require high speeds and good low-speed torque. The conceptual circuit shown in Fig. 5.17 uses a small MCU and IGBT to vary the motor's speed. This method chops or pulse-width-modulates the pulsating 120-Hz DC, which minimizes motor

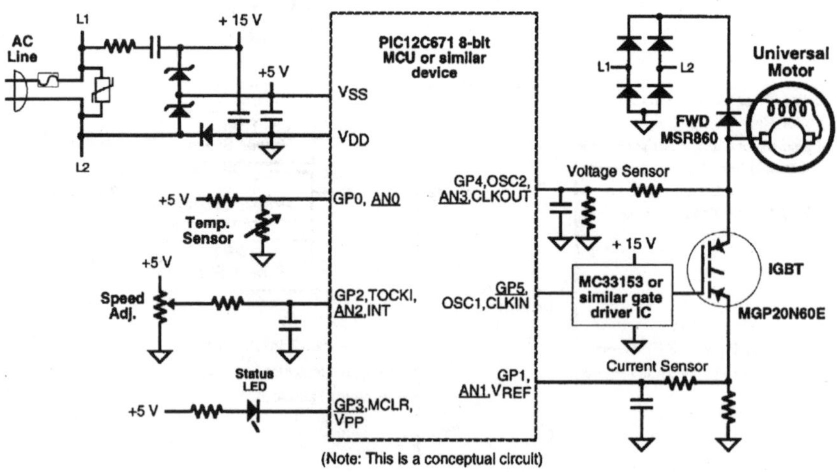

Figure 5.17 AC-powered lawn and shop motor speed control.

harmonics. The IGBT needs a gate buffer because the MCU can supply only 25 mA from a 5-V source. A standard 20-A 600-V IGBT is usually specified to operate from a 15-V gate supply and has a total gate charge of 20 to 75 nC, depending on the exact IGBT type. This means that the MCU cannot supply the necessary voltage or current to directly drive the IGBT, so a gate driver IC is required. This MCU, however, could directly drive a sensitive-gate triac or optocoupler device. If directional control is necessary, a field winding reversing switch can be added. The most interesting aspect of this circuit is that an 8-pin MCU can be employed for a basic motor control. Unlike a fixed analog IC design, the MCU allows programmability of the motor control strategies and also senses the motor's critical operating parameters.

5.7.2 Battery-powered lawn and shop tools

Portable lawn and shop tools use motors and control circuits that are designed to operate from low-voltage DC power, usually between 6 and 36 V. Permanent magnet motors are typically used, since they offer high efficiency and good low-speed torque. Power MOSFET devices are available with less than 0.01 Ω in resistance, making cost-effective speed control possible for battery-powered motors. (Refer to Chap. 3 for DC motor control methods.)

5.8 Other Appliance Motor Control Systems

Motor controls are utilized in many varied household equipment, such as ceiling fans, garage door openers, exercise equipment, and even a few types of toilets. In some cases, motor controls are simple ON/OFF devices or reversing elements, which can be accomplished with mechanical switches, triacs, or relays. If variable motor speed is required, an electronic control can be employed whose design would be similar to those that have been discussed for major appliances. There are some unique motor-powered appliances that can employ electronics to solve operational problems.

5.8.1 Water pump motor control

Sometimes an electronic motor control can be expanded to solve difficult system problems. Small water pump systems, which are commonly used in rural areas, often present the problem of water-pressure regulation. A typical small pump system must use a reservoir tank to ensure stable operation of the pump, especially at low water usage rates. (This assumes that the pump system can already meet the maximum usage rates.) A problem can occur if the pump's flow rate is significantly higher than the flow rate of the water usage. If the water reservoir tank is small, the pump's outlet pressure rises quickly as the tank fills to maximum and activates a pressure cutoff switch. As the pressure drops quickly because of its small tank size, the pump turns on once more, instigating the cycle all over again. Under certain conditions when the water flow rate is low, the pump may cycle several times a minute, which wastes power and overheats the pump motor. Adding a more complex pressure switch (one with a wide hysteresis) and a larger reservoir tank can help solve this problem. A water-level float switch in the reservoir tank is another solution but may present long-term maintenance problems. Going back to the basic problem, which is that the pump was selected for the maximum water usage rate and therefore operates poorly at low usage rates, the solution is a variable-speed pump. This allows a smaller reservoir tank to be used or even eliminated if the water supply is adequate. The pump's water output is matched to the water usage rate, thereby minimizing the pump motor's power cycling.

An MCU can be used, as shown, to control the pump. One of the analog input lines is connected to a pressure sensor device. The MCU can be programmed to allow hysteresis and also to calculate the amount of time by which the water pressure is varying. For example, under high water usage rates, the pressure will drop very quickly and the MCU program can then operate the pump motor at a high speed. If the pressure drop is gradual, the pump is operated at a low rate of speed. The program can be designed to minimize pump motor cycling in any situation, which will save power. This type of control would not be practical without the use of an electronic speed control.

Summary

Because of the need to achieve high reliability and long life at minimal cost in a hostile environment, designing appliance motor controls is a challenge. As we approach the twenty-first century, motors and electronics will continue to evolve in all types of appliances. Additional improvements in various systems will take advantage of more electronics, including those in clothes washers, refrigerators, and air-conditioning systems. Some of the significant points presented in this chapter are listed below.

- Two motor applications that could employ electronic motor controls to achieve high long-term power gains are HVAC equipment and refrigerators. This is mainly because these units operate at a high usage rate, and, in the case of a large, HVAC compressor motor, draw several kilowatts of power.

- Operating an AC induction motor from a 50-Hz AC power line instead of 60 Hz will decrease the motor's shaft speed by 16 percent and raise its power consumption by approximately 10 percent.
- The universal motor is commonly used in small appliances such as blenders, mixers, and vacuums because it is economical and has the ability to operate from a DC to 60-Hz power supply. Brush-commutator maintenance, noise, and poor speed performance are among its disadvantages. Phase-angle speed control with an SCR or triac is a cost-effective speed control method but allows high-energy harmonics to occur in universal motors.
- When a power failure occurs, the appliance's switch settings can be recalled by the motor electronics if it has an MCU that can save the critical operational data in nonvolatile memory (EEPROM). A special power supply that employs a battery or special high-value capacitor can also be utilized to supply a few microwatts of power to keep the MCU in a sleep state that will allow it to retain its memory.
- The AC power line is a source of transient voltages that can rise to over 1000 V. They are usually generated from the inductance of nearby motors or electrical equipment as they are switched off. Lightning strikes can generate 10,000 V or more and can travel long distances over power lines.
- An SCR device is used in simple universal motor speed controls and limits the motor's maximum voltage to one-half the AC power line. The triac can allow the full AC power line to be fed to the motor. Phase-angle control is used with both the SCR and triac to change the motor's speed and torque. When the SCR or triac is triggered, it will remain latched on until the AC voltage switches its polarity. The SCR and triac are turned on with a current pulse and, once switched on, require minimal gate current.
- An IGBT-based motor power stage uses pulse-width modulation and requires a high gate peak current. The IGBT's average gate power will be higher than that of an SCR or triac.
- To determine which type of MCU to use, first identify all the hardware functions, which include the number and type of input and outputs. You should also be aware of the throughput requirements, which will determine the MCU performance criteria. A final consideration is to decide how easy the MCU development tools are to use

and the level of technical support that will be available from the MCU manufacturer.

- One way to determine the load on a motor is by measuring its current—this usually indicates the degree to which the motor is loaded down. For example, in a clothes dryer, the drum motor current can be used to determine the weight of the wet clothes. An MCU can then compute what the weight of the clothes should be when they are dried, as well as the optimum airflow and temperature settings.
- Electronic control switches will usually be lighter weight and have smaller size density than will a mechanical switch. The electronic switch contacts conduct only a fraction of the current and operate at lower voltages. If a power failure occurs, the MCU will reset all switch functions to standby, unless a nonvolatile memory has been incorporated. Mechanical switches usually retain their settings during a power failure, which allows the appliance to complete its operation when the power is restored.
- The advantages of an electric motor in lawn equipment include instant starting, no flammable fuel storage, and no toxic emissions. The electric motor's power density is similar to that of a gas engine; its disadvantages are its power source and lack of integral speed adjustment. The use of extension cords to supply power to electric-powered lawn equipment is not without problems—tangles, danger of electrocution or fire from cutting the cord, and limited operating distance.
- The design of battery-powered lawn equipment is evolving in a manner similar to that of cordless power tools. Battery technology will remain crucial for electric-powered lawn equipment. Lithium or other highly reactive-type batteries offer very high power density and may eventually become cost-effective for cordless lawn or tool applications at some time in the future.

Acknowledgments

The author would like to thank Nigel Allison, Jean-Rene Bollier, Leos Chalupa, Charles Cordonnier, Bill Lucas, Tom Newenhouse, Khalid Shah, Ivan Skalka, and Warren Schultz for their assistance with this chapter on appliance motor controls.

Further Reading

Althouse, A., Turnquist, C., and Bracciano, A., *Modern Refrigeration and Air Conditioning,* 1992 edition.

Babyak, R., "Revolution in the Kitchen," "Big Brain," *Appliance Manufacturer,* May, June, 1995.

Emerson Motor Company, *Motor Technology News,* July 1996.

Janesurak, J., "In Their Opinion," "Major Appliance Dealers Study," *Appliance Manufacturer,* April 1995.

Microchip Technology Inc., PIC12C672 data sheet number DS30561A.

Rajamani, H., McMahon, R. A., "Induction Motor Drives for Domestic Appliances," *IEEE Industry Applications Magazine,* May-June 1997.

Sabin, D., and Sundaram, A., "Quality Enhances Reliability," *IEEE Spectrum,* Feb. 1996.

Chapter

6

Electric Car Drives

Richard J. Valentine, Peter Pinewski, and Thomas Huettl
Motorola Semiconductor Products

Introduction

Electric-powered vehicles use a variety of motor drive systems. The most common has been a field wound DC brushed motor. The recent introduction of purpose-built electric vehicles indicates that both AC induction and BLM motors are gaining in popularity for traction motor application. Hybrid vehicles also use electric motors with internal-combustion engines (ICE) and, in some cases, an ICE-powered alternator or flywheel mechanism.

This chapter reviews DC and AC motor drive designs for electric vehicles. These designs are based on motor drive research and a conversion test vehicle (a 1993 Dodge Dakota midsize truck). One of the main differences between a vehicle powered by an electric motor and a standard ICE is that the electric motor provides high torque at very low rpm. The series-wound DC motor produces very high starting torque, while an AC induction motor with vector control, in addition to providing high starting torque, can also continue to supply that torque level at higher rpm than can the DC series motor.

Figure 6.1 shows both the basic DC and AC motor drive systems. The DC brushed motor system is undoubtedly the easiest to implement. The AC motor system, on the other hand, is more complex but offers higher performance. The DC motor design can use linear control ICs, while the AC vector control motor requires a high-performance MCU or a custom ASIC device.

The EV motor and drive electronics are similar to those of large commercial or industrial electronic motor control systems. Some differences are worth noting—the EV application requires high energy efficiency, wide speed range, minimal weight, and operates from a high-voltage DC power supply. The EV's speed is controlled by large power transistor stages or modules that regulate the motor's power levels. These power modules incorporate very large power transistors and rectifiers that control the motor's electrical energy levels. The

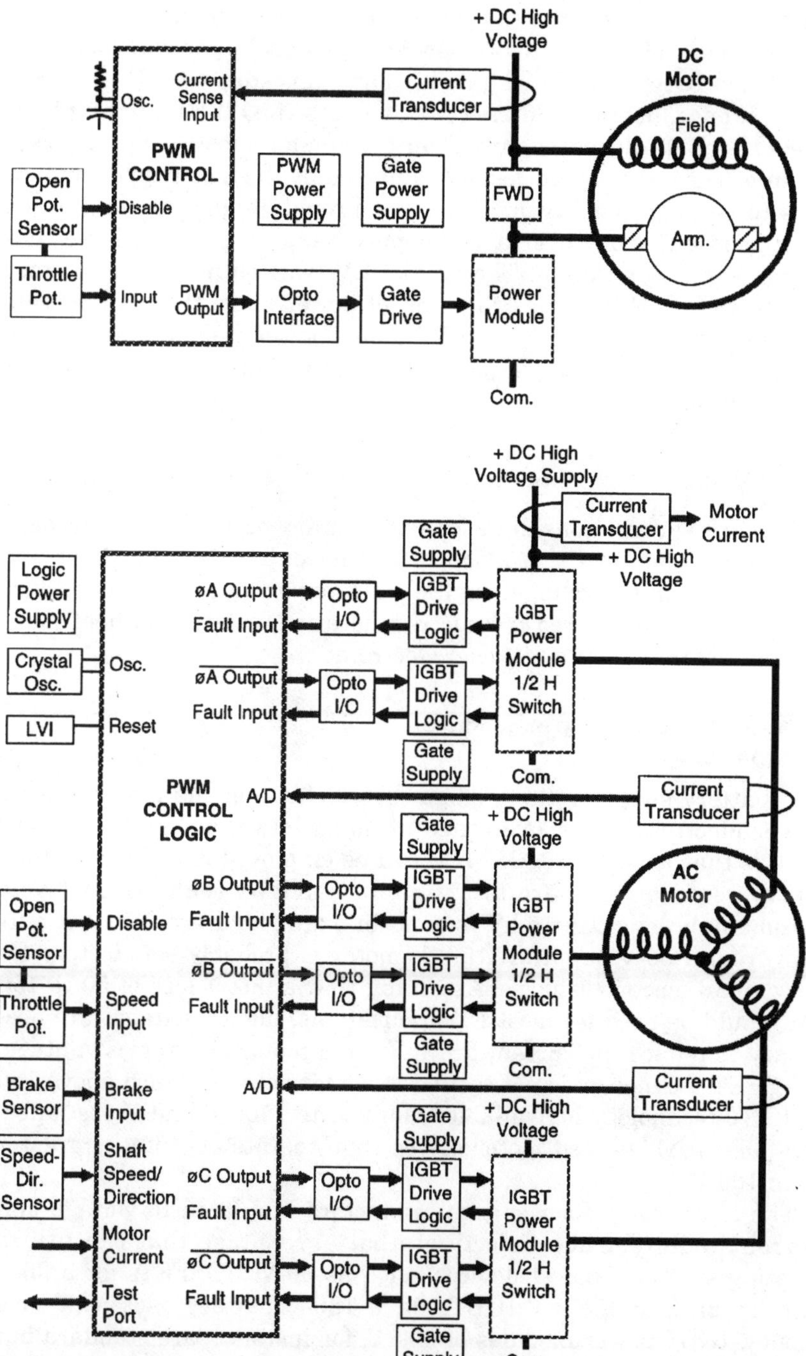

Figure 6.1 Electric vehicle motor system comparison.

management of the motor's speed and torque can utilize 8-, 16-, or 32-bit microcontrollers. The basic power stage usually consists of several paralleled N-type insulated gate bipolar transistors (IGBTs) or metal oxide semiconductor field-effect transistors (MOSFETs) with FWDs (free-wheeling diodes) mounted on a common substrate in a power module. The power stage design is affected by the motor type and how it is to be operated. EV motors must provide high starting torque, offer good efficiency over a wide speed range, and be reliable in a rather harsh application. The type of EV traction motor and battery supply voltage determines the kind of power stage that can operate the motor most effectively. The EV motor power stage requires large, reliable, and cost-effective power transistors and FWDs, no matter what type of motor is utilized.

6.1 EV Motor System Requirements

There are several considerations that must be taken into account when designing an EV traction motor drive. The type of motor and the battery supply voltage are the most important, but other factors that must be considered include regeneration, safety, and heat management of the motor and power electronics.

6.1.1 Voltage effects on motor drive electronics

The battery supply voltage has a direct effect on the motor and its power electronics. An advantage in using less than a 50-V battery pack is that low-$R_{DS,on}$ MOSFETs can be employed, but if the motor's power requirements are fairly high (>15 kW) the current levels become impractical. A 15-kW motor, for example, draws 250 A of current from a 60-V supply. A 15-kW motor is probably satisfactory for very small electric vehicles. Assuming that a motor size of 50 to 150 kW would be used for most EV applications, higher voltages are necessary to reduce the current draw. From a motor electronics vantage, MOSFET technology is not cost-effective for voltages much above 150 V. IGBT technology is available from 250 to 1700 V and is cost-effective for variable-speed motor drives that operate at voltages greater than 100 V.

The size (cost) of power semiconductors is related to voltage and current requirements. Selecting a battery voltage that can utilize existing standard power devices, such as 60-V MOSFETs for a 36-V battery pack and 250-V IGBTs for a 180-V battery pack, can save money. IGBT power modules of 600 V, for instance, are standard but could be rated at 500 V for use in a 300-V battery pack. Unless purchased in high volumes, a 500-V IGBT module would probably cost the same as a 600-V IGBT. The designs shown in this chapter will use standard IGBT power devices because of their cost/performance advantage, temperature stability, and high-voltage capability. If the battery pack voltages are lower than 50 V, the substitution of MOSFETs should be considered.

6.1.2 Regeneration

Since most electric vehicles are driven in a stop-and-go fashion, it makes sense to recapture some of the stopping energy. This increases the vehicle's driving range by up to 20 percent and may extend brake life. Regeneration requires that the EV traction motor switches from driving the wheels to being driven from the wheels. In effect, the motor must act like a generator. Changing a motor into a generator is possible when the motor exhibits certain attributes, and the control electronics is designed for both motor and generator operation.

Some form of field control is required for a motor to operate as a generator. This can be difficult with a series-wound DC motor, since the field winding is normally connected in series to the armature. Breaking this connection and electronically controlling the field winding can allow regeneration. Motors with built-in magnets, such as PM or BLM, will automatically act as a generator but are somewhat limited in versatility because the magnets always present a fixed-field strength. AC and SR motors can operate as generators and allow a wide range of operation because the field strength is controlled by electronics. The AC induction motor can provide regeneration with no extra circuitry—it is accomplished simply by commanding a negative slip. In a vector control strategy, this would be accomplished by commanding a negative torque current. In a slip control algorithm, it is done by controlling current magnitude and commanding a negative slip.

In some cases, more energy can be obtained during regeneration than in the driving mode. This occurs because a moving vehicle's inertia is quite substantial and can put a significant amount of force back into the motor, especially at high vehicle speeds. Remember that the level of energy is equal to the vehicle's speed squared times its weight. Therefore, regeneration at high vehicle speeds must be carefully controlled or too much energy may be incurred, which can damage the power electronics or battery pack. If the motor has sufficient torque to break the tires loose during wide-open throttle, the tires can also lose traction and contribute to vehicle instability during maximum generator operation or regeneration.

A brake pressure sensor or pedal travel sensor and antilock brake system (ABS) may be necessary to gain the largest amount of energy during regeneration while maintaining a stable braking condition. The data from these sensors, the motor's rpm, and the vehicle's speed would be used to determine the optimum level of regeneration. Controlling the level of regeneration based just on vehicle speed and its rate of deceleration is also possible. This could be as simple as adjusting the regeneration energy to the inverted square of the vehicle's speed with a regeneration program in the MCU motor controller. The state of the battery pack charge would also be considered. A fully charged battery pack would be damaged if too much regeneration were allowed, which could happen when coasting and braking down a long mountain road.

6.1.3 Safety issues

A runaway traction motor presents the same safety concern as a stuck wide-open throttle in an ICE. Since the speed control or the torque command control is a potentiometer that is hooked to the throttle pedal, it is easy to imagine an electrical failure condition that might allow the motor to operate in a full-speed mode. A broken wire or potentiometer, for example, might cause this to happen. (See Chap. 3 for methods to detect and correct this problem.) In addition to continuously testing the integrity of the throttle control elements, it may also be wise to ensure that the motor is not in a normal driving mode whenever the brake pedal is depressed. This is similar to a cruise control system—the cruise control automatically disengages whenever the brake pedal is engaged.

Another safety concern is the reliable containment of a traction motor short circuit. Since batteries can supply over 1000 A of current and the battery pack voltage may be 300 V, over 300 kW of energy is on hand, about 6 times more energy than the average home's electrical service can safely control. This means that fusing and short-circuit protective methods are an absolute necessity for the EV. The battery pack should have some form of disconnect and fusing right at its terminals to protect against wiring harness mishaps. A floating common can also be used in conjunction with a chassis-fault interrupter (CFI) to protect against electrical shocks. This would detect motor current levels flowing through the chassis, which might occur if the battery pack made contact with the frame and someone came into contact with the battery pack bus. The CFI would trigger a circuit breaker or relay, turning off the high voltage. However, designing this type of CFI is difficult because of the high peak currents and induced electrical noise into the chassis.

Another aspect of safety and reliability involves the temperature of the motor and its power stage. Their temperatures must be constantly monitored because of their high power dissipation, which can lead to damage or a fire if not properly managed. There are several factors to consider when dealing with the temperature sensor data. One is to feed the sensor data to the motor control MCU, which can evaluate the temperature and its rate of change. If the temperature is computed to be in a marginal zone, a warning signal is activated and invokes auxiliary cooling fans to maximum. If the temperature enters a critical zone, more drastic action is taken, such as reducing the motor power consumption. A simple thermal cutoff switch, which can be wired directly to the main power relay and does not depend on the MCU for activation, may also be employed to prevent catastrophic failures.

6.1.4 Cooling

The traction motor power electronics will dissipate a few hundred watts to over 3000 W, depending on the vehicle's driving conditions and the power rating of the motor. There are three main choices to consider for cooling an EV's motor and power stage: natural-air-convection-cooled heatsinks, forced-air-convection heatsinks, and liquid-cooled heatsinks. (Refer to Chap. 15 for heat management design.) Because of the power stage's high power dissipation during maximum torque, a very large natural convection heatsink is required, in the range of 0.02°C/W for a large 150-kW motor and a power stage that dissipates up to 3000 W. A 0.02°C/W natural-convection heatsink is not practical for use in a typical EV. Large bonded fin heatsinks are available in 0.2°C/W sizes. Adding forced-air fan cooling would increase the bonded fin heatsink's thermal performance by a factor of about 7 or more, depending on the fan's speed and airflow characteristics. A liquid-cooled heatsink can obtain very high thermal efficiencies and is dependent on the coolant surface area, flow rate, and radiator size. The main disadvantage of liquid cooling is cost and complexity, but, for EV applications, this approach is similar to the ICE cooling system, which has been successfully employed for many decades.

Cooling design affects the reliability of the power stage transistors because it is directly related to junction temperatures and thermal cycles. Excessive junction temperatures will cause degradation of the semiconductor materials or latchup of IGBT-type devices. Lab tests of electronic motor controls reveal that most power transistors will tend to fail in a sudden manner when peak junction temperatures reach above 225°C.

Thermal cycling is a more insidious failure mode. The power transistor and FWD dies expand at different rates than the die's attaching materials, which include the die attachment solder, the substrate, and the copper baseplate. Each of these layers has a different coefficient of thermal expansion (CTE). After several hundreds or thousands of thermal cycles, caused by fluctuating motor power levels, separations or cracks in the semiconductor's die or attaching material may occur, leading to a catastrophic failure of the power stage. The connections from the power dies to the module's leadframe are made with wires that are ultrasonically welded. Reliability tests suggest that this is also another area of concern—the wire bond on the die heats up and expands at a CTE different from that of the silicon die and may eventually fail. The wire bond failure rate is controlled by the range of temperature variations and their repetition. The thermal cycling issue is particularly serious for EV systems because of stop-and-go motor operation, 10-year (or longer) vehicle operation, and wide extremes in the vehicle's ambient temperature.

6.2 DC Traction Motor Controller Design

The DC series wound motor power stages can be very simple, as shown in Fig. 6.2, which is a conceptual design. This DC series wound motor controller needs few control elements, and low cost is obtained partially because only one power switch and FWD are required. The control and interface logic use standard devices. The power module design is a large single switch with a separate FWD across the motor. Most motor controls use PWM techniques to chop the motor's voltage, which changes the motor's speed and torque.

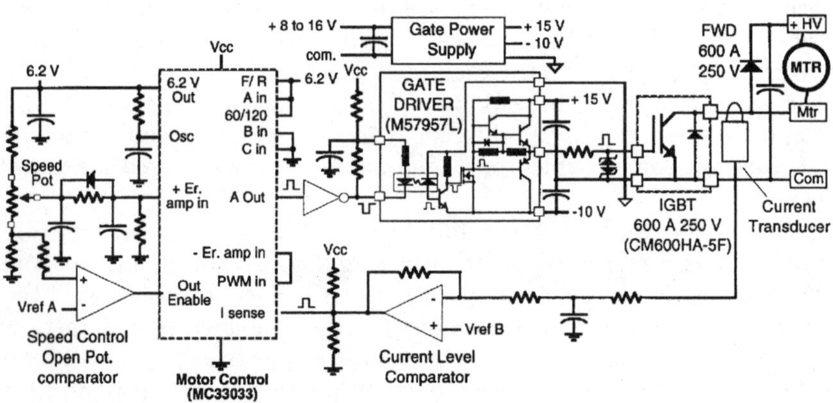

Figure 6.2 DC motor control design concept.

The logic control generates the PWM signal that drives the power stage. While the series motor design is simple in nature and offers excellent low-speed torque, regeneration is difficult to implement. The simplicity and low cost of this power stage and the high starting torque of the DC series wound motor make this design popular for many types of traction systems.

When compared to AC drive systems or bidirectional controls, a unidirectional brushed DC motor system offers higher reliability. A glitch in the DC motor drive signal may cause a slight roughness in the motor's speed, while the same glitch could permanently destroy an AC drive or bidirectional motor system by turning on both top and bottom devices of the bridge power stage at the same time. The series wound DC motor and control is less complex and more cost-effective than AC variable-speed motor systems, but it does not allow wide-speed-range performance. The DC series-wound motor offers formidable torque in the 0- to 2000-rpm range. The motor's speed is controlled by altering the average voltage to the motor by PWM at a fixed frequency. This speed control method (also called "chopping") relies on the motor's inductance to integrate or smooth out the PWM pulses and requires that the entire motor current be switched; in the case of a 100-kW rated motor, current levels in the 1000-A range are needed.

6.2.1 DC control design

The DC motor control design regulates the motor's speed and limits the motor's maximum current level. The motor speed range can vary from zero to wide-open throttle (WOT), which requires the controller's power transistors to sustain high peak currents and provide good efficiency at nominal cruise speeds. The controller must also be self-protecting against electrical disturbances, such as an intermittent battery cable or a faulty throttle position sensor. A large heatsink with auxiliary-fan cooling or a liquid-cooled heatsink that may also be tied in with the electric motor cooling system is needed because of the significant amount of heat dissipation from the controller's power transistor modules. Other important aspects of the motor controller design include detection of excessive heatsink or motor temperatures, and diagnostic indicators.

The experimental DC motor controller about to be reviewed uses IGBT power transistor modules. By choosing IGBT (insulated gate bipolar transistor)-type modules, the gate drive power requirements can be kept under 5 W, and the battery pack voltage can be well above 150 V. In this particular design, the motor voltage ratings were limited to under 150 V. Other types of power devices were considered,

TABLE 6.1 Electric Vehicle Motor Power Transistor Comparison

Device type	Die area, mm^2	ON voltages		Switching times			Driver stage	
		100 A	400 A	T_{rise}	T_{delay}	T_{fall}	Type	P_d
250-V Darlington	376	0.9 V	2.0 V	0.3 μs	1.2 μs	0.15 μs	16 A	80 W
250-V IGBT	356	0.8 V	1.0 V	0.3 μs	1.0 μs	0.5 μs	15 V	1 W
200-V MOSFET	384	1.0 V	3.9 V	0.3 μs	0.5 μs	0.3 μs	15 V	1 W
60-V MOSFET	378	0.05 V	0.21 V	0.2 μs	0.4 μs	0.2 μs	15 V	1W

NOTES: T_j = 125°C, inductive load; die area is estimated.

such as bipolar Darlingtons and MOSFETs, but, as shown in Table 6.1, the IGBT technology offers the best overall high current performance on the basis of switching speed, gate drive power, and ON voltages. The high-voltage Darlington requires substantial base drive current (2 to 25 A) to maintain a low $V_{CE,on}$ when conducting 100 to 600 A and usually has poor switching times. High-voltage MOSFETs exhibit high $V_{CE,on}$, especially at elevated junction temperatures, and would require a number of devices to be connected in parallel.

The design of an EV (electric vehicle) traction motor controller that manages 100,000-W power levels with 1400-W-rated power transistor modules presents many challenges. Operating frequencies beyond a few kilohertz are especially troublesome because of the motor's intrinsic inductance and the controller's high *Dv*/*Dt* values. The power transistor module's heat dissipation requires the use of large air- or liquid-cooled heatsinks. Design criteria and a working circuit for a DC motor controller are demonstrated in Fig. 6.3. (Possible improvements to this design are also reviewed later in this chapter.)

The circuit shown was developed to control a 9-in-diameter DC series-wound motor (Advanced DC Motor Model FB-4001). The PWM control is based on a MC33033 bipolar linear IC that was originally designed for controlling 3ø (three-phase) brushless motors. This part, which is specifically designed for motor control, was chosen primarily because of its ability to operate from 0 to 100 percent PWM while maintaining an overcurrent function even at 100 percent duty cycle. The circuit operation can be broken down into a number of functions: the PWM control, speed control fault detector, PWM output–gate driver opto interface, gate drivers, IGBT $V_{CE,on}$ sense opto interface, power IGBT control, heatsink temperature sensor, gate drive power supply, and 12-V regulator.

6.2.1.1 PWM speed control. Selecting a PWM frequency of 16 kHz ensures minimal audible noise since this frequency is usually just

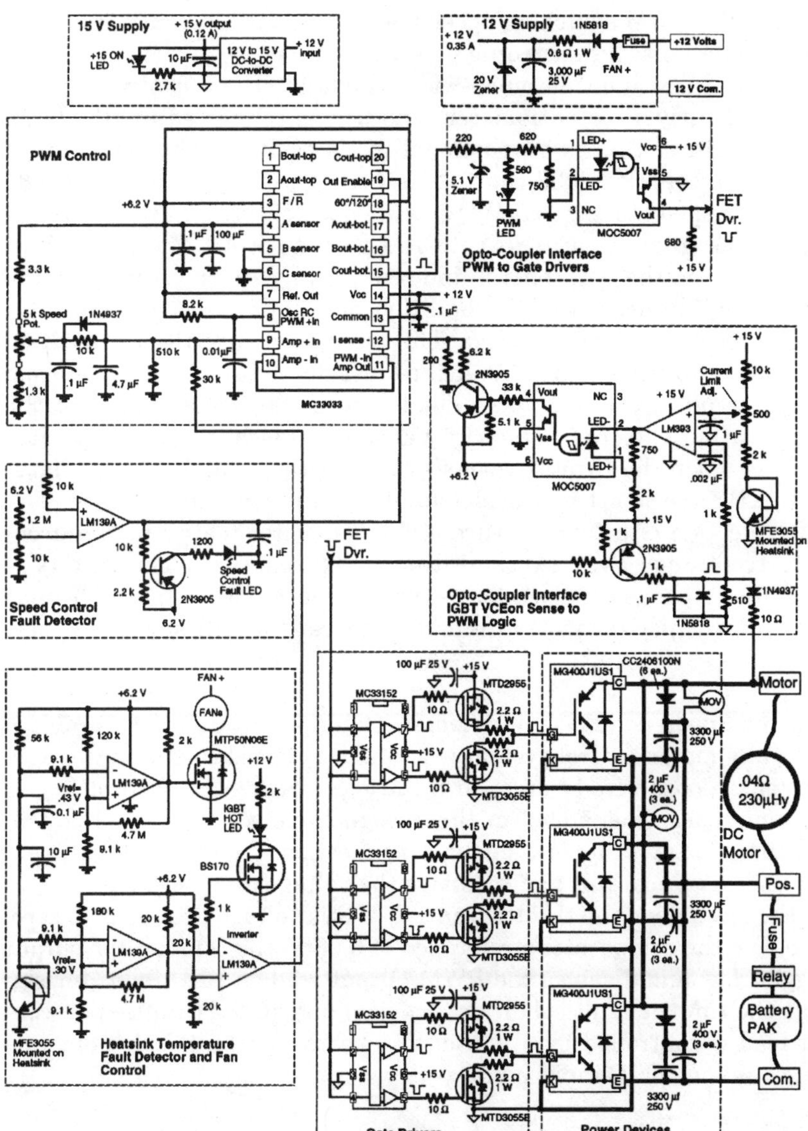

Figure 6.3 EV DC motor control design.

above normal human hearing. Extreme PWM switching (<100 ns) will reduce switching edge power loss but create another set of problems due to the inherent inductance in the wire and component leads. Fast-switching edges can also generate significant radio-frequency interference (RFI) that radiates into the vehicle's electrical system.

Setting the PWM switching speeds or transitions to about 250 ns is a good compromise for switching power loss.

The 16-kHz signal is pulse-width-modulated and controlled with a throttle-driven potentiometer. The pot. is connected to supply an analog voltage that is proportional to its position. The pot's output voltage to the MC33033 controller IC input varies from about 0.85 V at zero speed to 4.1 V at WOT (wide-open throttle). The PWM signal varies from 0% at 0.85 V to 100 percent at 4.1 V input. A low-pass filter on the pot. control line reduces any spurious noise on the speed signal and allows the addition of a simple resistor-capacitor (*RC*) acceleration rate limiter. The maximum throttle acceleration rate is set to about 0.3 s. To ensure fast deceleration response from the controller, a diode was added across the resistor part of the acceleration rate *RC* network, which will discharge the capacitor quickly when the throttle goes to zero position. The PWM oscillator is 16 kHz and is set by the *RC* values tied to the controller OSC pin. The selection of a temperature stable capacitor maintains an accurate frequency over extreme temperatures. Most ceramic-type decoupling capacitors, while inexpensive, are not adequate; a mica or similar capacitor with minimal temperature drift should be used instead.

6.2.1.2 Speed control fault detector. One potential safety problem with any motor control system is a runaway speed condition due to a broken or loose wire or an open-throttle control device. A throttle pot. current sensor was added in this design to detect an open circuit in the throttle pot. or its low-side interconnection. The throttle fault sensor uses a comparator and monitors the voltage drop across a resistor on the low side of the throttle control. If this voltage drop goes to zero, or below the comparator's reference of 0.05 V, the comparator output toggles low and disables the PWM output while a LED indicator signals the throttle fault condition. An open line on the throttle pot. center lead is controlled by a pull-down resistor on the PWM input line near the control IC input pin.

6.2.1.3 Gate driver opto interface. The gate driver interface is one area in which problems can occur. If the MC33033 PWM output signal is directly connected to an amplifier that drives the IGBT power devices, severe instabilities can result because the common-mode noise that is present in the power stage will disrupt the normal operation of the control IC. Adding an optoisolator solves this problem and also provides protection for the control IC in the event of a power stage high-voltage short.

The PWM control IC output signal, pin 15, is connected to the input side of an optocoupler device. A 5.1-V zener shunt regulator was added to maintain a constant current to the optocoupler. This ensures that the optocoupler's output signal remains stable during varying power supply conditions, for instance, when the 12-V battery, with its charge nearly exhausted, drops below 11 V. A 10-V undervoltage lockout (UVLO) circuit built into the control IC disables its output if the 12-V battery goes below 10 V. The optocoupler shown is a Schmitt logic type, which tends to minimize noisy signals at the expense of reduced response time. A delay time and minimal pulse width of about 1.5 μs was observed. This means that the minimal PWM signal can be only 1.5 μs, which is a good number to work with, since it matches the switching speed response of the IGBT output stage. A narrow 0.5-μs pulse would not allow the IGBT to switch fully ON and would stress the IGBT in a forward-bias safe operating mode (FBSOA). *Even though these IGBTs are rated at 400 A, they have limited power capability in the linear mode of operation, less than 1 A at 100 V DC.*

6.2.1.4 Gate drivers. The optocoupler inverted output signal is connected to each IGBT gate driver printed-circuit board (PCB). The gate driver PCB is mounted directly on top of each power IGBT module.

The gate driver uses a MOSFET driver IC, an MC33052, to drive a complementary MOSFET amplifier stage. This stage is capable of switching the 5- to 7-A peak gate current levels required to charge and discharge the large 40,000- to 80,000-pF IGBT gate capacitance in less than 500 ns. The series gate resistor values largely determine the switching speed of the gate voltage. It is possible to slow down the turnon time by increasing the upper P-MOSFET resistor value and to increase the turn-off time by lowering the N-MOSFET resistor value. The gate resistors should be flameproof types to minimize PCB damage in the event of an IGBT collector-to-gate short.

6.2.1.5 IGBT $V_{CE,on}$ sense opto interface. The IGBT module's current level is sensed by measuring the IGBT transistor's forward ON voltage. This method does have some inherent drawbacks. The collector-to-emitter voltage measurement circuit has to be synchronized with the gate drive voltage, and the $V_{CE,on}$ will vary somewhat with temperature fluctuations. In addition to these drawbacks, the IGBT's $V_{CE,on}$ versus current does not have a nice linear curve, and therefore places a performance limit on this type of current limiter. (In hindsight, a better current-limiting method would be to use a fast noncontact type of current sensor.) A temperature compensator added into the current limit detection circuit minimizes the temperature drift problem, and a simple transistor stage synchronizes the current sensor stage. A closer look at the current sensor design shows that the IGBT's ON voltage value plus a diode drop appears at the comparator's negative input. Normally, this voltage will vary from 1.5 V during minimal motor load to over 3.5 V during WOT or a stalled motor condition. The comparator's reference positive input determines the trip point. A current limit range of 300 to 900 A is obtained by varying this reference input voltage from about 2.6 to 3.6 V. An NPN bipolar transistor mounted near the IGBT modules is used for temperature compensation. This is required because the IGBT ON voltage (with the current held constant) will vary with a slightly negative coefficient. Unfortunately, the coefficient factor varies with different IGBT brands, as shown in Fig. 6.4. If no temperature compensation is used, the maximum current level will increase as the IGBT modules heat up. This causes a regenerative action—more current will flow, adding more heat to the modules, leading to more current flow, which eventually causes a motor to burn out or an IGBT module to fail. It is possible to increase the temperature compensation factor to purposely reduce the motor current as the modules heat up, thereby decreasing the chance of a heat-induced failure.

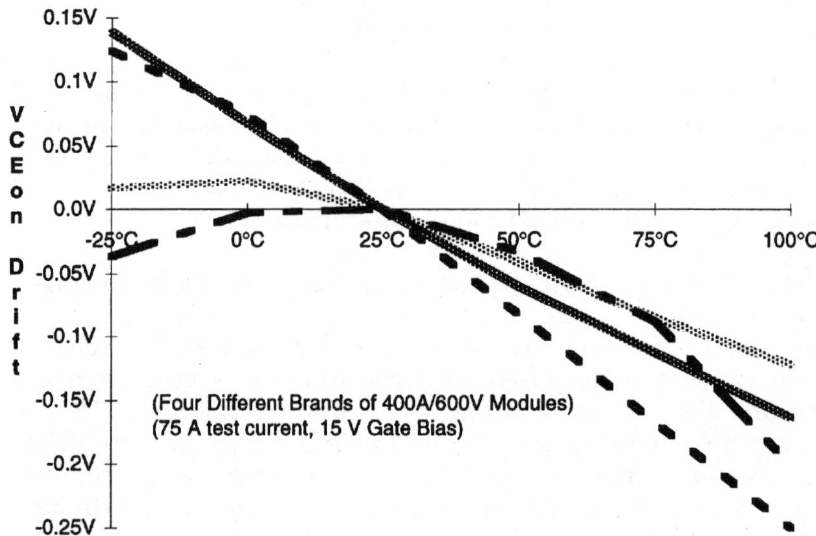

Figure 6.4 IGBT $V_{CE,on}$ temperature drift.

6.2.1.6 Power IGBT control. Three 400-A 600-V fast IGBT power modules are connected in parallel with large copper or aluminum bus bars. Several 100-ns "soft" fast-recovery rectifiers are connected in parallel in close proximity to the IGBT's collector terminals and the battery-motor positive bus bar. The high-frequency 2 μF 400-V metalized film capacitors, and the large 3300-μF 250-V low-ESR (equivalent series resistance) electrolytic capacitors are also near the battery-motor positive and battery-common bus bars. It is a good idea to match the IGBT devices for similar $V_{CE,on}$ values over their 100 to 300 A, equalizing current division among all three units.

Proper mounting of the IGBT modules to the heatsink is important. The thermal interface between the module's case and the heatsink must be kept as low as possible or excessive junction temperatures will occur under WOT. The modules should be mounted onto the heatsink with either nondrying silicone thermal grease or a metalized silicone pad. (Silicone pad and other interface materials are available from several companies such as Berquist Co. or Power Devices Inc.) The heatsink mounting interface thickness should be kept to a minimum, and the module fasteners should be tightened with a torque wrench.

It is a good idea to protect the IGBTs against voltage spikes. Voltage-clamping networks, such as metal oxide varistors (MOVs) across the IGBTs and free-wheeling rectifiers, ensure that these devices will not be destroyed by excessive voltage spikes that may occur due to intermittent wire connections.

When designing the power stage, you must consider the effects of long power cables to the battery pack. These power cables will exhibit a few microhenries of inductance that will cause severe problems unless filter capacitors are used in the controller's power stage. When employing fast switching edges, a capacitor filter network is mandatory near the motor and controller power supply lines. This filter network smooths out both the PWM frequencies, as well as the much higher frequencies associated with the 200- to 1000-ns PWM switching edges. The filter consists of capacitors that are able and specifically designed to operate in electronic motor control applications. The interconnections between the large electrolytic-type capacitors, smaller metalized film units, power modules, and FWDs should be constructed to minimize inductance. Figure 6.5 shows a good-versus-poor layout for the capacitors and other associated devices in the controller. Note that the desirable layout is also the most compact.

6.2.1.7 Free-wheeling rectifier. The motor winding's inductance also plays a role in the design of the motor controller. Besides being an

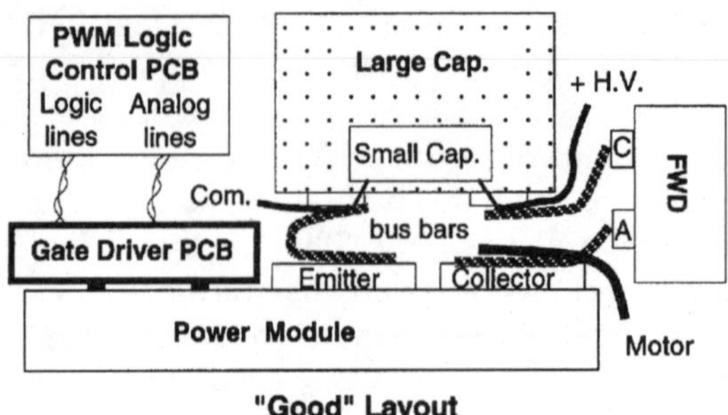

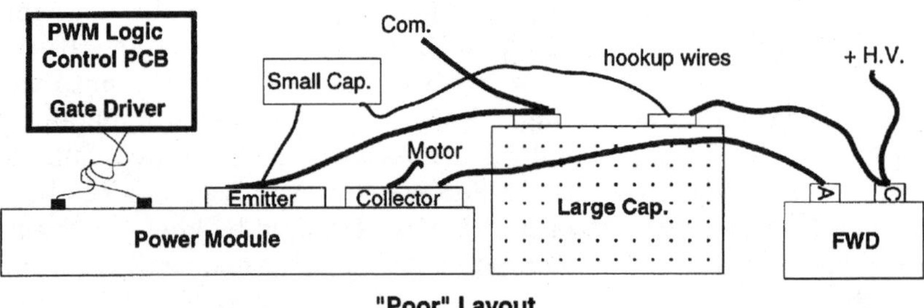

Figure 6.5 EV controller layout.

integral part of the motor design, the motor's winding inductance value must be considered when the power transistor switches off and interrupts the motor winding's current flow. This action generates a kickback voltage that is mostly determined by the motor's current, the motor's internal and external wire inductance value, and the rate at which the current is switched off. Normally, a high-current FWD network is connected across the controller's battery and motor terminals to clamp the motor's inductive kickback voltage spike. The reverse recovery time of the silicon FWD does affect the power module's switching performance.

Silicon junction diodes exhibit varying degrees of reverse recovery time (the time it takes the diode to stop conducting and revert to a blocking state). In electronic motor controls, the motor's inductance will store enough energy to keep the FWD in conduction until the next PWM pulse occurs. The diode continues to conduct for a few tens of nanoseconds while the power transistor is switching on. This creates a momentarily shorted load and results in high shoot-through currents in the diode and power transistor stage. (Chapter 2 reviews this effect.) To help minimize noise spikes, you should select fast reverse recovery rectifiers for the FWD that also exhibit a soft knee,

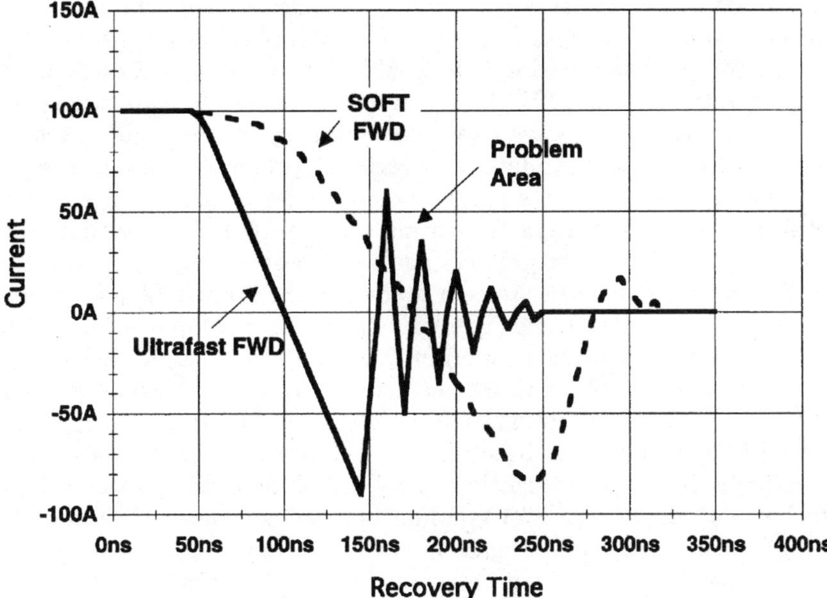

Figure 6.6 FWD reverse recovery.

as shown in Fig. 6.6. Snubber networks across the power transistor and FWD can also help minimize noise spikes. Some new rectifier products based on gallium arsenide (GaAs) have less than 15 ns reverse recovery time (for example, a MGR1018 rectifier from Motorola).

6.2.1.8 Heatsink temperature sensor. Protection against excessive heatsink temperatures is accomplished by using a temperature-sensing network. An NPN bipolar transistor is connected as a base-emitter diode mounted to the heatsink. The output of this diode is connected to two comparators. One comparator is set to trip at a voltage that is equal to +20°C and controls the heatsink's cooling fans. A feedback resistor in the comparator circuit allows a 5°C hysteresis factor, which means that once the fans are on, the heatsink must cool to 5°C below the initial setpoint temperature. The second comparator will trip if the heatsink temperature rises above 75°C. If this occurs, the PWM control input voltage is reduced to minimize the PWM, thus lowering the heatsink's excessive temperature. A resistor from the comparator's output is used to pull down the PWM voltage input level, thereby limiting the PWM. The value shown curtails the PWM to 50 percent at WOT.

6.2.1.9 Gate drive power supply and 12-V regulator. The vehicle's 12-V battery supply was chosen to power the controller's analog functions mainly for the sake of simplicity. As with any electronic equipment that operates from a 12-V battery system, protection against reverse battery hookups, excessive charging voltages, and intermittent battery connections is mandatory. A rectifier protects against reverse battery hookups, a series resistor establishes a maximum current level, and a 3000-μF capacitor forms a large *RC* filter network. The addition of a 20-V zener protects against excessive voltage on the 12-V filtered output. Another reason to include the large 3000-μF capacitor is to ensure that the controller's operation remains stable when the 12-V line is switched on-off-on and so on in a rapid fashion. A small internal 12- to 15-V-DC converter is used for the gate driver's +15-V supply. It is necessary to provide an isolated power source to the IGBT power control because of its electrical noise levels and to electrically isolate the throttle control and diagnostic functions from the battery supply. The +15-V-DC output matches most IGBT specification sheets for a nominal gate bias voltage. Lower $V_{CE,on}$ values can be obtained with a 20-V gate bias level but with a reduction in short-circuit capability of the IGBT. In retrospect, a better method than using the 12-V supply would have been to use two internal high-voltage (50- 200-V-DC and +15-V-DC) converters—one for the analog functions and the other for the IGBT drivers. This would guarantee that the motor controller would operate as long as the main battery supply was operational. When powering the motor controller from a high-voltage battery supply, you must design the logic control circuits so that they always operate in a stable manner, even under wildly varying battery supply voltages.

6.2.2 DC motor current control considerations

The DC series wound motor can draw significant current levels, especially during a locked-rotor or start-up operating condition. Using the 9-in DC EV motor (previously shown in Fig. 6.3) as the load, the maximum stalled or locked-rotor current can be calculated as shown:

$$I_{max} = E_{max} \div R_{mtr}$$

$$2400 = 120 \div 0.05 \qquad (6.1)$$

where E_{max} is the maximum power supply voltage (120 V) and R_{mtr} is the minimum motor (0.04-Ω) and hookup wire (0.01-Ω) resistance.

Even though this is basic Ohm's law, it is often overlooked by motor control engineers who are accustomed to dealing with normal AC power sources that are protected with circuit breakers and have a limited current source. You don't usually expect a 60-A power source to have the capability to supply well over 2000 A, as can happen with an EV battery pack. *EV battery systems can supply a very high current level, such as the calculated 2400 A, which is well beyond the EV motor's safe limit of operation and will result in brush assembly destruction and/or severe winding overheating.*

Since it is not practical to select power modules that can sustain the maximum energy levels that occur during a catastrophic motor failure (e.g., a dead short circuit), a current-limiting design becomes a necessity. The current-limiting design can also limit the motor's peak current level during maximum torque command or WOT and a locked-rotor condition.

Another factor to consider is the driving range, which diminishes as the maximum motor current level is increased. The nature of the driving cycle will greatly affect the vehicle's range; wide-open throttle during stop-and-go traffic will diminish it, while a steady speed in open country will increase it. On the basis of DC motor experimental design, the maximum motor current was adjusted from 300 to over 900 A, thereby allowing the driver to trade performance for distance. Current levels of 300 A give fair EV performance; acceleration times from 0 to 40 mi/h (mph) (0 to 64 km/h) are a sluggish 15 plus seconds (≥15 s), while 900-A levels allow the EV to approach the performance levels of an average four-cylinder gasoline engine.

Motor current can be measured by several methods, as shown in Fig. 6.7. A simple series resistor element is, in actuality, not that simple because the current-sensing resistor must have low inductance, offer good accuracy over a wide temperature range, and be able to sustain hundreds of amperes. Measuring the forward ON voltage of the power switch can give a rough motor current measurement. This method does have some inherent drawbacks and is best for detecting gross overcurrent conditions, such as shorted or locked rotor conditions. The noncontact current-sensing transducer offers ease of implementation, voltage isolation, and is a good choice for motor current control loops. Generally, any current measurement circuit must be synchronized with the gate drive voltage. The current sensor amplifier should be gated in a manner that negates the switching edges where high levels of noise will be encountered—for example, the current amplifier is turned on 1 or 2 μs after the power stage is switched on and 1 μs before the power stage is turned off. For shorted loads, it is important to protect the power stage. IGBT power transistors will

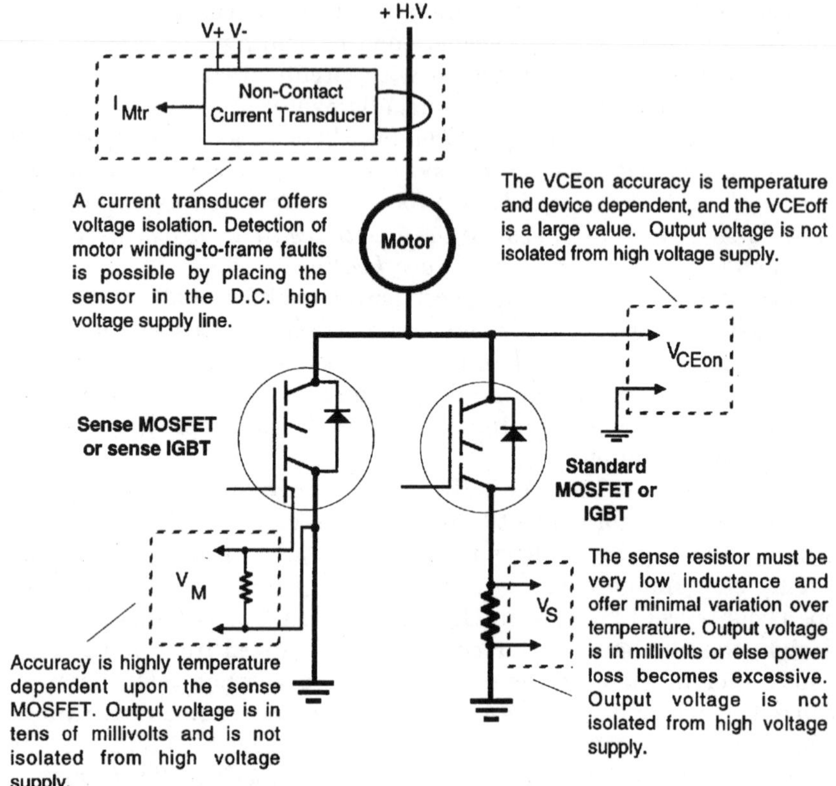

Figure 6.7 Current sensing methods.

succumb quickly to excessive junction temperatures produced by abnormal current levels.

The motor's running current can be limited by controlling the PWM. For example, if a maximum motor peak current of 400 A is desired, a feedback loop can be designed that shuts off the PWM cycle when the 400-A level is detected. This sounds fairly straightforward, but because of the inherent common-mode noise in whatever current sensor element is used, some care is required to obtain a stable cycle-by-cycle current-limiting design. An optoisolator or another means of voltage isolation between the current sensing element is highly recommended, not only to minimize common-mode noise problems, but for safety reasons as well. Without isolation, the operator controls, such as the speed potentiometer, would be subject to high voltages in the event of a short circuit in the power stage.

6.2.3 Power dissipation considerations

A prime consideration for reliable power transistor module operation in the EV application is heat dissipation. Obviously, the lower the voltage drop across the power stage, the lower its power dissipation will be. A decision must be made between silicon size (high cost) or heatsink size (medium cost). One important difference between MOSFETs and IGBTs is that the MOSFET's ON resistance almost doubles for a 125°C rise in junction temperature, while the IGBT's ON voltage remains fairly stable. The worst-case scenario would include maximum values for the ambient temperature, heatsink thermal resistance, contact thermal resistance from heatsink to transistor module, and the transistor module's case-to-junction thermal resistance. Equation (6.2) calculates the power transistor's maximum allowable heat dissipation during a WOT condition. The power stage is modulated at 100 percent, so its forward voltage drop will be the prime heat factor rather than switching losses. A maximum junction temperature of 150°C for the power transistors was chosen. It should be noted that the reliability will increase by about one order of magnitude for each 10°C drop in the power transistor's junction temperature. The heatsink's 0.03°C/W thermal resistance value represents either a large heatsink cooled by forced air or a smaller heatsink unit cooled by liquid.

$$\begin{aligned} PD_{max} &= (T_{j,max} - T_{a,max}) \div (R_{\text{øjc}} + R_{\text{øcs}} + R_{\text{øsa}}) \\ &= (150 - 40) \div (0.08 + 0.025 + 0.03) \\ &= 815 \text{ W per module} \end{aligned} \tag{6.2}$$

where $T_{j,max}$ = maximum allowable junction temperature
$T_{a,max}$ = maximum ambient temperature
$R_{\text{øjc}}$ = junction-to-case thermal resistance
$R_{\text{øcs}}$ = case to heatsink interface thermal resistance
$R_{\text{øsa}}$ = heatsink to ambient thermal resistance

After the maximum module power dissipation allowance is known (which is about 800 or 2400 W for three modules), each module's forward voltage drop, $V_{CE,on}$, can be calculated [Eq. (6.3)]. Three 400-A IGBT modules connected in parallel easily meet the 900-A current requirement, provided they are selected for good matching and correctly mounted onto a high-performance thermally efficient heatsink. A 2.67-V forward ON voltage specification at a 300-A nominal current level can be met by several commercially available 400-A IGBT modules. One note about the PWM modulation—these power loss calculations are for a 100 percent PWM mode or a full-ON condition. A higher power loss will be incurred if the PWM modulation is set to 95 percent. This is because, at 95 percent, most of the forward voltage power loss is still occurring, in addition to the power loss caused from the additional switching.

$$V_{\text{CE,max}} = \text{PD}_{\text{max}} \div I_{\text{max}}$$

$$2.67\ \text{V (at } 150°\text{C } T_j) = 800 \div 300 \tag{6.3}$$

It is interesting to note that the IGBT's ON resistance is calculated to be less than 0.01 Ω [Eq. (6.4)], which makes it highly cost-effective when compared to a similar-rated MOSFET. This is why IGBTs are preferred for high-voltage and high-current controls.

$$R_{\text{CE,max}} = V_{\text{CE,max}} \div I_{\text{max}}$$

$$0.009\ \Omega\ \text{(at } 150°\text{C } T_j) = 2.67 \div 300 \tag{6.4}$$

Each time the power stage is switched from off to on or on to off, a power loss occurs. Selecting the best PWM operating frequency comes down to a choice between audible noise and power module reliability factors. Using frequencies above 16 kHz (usually nonaudible) affects the power module's reliability and is related to the power loss occurring during the PWM switching edges. The power loss goes up with frequency.

6.2.4 Paralleled IGBT drive considerations

Driving one IGBT power module is simple when compared to driving several in parallel. The main difficulty lies in the gate drive connections since they allow a second high-current path to occur when the emitter-Kelvin connections are all tied in parallel. Through experimentation with the emitter-Kelvin contacts and the gate driver circuits, it was found that a stable turnon waveform would be obtained when a large copper plate was tied across all three emitter-Kelvin contacts and the high-current emitter contacts. Selecting an IGBT module that has minimal internal emitter lead inductance will help avoid oscillations that occur during the turnon time of the IGBT. The root cause of the oscillations can be traced to the reverse recovery time of the free-wheeling rectifier across the motor that is used to clamp the turnoff voltage spike generated by the motor's internal winding inductance. Because of the motor's inductance value in respect to the 16-kHz switching frequency, the free-wheeling rectifier is in conduction during the OFF time of the power IGBT modules.

When the power IGBT modules are switching on, the freewheeling rectifier must clear, which, in effect, shorts out the main battery rail to common for a brief moment, usually less than 100 ns. This very high current level causes the emitter lead inductance to create a voltage spike that opposes the gate drive signal and begins to turn off the device. This event exhibits a 10- to 30-ns repetition rate and lasts for about 50 to 100 ns or until the rectifier has finally cleared.

The selection of a free-wheeling rectifier with a recovery time specification similar to the IGBT turnon time appears to minimize the unwanted turnon oscillation problem. The use of 20- to 30-ns reverse recovery rectifiers was found to be especially troublesome with IGBTs that have high internal emitter lead inductance. This problem intensifies when the IGBTs are connected in parallel.

Adding series resistors with a small value (1 to 10 Ω, for example) into the Kelvin IGBT emitter leads may also help minimize oscillation problems when paralleling IGBT or MOSFET power devices.

6.2.5 DC motor control enhancements

A single chip microcontroller (MCU) could greatly improve the operation of this DC motor control design by allowing more functionality. For example, a stalled motor condition, which might occur if the EV driver had left the parking brake engaged, could be detected, and a previously set shutdown mode invoked to protect the motor against burnout. Other types of predetermined modes and functions could easily be accommodated with an 8-bit MCU (such as a MC68HC705MC4), as shown in Fig. 6.8. The MCU could monitor the motor-battery voltage and current levels for use in calculating the average energy consumption rates and predict the remaining range of the vehicle. The main difficulty in using an MCU is its sensitivity to the electrical noise present in the motor controller. Careful printed-circuit-board (PCB) layout and high-frequency filters on the MCU's I/O lines are necessary, as well as metal shielding around the MCU PCB area. The metal shielding will also help minimize the effects of mutual coupling from the very high magnetic fields created when the controller is operating at WOT. Any connection from the MCU to the motor's power stage or battery pack must be isolated with an optocoupler or similar device to minimize EMI coupling into the MCU.

The gate driver must function in a stable manner under a wide variety of conditions: shorted loads, reduced power supply voltages, and variable pulse-width switching requirements. Isolation from the user control and gate driver is by optocouplers or pulse transformers. Adding a negative voltage OFF bias will increase the reliability of the power stage. Noise spikes generated by the silicon FWD or other

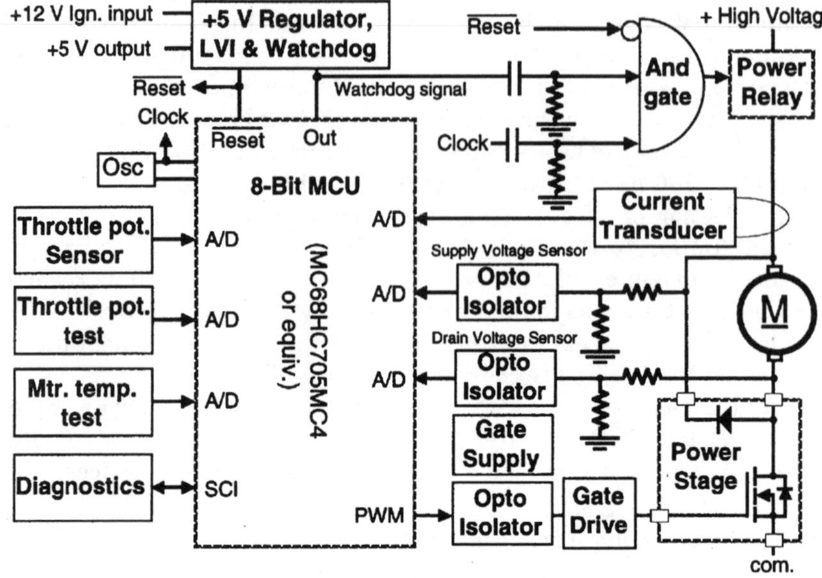

Figure 6.8 MCU-based DC EV motor control system.

noise spikes from external electromagnetic interference (EMI) must overcome the negative voltage bias as well as the gate's threshold voltage before the power stage can turn on and cause more problems.

It is possible to modify the series DC motor design to allow more control. This is done by switching both the field and armature, as shown in Fig. 6.9, which is a conceptual design, and requires the use of a DC motor that has separate field and armature connections. Note that the complexity of this system approaches that of an AC motor design.

6.2.6 EV PM motor control

Electric vehicles can use permanent magnet brush motors and are usually limited to 5 kW for smaller traction systems. If no regeneration control is required, a simple PM motor control would utilize the same power stage as the DC series motor. Regeneration occurs with a PM motor if its shaft speed exceeds its nominal speed. The PM motor acts like a DC generator whose output voltage is directly proportional to the motor's shaft speed, which is primarily set by the power-train gear ratios and the vehicle's coasting speed. Controlling the level of regeneration is important to prevent battery overcharging. The PM motor control requires two power stages to regulate speed and regeneration, as shown in Fig. 6.10.

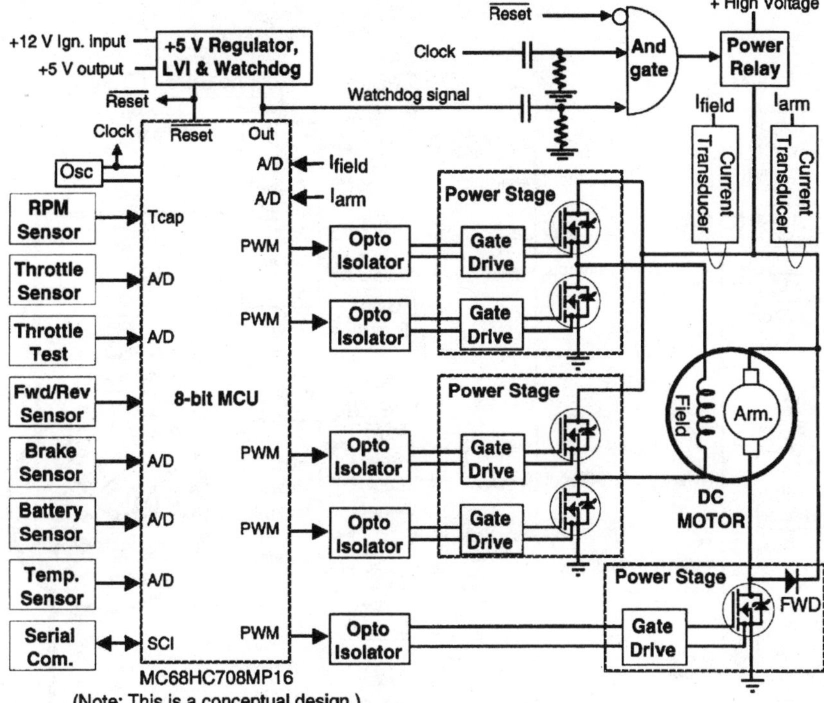

Figure 6.9 Regeneration for series DC motor.

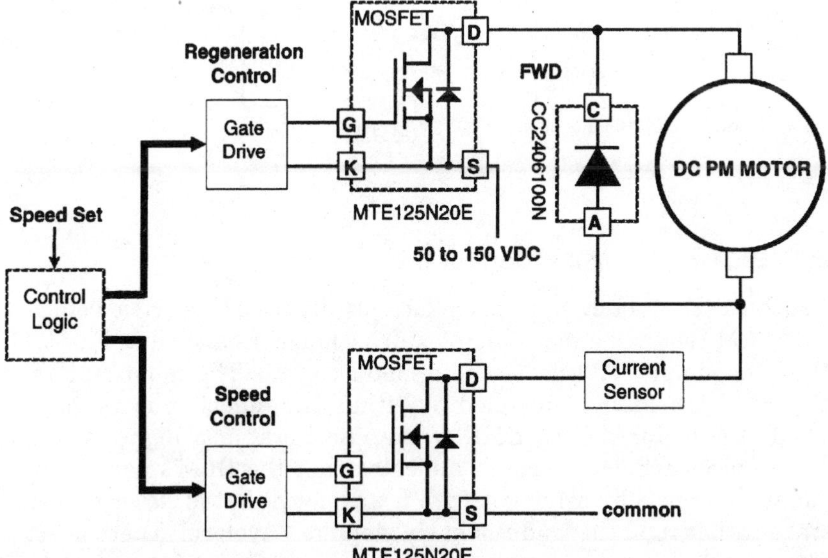

Figure 6.10 EV PM motor control system.

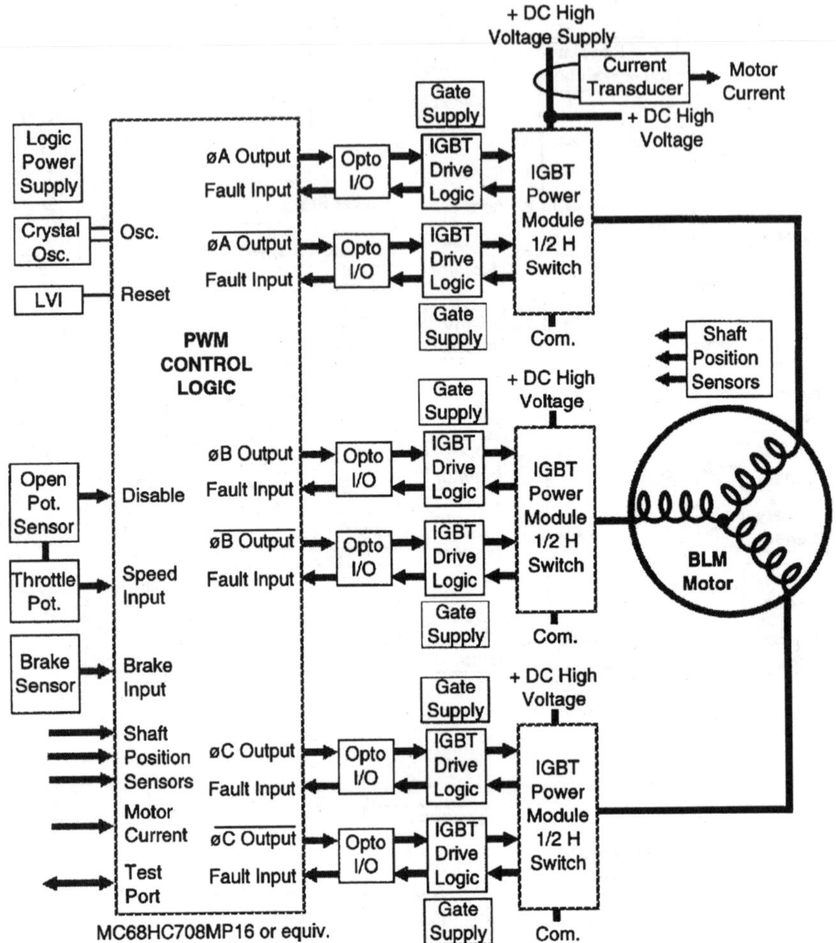

Figure 6.11 BLM control design concept.

6.3 BLM and SR Motor Control

Brushless DC motors have been successfully used in electric vehicles. The BLM requires a more complex drive stage, as shown in Fig. 6.11, than the simple DC series-wound motor control. The commutation of the BLM is usually controlled by either Hall-effect sensors located inside the motor, a back EMF sensing network, or a high-resolution shaft encoder. Some research indicates that the BLM's performance can be improved by driving it with a waveform that is closer to a sine wave rather than the traditional six steplike waveform. There is actually no difference in the power stage topology of the BLM and a standard AC induction motor; both require half H-bridge power devices for each phase. In some cases, all three half H-bridge stages can be combined into one large module, nicknamed a "six-pack." The BLM's power stage maximum current level is usually constrained to a level that will not weaken or damage the BLM's internal magnets. Like the PM motor, a BLM allows regeneration due to its permanent magnet construction. (See Chap. 3 for BLM design data.)

6.3.1 Switched reluctance motor

A switched reluctance (SR) motor is similar to an AC induction motor in that it requires an electronic control to generate the correct drive waveforms when operated from a battery pack. The SR motor power stage is more complex because both lines of each phase must be controlled, as shown in Fig. 6.12. SR motors are claimed to be the least expensive kind of motor to manufacture but suffer from high ripple torque, high audible noise, and require a shaft position encoder. The SR motor's windings are driven from single-ended power switches and require separate free-wheeling diodes, which is a similar requirement for the DC motor power stage, as previously shown in Fig. 6.2. The SR power stage includes a noteworthy design feature—it does not suffer from the potential short-circuit problem that is inherent in the BLM or AC induction half H-bridge power stage.

6.4 AC Traction Motor Controller Design

Speed and torque control electronics for vehicular motor applications requires high-performance/cost designs. This is especially true when designing an AC induction traction motor system. The induction motor has become popular for the traction drive, power steering, and HVAC systems of the electric vehicle. This popularity comes from the AC induction motor's chief attributes: low cost, rugged design, high torque capability, and wide rpm operation. Although the complexity of an AC motor drive and the cost of the electronics are greater than those of a DC brushed motor drive, digital control techniques and integration in electronics are reducing the cost and complexity of the AC motor drive hardware and making the induction motor more acceptable for variable-speed applications. However, even with these advances, complexity is still evident in the AC induction motor controller: The AC motor system must be able to create AC from DC, and it must be able to control both the amplitude and the frequency of these AC signals. While digital techniques shift these tasks to software, the electronics still play a major role by determining the limits of the software.

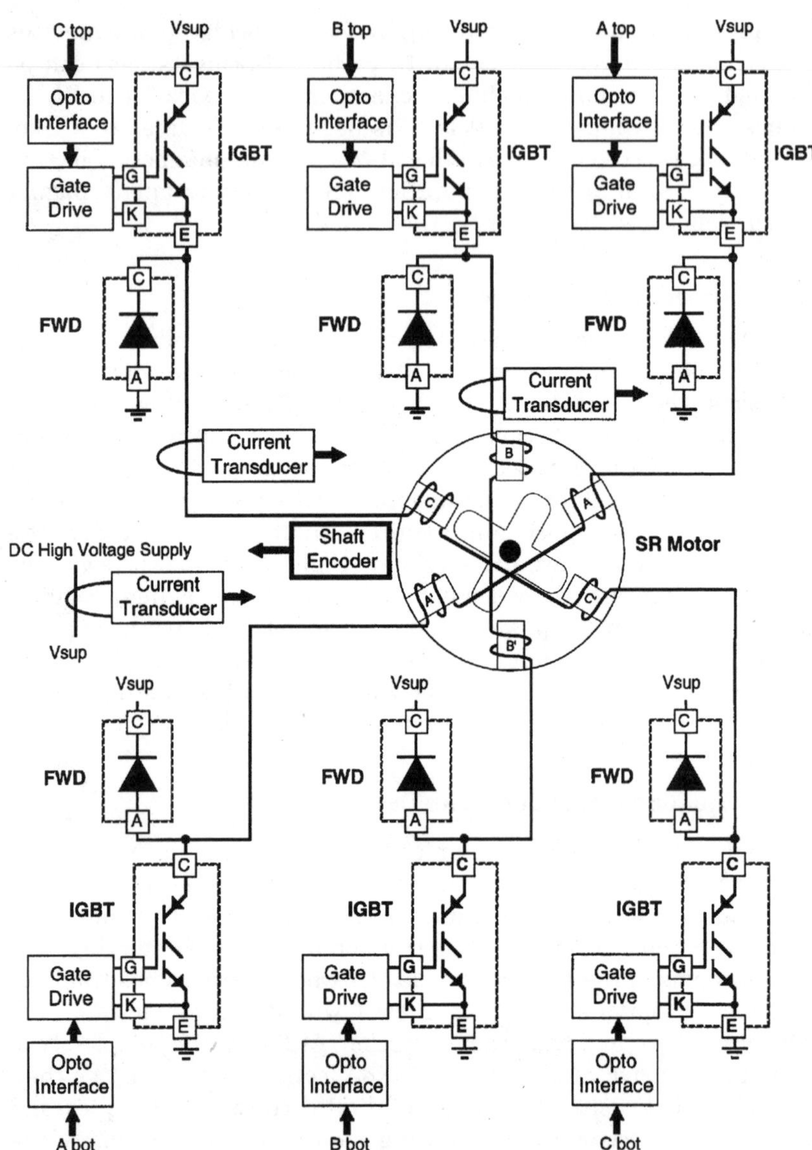

Figure 6.12 SR motor control design concept.

The electric vehicle's AC traction motor requires a torque-type controller that is capable of operating over a wide rpm range. Power efficiency gain in any EV motor control design means longer driving range and is affected by motor control strategies. Vector control strategy and modulation schemes such as space vector modulation (SVM) can boost control efficiency but require a higher-performance MCU.

The AC propulsion system's main control input is the driver's throttle control. This input is converted to a torque command whose final value is determined by several parameters besides the throttle position. The vehicle speed, battery voltage, and motor current values are used to calculate the most optimum torque command level. A regeneration sensor monitors the brake pressure and, if necessary, the ABS system. The amount or degree of regeneration is calculated that will give the most regeneration for a given situation. In a braking mode and low state-of-charge battery pack, the regeneration can be much higher (as determined by the ABS) than in a coastdown mode.

AC induction motors for EVs require a motor controller that generates AC voltages from the DC battery supply. This is readily accomplished with digital control techniques using microcontrollers (MC68HC16, MC68332 or equivalent) that have a built-in timer processor unit (TPU), as shown in Fig. 6.13. The power stage is similar to the BLM controller. The main difference is that a current sensor is required in two of the AC motor's phases, and a fairly precise shaft encoder may also be needed for regular or vector control strategies. AC motors can operate at higher voltages than most DC brushed motors, thus reducing the current rating of their power stage. The AC induction also allows more efficiency, since both its field and rotor flux levels can be adjusted by the control logic.

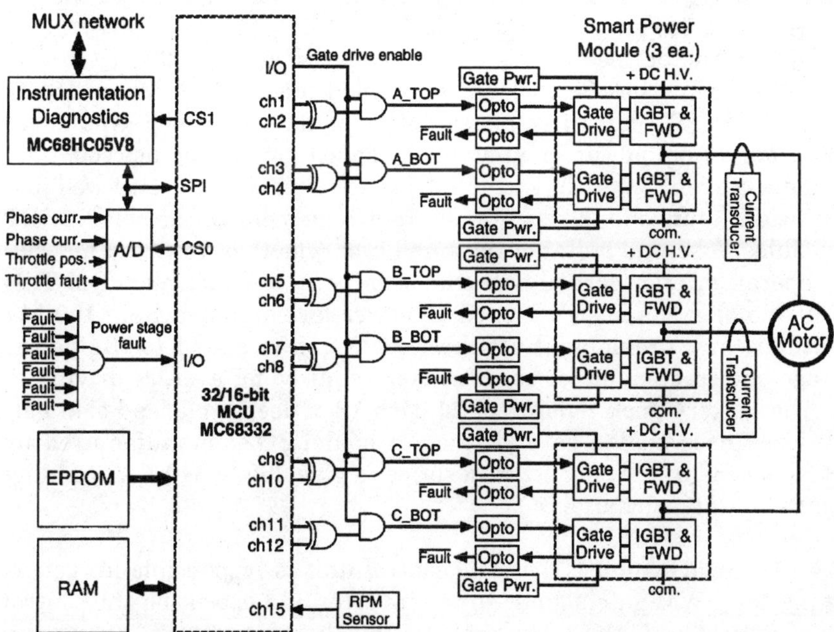

Figure 6.13 AC motor control EV system.

6.4.1 AC motor control design

An EV traction drive requires direct control of the torque and may implement more complex algorithms such as slip control or vector control. In our test EV, several different control algorithms were implemented on both the MC68HC16Y1 and MC68332 with the exception of sensorless vector control. A *V/F* control strategy was employed first to verify the hardware. A slip-type control strategy was also used and resulted in good performance but required complex slip tables. A vector control algorithm was found to be superior to the other control strategies. (The various motor control algorithms for the AC-powered EV are described in more detail in Chap. 8.)

The AC motor hardware is divided into two main parts—the logic control and the PWM inverter. The logic control part consists of the motor control MCU, auxiliary MCU for MUX, A/D converters, voltage regulators, and other miscellaneous MCU-related devices. The PWM inverter includes the opto interfaces, gate drivers, DC-DC gate bias converters, IGBT power modules, bus bars, and filter capacitors. A motor speed sensor with high-resolution motor is also required to provide data to the MCU.

Implementing an AC motor drive involves many considerations. They include sine-wave generation, algorithm execution time, speed sensing accuracy, A/D resolution, and A/D speed. The system requirements or limitations become apparent as various aspects of the AC motor drive design are investigated.

The AC motor used in this test EV conversion was a 20-hp (15-kW) size 256 frame totally enclosed fan-cooled model (U.S. Motor A983). The 235-lb (518-kg) three-phase 230-V motor was first rewound for 30-V operation but, unfortunately, the smart power modules could not handle current levels in the 450-A range. This current level was just sufficient to invoke the current shutdown element in the smart IGBT modules. A second motor of the same type was then rewound for 150-V operation and worked with the 400-A-rated smart modules. This 150-V motor has been operated at power levels approaching 100 kW without any problems. Extra fan cooling was added to the power stage and traction motor but is seldom required for average driving.

The battery pack consists of 24 each 12-V deep-cycle lead-acid batteries. These sealed gel batteries (a model 8G27 manufactured by Sonnenschein) are connected in series to give 260 to 340 V, depending on their state of charge.

6.4.1.1 Logic control. The logic control unit is responsible for generating the PWM gate signals to the inverter and executing the control algorithm. Of all the tasks executed by the control unit, generating the AC waveforms is the fundamental action. The MCU continuously updates the generation of the AC waveforms on the basis of the motor's phase current, rpm, and throttle input. The AC induction motor logic design (Fig. 6.14) is based on a 32-bit MCU (a MC68332). (Note that a MC68HC16Y1 MCU could also be used.) A special module called a "business card computer" (BCC) that contains the MC68332, 64 kbytes of RAM, and 128 kbytes of EPROM was used as a daughterboard to the main logic control printed-circuit board. The

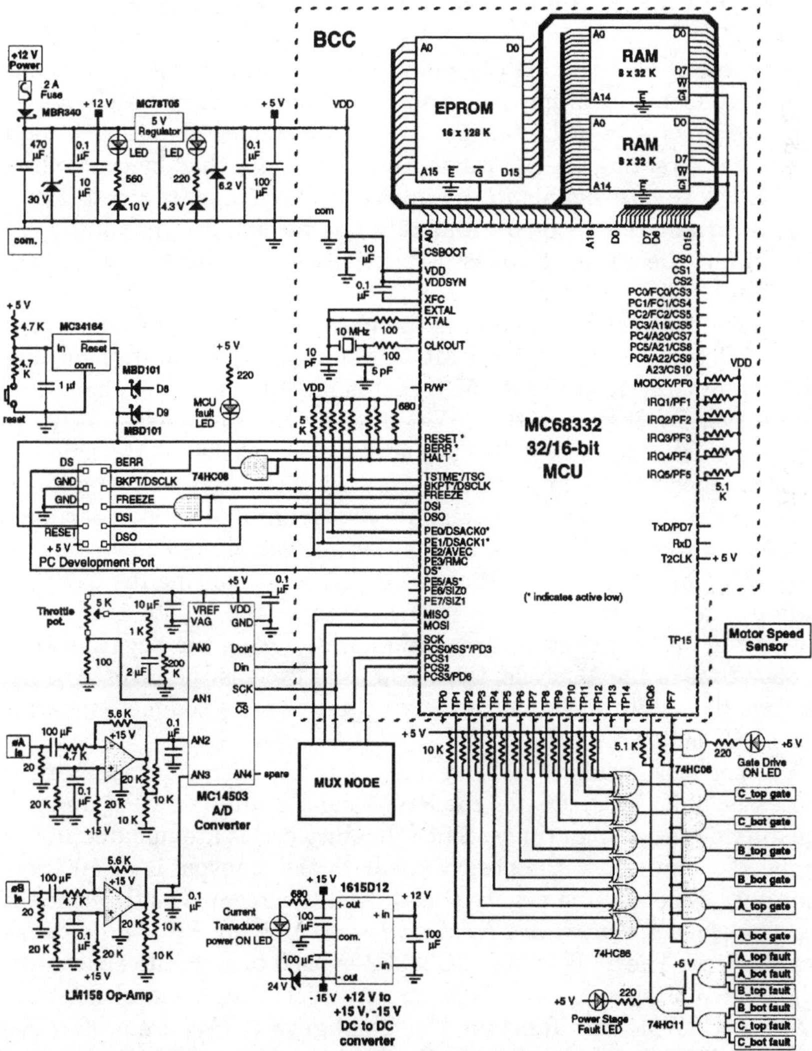

Figure 6.14 AC motor control logic design using MC68332.

BCC was selected because it was easy to implement since it contained the necessary external memory elements. A better approach would be to use the MC68HC16Y1 because it is a single-chip MCU. The main logic control board also contains a MC68HC705V8 MCU for instrumentation. This particular MCU contains a MUX network port. (See Chap. 11 on networking.) A MC34164 is used as a reset control. Exclusive-OR gates, MC74H08, are required to create the center-aligned PWM gate drive output signals. The output of these exclusive OR gates then drive AND gates. The AND gates provide a means to totally shut off the PWM output signals in the event of a serious fault.

A review of how the MCU generates the necessary AC waveforms might be useful. It is important to understand that the MCU's outputs are always in a digital format, not an analog signal such as a sine wave. Creating the AC waveforms is accomplished by controlling the six PWM gate signals to the inverter. These signals are still in a digital format (i.e., 1 or 0), and rely on the motor's inductance to smooth out their square edges. Essentially, the motor acts as a low-pass filter.

Sinusoidal PWM modulation is the most basic method used to generate AC waveforms with an MCU. Sinusoidal modulation means that the duty cycle of the PWM signal is modulated in the shape of a sine wave. By changing the PWM duty cycle, the phase voltage is altered. Most reports describe this sinusoidal creation as the sine-triangle method that is implemented in analog circuitry using op amps and comparators. The PWM signals are generated by comparing a sinusoidal waveform with a triangular reference, as shown in Fig. 6.15. It is important to note that the PWM output during the 0° to 90° time of the positive AC cycle is the same as the 90° to 180° time, except the pulses are in reverse order, and, during the negative 180° to 360° time of the AC cycle, the PWM duty cycle is the inverse of the positive 0° to 180° PWM output. (See Chap. 8 for a complete description of AC motor control.)

A closer look inside the 332 reveals that his microcontroller's time processor unit (TPU) creates the PWM signals while its CPU updates the duty cycles or high times. Since the duty cycles are updated in the shape of a sine wave, the positive peak of the sinusoid is at 100 percent duty cycle and the negative peak is at 0 percent. *The frequency of the sine wave is determined by the rate at which the PWM duty cycles are changed.* The TPU in the MC68332 is used to generate six center-aligned PWM signals to the inverter by executing the multichannel PWM (MCPWM) time function. Center-aligned PWMs are preferable; they result in signals which are exactly the same as those in the sine-

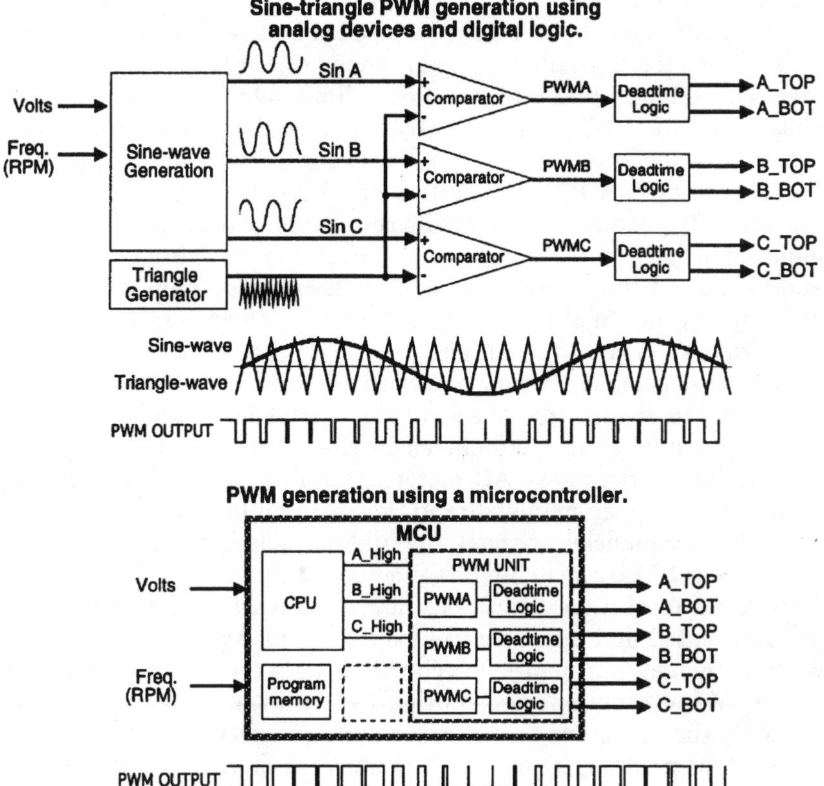

Figure 6.15 Sinusoidal modulation.

triangle method, reducing the likelihood that the three phases will switch at the same time, and producing a more symmetric waveform. In addition to generating center-aligned PWMs, the MCPWM time function generates the deadtime between top and bottom PWM signals. The deadtime prevents both top and bottom power devices from conducting high-current spikes produced by switching both devices at the same instant. Having this as part of the PWM unit reduces external deadtime generation circuitry. In addition, it is programmable via software.

Determining the PWM switching frequency is, for the most part, straightforward. There are, however, other effects to consider when selecting the switching frequency. Switching losses and audible noise are the most common factors in determining PWM switching frequencies; deadtime distortion and PWM bit resolution are among the others. PWM resolution is an important issue because it determines how accurately the sine wave can be represented and can affect control loop stability. The bits of resolution needed is determined by the fineness of the step desired. The trend is toward PWM bit resolutions in excess of 10 bits. PWM bit resolution is limited by the timer clock resolution and the PWM switching frequency. In this design, the TPU is driven by 4-MHz maximum clock. Choosing a switching frequency of 8 kHz results in a PWM resolution of 9 bits. A lower switching frequency will provide higher PWM resolution. Although the factors mentioned here point to a need to lower the switching frequency, audible noise usually precludes it. Because of this, there is a push in the industry for PWM units, which can be driven with faster clocks to increase the PWM bit resolutions.

Audible noise is a primary concern in determining PWM switching frequency. To ensure that the switching frequency is above the audible range, switching frequencies in the 16- to 20-kHz range are needed. The three-phase AC motor controller, however, is different than a DC chopper in that the AC motor may be operated at lower switching frequencies without the audible noise. Tests have shown that an 8 or 10 kHz switching frequency is acceptable.

PWM bit resolution determines how accurately the sine waves can be represented and affect control loop stability. An impure sine wave will create more harmonics, which will disturb the motor's power efficiency. Motor frequency resolution determines how accurately or smoothly the motor can be controlled and can affect loop stability.

One important aspect of the AC motor control system is the overall loop execution time. This execution time is an important factor in determining control loop stability. In addition, the PWM update routine is normally executed at the same rate as the control loop. Since the frequency of the sine wave is determined by the update rate and the number of sample points, the control loop time affects the maximum sine frequency that can be achieved. Another way to consider this is that the loop execution time or update rate determines the number of sample points used to represent the upper frequency. The faster the update rate, the more samples used to create this upper frequency (see Chap. 8).

The factor that affects update rate is the MCU's processing capability. The MCU must be able to perform the control algorithm calculations necessary within the update rate. If it is desired to produce frequencies up to 500 Hz with a minimum sample points of 16, an update rate of 125 μs is required. However, if the calculations needed take longer than 125 μs, the update rate must be extended at the expense of less sample points to create the upper frequency. The minimum number of sample points tolerable is difficult to specify.

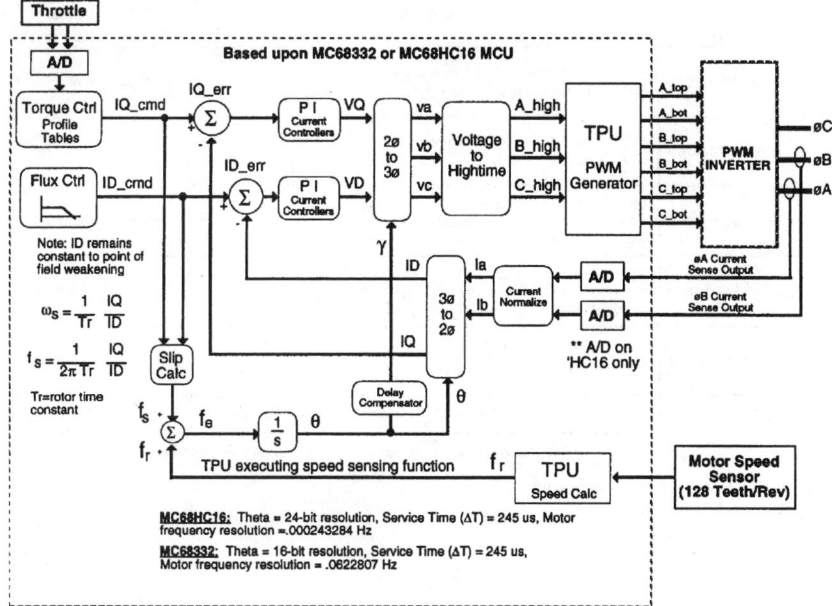

Figure 6.16 Vector control algorithm.

However, it is partly dependent on the stator time constant. In general, the number of sample points can be as low as eight.

The MC68332 and MC68HC16 designs are both limited by the execution of the control loop, which is shown in Fig. 6.16. The loop time is set at 245 μs, synchronized to every second PWM period (8-kHz switching). The loop time was set by the control algorithm execution. The control loop required 180 μs of processing, including the necessary loop equations and software for fault detection and protection.

The processing times required for different portions of the control algorithm for both the MC68HC16 and MC68332 are outlined in Fig. 6.17. For better control stability and higher-frequency operation, the control loop should preferably execute every PWM period (122 μs) but CPU limitations prevented this. (The MCU's internal-bus speed is the limiting factor.)

6.4.1.2 PWM power inverter. The PWM inverter is the basic component of the three-phase AC motor drive. It consists of the six power switches, the gate drive circuitry, and the power device protection circuit, as seen in Fig. 6.18. The power devices were chosen to meet the kilowatt rating of the AC motor drive and also contribute to the

Algorithm Function	Equation(s)	HC16 (16.7 MHz)	332 (16.7 MHz)
3ø-2ø Transformation	IQ = (ia * cosθ) + ((.577 * ia + 1.155 * ib) * sinθ) ID = (ia * sinθ) - ((.577 * ia + 1.155 * ib) * cosθ)	39.6 μs	32.2 μs
PI Current Controllers	VQ = (VQ * z^{-1}) + (Kp * (IQ_err - IQ_err * z^{-1}) + Ki * IQ_err)) VD = (VD * z^{-1}) + (Kp * (ID_err - ID_err * z^{-1}) + (Ki * ID_err))	21.7 μs	30.2 μs
2ø-3ø Transformation	va = VQcosθ + VDsinθ vb = (-.5va) - (.866 * (VDcosθ - VQsinθ)) vc = - (va + vb)	41.7 μs	38.2 μs
Voltage to High Times	A_HIGH = (va * (PERIOD / 2)) + (PERIOD / 2) B_HIGH = (vb * (PERIOD / 2)) + (PERIOD / 2) C_HIGH = (vc * (PERIOD / 2)) + (PERIOD / 2)	14.3 μs	19.0 μs
Slip Calc. + Integrator	fs = (l / (2π Tr)) * (IQ / ID) fe = fs + fr θ = (θ * z^{-1}) + fe where rotor time constant, Tr = Lr / Rr	10.4 μs	12.1 μs
	TOTAL MOTOR CALCULATION LOOP-TIME* =	127.7 μs	131.7 μs

* Totals do not include current normalization or fault detection and handling.

Figure 6.17 Loop execution times.

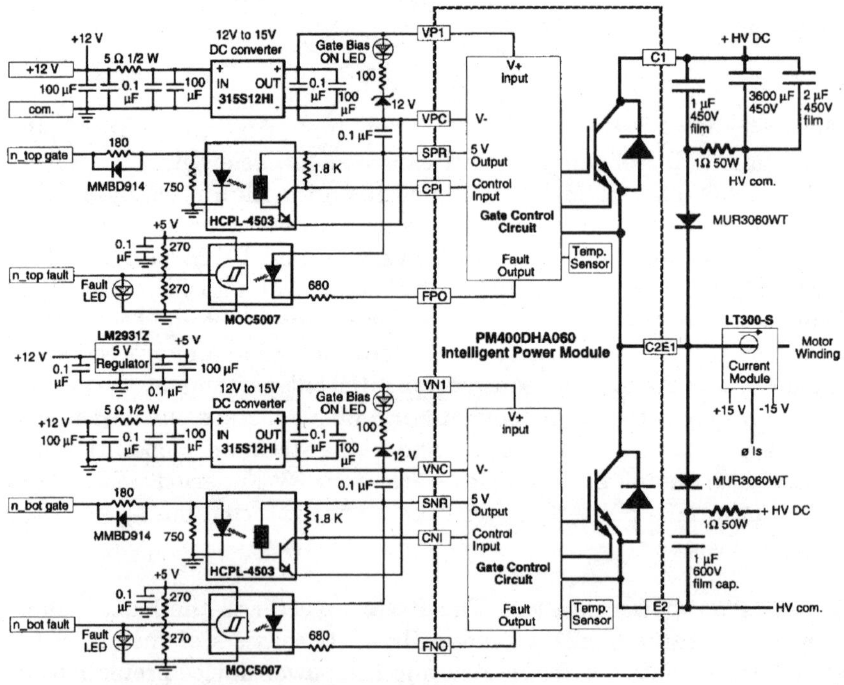

Figure 6.18 AC motor PWM power inverter.

majority of the AC controller's cost. The same power circuit design considerations used in DC motor controller designs are followed for three-phase AC PWM inverters—the power modules are laid out with a low inductance bus bar design, the gate drivers are designed to minimize switching losses and reduce switching noise, and thermal and current protection are employed. The greatest complexity in the inverter design is in the gate drive circuitry and fault handling. In this example, smart power modules (Powerex part number PM400HA060) were employed, simplifying the gate driver design, since these smart half H-bridge IGBT modules contain the gate drive and protection circuitry. An interface PCB, however, was still required. Each interface PCB contains two channels, one for the upper and one for the lower IGBT power stages. Each channel includes a gate drive optoisolator, fault output optoisolator, and DC-DC inverter.

Optocoupler devices are used to isolate the power stage from the motor control logic and need to be rated for high-common-mode-noise operation. The devices in this experimental circuit work well. The HCPL4503, used for the gate drive signals, is rated to operate in applications that have voltage transients of 15,000 V/μs. Chosen for its low power requirement, the MOC5007 is used for the fault signal. In retrospect, another HCPL4503 device would have been a better choice for the fault output optoisolator if the DC-DC converter could supply the extra current. It is important that high-Dv/Dt-rated optocoupler devices are employed or the power stage's noise spikes may feed back into the control logic.

One way to minimize noise problems in the gate driver circuit is to utilize low-impedance signal lines. In this design, 74HC08 AND gates in the logic control section buffer the PWM output signals as well as provide a total fault shutoff of all the PWM output lines. It was found that the PCB layout is also critically important to minimize common-mode coupling of the IGBT noise spikes back into the driver circuits or even into the microcontroller. All crucial control lines should be routed for minimal distances, and the input traces of the interface optoisolators should never be located close to the output traces.

A DC-DC inverter was chosen to provide the gate bias voltage to each IGBT power switch. This required a total of six inverters but was worth the extra cost since no problems were encountered with common-mode noise. Other methods that could be employed to supply isolated gate driver power include a "bootstrap" from the high-side IGBT collector or a photovoltaic device with a light source and solar cells. The bootstrap method requires that the high voltage be applied to activate the control circuit. This does not allow much time for the gate driver circuits to stabilize and basic diagnostics to be performed before the high voltage is switched on. In a 100,000-W control system the logic and interface circuits should always be "first on" and "last off" to ensure reliable and safe operation. (Some systems actually lock out the high voltage until the control logic and power interface is stable.) A photovoltaic (solar cell) device might be practical if the gate driver power requirements could be significantly reduced. A solar cell is about 12 percent efficient; to supply about 2 W average power to each gate driver would require employing a 16-W light source. The power stage circuits need a high level of isolation between the logic control circuit and the high-voltage power stage. The isolation is necessary for both safety and operational requirements. Generally, an isolation barrier of 2500 V will meet most regulatory standards.

6.4.1.3 Current sensing (A/D conversion). For the vector control strategy to be successful, the AC motor's phase currents must be sensed, accomplished, in this case, with a current transducer (part no. LT300-S, manufactured by LEM USA Inc.). The current output of the current transducer is fed to a resistor that develops an analog signal that can vary from about −6 to +6 V. Its voltage output sensitivity is 1 V per 100 A or 1 mV per ampere. Unfortunately, the MC145050 A/D converter input works only with a positive input. This required the use of an LM158 op amp to level-shift the current sense signal to a 0- to 15-V output with a center or zero current at about 7.5 V. The 0- to 15-V signal is then divided by 3 with a resistor divider that gives a 0- to 5-V output with a zero current or center of 2.5 V. The output voltage is set to about 0.396 mV/A, which means that a 10-bit A/D will resolve to only 12 A (5 mV per bit). This demonstrates why precise A/D converters are a critical part of the vector motor control strategy. Another limitation in the MC145050 A/Ds in this MC68332 design is their execution time of 20 μs. In an MC68HC16 implementation, its 9-μs onboard converters would be used rather than the MC145050. Even its 9-μs conversion time introduces a phase delay which affects stability, in addition to another delay between the sampling of the two-phase currents. This introduces an error, which also affects stability. There are better A/Ds available that provide simultaneous sample-and-hold circuits, 2 μs conversion, and 12 bits of resolution. (One example is an AD7890 device from Analog Devices.) For control accuracy, these A/Ds would be preferable. Future motor control MCUs with enhanced internal A/Ds may approach this level of refinement.

6.4.1.4 Motor speed sensor. A motor speed sensor is necessary for the electric vehicle AC motor drive system. An off-the-shelf unit that would fit into the space allocated could not be found, so a custom speed sensor needed to be developed. A 128-slot encoder disk was designed, which gives a 0.5-Hz resolution by counting every edge of the 128-tooth disk (256 edges). Speed sensing is done with the TPU. A custom time function was written to count transitions from the speed encoder at a periodic rate, smoothing out pulse jitter. The speed sensing function samples the encoder every 15 ms. Since frequency resolution is a function of sample time, speeding up the encoder sample time would require a higher-resolution encoder disk, specifically, one with 256 slots. Another sensing method would count transitions at a higher sampling rate when at high rpm levels and switch to a pulse measurement technique at low rpm levels.

The speed sensor design shown in Fig. 6.19 has given satisfactory performance in the AC motor EV system. It is also compact enough, about 16 cm^3 (1 in^3), to fit on the motor end housing of large motors.

The differential line drivers and receivers incorporated make this design suitable for the EV's electrical noisy environment. The speed sensor transmitter circuit uses a slotted optical switch, a high-performance line driver IC, and a TO-220-sized +5-V regulator. A small heatsink is required on the +5 regulator because of its 1-W heat dissipation when connected to a 12- to 15-V input voltage. An LED, in series with the optical switch's internal LED, provides an indication that the optical switch's LED has not failed open. The twisted-wire pair is connected to the receiving circuit. One of the receivers is linked to the motor drive's logic control, while the other drives an

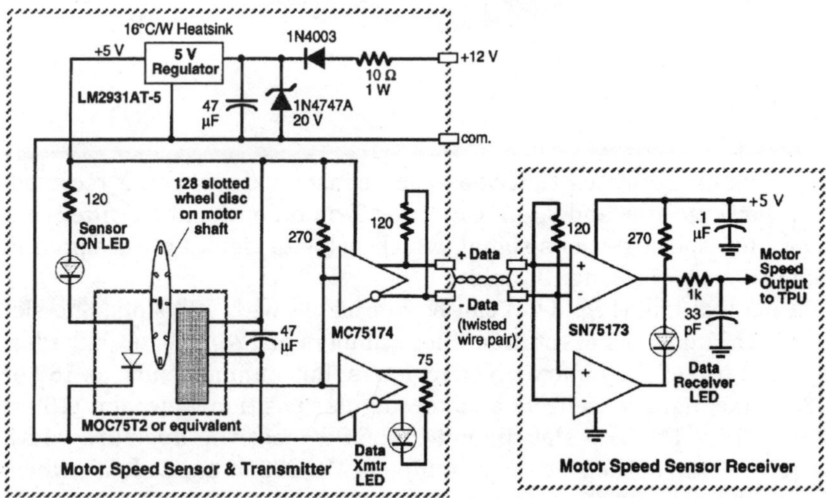

Figure 6.19 Speed sensor circuit.

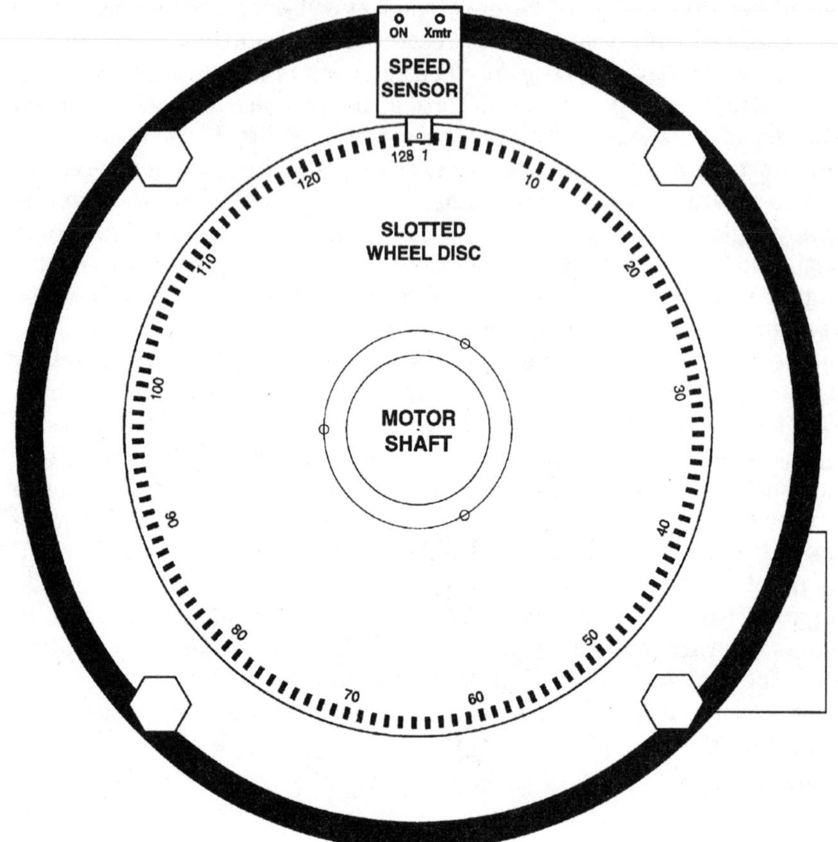

Figure 6.20 Speed sensor layout.

LED which indicates that the speed sensor data is being received. The 1-kΩ resistor and 33-pF capacitor form an optional RC filter that slows the speed sensor's signal, which helps to clean up the signal in the presence of strong RF fields.

A slotted disk (Fig. 6.20) can be fabricated with 128 slots, one slot per 2.8125 degrees of rotation. Slot numbers of 128, 256, or 512 allow optimal use of the binary timer registers, but numbers such as 180 or 360 make more sense from a motor angular position viewpoint (180 = 2.0°, 360 = 1°). The slot diameter should meet the aperture width requirements of the slotted optocoupler (MOC75 or equivalent), which is about 1 mm (0.04 in), and can be about 2.5 mm (0.1 in) or longer to allow for some up/down error in the wheel and sensor mounting.

6.4.2 AC motor MCU

The AC motor MCU relies on a fairly complex software program. The software can be written at a high level, such as C, or at the assembly level. High-level languages, such as C, have the advantage of being somewhat MCU independent, but usually require more code. Assembly language is more difficult to incorporate but is usually more compact. Both approaches have been used for AC motor controls described in this book (see Chap. 8). (The electric vehicle MC68332 code was written in assembly language and is listed in App. E.)

The AC motor control algorithms used in this electric vehicle traction system include slip control and vector control. The source code shows the implementation of motor control algorithms but does not reveal how they affect both the MCU and EV motor application. The control algorithms that are reviewed next determine much of the performance of the AC drive and determine some of the system requirements in terms of sensing elements and processor performance.

6.4.2.1 Slip control. AC motor drives can be either speed controllers or torque controllers. The slip control algorithm is for torque control, which is achieved by controlling both the phase current and the slip frequency, as shown in Fig. 6.21. The voltage is adjusted to maintain the desired current magnitude, and the frequency is adjusted to maintain desired slip frequency. Slip control in the EV application provides good performance because current feedback from the motor allows the software to compensate for variable driving conditions.

Current and slip tables are generated to provide the appropriate amount of current and slip on the basis of a desired torque. Given a

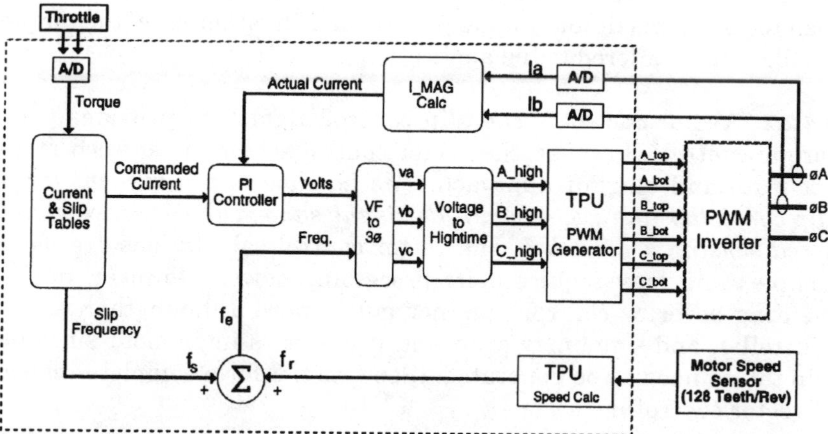

Figure 6.21 Slip controller algorithm.

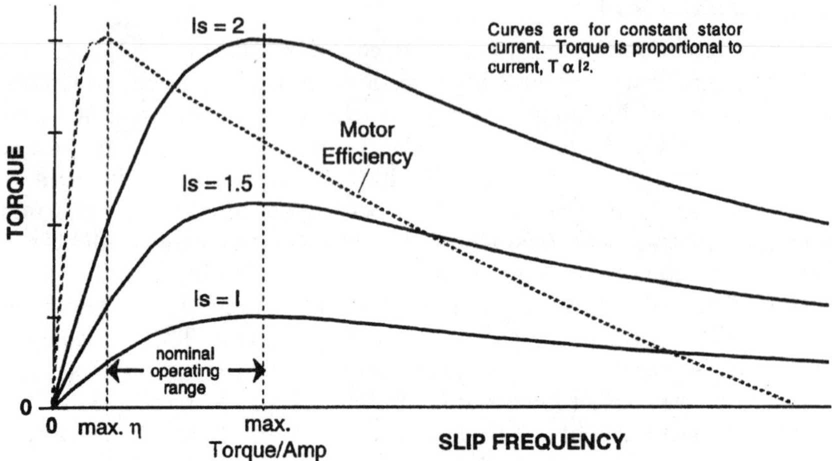

Figure 6.22 Torque versus slip, and current characteristics.

fixed slip frequency, increasing current increases torque. Similarly, given constant current, increasing slip frequency will increase torque, that is, to the point of breakdown, as seen in Fig. 6.22.

The torque/slip curves are dependent on the motor. Therefore, the slip tables need to be customized to the motor. Although the slip control algorithm is not difficult to implement, generating the slip tables may not be so easily accomplished. The difficulty is determining the value of slip for a given current. Since maximum motor efficiency does not occur at maximum torque, it is tempting to choose low slip frequencies to maximize motor efficiency. However, since the controller's power loss is related to current, it may be more beneficial to lean toward a maximum torque per ampere. In either case, the torque profile can be tailored to the application.

6.4.2.2 Vector control. The slip control algorithm provides good torque control. However, the vector control scheme is a much more accurate and elegant approach. *The hardware requirement for a slip controller and vector controller is the same; current sensing and speed sensing are needed.* The vector control calculations are more complex and may require more processing power. However, implementing a vector controller is not much more difficult than a slip controller, and some may even find it easier to implement since no slip tables need to be generated. (See Chap. 8 for a complete review of vector control.)

6.4.2.3 Other PWM modulation techniques. Besides different motor control strategies such as slip and vector, there are also different

PWM modulation techniques. Sinusoidal modulation is the standard technique, as shown previously in Fig. 6.15. However, other modulation methods have been employed in AC motor drives. These alternate modulation techniques, including third-harmonic modulation, 60° modulation, and space vector modulation, allow the drive to obtain more power from the battery pack by increasing the voltage to the motor. Third-harmonic and 60° modulation are variations of the standard sine modulation technique and could be implemented very easily in the MCU with a simple table change.

Third-harmonic modulation consists of a waveform containing the fundamental and a third-harmonic component. The characteristics of the three-phase motor filter the third-harmonic component, leaving only the fundamental. The peak-to-peak amplitude of the function containing the fundamental and third harmonic is limited by the available DC supply voltage, but the fundamental component is around 15 percent higher. Therefore, the third-harmonic method allows more power to be delivered to the motor by the increased fundamental. This PWM modulation method performed well in the EV application.

Space vector modulation (SVM) allows greater bus (voltage) utilization such as third-harmonic modulation. Space vector requires different calculations and is reported to reduce harmonics. In the EV AC traction motor application, SVM has been implemented and also performed well.

6.4.3 AC motor control enhancements

The AC motor control electronics previously described could be enhanced in several areas. If the smart power modules were replaced with either a higher-current version (600 A) or a standard IGBT module and custom gate drive stage, the 150-V AC motor could be rewound with lower voltage versions such as a 30- or 75-V ratings, providing more power from the motor. The design of the optimal AC motor for an EV traction system is beyond the scope of this book, but controller electronics will play a factor in obtaining the most performance from a given motor. It is important to note that IGBT smart modules, with their built-in current-limiting circuit, would shut down the module during a normal motor overload. Motors inherently can draw high currents for short periods of time. A better method is just to detect the overcurrent condition and tell the MCU that this condition is occurring. The MCU can then figure out what to do. An overtemperature shutdown and shorted motor shutdown should be included in the smart module, but an overcurrent condition is not that serious, unless sustained for several minutes.

Another area of improvement would be to replace the large bonded-fin heatsink with a liquid-cooled version that would allow better packaging of the power stage. You would need to add a pump, radiator, fan, and plumbing, but the power stage could be more compact and mounted closer to the motor. A purpose-built motor with liquid cooling would also allow a smaller motor to be utilized, again reducing the size of the overall traction system. The AC motor control MCU could be improved with a more powerful PWM unit and faster computational power. (Some other MCUs that offer higher motor control performance are listed in App. C.) The speed sensor could be improved with a quadrature-type encoding method to detect rotation direction and improve accuracy. The quadrature approach does, however, require the use of two slotted optocouplers.

6.4.3.1 Motor control power device improvements. Most of the cost of the motor controller electronics is divided between the power switch modules, power line filter capacitors, heatsink system, interface circuits, and control logic. The power switch modules contain several power transistors and FWD dies that are connected in parallel and configured as either a single switch or half-bridge. Using smaller individually packaged devices requires careful mechanical layout, device matching, bus bar layout, and filter network placement. Designing a stable gate driver and interface can be a challenge and has led to the development of smart power modules. The smart module design places the interface circuits, which include the gate driver and protective functions, on a circuit board that rests on top of the power device substrate. This approach minimizes lead lengths and ensures that the power stage works properly with the interface circuits. The cost of including the gate driver and protective elements inside the power module is a significant issue for manufacturers.

The cost of the smart module will always be more than the individual module and gate driver board. There are two main reasons for this cost increase: (1) completed smart modules cannot be reworked if a defect occurs (the gate drive circuit consists of dozens of parts and will statistically have more defects than the power devices) and (2) the module assembly and testing are more complex. Besides the cost concerns of manufacturing smart modules, there are application issues that arise when employing them.

The smart modules internal circuit parameters cannot be adjusted or changed for specific application needs. For example, some smart module protective designs include both shorted load and overcurrent sensing functions that shut down the power stage in the event of either an intermittently loaded-down motor or a shorted motor. Protecting the power module against shorted loads by shutting it off is a good plan, but shutting off the module for a simple short-term motor overload limits the motor's application. Shutting down the power stage for motors that operate intermittently causes the motor system to be unusable. By just detecting motor overloads, the motor control logic can track how long and how many times the overcurrent conditions occur and can take any necessary action. A temperature sensor inside the power module can also ensure that the motor overloads do not overheat the power modules. The advantage of smart modules is that they are easy to use since the sensitive analog gate driver circuits are embedded; if the application requirements, however, are slightly different from the smart module's internal operation, the motor control designer must either change the motor system or redesign its gate driver or other components in the smart module. It should also be noted that most industry standard smart modules still require an interface board.

6.4.3.2 Interface improvements. The gate control and optoisolator products could be improved on to simplify the motor's power stage. A gate driver IC, which integrated the current sensing amplifiers and logic, undervoltage detection, +15- and −5-V gate drive operation, temperature detection, and a coded fault output signal, would eliminate the need for extra op amps and logic gates. Adding user programmable functions to the gate driver design, such as current-limiting trip points and shutdown time selection, would allow the module to match each application. The optoisolator devices should include a built-in logic level input, a push/pull or totem-pole output, and matched positive and negative switching times faster than 1 μs. Existing optoisolator products use an open collector output that exhibits a high OFF-state output impedance. This makes the output line more sensitive to induced or capacitance-coupled noise, generated by the high *Di/Dt* and *Dv/Dt* values from the power IGBT stages. The temperature sensor function could be integrated into one standalone IC with either internal or external diode strings or integrated into the gate driver IC. The cost of embedding a complex printed-circuit board in the power module is difficult to justify in high-volume production. If the control circuits need to be in the power module, the control circuit devices should be further integrated and directly attached to the power substrate, possibly eliminating the internal PCB and its associated drawbacks.

Figure 6.23 shows what an optimized optocoupler for vehicular motor control applications might look like. This conceptual device minimizes external parts and uses an H-bridge amplifier with a current source. (An optocoupler device, an HCPL-3150 from Hewlett-Packard, offers a direct gate drive output but still requires an input amplifier.)

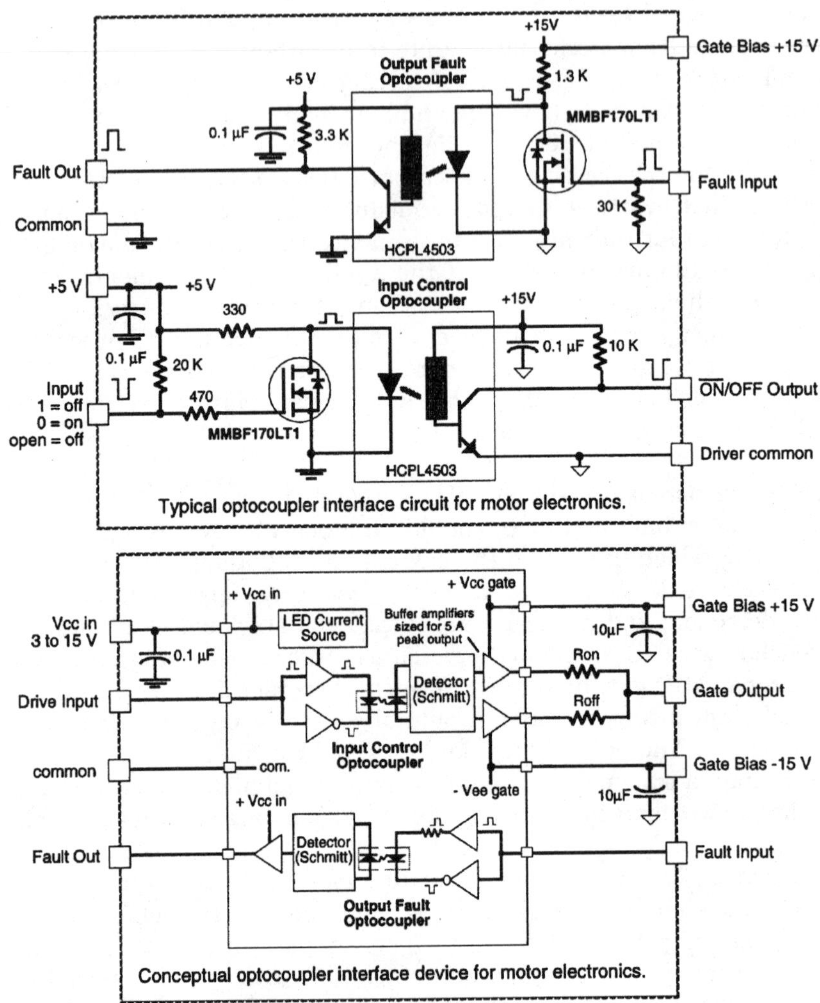

Figure 6.23 Conceptual optocoupler design.

6.5 EV Power Transistor Technology

The recent availability of power transistors that can handle high current and high voltage while offering reasonable cost-performance has opened the door for EV motor power electronics. The EV propulsion motor's power levels require power transistors that can operate at 100 to over 800 A current levels, switch in less than 500 ns, and block up to 600 V. In addition to these prodigious specifications, the transistors must also be power-efficient by having a low forward voltage drop. Since most devices with these types of ratings are not perfect switches, an excellent thermal package, as well as exemplary long-term thermal cycling ability, are also required. A review of the EV power modules and device technology reveals that a number of compromises have been made.

6.5.1 Power transistor analysis

Advances in power transistor technology, such as high-density MOSFETs and large (≥100-A) IGBTs, provide the means to control fairly large motors. Choosing the best power transistor technology usually comes down to price versus performance. Heat dissipation reduction is also a consideration for reliable power transistor module operation. Obviously, the lower the voltage drop across the power stage, the lower its power dissipation will be. The tradeoff is silicon size (high cost) for heatsink size (medium cost). As stated at the beginning of this chapter, selecting the appropriate silicon technology is somewhat dependent on the EV's battery pack voltage. Higher-voltage systems allow lower current levels for a given-size motor. This is a significant consideration when designing the most efficient EV power system and is one reason why most large EV designs are using 300- to 400-V battery supplies. Heat dissipation from any system that operates from the EV's battery supply means reduced driving range. The goal is to reduce heat dissipation whenever possible, which means that the power stage must be designed with highly power-efficient transistors.

6.5.2 Die size comparison

Choosing the best power transistor technology directly relates to die current density, namely, how much current a given die area will conduct under a given set of test conditions. For EV power stage applications, IGBTs have a definite advantage when used with 100-V or higher battery pack voltages. The IGBTs have about 2 to 3 times the current die density than MOSFETs, as shown in Fig. 6.24, from 100 V and higher. This means that the IGBT's die size will be significantly smaller than a MOSFET's for the same voltage and current rating. Small one- or two-person low-speed electric vehicles usually operate at less than 50 V, which means high-density MOSFETs would be preferred.

One main difference between IGBTs and MOSFETs is that IGBTs always exhibit about a 1-V drop that is independent of the IGBTs voltage or current rating. Figure 6.25 compares the forward ON voltages for similar IGBT and MOSFET die areas at 125°C T_j. Note that the 250-V IGBT device performs better than a 100-V MOSFET when operating above 300 A. This is because the MOSFET's ON resistance almost doubles for a 125°C rise in junction temperature, while the IGBT's ON voltage remains fairly stable. MOSFETs can give better switching performance, since IGBTs exhibit slower turnoff times due

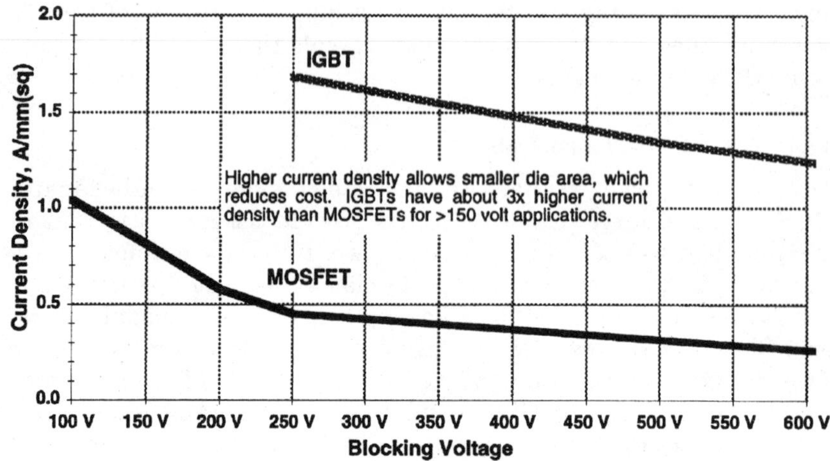

Figure 6.24 IGBT and MOSFET current die density.

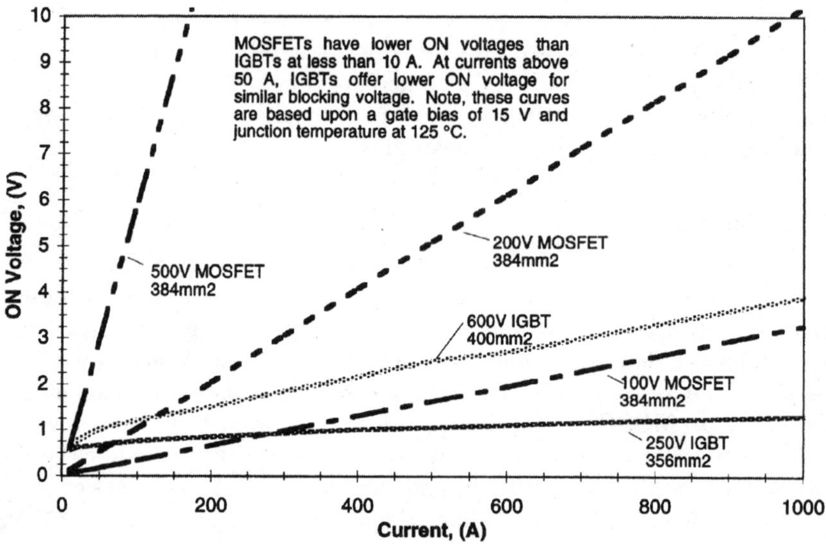

Figure 6.25 Comparison of IGBT and MOSFET forward ON voltages.

to their partial internal bipolar design. IGBTs, however, can operate satisfactorily when the PWM operating frequency of motor power stages is limited to 20 kHz with 250- to 500-ns rise and fall times. A significant disadvantage of IGBTs for EV traction applications is their inherent 0.7-V drop at low motor current levels, which causes a loss of power efficiency when driving at slower speeds.

6.5.3 Transconductance

Another aspect to consider when using large MOSFETs or IGBTs in EV traction motor applications is their high transconductance. For example, a 400-A-rated power module that typically consists of four 100-A transistors wired in parallel can usually conduct 800 A or more if the motor's load and control system does not employ current limiting. This is an important design aspect since the IGBT or MOSFET power stage will allow 2 or 3 times higher load current levels to flow, which can be utilized for momentary motor loads, if care is taken to ensure that no damage to the motor or power stage occurs.

6.5.4 Package constraints

Lead lengths and interconnections in power control circuits usually are a source of trouble. If the lengths are too long, as mentioned in Chap. 2, their inductance will cause problems. Since stray inductance is a fact of life when designing motor electronics, it is important to consider the package layouts of the power stage. There are definite advantages in locating the gate control circuits as close as possible to the IGBT power devices: increased response times, thermal sensing, and short lead lengths. However, mounting the gate control circuits inside the power module is not without problems. It is difficult to combine a multichip wire–die bond assembly with a PCB attachment. These two types of manufacturing operations are quite different and require extra manufacturing steps. The smart module manufacturing costs will be higher than a standard module and external control PCB, partly because completed modules which fail final testing are not readily repairable.

There is, however, an advantage for the smart power module user that can offset the extra cost, especially in low volumes or prototype situations. The interfacing from an MCU control system to the smart power modules is highly simplified when the interface control circuits are already inside or attached directly to the module. One disadvantage of incorporating control circuits inside the power module is that some applications may operate beyond the control circuit's design parameters, and the internal gate driver circuit values cannot be adjusted to perfectly match the application. Also, the gate driver circuit cannot be repaired, which means that a power module, whose main cost is in the power devices, must be discarded if the gate driver fails.

6.5.5 Power module interfaces

One method to simplify the MCU interface to the power module is to construct a PCB that attaches directly to the IGBT power module. This interface module should be designed to match the power module's parametrics and allow some user adjustments to match the spe-

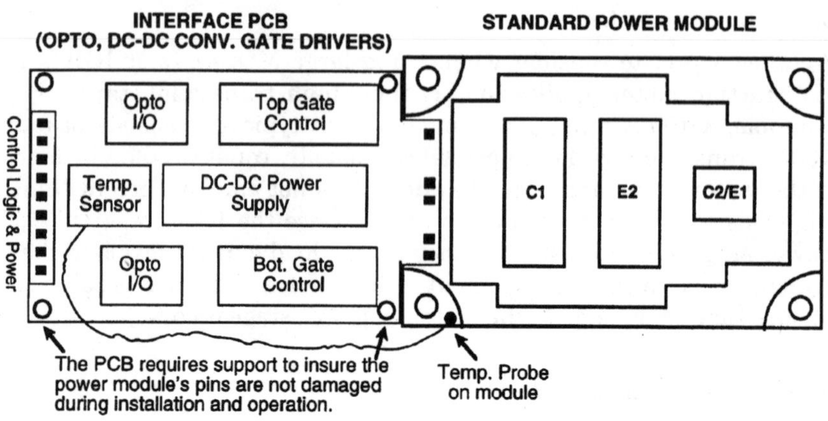

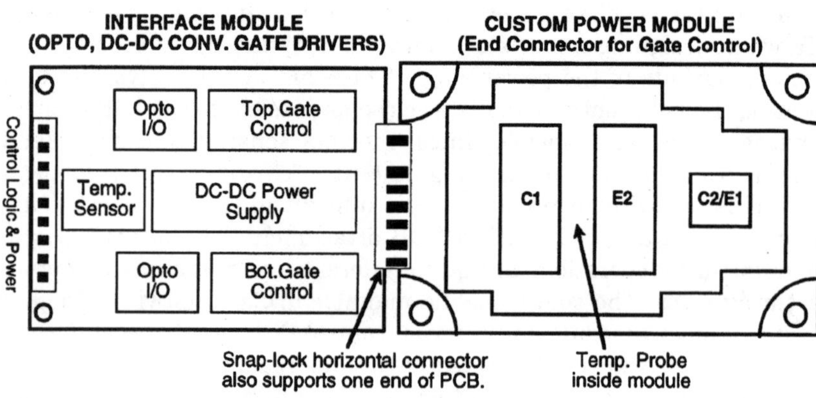

Figure 6.26 Power module interface concept.

cific application. The interface module's physical layout could be designed to allow a complementary mechanical latch to the power module. A mechanical latch of some type or screw fasteners would be used to attach the interface module to the power module. This method is illustrated in Fig. 6.26.

6.6 Other EV Motor Systems

Aside from the traction control system, an electric vehicle may use other electric motor powered systems, such as direct steering motor assist or hydraulic steering motor assist and a heat pump for air conditioning and heating. AC induction motors are often chosen for these motor designs because they are more cost-effective. Table 6.2 shows

TABLE 6.2 Other Electric Vehicle Motor Controller Requirements

Motor type	EV application	Control strategy	Power efficiency,%*	System cost*	MCU requirement†	Power stage‡
PM brush	Electrohydraulic power steering Fans for HVAC	Variable-speed, open-loop	85	Lowest	8-bit low performance	1 single switch 1 FWD
Three-phase PM brushless	Electric power steering Compressor for HVAC (Fans for HVAC)	Variable-speed, variable-torque, closed-loop	95	Highest	8-bit high performance	3 dual switch
Three-phase AC induction	Electric power steering Compressor for HVAC (Fans for HVAC)	Variable-speed, variable-torque, closed-loop vector control	90	Medium	16-bit high performance	3 dual switch
Three-phase switch reluctance	Electric power steering Compressor for HVAC	Variable-speed, variable-torque, closed-loop	90	High	16-bit high performance	6 single switch 6 FWD

*Power efficiency and system cost is estimated.
†MCU requirements relates to speed, timer complexity, and A/D performance.
‡The number of power IGBT switches and external free-wheeling diodes.

the essential specifications for these applications. The choice of software algorithms is dependent on the application. For example, a known load application in which precise speed control is not needed, such as an hydraulic steering motor assist unit or HVAC system, may use a *V*/*F* algorithm.

6.6.1 Electric power steering

As discussed in Chap. 4, electric steering can use either an electric motor driving an hydraulic pump or a direct-assist method that drives the steering rack directly with an electronically controlled motor. The EV's high voltage battery pack allows a more cost-effective implementation of either system than when powered from a 12-V source. A 1- to 2-kW motor power stage can use a six-pack power IGBT module that contains all three half H-switches. Unlike the standard internal-combustion engine (ICE) 12-V designs, the steering system would be powered from the same power source as the traction motor. This means that, as the battery pack is drained, the power steering assist will diminish. Some type of load control may be necessary to determine which system is given priority for consuming the remaining energy from the battery pack. The traction motor and high voltage to 12-V DC inverter may be given first priorities, while other less essential systems, such as HVAC, would be given low priority and would be turned off first.

6.6.2 Heating and air conditioning

The HVAC system for an electric vehicle can use a self-contained motor compressor that is very similar to a residential heat pump. A standard three-phase AC induction motor is utilized, as shown in the conceptual design in Fig. 6.27. A 1- to 5-kW rated motor would be required with the size dependent on the type of EV. The compressor's speed is set by the climatic load requirements. The MCU requirements for the compressor motor are similar to those of the hydraulic steering motor assist system. A *V*/*F*-type control strategy can be utilized. The ventilation fans can also employ variable speed AC motors. The main input is the user's temperature setting. More efficient control of the compressor is possible by sensing the outside temperature. For example, if the vehicle has been parked in the sun on an 85°F day, and the driver sets the temperature control for 78°F, the compressor can be operated at a minimal speed for efficiency. Conversely, for a 115°F day, the compressor would run at maximum.

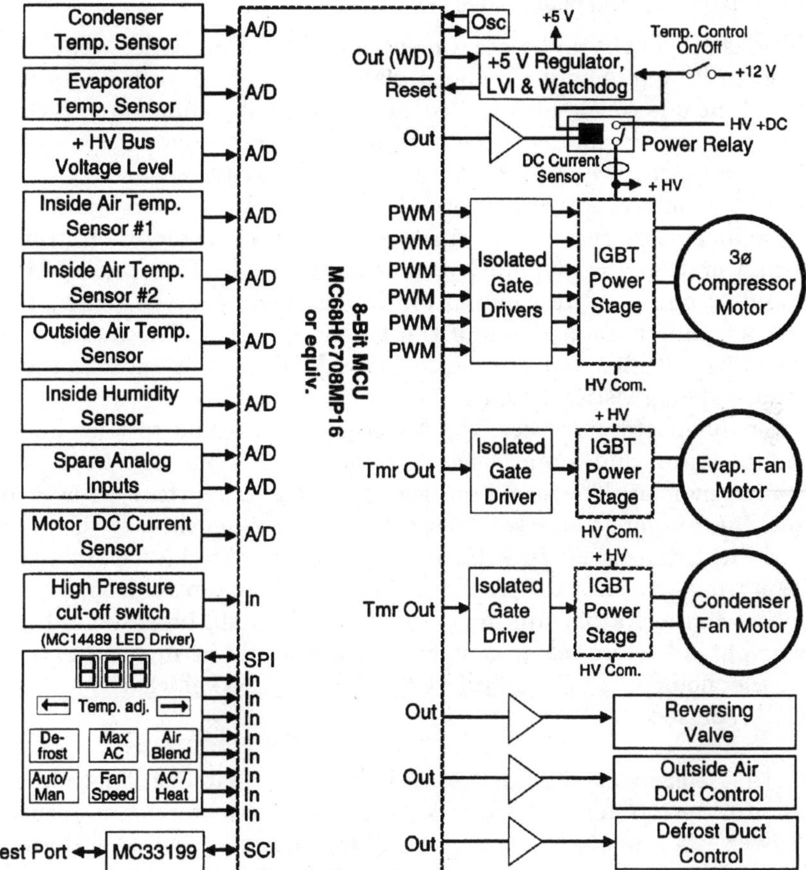

Figure 6.27 HVAC system for EV.

6.7 EV Motor Control Diagnostics

The term *electric vehicle* implies that most, if not all, of the vehicle's mechanical functions are under electronic control. The instrument panel (IP) would be similar to a standard ICE-powered vehicle, except that the gauges would give information about the EV's traction system. Instead of an oil-pressure gauge, the traction motor's load-bearing temperature may be displayed. The water-temperature gauge could indicate the power stage coolant temperature. The fuel gauge would display the amount of kilowatts left in the battery pack.

6.7.1 Motor system diagnostics

Besides IP gauges and displays, some form of diagnostics would also need to be implemented. These traction motor control diagnostics would include continuous testing of critical control functions, such as the rpm sensor, motor current sensors, power stage temperature, motor temperature, speed potentiometer integrity, power stage fault response, microcontroller watchdog, and so forth. Development of these diagnostic systems are an integral part of the motor electronics design and may be supplied in a separate MCU. This second MCU would be dedicated to running continuous testing of the critical sensors and other components of the traction system. The information would be collected and sent out over an MUX network port to the IP or special diagnostics display.

For diagnostic purposes, the MCU program would test for out-of-bound sensor values and, once detected, place a message in its non-volatile memory. The saved message would indicate the type of error, time, and number of error occurrences. The diagnostic MCU would work in close association with the main motor control MCU. In a situation where a sensor failure occurred in a critical control loop, such as phase current, a backup mode of operation could be activated. For example, a backup motor control program would be invoked to operate the motor in a *V/F* control mode, allowing the vehicle to be driven until repaired.

6.7.1.1 Temperature diagnostics. A key to the reliability of a system is monitoring the power devices and the electric motor temperatures. Service life and performance of a semiconductor device is directly affected by its junction temperature. If the junction temperature is sensed and found to be extreme, the software can reduce the current level to protect the device. Hopefully, this will not occur if the thermal management system was designed for worst-case situations. Traction motor temperature is another parameter that requires monitoring. At 180°C the motor's insulation may start to deteriorate. A temperature sensor should be located in the windings and another near the motor's drive bearings. Again, when temperature approaches the critical value, the software in the microcontroller should limit the amount of energy in order to maintain the reliability of the motor. The diagnostic MCU can also predict how fast the temperature is climbing and can make a better judgment as to what action is required. For example, if the motor or power stage temperature climbs 20°C in less than 1 min, a critical event is occurring, and immediate action is necessary. On the other hand, if the 20°C climb occurred over 10 min, this is probably a normal mode of operation, and no immediate action is necessary.

6.7.1.2 MCU diagnostics. Safe operation of the traction motor system requires monitoring of the main motor control MCU. In the environment of an EV, all of the motor control system components are subject to erratic operation instigated by either capacitive or magnetic coupled noise (EMI), usually from the power stage. If the MCU program crashes or locks up, a monitoring device must be ready to shut down the system. Some other factors besides EMI that would cause the software to fail are low-voltage power conditions, undetected software bugs, a poor RESET circuit, and a defective MCU. Usually, undervoltage conditions are detected by external ICs which cause the MCU to shut down in a stable manner. However, the EMI created by switching the motor is still a factor. These transients affect not only the MCU but other remote-sensing devices as well. It is important to ensure that any remote sensor line or any input line be carefully designed to handle EMI or the sensor or input data will be corrupted. (See Chap. 12 for MCU protection methods.)

6.7.1.3 Data sharing. In electric vehicles, the various motor systems may be somewhat interlinked, allowing data from sensors, MCU computations, and diagnostic messages to be shared. These systems can be linked together through a simple bus or multiplex (MUX) system. From a performance standpoint, the data sharing and information management will help in overall system safety. Figure 6.28 shows the instrumentation design used in the EV test platform.

6.8 Futuristic EV Motor Control

Both electric- and ICE-powered vehicles have been around since the 1900s. In the last decade, the use of electronics has vastly improved the ICE-powered vehicles with computer-controlled fuel injection and spark timing. Electric vehicle motor control systems are also improving, thanks to the recent introduction of high-performance motor control specific MCUs and large cost-effective power transistors. In fact, a modern EV's traction motor system performs in a manner that is similar to that of a normal ICE-powered vehicle. The first generation of purpose-built EVs were introduced in late 1996 (General Motors EV1). The next generation of EVs will offer even higher performance traction motor systems. The cost of the electronic motor control devices may decrease as a result of further integration and finer geometry used in the manufacturing of MCUs. The motor system will still remain somewhat partitioned, due to the safety and EMI requirements of isolating the power stage from user controls.

Motor designs are currently being reengineered to accommodate power electronics. The industry-standard 3-phase motor topology may

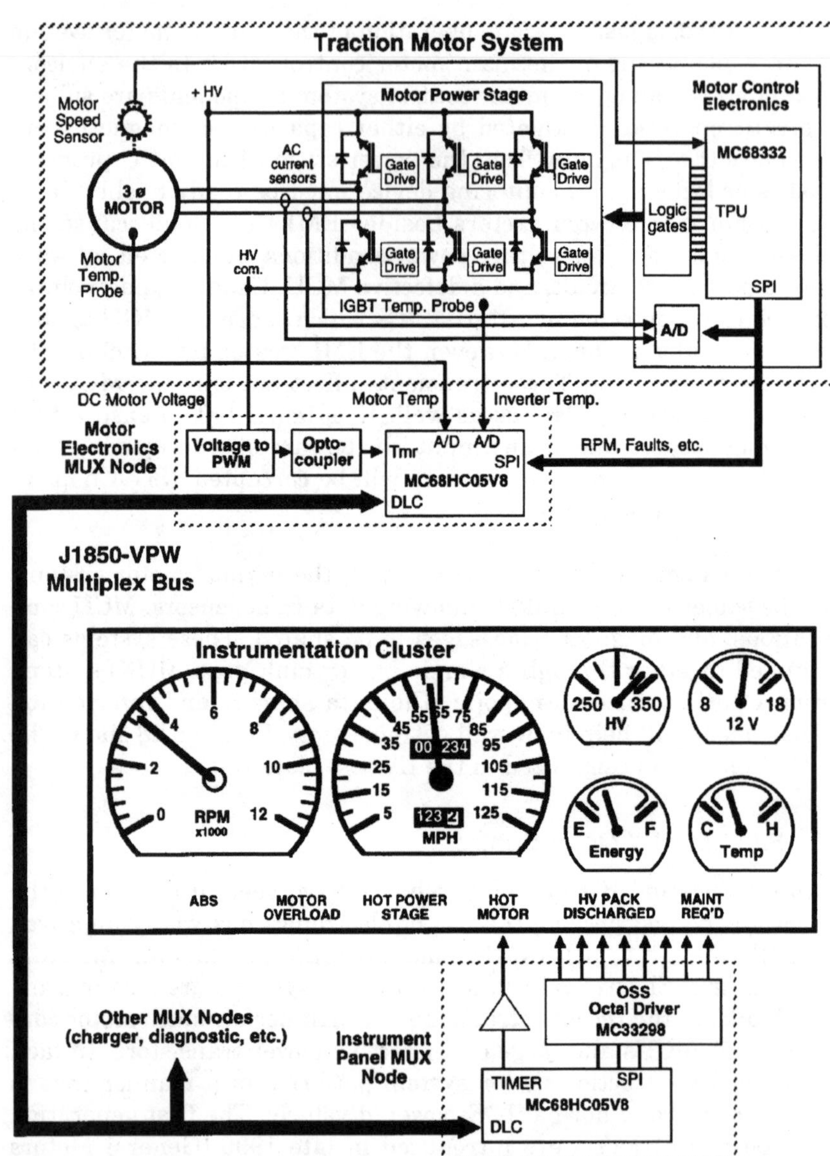

Figure 6.28 Instrumentation functions.

change over to 4-, 6-, or higher-phase numbers. The motor control electronics can probably drive up to 16 phases, which would greatly reduce ripple torque at low rpm levels.

EV motor speed controls require the use of automotive rated components that traditionally represent high reliability, good cost-to-performance value, and are designed for long-term use. The present commercial-grade power modules, however, are not optimized for the automotive environment. Power switch module technology improvements will improve their price performance with higher-current-density power devices and new module packaging materials. Microcontroller architectures will be tailored not only for motor controllers but also for specific motor types. Diagnostics of the electric vehicle motor control electronics at the factory and service shop will be similar to the test instrumentation used for ICE controllers. The service technician will probably use a general-purpose PC connected to an EV test port. The PC would run special diagnostic software that would interrogate the EV electronics and help pinpoint faulty components.

6.8.1 Futuristic motor control power transistors

One of the more important improvements in power transistors has been the development of IGBT technologies. The IGBT power device technology allows low-energy drive signals, while also allowing the die sizes to be much less than a comparably rated high-voltage MOSFET. The evolution of high-voltage power transistors will continue, but improvements may not be as dramatic as in the world of MCUs because the basic power transistor structures must block fairly high voltages. Look for future improvements in the semiconductor materials, die attachment design, wire bonds, substrate materials, and packages. There is also research under way to develop high-power diodes (FWDs) for motor controls that minimize reverse recovery time.

6.8.1.1 New packaging materials. A new power module package material is under development that should reduce the thermal cycling reliability concerns for the EV's stop-and-go motor. This new module material uses a metal matrix composite (MMC) that is based on silicon carbide. The coefficient of temperature expansion (CTE) of this new MMC package is about one-third that of copper, which means greatly improved thermal cycling reliability, since it closely matches the silicon die's CTE.

6.8.2 MCU evolution

Further integration in motor specific MCUs is removing much of the complexity from the AC motor control circuit design and shifting it to the software design. The PWM unit would include six center-aligned PWMs with deadtime control as well as a fault input. More importantly, the PWM unit should be run from a high-speed clock (>16

MHz). A 16-MHz clock (with 60 ns resolution) would increase the number of PWM bits of resolution, allowing the software to choose a high switching frequency without sacrificing resolution.

The A/D converters should be very fast (<2 μs) and more accurate (>12 bits). These parts are available as standalone devices but are not available on the microcontroller unit. Integrating these into the microcontroller reduces the hardware complexity and speeds up communication time.

The CPU is probably the biggest limitation to achieving higher throughput because the PWM unit and A/Ds can be accomplished externally. The CPU must be able to execute the control loop expeditiously. The 245-μs loop time as in the MC68332 design may not be acceptable for some motor control applications. It would be advantageous to perform PWM cycle-by-cycle control, which would require very fast control loop times at the higher switching frequencies. DSP processors are fast but typically do not offer microcontroller functions, such as on board A/Ds and PWM units. A RISC microcontroller or coprocessor solution may provide a better solution.

Advances in electronics will continue to reduce the cost and complexity of the AC drive hardware. In addition, electronics suppliers will reduce the software complexity by providing application firmware or downloadable application code. As this happens, the induction motor will become even more popular for all types of variable-speed applications. (See Chap. 10 for a description of MCU motor development tools that include motor control software.)

Summary

This chapter has provided application examples of both DC and AC traction motor drives. The design of the control electronics for these motors is an ambitious endeavor because of their large size and reliability requirements. Key points about this chapter are listed below.

- The series-wound field DC brushed motor offers high starting torque and can use a single electronic power switch. Its main disadvantages are poor high-speed torque, difficult regeneration, and only unidirectional control when just one electronic power switch is used.
- An AC induction motor has no brush or commutator assembly that can wear or create RFI. The AC motor requires complex control electronics but can offer high torque over a wide speed range. Regeneration and bidirectional control is fairly easy to implement with the AC motor control system.

- Brushless DC motors and PM motors use fixed magnets to generate the field flux and are therefore power-limited by the field's magnet size. Large BLM motors cost more than other motor types but can offer higher power efficiency at their rated power output.
- The two top design considerations for an EV traction drive system are the motor type, which determines the complexity of the control electronics; and the battery voltage, which has a direct effect on the power stage design.
- MOSFETs will usually be more cost-effective for motor voltages of less than 50 V. IGBTs may be considered when motor voltages are between 50 and 100 V, while IGBTs will usually be the preferred choice for motor voltages above 100 V.
- Reliability is the top design issue for the electric vehicle's traction control electronics. A defective EV motor system means a tow truck and a dissatisfied EV owner. Reliability has been addressed in several areas: the operating temperatures of the power stage electronics, bug-free MCU software, EMI-resistant designs, and smart diagnostics, which may help alert the user to impending problems.
- Regeneration is implemented for electric vehicles by electronically switching the traction motor to operate as a generator, easily accomplished with an AC motor drive. The degree of regeneration is dependent on the braking effort and may need to be limited if the vehicle enters an ABS mode.
- The root cause of many problems with EV motor control electronic designs is the EMI generated by the power stage, motor, and associated cables. High current levels, fast-switching edges, and stray inductance cause the EMI. A FWD's reverse recovery time also contributes to EMI, since it allows a high-shoot-through current to occur.
- In a three-phase AC motor control the MCU generates the sinusoidal waveforms by the use of a complex timer to generate a PWM digital signal whose duty cycle varies, allowing the motor's current to look like a sine wave. The motor's rpm is varied by changing the fundamental PWM frequency.
- A vector-type electric vehicle motor drive requires motor phase current sensors, shaft speed and direction, and a high-performance motor-specific MCU.
- Smart power modules offer fast implementation but cost more than standard modules. Smart modules also cannot be adjusted to perfectly match each application, since their internal control circuits are sealed.

- The greatest improvement in motor control electronics will be in the MCU and software. Motor control specific MCUs will offer improved cost/performance ratio, and motor control software libraries will become commonplace for these MCUs.

Acknowledgments

The author would like to thank Jeff Baum, Ken Berringer, Y. H. Cheng, John Forsight, Loretta Garcia, Marko Koski, Dan Mason, Jeff Marvin, Chuck Powers, Mark Torfeh, and Dave Wilson for their contributions that led to several of the successful EV motor control designs described in this chapter.

Further Reading

Bates, B., *Electric Vehicles: A Decade of Transition,* SAE PT-40, 1992.

Behr, M., U.S. Patent 4660671, *Electric Steering Gear,* TRW Inc., April 28, 1987.

Berringer, K., Baum, J., "The Right μp Simplifies Using Induction Motors to Propel Electric Cars," *EDN,* March 1994.

Brant, B., *Build Your Own Electric Vehicle,* Tab Books, 1994.

Buckley, G., "Making the Switch," *Electric and Smart Vehicle Technology '95,* UK and International Press.

Chokhawala, R., Catt, J., and Kiraly, L., "A Discussion on IGBT Short Circuit Behavior and Fault Protection Schemes," *International Rectifier,* TPAP-3, 1993.

Earle, M., "Propulsion Technology: An Overview," *Automotive Engineering,* July 1992, vol. 100, no. 7.

Fitzgerald, A. E., Kingsley, C., and Umans, S., *Electric Machinery,* 5th ed., McGraw-Hill, 1990.

Flett, F. P., "Vector Controller IC Simplifies Induction Motor and PM DC Motor Systems," *Powerconversion Intelligent Motion,* June 1993.

Fukino, M., Irie, N., and Ito, H., *Development of an Electric Concept Vehicle with a Super Quick Charging System,* SAE 920442.

Habetler, T., "Direct Torque Control of Induction Machines Using Space Vector Modulation," *IEEE Transactions on Industry Applications,* vol. 28, no. 5, Sept./Oct. 1992.

Ishida, T., "Making Sense Out of Current Sensors," *Powerconversion Intelligent Motion,* April 1994.

Jönsson, R., "Natural Field Orientation for AC Induction Motor Control," *Power conversion Intelligent Motion,* May/June 1994.

Jurgen, R., "Electric and Smart Vehicles," in *Automotive Electronics Handbook,* McGraw-Hill, 1995, chap. 30.

Kobe, G., "Point of Impact," *Automotive Industries,* July 1992.

Motto, E., *Power Circuit Design for IGBT Modules,* Powerex Application Note, 1993.

Pinewski, P., *Understanding Space Vector Modulation,* EDN Products ed., March 7, 1996.

Powerex Inc., *IGBT Module Applications and Technical Data Book,* 3d ed., 1993.

Powerex Inc., PM400DSA060 module data sheet, 1994.

Powers, C., and Huettl, T., "MUX for Electric Vehicle Systems," *Future Transportation Technology Conference Transcripts,* Aug. 1995.

Rashid, M., *Power Electronics Circuits, Devices, and Applications,* Prentice-Hall, 1988.

Romero, G., "Metal Matrix Power Modules: Improvements in Reliability and Package Integration," *IEEE Industry Applications Conference Record,* 1995.

SAE International, *Electric Vehicle Design and Development,* SP-862, 1991.

SAE International, *Electric Vehicle R&D,* SP-880, 1991.

SAE International, *Electric and Smart Vehicle Technology,* SP-915, 1992.

Smith, M., Sahm, W., and Babu, S., *Insulated-Gate Transistors Simplify AC-Motor Speed Control,* Application Note AN9318, Harris Semiconductor, Sept. 1993.

Stokely, G., "Absolute Encoders Provide Precise Position Information," *Powerconversion Intelligent Motion,* April 1994.

Valentine, R., Pinewski, P., and Huettl, T., *Electronics for Electric Vehicles Motor Systems,* SAE 951888.

Wright, J., *Multi-Channel PWM TPU Function,* Motorola Application Note TPUPN05-D.

Part

3

Digital Controls

Chapter

7

Motor Drive Electrical Noise Management

Warren Schultz
Motorola Semiconductor Products

Introduction

In motion control systems much design time is typically devoted to electrical noise management. In this chapter, we will discuss various techniques that make the nuts and bolts of noise management easier, reducing the need to redesign and debug a motor drive system. Many of these techniques provide electrical noise robustness at the expense of some additional component cost. (*Editor's note:* Power control engineers use the term *robustness* to indicate ability to withstand stress from outside sources, such as voltage transients, electrical noise, thermal cycling, and current peaks). The benefits of increasing electrical

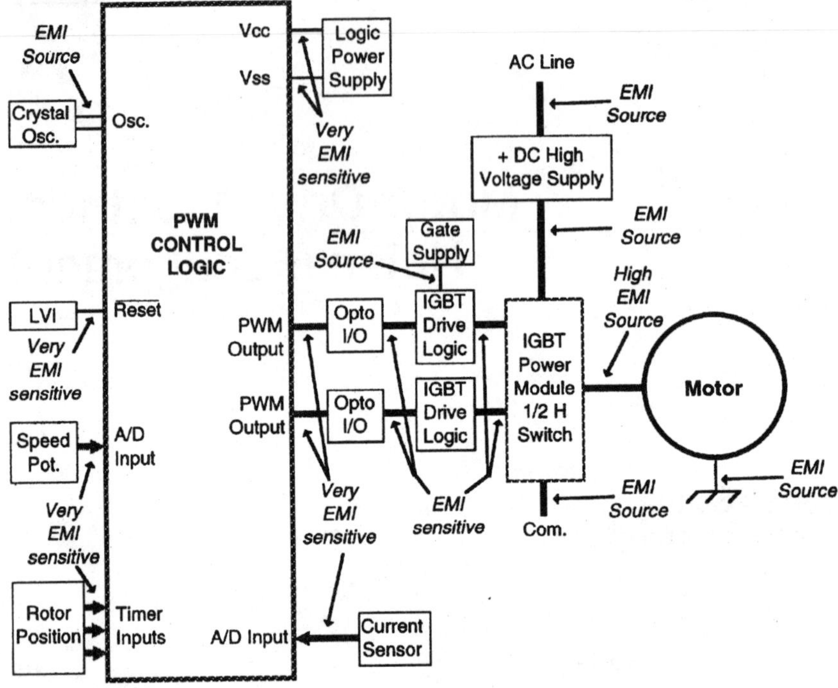

Figure 7.1 Electrical noise problem areas in a motor drive system.

noise robustness include reduced development cost, faster time to market, and a higher likelihood of trouble-free operation in the field.

7.1 Circuit Techniques

Circuit design can have a profound influence on both the amount of electrical noise produced and the susceptibility of motor drive circuits to the noisy environments in which they operate. Figure 7.1 is a block diagram showing some of the most troublesome noise areas in a motor drive. Most of the electrical noise is generated from the motor's electronic power stage. In the following discussion we will examine N-channel output stages, complementary output stages, and controllers.

7.1.1 N-channel output stages

In a high-voltage (+150-V-DC) motor drive that uses N-channel IGBTs, an illustration of a noise robust circuit design is provided by comparing Figs. 7.2 and 7.3. Figure 7.2 shows a minimal circuit topology for one phase output. Figure 7.3 adds the components necessary

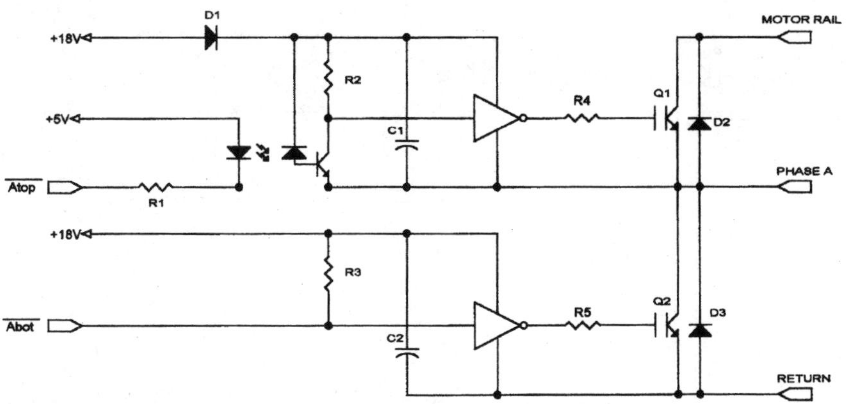

Figure 7.2 Minimal N-channel IGBT power stage.

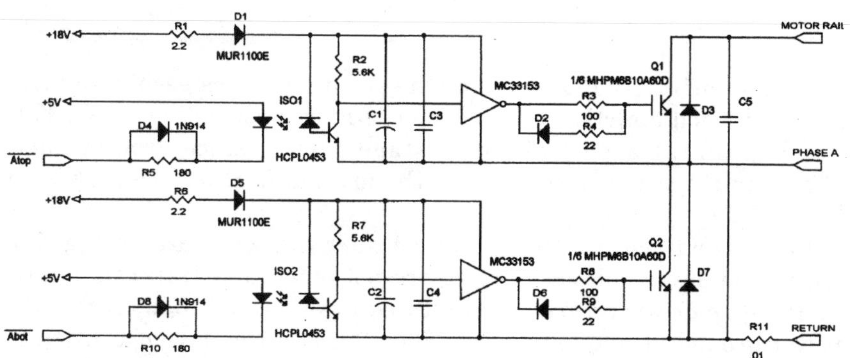

Figure 7.3 Robust N-channel IGBT power stage.

to make this topology achieve enough noise robustness for a 740-W inverter.

Perhaps the most important circuit design influence is the use of optocouplers. Optocouplers are widely employed in upper half-bridge IGBT drives for level shifting. When they are used in the lower half-bridge gate drives as well, noise robustness is significantly improved.

Consider the robust design in Fig. 7.3. Both top and bottom inputs utilize optocouplers that are connected to inverting gate drivers. This arrangement provides level shifting and also isolates the inputs from the high-voltage power stage. The isolation provided by this circuit topology significantly improves noise immunity for two reasons: (1) the optocouplers are very effective at keeping conducted noise away from microcontrollers—the noise isolation they provide adds a degree of robustness that can make controller layout and debugging much simpler; and (2) the use of optocouplers in the lower half-bridge facili-

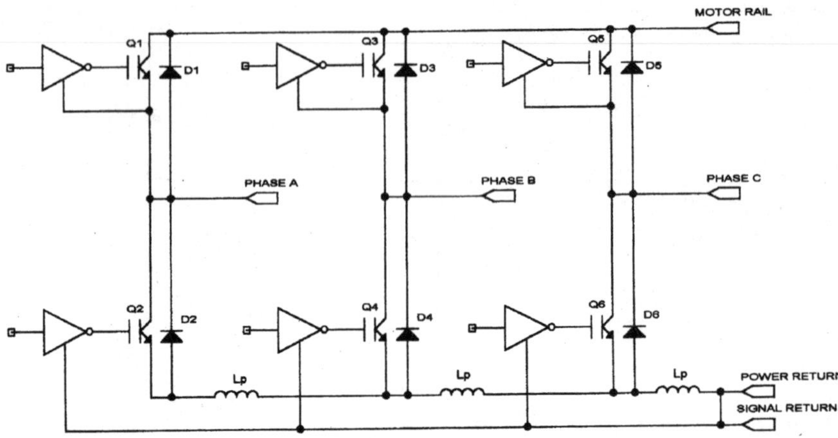

Figure 7.4 Ground inductance.

tates gate drive grounding, since the isolation allows each gate drive to be returned directly to the emitter of its corresponding IGBT. This is a significant aspect of power stage design, since ground noise effects on the gate drives is one of the more difficult issues that must be resolved.

To further illustrate this point, consider the schematic in Fig. 7.4, where the lower half-gate drives are not optocoupled. In this figure, all the gate driver returns are first tied together and then contact the power ground at only one point. The effects of this layout constraint are illustrated by showing parasitic ground inductance between phases as inductor L_p. When a switching transient occurs, the voltage drop across L_p shows up between the gate drivers and their respective IGBT's emitters. The result is unwanted gate-emitter voltage spikes that can cause turnon or turnoff at inappropriate times. In contrast, the use of optocouplers permits driver returns to be connected directly to their respective emitters and eliminates the effects of phase-to-phase parasitic inductance.

To put this issue in perspective, let's assume that 10 A is switched in 25 ns and that $L_p = 25$ nH. The resulting voltage transient across L_p is then $25E - 9(10/25E - 9) = 10$ V. Ten volt spikes can easily cause shoot-through currents and can add considerably to switching losses. For drives that are less than 2 kW, careful layout can yield acceptable results. However, using optocouplers in both top and bottom power stages provides a much more noise-robust design.

To help with gate transients, the MHPM6B10A60D IGBTs that are shown in Fig. 7.3 have a 6-V gate threshold, as opposed to the standard 3.5-V threshold for MOS gated power devices. The higher threshold provides an additional 2.5 V of noise margin for an IGBT. The MC33153 gate drivers in this circuit have an undervoltage lockout that is designed for the 6-V threshold.

Referring again to Fig. 7.3, diodes D4 and D8 reduce input impedance to the optocouplers when the inputs are high. The lower impedance provides higher noise immunity by ensuring that the optocouplers remain off when they are supposed to be off, given an environment that includes high *dv/dt*. Resistors R1 and R6 protect the supply voltage to both upper and lower gate drivers from *di/dt*-induced voltage transients. Without these resistors, it is very difficult for an all N-channel power stage to work properly. They, in effect, act as shock absorbers, isolating the gate driver's bias voltage from *L(di/dt)* voltage spikes that are produced by switching the power devices. Although unnecessary for rectification, diode D5 serves the same function on the lower gate drive. Capacitors C3 and C4 are ceramic components that provide an improved high-frequency return path for the gate drive during switching transients. The addition of these capacitors helps provide cleaner signals to the IGBT gates and to facilitate proper operation of the gate driver ICs. A short high-frequency return for the power devices is provided with capacitor C5. Polypropylene film capacitors (e.g., a WIMA MKP 10 type) work well for this purpose.

Between the driver output and IGBT gates, two resistors and a diode are used instead of a single gate drive resistor. The additional components allow turning the IGBTs on more slowly than they are turned off. Careful selection of turnon time is important, since peak reverse recovery *di/dt* of the opposing transistor's free-wheeling diode is dependent on turnon time. Since the most troublesome noise in a typical power stage is generated during reverse recovery, the two-resistor topology and careful choice of values are important aspects of the design.

7.1.2 Complementary output stages

Complementary p-channel/N-channel MOSFET output stages are generally somewhat simpler than all-N-channel output stages, since voltages and power levels are lower. A typical circuit is illustrated in Fig. 7.5. It shows a complementary output stage that is capable of operating at 5 A and 48 V. The design challenges in this type of circuit arise from the reverse recovery characteristics of the power MOSFET's drain-source diodes. Unlike IGBTs with discrete diodes that are designed for softness, power MOSFET drain-source diodes tend to be very snappy during reverse recovery. (*Editor's note:* The

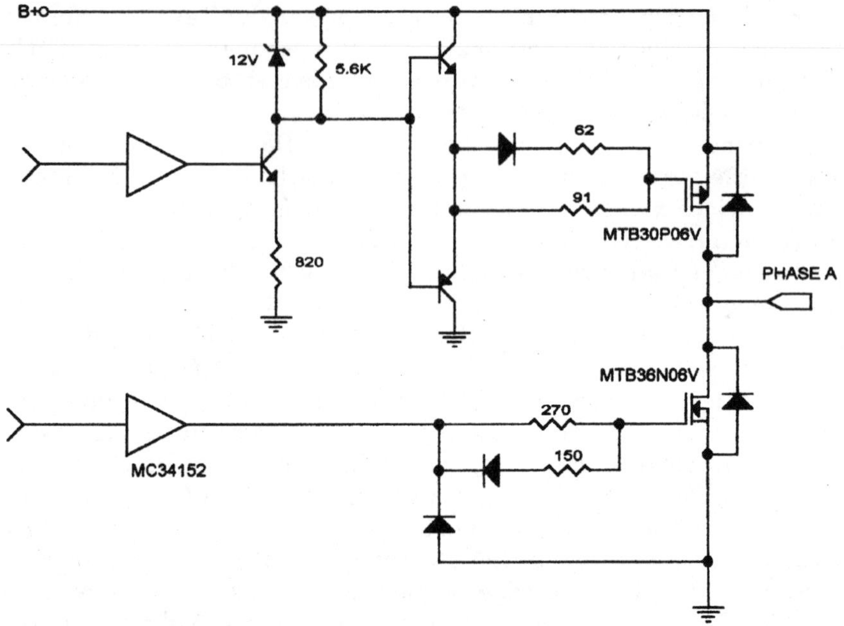

Figure 7.5 Complementary power stage.

term "snappy" in power control designs refers to diodes or rectifiers that snap off. A diode whose negative-to-zero transition occurs faster than its positive-to-negative transition is usually considered to be snappy.) Figure 7.6 illustrates this point. It shows a typical power MOSFET's drain-source diode recovery characteristics. After reaching a negative peak at turnoff, the diode's current rapidly returns to zero. This behavior can produce *di*/*dt* values on the order of 1 A/ns.

At this rate, even 10 nH of parasitic inductance can produce troubling transients. Fortunately, diode snap can be slowed down somewhat with the appropriate choice of gate drive resistors. Doing so requires different values for turnon and turnoff. Figure 7.5 shows how to do just that, with two resistors and a diode for the gate of each MOSFET. In addition to using two values, the switching times that they target are very important. The values shown in Fig. 7.5 produce rise and fall times of approximately 200 ns. With 200 ns switching times, the resulting gate drive impedances are high enough to permit a small amount of *dv*/*dt*-induced shoot-through current to flow, which considerably softens diode snap.

This works in the following manner. In addition to the reverse recovery current that you would normally expect, there is another current generated by switching transitions that can be called a *dv*/*dt*-

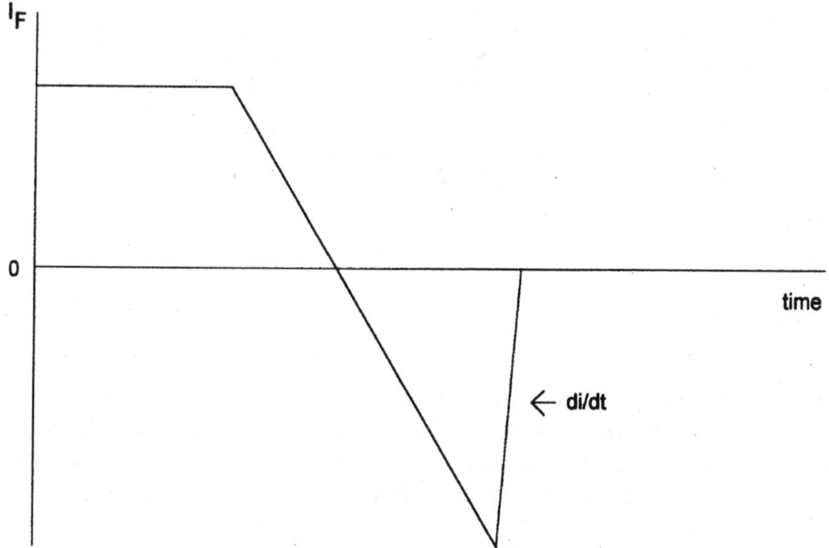

Figure 7.6 Drain-source diode snap.

induced shoot-through current. When one transistor in the bridge turns on, its opposing transistor's drain is pulled rapidly to the opposite rail. The *dv/dt* impressed at the drain causes a current to flow through the gate to drain capacitance and shows up at the gate as input current. This current returns to the source potential through the gate drive's OFF-state impedance. It forward-biases the gate by drive impedance times *dv/dt* current. If this voltage exceeds the OFF transistor's turnon threshold, *dv/dt*-induced shoot-through current is produced.

Although shoot-through current is something that one normally strives to minimize, a small amount, in this case, is a good tradeoff for the resulting reductions in *di/dt.* During reverse recovery, *dv/dt*-induced shoot-through current adds to the diode's current waveform in a way that significantly reduces negative peak to zero *di/dt.* The correct choice of gate drive resistors can easily produce a three- to fourfold improvement.

7.1.3 Brushless DC motor controllers

Brushless DC motor controllers that use Hall sensors pose noise immunity challenges related to the sensor inputs. The sensors are in an inherently noisy environment, since they are located in the motor close to PWM noise that is present in the windings. Given this situa-

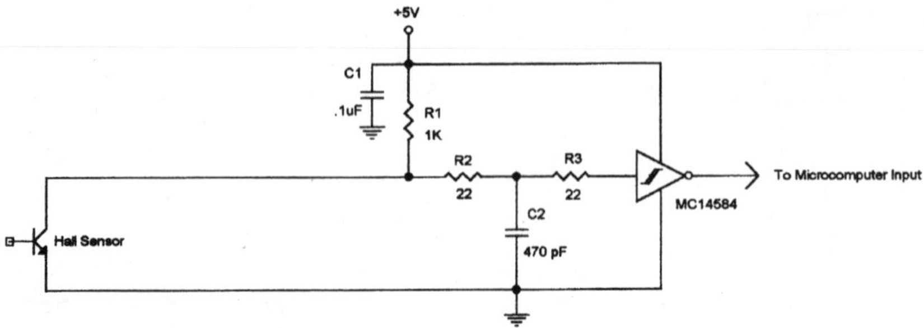

Figure 7.7 Hall sensor input.

tion, some motors do a much better job than others of presenting clean signals to the controller. In general, it is necessary to build some noise immunity into the way a brushless controller receives Hall sensor inputs. An example of how this can be done is shown in Fig. 7.7.

In this illustration, two techniques are used for isolating noise transmitted by Hall sensors. The first is a 100-ns filter consisting of two resistors and one capacitor: R2, C2, and R3. Since the rise times of Hall sensors are typically on the order of 500 ns, the 100-ns time constant does not significantly affect the timing of the Hall signals, yet it effectively suppresses spikes that occur on Hall sensor lines. Once the signal is filtered, it is also a good idea to run it through a relatively slow 14000-series Schmitt trigger. The Schmitt trigger improves noise immunity by virtue of its hysteresis and because 14000-series CMOS parts are inherently slower than most modern microcomputers. This is a case in which it is very helpful to avoid using devices that are any faster than they need to be.

In addition to circuitry, the way that microcomputer code is written also influences noise robustness. Since the sequence of commutation is known, it is relatively easy to detect an out-of-sequence Hall sensor input. Generally speaking, it is desirable, when this occurs, to turn off all the power transistors until a valid Hall code is received. In other words, it is better to let the motor coast in the presence of an incorrect Hall input than to commutate to the wrong state.

7.2 Layout

In a motor drive, layout is a critical part of the total design. Often, a system that works properly is actually more the result of layout than of circuit design. The following discussion covers some general principles, power stage layouts, and controller layouts.

7.2.1 General principles

Several general layout principles are important in motor drive design. They can be summarized in the following five rules:

Rule 1. Minimize loop areas: This is a general principle that applies to both power stages and noise-sensitive inputs. Loops are antennas. At noise-sensitive inputs, the area enclosed by an incoming signal path and its return is proportional to the amount of noise picked up by the input. At power stage outputs, the amount of noise that is radiated is also proportional to loop area.

Rule 2. Cancel fields by running equal currents that flow in opposite directions as close as possible to each other: If two equal currents flow in opposite directions, the resulting electromagnetic fields will cancel each other out as the two currents are brought infinitely close together. In printed-circuit-board (PCB) layout, this situation can be approximated by running signals and their returns along the same path but on different layers. Field cancellation is not perfect because of the finite physical separation, but it is sufficient to warrant serious attention in motor drive layouts. From a different perspective, this is another way of looking at rule 1, namely, minimize loop areas.

Rule 3. On traces that carry high-speed signals, avoid 90° angles, including T connections: If you think of high-speed signals in terms of wavefronts moving down a trace, the reason for avoiding 90° angles is straightforward. To a high-speed wavefront, a 90° angle is a discontinuity that produces unwanted reflections. From a practical point of view, 90° turns on a single trace are easy to avoid by using two 45° angles or a curve. Where two traces come together to form a T connection, adding some material to cut across the right angles can be a bit of a challenge with some software packages. However, it can be done, and it is an important part of optimizing motor drive design.

Rule 4. Connect signal circuit grounds to power grounds at only one point: The reason for this constraint is that transient voltage drops along power grounds can be substantial, because of high values of *di/dt* flowing through finite inductance. If signal processing circuit returns are connected to power ground at multiple points, these transients will show up as return voltage differences at different points in the signal processing circuitry. Since signal processing circuitry seldom has the noise immunity to handle what can be several tens of volts of power ground transients, it is generally necessary to tie signal ground to power ground at only one point.

Rule 5. Use ground planes selectively: Although ground planes are highly beneficial when used with digital circuitry, in power control systems it is better to use them selectively. A single ground plane in a motor drive would violate rule 4 by mixing power and signal grounds at multiple points. In addition, ground planes tend to make large antennas for radiating noise. In motor drives, a good approach is to use ground planes for digital circuitry and use ground traces in the power stages and for analog circuitry.

7.2.2 Power stages

There are two overriding objectives pertaining to power stage layout: (1) it is necessary to control noise at the gate drives sufficiently so that the power devices are not turned on when they are supposed to be off or vice versa; and (2) it is highly desirable to minimize radiated noise with layout, where tight loops and field cancellation can reduce the cost of filters and enclosures.

Looking first at gate drive, noise management is greatly facilitated by using the source or emitter connection for each power device as a miniature ground plane for that device's gate drive. This is particularly important for high-side N-channel gate drives, where the gate drivers have high *dv/dt* displacements with respect to power ground. If the power device's source or emitter connection is used like a ground plane, parasitic capacitive coupling back to power ground is minimized, thereby increasing the *dv/dt* immunity of the gate drive.

Consider the circuit shown in Fig. 7.8. Let's assume that the phase output swings 300 V in 100 ns and that the parasitic capacitance to power ground C_p is only 1 pF. A simple $i = C(dv/dt)$ calculation sug-

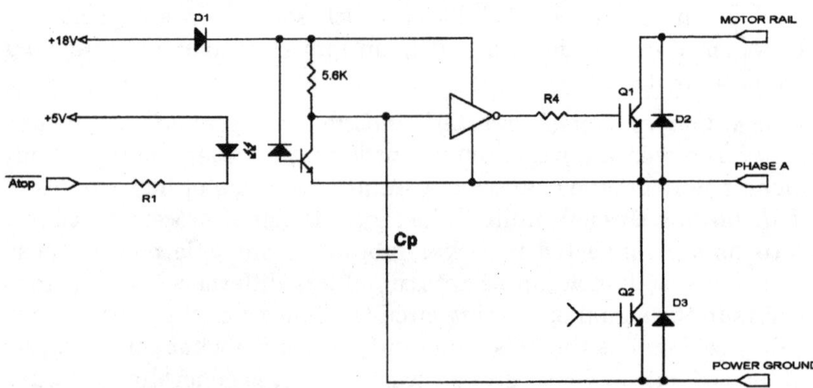

Figure 7.8 Parasitic capacitance.

gests that 3 mA of charging current will flow through C_p. This 3 mA into 5.6 kΩ of node impedance is much more than enough to cause false transitions. These numbers illustrate a very high sensitivity to parasitic coupling, which makes layout of this part of the circuit a primary design consideration.

In addition to viewing source or emitter connections as miniature ground planes, it is also important to keep any signals referenced to ground at least 3 mm (⅛ in) away from high-side gate driver inputs. Given this constraint, it is easy to see why use of a single ground plane in power stages is seldom a good design practice.

Gate drive noise immunity is also facilitated by minimizing the loop area that contains the gate drive decoupling cap, gate driver, gate, and source or emitter of the power device. One way to do this is to route the gate drive signal either directly above or beneath its return. If the return is relatively wide, 2.5 mm (1⁄10 in) or greater, it forms the miniature ground plane discussed previously. The resulting minimum loop area decreases capacitive coupling, as well as antenna effects that inject noise at the input of the gate driver. In addition, relatively high peak gate drive currents exhibit some field cancellation, thus reducing radiated noise.

The other major source of gate drive noise that causes false transitions is nonzero voltage drops in power grounds. Using optocouplers and a separate gate drive IC for each power transistor, as well as routing each gate drive return directly to the emitter of its corresponding power device, is the cleanest way to provide noise immunity. For fractional-horsepower drives where optocouplers are not practical, taking care to minimize the inductance between power device emitters or sources is a viable alternative.

Minimizing loop areas is a key consideration when attempting to reduce the amount of noise that is produced by power stages. The most important loop is the one that includes the upper half-bridge IGBT drain, lower half-bridge IGBT source, and high-frequency bus decoupling cap. This loop contains the high *di*/*dt* that is produced during diode reverse recovery and should enclose an area that is as small as possible. Running traces to and from the decoupling capacitor directly over each other is one way to do that. Since the currents in the two traces are equal and opposite, there are benefits due to field cancellation and minimizing inductance. There are also benefits from adding distributed capacitance.

Figure 7.9 illustrates the difference between a loop that has been routed correctly and one that has not. In this figure, the solid circles represent pads, the schematic symbols show the components that are connected to the pads, and two routing layers are shown with cross-

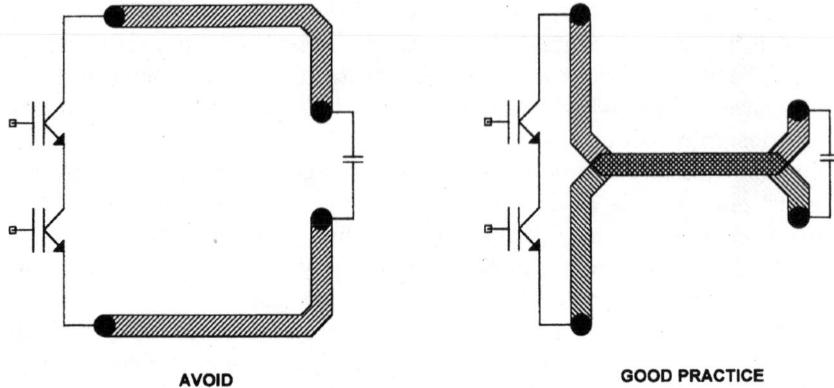

Figure 7.9 Minimizing loop areas.

hatching that run in opposite directions. By routing the two traces one over the other, the critical loop area is minimized.

For similar reasons it is desirable to run power and return traces directly on top of each other. In addition, if a current sampling resistor is used in the return, using a surface mount resistor is preferable because of its lower inductance. It also can be placed directly over the power trace, providing uninterrupted field cancellation. Again, for field cancellation, it is desirable to run phase outputs parallel and as close as possible to each other.

In the power stage, it's important to avoid right angles since they can produce unwanted reflections and current crowding at the inside corners. Single traces are easy—two 45° angles or a curve easily accomplish a 90° turn. It is just as critical to avoid 90° angles in T connections. Figure 7.10 illustrates correct versus incorrect routing for both cases.

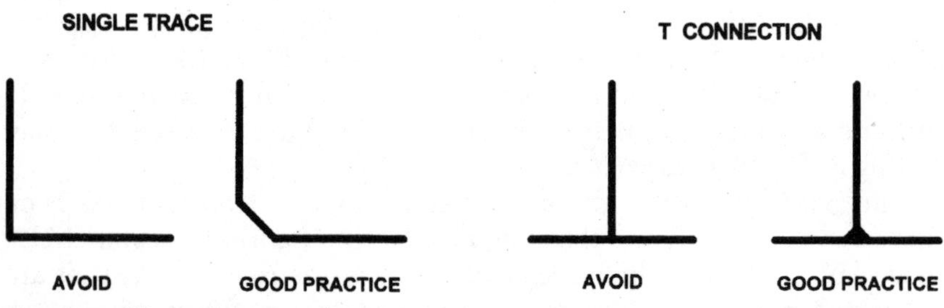

Figure 7.10 Ninety-degree angles.

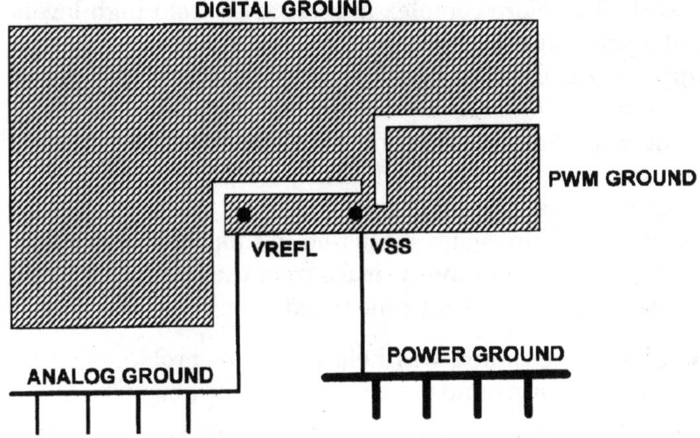

Figure 7.11 Ground architecture.

7.2.3 Controllers

The primary consideration when laying out controllers is ground partitioning. A good place to start is with the architecture shown in Fig. 7.11. This architecture has several key attributes—analog ground and power ground are both separate and distinct from digital ground, and both contact the digital ground at only one point. For the analog ground, it is preferable to make the one point as close as possible to the analog-to-digital converter's ground reference (V_{refl}). For the power ground, the connection should be as close as possible to the microcomputer's power supply return (V_{ss}). Note that the path from V_{refl} to V_{ss} is isolated from the rest of digital ground until it approaches V_{ss}.

PWM ground is also isolated as a separate ground-plane section until it approaches V_{ss}. This is most crucial in systems that use optocouplers, since the current that flows through the PWM ground return will be higher than other digital return currents. If a two-layer board is used, traces replace the ground planes that are shown in Fig. 7.11. The partitioning, however, remains the same.

In addition to grounding, controllers benefit from attention to avoiding 90° angles, since there are generally many high-speed signals on the digital portion of the board. Routing with 45° angles or curves minimizes unwanted reflections, increasing noise immunity.

Summary

For the most part, the functional architecture of motor drives is much more straightforward than some of the techniques that are required to get them to work. The more complex aspects arise from high levels of both *di/dt* and *dv/dt* that are present, producing a number of design issues that must be resolved in order to achieve acceptable noise management. As we have seen, a fraction of a picofarad of stray capacitance in the wrong place, a ground connection that is not carefully routed, or the absence of a functionally not-so-obvious series bootstrap resistor will cause improper operation.

The most important design issues discussed in this chapter are listed below. If you pay attention to these issues from the beginning, your design will operate properly the first time, right from the start.

- Careful attention to grounding will help minimize problems when mixing signal and power grounds.
- Minimizing critical loop areas causes less noise to be received at sensitive inputs and less noise to be radiated from the power stage.
- Careful attention to power transistor transition times can decrease electrical noise in the motor's power stage.
- Filtering sensor inputs improves the stability of the motor drive system.
- Avoiding 90° angles in PCB layouts reduces unwanted reflections and current crowding.
- Canceling fields by routing equal and opposite current flows as close as possible to each other is a good method to augment noise robustness.
- The use of optocouplers in the gate drive circuits provides high electrical isolation between the logic and power stage.
- The motor's high voltage and fast switching edges (*dv/dt*) directly and indirectly generate most of the electrical noise. Other factors include lead inductance, snappy free-wheeling diodes, and PCB layout.
- There is always some capacitance between conductors in an electronic circuit. Under certain conditions this parasitic, or stray, capacitance can couple unwanted signals into sensitive control circuits, which can produce instability in the motor system.
- Ground noise may reach several volts or several tens of volts in a large motor drive, resulting in erratic operation in the gate drive circuits and leading to a power stage failure.
- A series resistor helps to isolate the gate driver power supply from *L*(*di/dt*) voltage spikes that are produced by switching the power devices.

- Software can affect the motor drive's performance in its design and operation: 1. During design, a printed circuit layout program can automatically disallow sharp 90° traces. 2. When operating, the motor control software can be programmed to look for and ignore false sensor input signals.

- Most motor control IGBTs are more noise robust because they have a higher gate threshold voltage than MOSFETs and generally use softer drain-source free-wheeling diodes than MOSFETs.

- A snappy FWD is one that exhibits a very fast negative peak-to-zero current transition. It is this transition's speed and the small amount of inductance in the FWD circuit that can cause excessive voltage ringing.

FURTHER READING

Sabin, D., and Sundaram, A., "Quality Enhances Reliability," *IEEE Spectrum,* Feb. 1996.

Janesurak, J., "In Their Opinion," "Major Appliance Dealers Study," *Appliance Manufacturer,* April 1995.

Emerson Motor Company, "Motor Technology News," July 1996.

Chapter

8

AC Induction Motor Control

Peter Pinewski
Motorola Semiconductor Products

In this chapter we will describe various types of speed and torque control methods used with AC induction motors. Although not all AC induction motor control techniques will be discussed, we will give the reader a solid foundation of AC induction motor control basics. Some of the topics include implementing common modulation techniques, such as third-harmonic PWM and space vector modulation, and control algorithms, such as volts per hertz [voltage/frequency (V/F)] and vector control. Implementation issues will accompany application examples. We will not, however, discuss AC induction motor construction or motor differences in this chapter.

The AC induction motor is widely used in a number of variable-speed or torque control applications. But many engineers think that speed or torque control of the AC induction motor is a complex and costly task as compared to a DC motor. The AC motor drive not only needs to create AC waveforms for three phases but must also control the amplitude and frequency of these AC waveforms. While improvements in electronics such as power semiconductor devices and microcontrollers have made implementation of an AC motor drive hardware simpler and more cost-effective, digital control techniques have shifted the implementation burden to software. However, digital control techniques allow greater flexibility in control algorithm execution and allow new advancements in control techniques to be easily adapted.

8.1 AC Motor Drive

Since the normal line frequency and voltage are fixed, AC induction motor drives are based on the fact that variable-frequency, variable-voltage AC waveforms can be generated. Although there are different types of AC induction motor drives, the most common drive utilizes a converter-inverter structure as shown in Fig. 8.1.

The converter creates DC from the fixed AC line while the inverter develops the variable-frequency, variable-voltage AC from the DC.

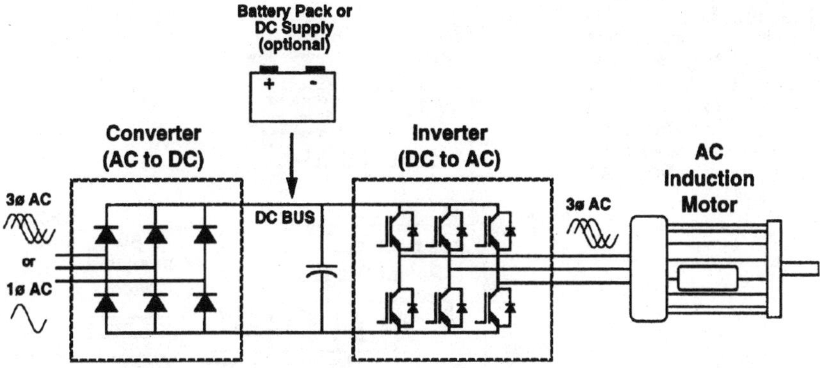

Figure 8.1 Three-phase AC motor drive.

Because of the inverter structure, the three-phase AC induction motor can be easily adapted to run off of single-phase power, three-phase power, or a DC supply. For example, some localities that only have single-phase power available can still use larger and more power-efficient three-phase motors by rectifying the single-phase power line and generating the necessary three-phase power. In other applications, such as an electric vehicle, the high-voltage battery pack is the DC source, and the inverter is used to generate the three-phase power.

Many refer to AC motor drives as AC *inverters,* since the inverter is the primary component of the motor drive. As seen in Fig. 8.2, a typical voltage-source inverter is comprised of a power stage, control circuitry, and a large capacitor or capacitor bank (2000 to 20,000 μF).

8.1.1 Power stage

The power stage consists of three half H-bridge drivers, which are used to switch the high-voltage DC bus and allow current to be driven into and out of each phase of the motor. These half H-bridge drivers are controlled by an algorithm, which is executed in the control circuitry, to create the proper AC waveforms. Because the power stage determines the current that drives the motor, the power stage establishes the kilowatt rating of the AC motor drive. In addition, the power stage becomes the dominating cost factor for increasingly large drives.

The power stage, as shown in Fig. 8.3, is critical to overall motor drive performance. IGBT technology and gate drive design play major

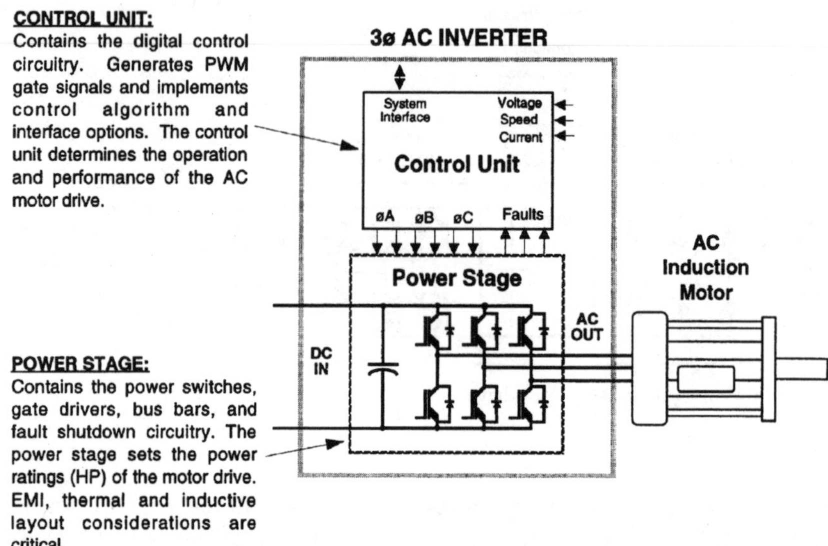
CONTROL UNIT:
Contains the digital control circuitry. Generates PWM gate signals and implements control algorithm and interface options. The control unit determines the operation and performance of the AC motor drive.
3ø AC INVERTER
System Interface
Voltage
Speed
Current
Control Unit
øA øB øC Faults
Power Stage
AC OUT
DC IN
AC Induction Motor
POWER STAGE:
Contains the power switches, gate drivers, bus bars, and fault shutdown circuitry. The power stage sets the power ratings (HP) of the motor drive. EMI, thermal and inductive layout considerations are critical.

Figure 8.2 Three-phase voltage-source inverter.

roles in switching performance and thermal management. In addition, power stage layout is important in minimizing inductive overshoot and EMI, which can affect control unit operation. The power stage usually incorporates some type of fault shutdown circuitry which is used to protect the IGBTs and gate drive circuitry. This fault detection typically includes overcurrent detection, short-circuit shutdown, and temperature faults. Catastrophic faults, such as short circuits, usually disable the power stage automatically until the fault is removed in order to prevent serious damage to the motor drive. Other faults, such as overtemperature or overcurrent, are normally fed back to the control unit, and the control unit is responsible for taking the corrective action. To reduce the number of fault signal lines, individual fault signals are often combined. In this case, faults can be categorized by the action required, such as warning faults or catastrophic faults.

8.1.2 Control unit

Although the power ratings of the AC motor drive are established by the power stage, the control unit and its algorithm are important factors in determining the speed or torque response and other motor drive functionality. The control unit executes the control algorithm, implements the interface options, and supplies the switching signals

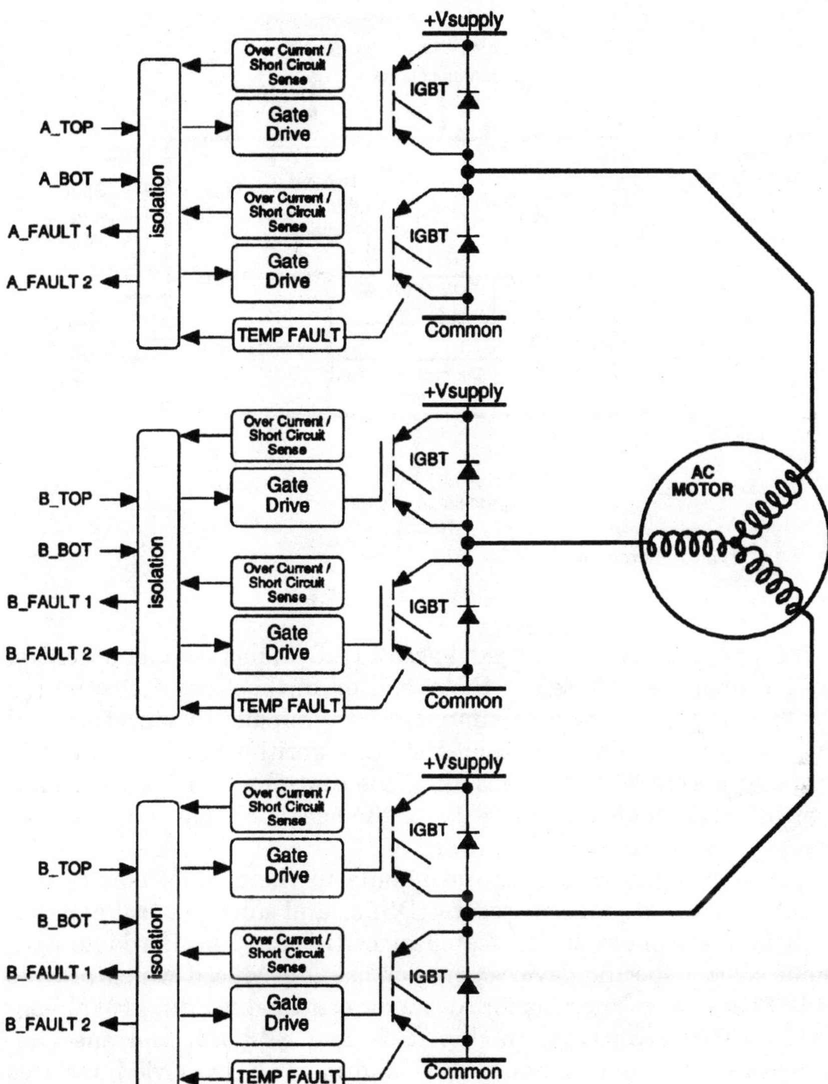

Figure 8.3 Three-phase power stage.

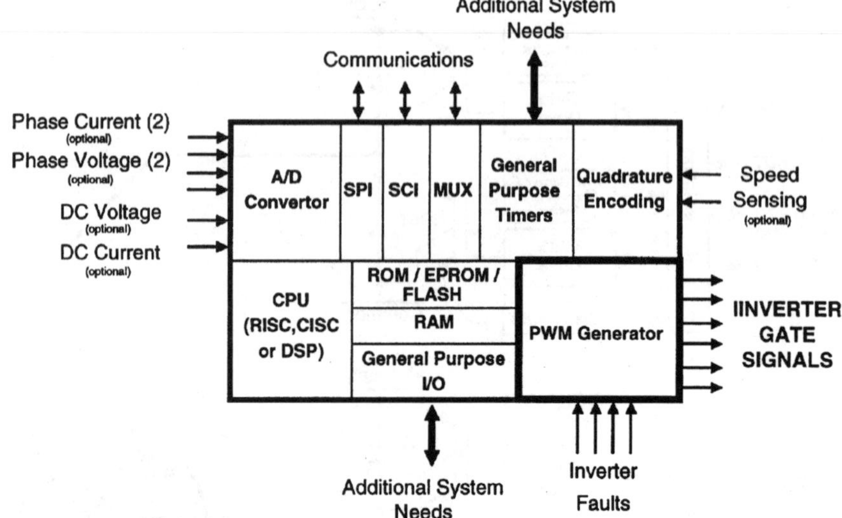

Figure 8.4 Typical motor control MCU.

to the power stage. Recent advances in AC induction motor control can be largely attributed to the evolution of the control electronics. Digital control techniques simplify the implementation of control algorithms and make more sophisticated algorithms feasible. In addition, digital control techniques allow the control algorithm to be easily modified through software without the need for component changes or costly board redesign.

Speed and integration improvements in the control electronics, mainly digital signal processors (DSPs) and microcontroller units (MCUs), allow more sophisticated algorithms at an affordable cost. Motor-control-specific devices are now being produced which provide all the necessary functions for AC motor control, including PWM generation, A/D conversion, on-chip RAM and EPROM, and miscellaneous interface components, such as multiplex (MUX) style interfaces and extra timers. Figure 8.4 shows a motor-control-specific MCU and its interface needs. (See Chap. 10 for more details on motor control MCU devices.)

8.2 Modulation Methods (DC-to-AC Conversion)

There are two parts to controlling the AC induction motor, both of which are accomplished in software: (1) variable AC waveforms need

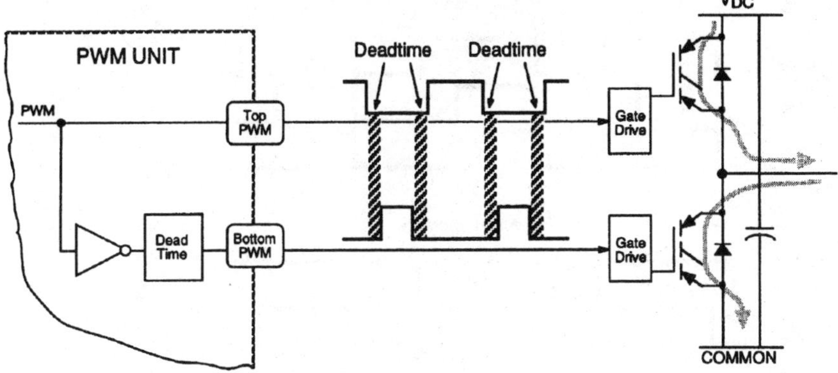

Figure 8.5 Inverse PWMs with deadtime.

to be created; and (2) the frequency and the voltage of the AC waveforms must be controlled. Waveform creation is accomplished by the modulation strategy and is addressed here. Methods of controlling the frequency and voltage (control algorithms) are presented later in this chapter.

Pulse-width-modulation (PWM) techniques are the standard for AC waveform creation. However, these techniques are different from those applied to DC motor control. Each phase of the AC motor requires two PWM signals, one for the top switch and one for the bottom switch. The two signals, as seen in Fig. 8.5, are derived from the same PWM where one is the inverse of the other with the inclusion of deadtime. These requirements include a deadtime period to ensure that the top and bottom switches do not conduct at the same time.

In addition to the PWMs shown in Fig. 8.5, PWMs for AC motor control are typically center-aligned as illustrated in Fig. 8.6. Center alignment not only reduces the likelihood that more than one power device will switch at any given time, thereby minimizing noise in the system, but also causes a frequency doubling effect in the motor, which reduces current ripple and acoustic noise.

Control of the PWM signals for AC motor control requires constant updating of the PWM duty cycles, creating the AC waveforms at the motor phases. The method by which the PWMs are updated is termed the *modulation technique*. Some of the more popular techniques are outlined here. Sinusoidal PWM is the most basic. However, other modulation techniques, such as third-harmonic PWM, 60° PWM, and space vector modulation (SVM), have gained popularity because they utilize more of the available DC bus. Sinusoidal PWM, third-harmonic PWM, and 60° PWM share a common implementation scheme.

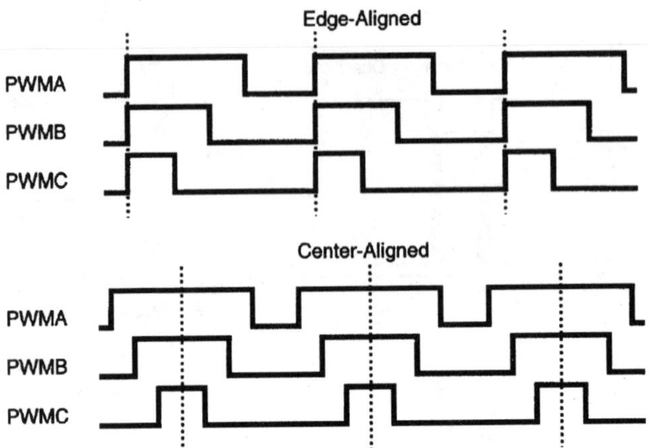

Figure 8.6 Edge alignment versus center alignment.

However, other techniques, which include SVM and six-step modulation, have different calculation and implementation requirements.

8.2.1 Sinusoidal PWM

Sinusoidal pulse-width modulation is the easiest modulation scheme to understand. It is used here to illustrate the implementation scheme shared by all PWM techniques. The sinusoidal AC waveforms are created by changing the PWM duty cycles to the shape of a sine wave. Early designs often referred to sinusoidal modulation as *sine-triangle-modulation* because a triangle carrier signal was compared with a sinusoidal reference to create the modulated PWMs as shown in Fig. 8.7.

The majority of today's AC motor drives utilize an MCU with a PWM unit designed specifically for motor control. There are many of

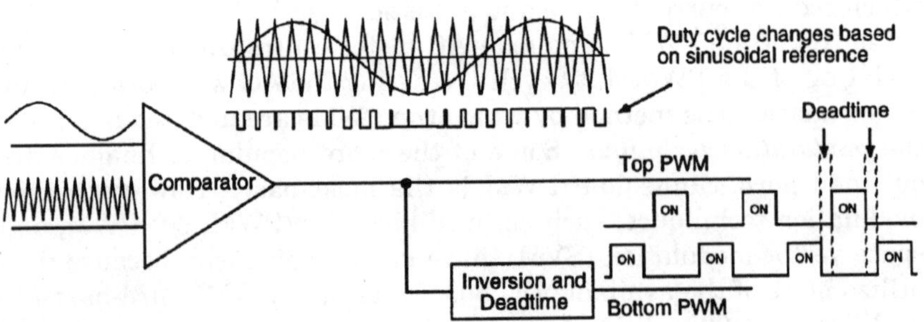

Figure 8.7 Sine-triangle modulation.

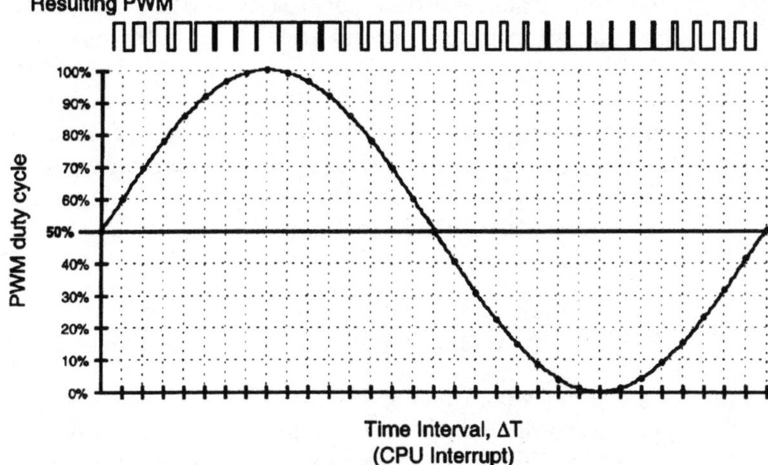

Figure 8.8 Sine-wave creation with MCU.

these devices available on the market. In the MCU implementation, the software is responsible for continuously updating the PWM duty cycles.

The PWM duty cycle calculations are based on a point in the sine wave. At full modulation (maximum voltage), 100 percent duty cycle corresponds to the positive peak of the sine wave, and 0 percent duty cycle is equivalent to the negative peak. The zero crossover point is represented by 50 percent duty cycle, as can be observed in Fig. 8.8. The math used for sinusoidal modulation is as follows:

$$\text{A\_HIGH} = m \cdot \sin(\phi) \cdot 50\% + 50\%$$

$$\text{B\_HIGH} = m \cdot \sin\left(\phi - \frac{\pi}{3}\right) \cdot 50\% + 50\%$$

$$\text{C\_HIGH} = m \cdot \sin\left(\phi + \frac{\pi}{3}\right) \cdot 50\% + 50\% \tag{8.1}$$

where m is the modulation index (0 to 1, 0 = zero voltage, 1 = full voltage).

Since two PWM signals are required for each phase, the output duty cycle at the phase winding corresponds with the duty cycle of the top device. As an example, if the desired output at the phase winding is 75 percent duty cycle, the top PWM would be 75 percent duty cycle and the bottom PWM would be at 25 percent. In actual practice, the PWM at the phase winding will be slightly different because of the necessity for deadtime. This deadtime creates an output waveform distortion, which causes torque pulsations. (See Chap. 9 for a review of deadtime.)

8.2.1.1 Changing output frequency. The PWM duty cycles are normally updated at a periodic rate, and the frequency of the sine wave is determined by this update rate (ΔT) and the number of sample points in a cycle. This is common to all PWM techniques. The relationship is given by

$$\text{Output frequency} = \frac{1}{\text{update\_rate} \cdot \text{number\_of\_samples}} \tag{8.2}$$

For example, a system which has a 250-μs sample rate ($\Delta T = 250$ μs) and 64 samples would produce an output frequency equal to 1/(250 μs · 64) = 62.5 Hz. To vary the frequency, either the number of sample points must change or the time between updates must change. Generally, the number of samples is changed, since the update rate is normally determined by the PWM period and therefore is kept constant. Because the sample points typically come from a table, the number of sample points are easily changed by varying the increment value through the table as given by Eq. (8.3). The resulting frequency can also be calculated as shown in Eq. (8.4).

$$\text{Number\_of\_samples} = \frac{\text{table size}}{\text{increment value}} \tag{8.3}$$

$$\text{Output frequency} = \frac{\text{increment value}}{\text{update\_rate} \cdot \text{table size}} \tag{8.4}$$

Using every point in the sine table (increment value = 1) represents the slowest frequency achievable and can also be referred to as the *frequency resolution*. By skipping table points (increment value ≠ 1), one can achieve a higher-frequency output at the expense of fewer samples. For example, if the increment value were doubled (increment_value = 2), the frequency would double. Figure 8.9 illustrates how the frequency is adjusted by changing the number of samples while keeping the interrupt rate constant.

8.2.1.2 Changing amplitude. The PWM duty cycles affect waveform amplitude. This is normally defined in terms of a modulation index m. A modulation index of one ($m = 1$) is equal to full amplitude and the PWM duty cycles swing from 100 percent to 0 percent. Half-voltage is achieved with $m = 0.5$ and the duty cycles swing from 75 to 25 percent. The zero crossover remains at 50 percent duty cycle. This means that the duty cycles are always centered about 50 percent. In

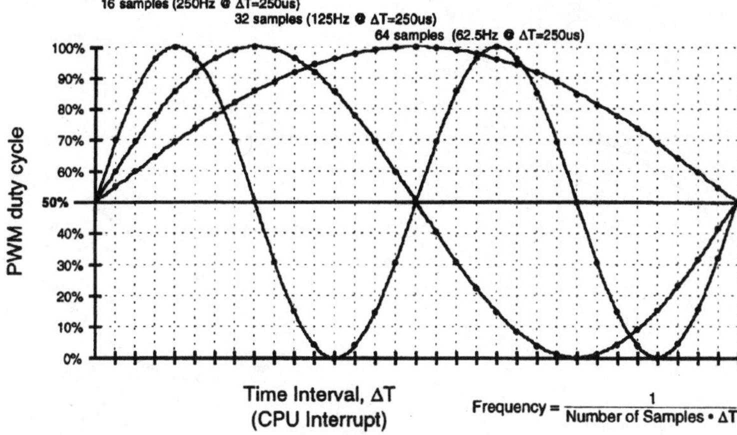

Figure 8.9 Varying frequency with MCU.

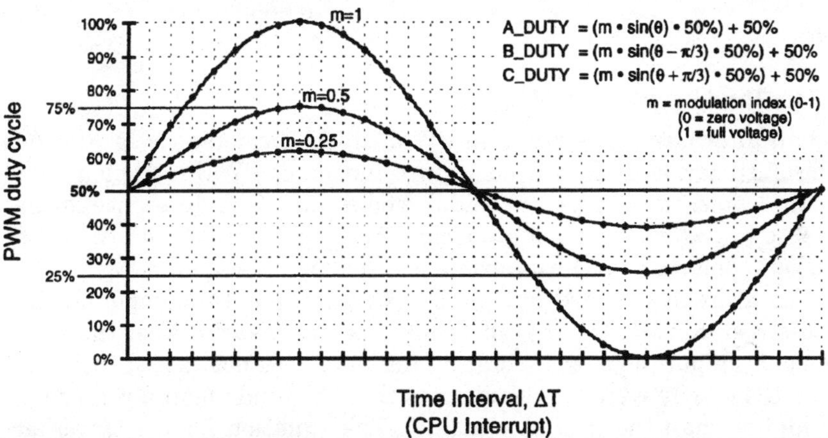

Figure 8.10 Varying voltage with MCU.

addition, a continuous duty cycle of 50 percent on all three phases ($m = 0$) results in zero voltage on all three phases as seen in Fig. 8.10.

With sinusoidal PWM, the neutral point of the motor floats to 0.5 V_{DC}. This relies on the fact the motor neutral is not connected—a necessary condition for AC waveform creation with a PWM inverter. The voltages from motor phase to common swing from $+V_{DC}$ to common, while the actual phase voltage across the motor winding ($V_{L\text{-}N}$) swings from $+0.5\ V_{DC}$ to $-0.5\ V_{DC}$, as observed in Fig. 8.11.

The maximum phase voltage for sinusoidal PWM is $0.5V_{DC}$. This means that the maximum phase-to-phase voltage as calculated here:

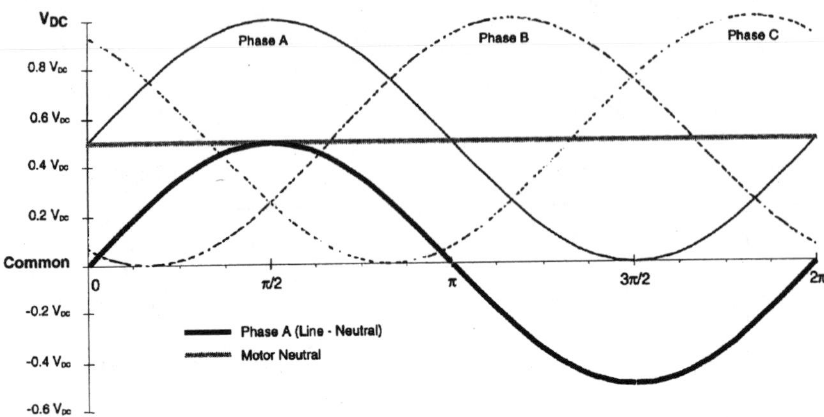

Figure 8.11 Phase A (line-to-neutral) for sine PWM.

$$V_{\text{L-L}} = \sqrt{3} \cdot V_{\text{L-N}} = \sqrt{3} \cdot 0.5V_{\text{DC}} = 0.8660254V_{\text{DC}} \tag{8.5}$$

8.2.2 Third-harmonic PWM

As an alternative to sinusoidal PWM, third-harmonic PWM provides more utilization of the DC bus. Third-harmonic PWM is implemented in the same manner as sinusoidal modulation. The difference is that the AC waveforms are not sinusoidal but consist of both a fundamental component and a third-harmonic component as seen in Fig. 8.12.

The amount of third harmonic and fundamental component are chosen so that the latter is maximized while ensuring that the peak-to-peak amplitude of the resulting function does not exceed the available DC supply voltage. The effect is that the fundamental component is higher than the available supply. The equation for the third-harmonic function is given in Fig. 8.12.

As was the case in the sinusoidal modulation method, the motor neutral is not connected but is floating. However, unlike sinusoidal modulation, where the neutral point floats to $0.5V_{\text{DC}}$, the motor neutral follows the third-harmonic component (which is exactly the same for each phase). This results in an effective cancellation of the third-harmonic component and the line-to-neutral phase voltages ($A_{\text{L-N}}$, $B_{\text{L-N}}$, $C_{\text{L-N}}$) are sinusoidal with a peak amplitude of $0.57735V_{\text{DC}}$ ($2/\sqrt{3} \cdot 0.5V_{\text{DC}}$). This is exactly the same as the fundamental. The maximum phase to phase voltage is determined by the equation

$$V_{\text{L-L}} = \sqrt{3} \cdot V_{\text{L-N}} = \sqrt{3} \cdot 0.57735V_{\text{DC}} = V_{\text{DC}} \tag{8.6}$$

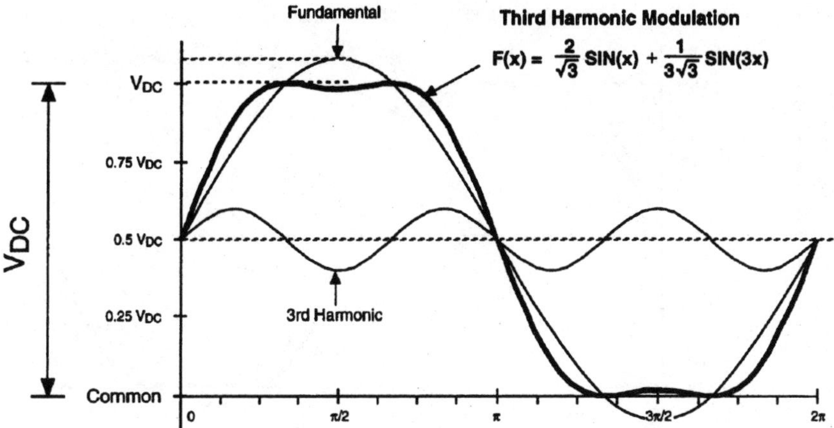

Figure 8.12 Third-harmonic PWM.

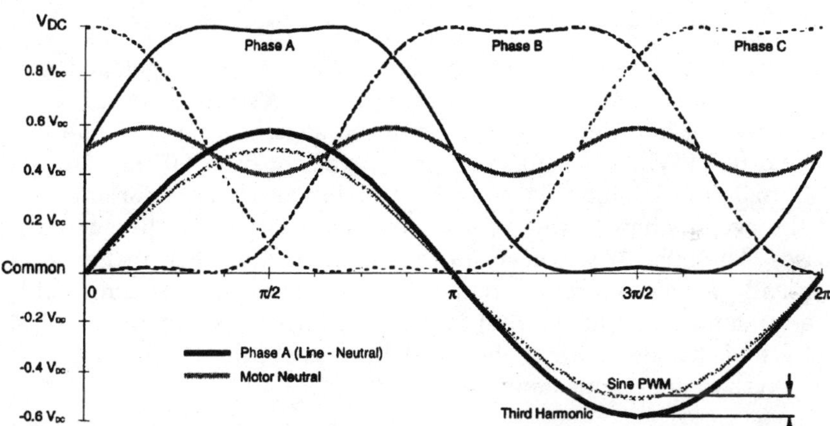

Figure 8.13 Phase A (line-to-neutral) for third-harmonic PWM.

This is approximately 15.5 percent higher in amplitude than that achieved by the sinusoidal method, which means that it is possible to get more energy into the motor using third-harmonic PWM. Figure 8.13 shows this effect.

8.2.3 60° PWM

As stated previously, the nature of the three-phase motor causes the neutral point to follow the third-harmonic component, thereby filtering out the third-harmonic component as seen by the motor winding. The same phenomenon occurs for all triple harmonics (3d, 9th, 15th,

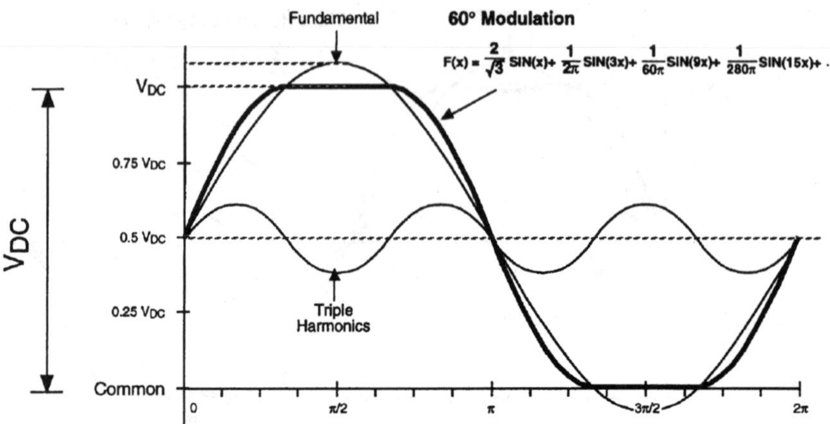

Figure 8.14 Sixty-degree PWM.

21st, 27th, etc.); as a consequence, 60° PWM has been developed. Like third-harmonic PWM, 60° PWM creates a larger fundamental (2/√3). However, its main advantage is reduced switching losses. The idea behind 60° PWM is to "flat-top" the waveform from 60° to 120° and 240° to 300°. In doing this, the power devices are held on for one third of the cycle (when at full voltage) and will not suffer the switching losses. The 60° PWM waveform can be defined as seen in Fig. 8.14. Typically, a close approximation of 60° modulation can be achieved by the fundamental and the first few terms. Like third-harmonic PWM, 60° PWM utilizes more of the available bus ($V_{L\text{-}N} = 0.57735V_{DC}$, $V_{L\text{-}L} = V_{DC}$) than does sinusoidal PWM.

8.2.4 Space vector modulation (SVM)

Space vector modulation is quite different from the PWM methods. With PWMs, the inverter can be thought of as three separate push/pull driver stages, which create each phase waveform independently. In contrast, SVM treats the inverter as a single unit; specifically, the inverter can be driven to eight unique states, as shown in Fig. 8.15. Modulation is accomplished by switching the state of the inverter. Like third-harmonic and 60° PWM, space vector modulation utilizes more of the available DC bus than does sine PWM. Space vector modulation has been gaining more attention in the industry. The reason for this is unknown but may be attributed to the claim of better harmonic performance and alignment to digital control techniques.

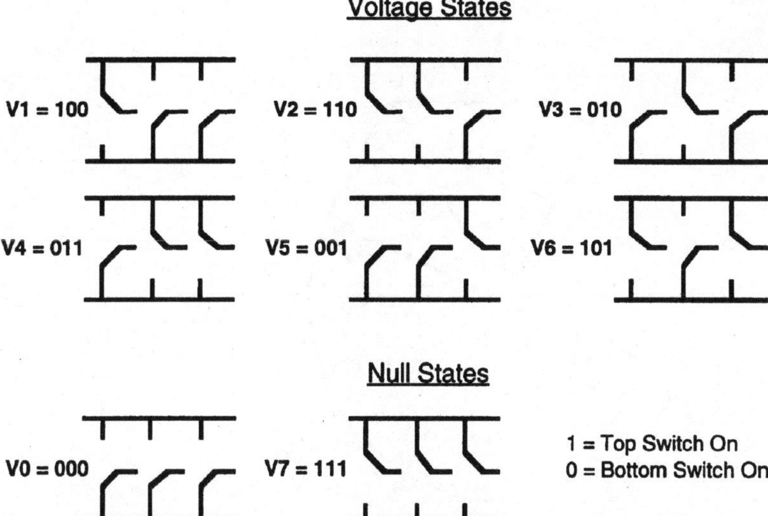

Figure 8.15 Space vector modulation inverter states.

8.2.4.1 How SVM works. As mentioned earlier, the inverter has eight possible states. If these states are mapped on a two-coordinate axis (*d-q* axis), they create a state map, as illustrated in Fig. 8.16. Each state creates a voltage vector. Two of the vectors, V_0 and V_7, short the motor load and are considered null or zero vectors because they result in zero voltage across the phase windings. Modulation is accomplished by rotating a reference vector around the state diagram. A circle can be drawn inside the state map to represent this and corresponds to sinusoidal operation; thus, the resulting voltages across the motor windings are sinusoidal.

The direction in which the vector rotates determines the direction in which the motor turns. The rate at which the vector rotates determines the frequency of the AC waveforms. The amplitude of the waveforms is determined by the radius of the circle inside the state diagram. The maximum sinusoidal phase voltage that can be achieved by space vector modulation is $V_{max} = 0.57735V_{DC}$, which is 15.5 percent higher than that achievable by sinusoidal PWM and is equivalent to third-harmonic or 60° PWM.

8.2.4.2 SVM switching. AC waveform creation is accomplished by rotating a reference vector around the state map. Any instance of the reference vector can be created by the combination of the two adjacent vectors and a null vector. This is done by time-modulating the

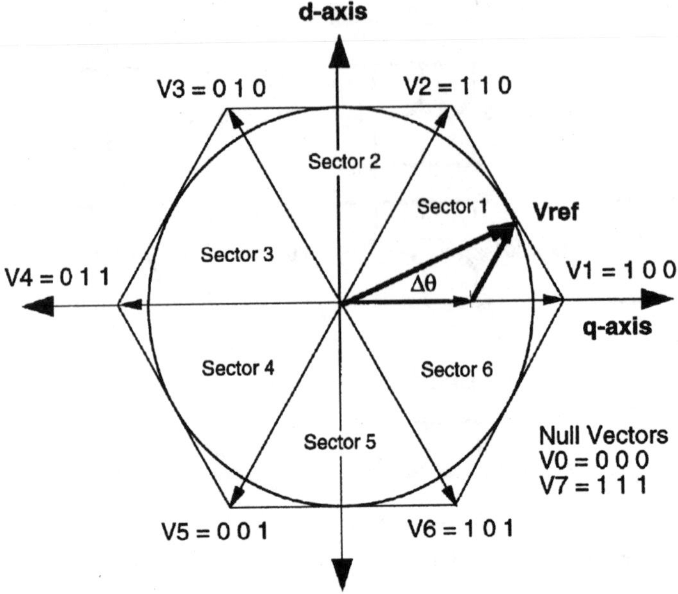

Figure 8.16 Space vector modulation state map.

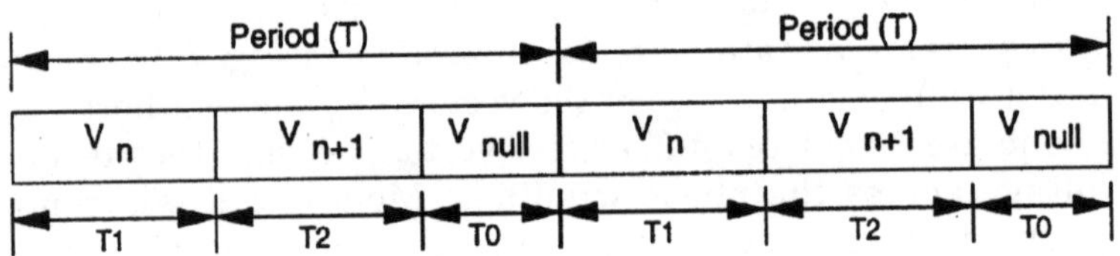

$$\text{Vref} = \text{V}_\text{n}\bullet\text{T1} + \text{V}_{\text{n+1}}\bullet\text{T2} + \text{V}_{\text{null}}\bullet\text{T0}$$

Figure 8.17 Space vector modulation state switching.

adjacent two vectors and a null vector, as shown in Fig. 8.17. As an example, a voltage vector, V_{ref}, in sector 1 can realized by the V_1 and V_2 vectors and one of the null vectors (V_0 or V_7).

To clarify, the V_1 state will be active for time T_1, V_2 will be active for T_2, and the null vector, V_0 or V_7, will be active for T_0. The only criterion is that the accumulated time of the states must equal the period, as illustrated in Fig. 8.17. The state times are calculated by

$$T_1 = T \cdot m \cdot \sin(60 - \Delta\theta)$$

$$T_2 = T \cdot m \cdot \sin(\Delta\theta)$$

$$T_0 = T - T_1 - T_2 \tag{8.7}$$

where T = time period
m = modulation index
$\Delta\theta$ = angle between V_{ref} and V_n

8.2.4.3 SVM variations. Variations of space vector modulation based on the choice of null vectors and the ordering may provide different switching performance, harmonic losses, and current ripple. Five variations are described here.

Null = V_0 **and null** = V_7. The two most obvious variations are to always use the V_0 null vector and a straightforward sequence $T_1 - T_2 - T_0$ or to always use V_7 with the same sequence, $T_1 - T_2 - T_0$. An example of the sequences for these two types of SVM is shown in Fig. 8.18.

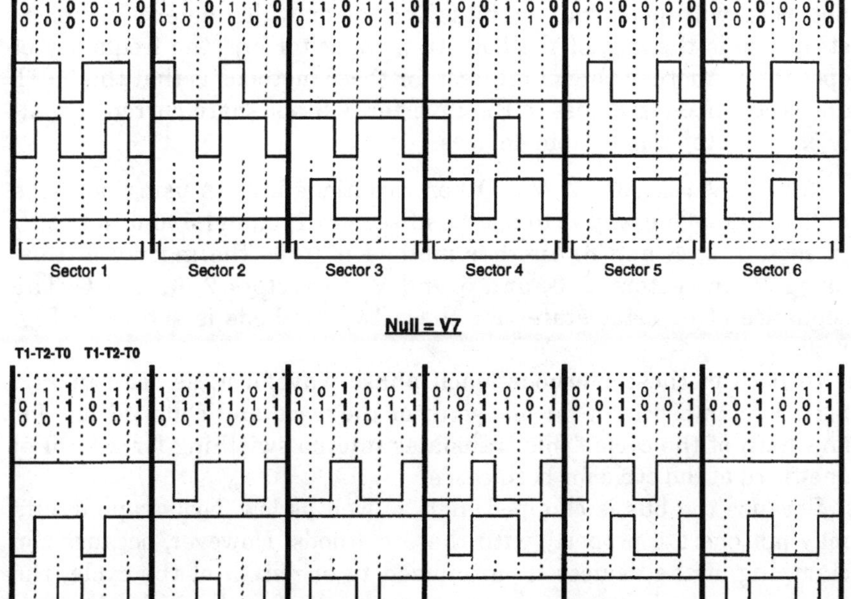

Figure 8.18 Switching sequences for null = V_0 and null = V_7.

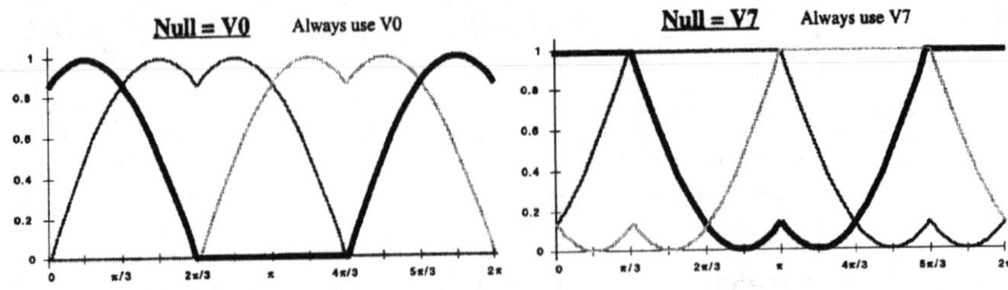

Figure 8.19 Phase waveforms (null = V_0 and null = V_7).

Note that, in both cases, each phase does not switch for one-third of its cycle. In one case, the phase winding is clamped low (null = V_0), and in the other the phase winding is clamped high (null = V_7). The resulting phase waveforms are shown in Fig. 8.19.

The resulting phase waveforms of these two methods are simply inverses of each other. However, the two methods differ in how they operate with bootstrap circuits. In the null = V_0 case, bootstrap circuits will still work, since the clamping keeps the bootstrap capacitors charged. In the null = V_7 case, the clamping occurs so that the bootstrap capacitors can discharge. Therefore, the null = V_7 method may not work with bootstrap circuits. The factor which determines bootstrap use is the size of the bootstrap capacitor and the frequency of operation. Another note of interest for these methods is that the IGBT and free-wheeling diodes of the inverter will not share current equally between top and bottom devices.

Null = V_7, V_0 and null = V_0, V_7. Other variations involve using both the null vectors. One way is to use V_7 in sectors 1, 3, and 5 and to use V_0 in sectors 2, 4, and 6. Another method is to do the inverse of this, using V_0 in sectors 1, 3, and 5 and V_7 in sectors 2, 4, and 6. The sequence of inverter states for these two methods is shown in Fig. 8.20.

In both of these methods, each phase waveform, as seen in Fig. 8.21, is clamped high for one-sixth of the cycle and clamped low for one-sixth of the cycle. This translates into no switching for a total of one-third of the cycle for both cases.

Because the bus is clamped high as well as low, bootstrap circuits may not operate properly with these methods. However, because the clamping of the devices is minimized to one-sixth of the cycle, the demand on the bootstrap capacitor is less than in the null = V_7 case. Although these two methods are very similar, their switching loss performances can be significantly different. Since the induction motor generally has a lagging power factor during motoring, the null = V_7, V_0 method will have lower switching losses because the bus clamping

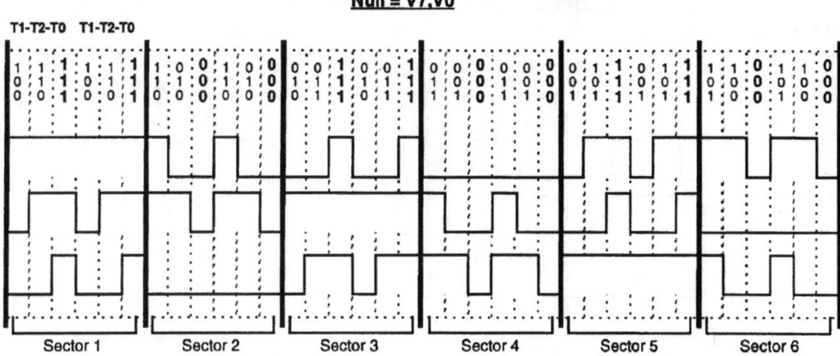

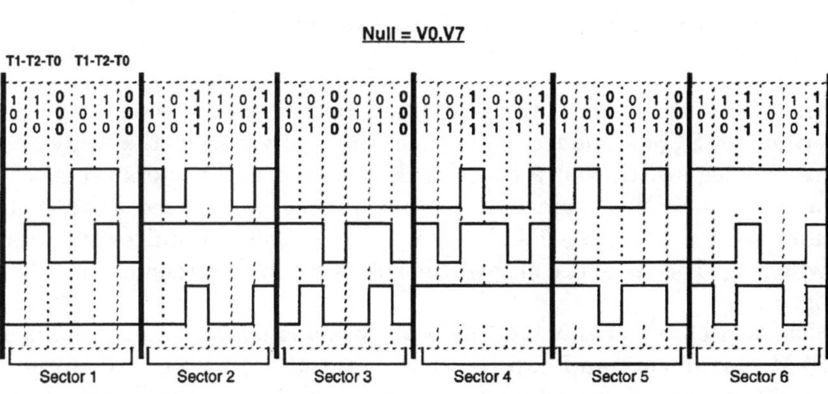

Figure 8.20 Switching sequence for null = V_7, V_0 and null = V_0, V_7.

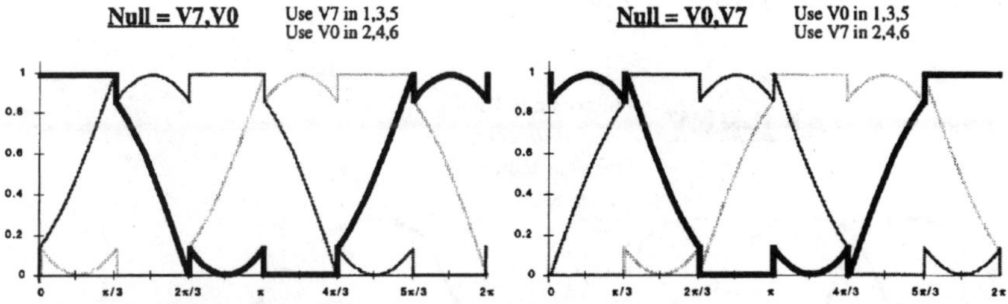

Figure 8.21 Phase waveforms (null = V_7, V_0 and null = V_0, V_7).

occurs during the points of highest current. The null = V_0, V_7 method would be better for generator operation for the same reasons.

8.2.4.4 Alternating-reversing sequence. The problem with the techniques described above is that for each sequence, four commutations occur and two occur at the same time. This means that two inverter

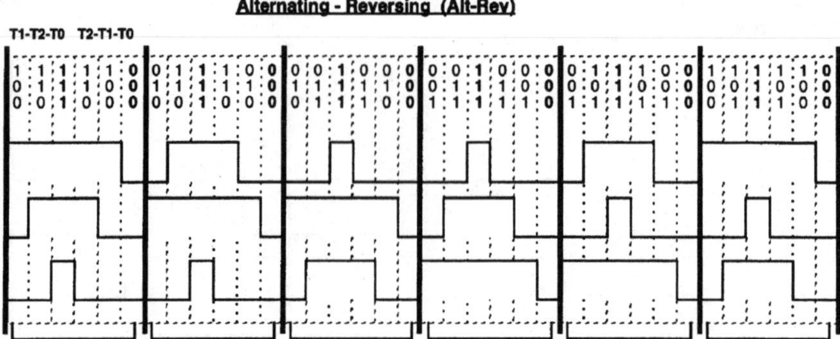

Figure 8.22 Switching sequence for alternating-reversing method.

legs switch at the same time. Another method, which seemingly has become an SVM standard, is to select the null vector so that only one inverter leg switches at a time. To do this, the null vectors are alternated each SVM sequence and the sequence is reversed after each null vector (i.e., T_1–T_2–T_0–T_0–T_2–T_1). Doing this minimizes inverter switching. The sequence of inverter states for this method is shown in Fig. 8.22.

The switching for this type of SVM closely resembles center-aligned PWM waveforms and can be implemented with a center-aligned PWM unit, as illustrated in the following section. The technique benefits from only three commutations per switching sequence, and only one inverter leg switches at a time. The phase waveforms for this method are shown in Fig. 8.23.

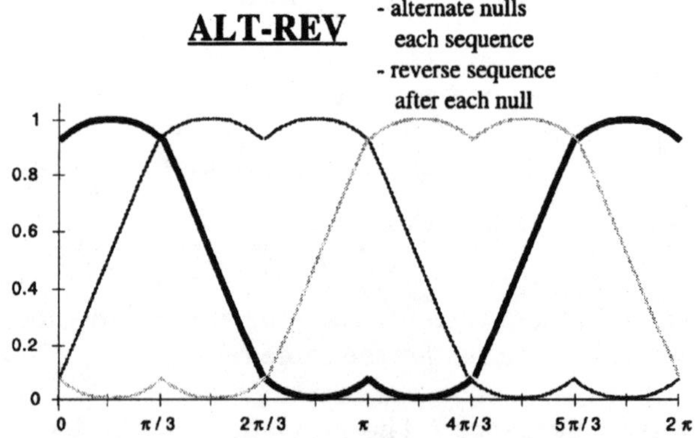

Figure 8.23 Phase waveform (alternating-reversing method).

8.2.4.5 SVM techniques with center-aligned PWMs. The calculations and switching for the SVM techniques are different from the PWM techniques. Instead of adjusting PWM high times, the SVM changes inverter states. Because of this fact, implementing SVM techniques has typically been done with specialized hardware that generates the state timing. However, in one case (as shown in Fig. 8.22) the resulting switching patterns are identical to center-aligned PWMs. This implies that center-aligned PWMs could be used to implement this type of SVM. The difference, as far as the MCU is concerned, is in the way the high times are calculated and the fact that two SVM sequences are equal to a single PWM period. As for the motor and inverter, there is no difference. In addition, a PWM unit with half-cycle reload capability could implement single sequence updates. The calculations for space vector modulation require the software to check the sector boundary of the reference vector, calculate T_1, T_2, and T_0, and update the high times of the PWMs. The high times are calculated as summations of the state times.

The other SVM techniques can be analyzed in a similar manner to produce center-aligned PWMs. Table 8.1 outlines the PWM high time calculations for implementing the SVM techniques with center-aligned PWMs.

In fact, all SVM techniques can be implemented with center-aligned PWMs. The secret is in reordering the states in a manner to achieve center alignment while not changing the effective SVM voltage. As an example, the normal switching sequence for the null = V_0 case (Fig. 8.18) can be reordered so that the sequence starts with the null state (T_0–T_1–T_2). The next sequence reverses the sequence (T_2–T_1–T_0). This is also true for sectors 3 and 5. In sectors 2, 4, and 6, the sequence would be slightly different (T_0–T_2–T_1–T_1–T_2–T_0). Doing this to achieve center alignment is valid as long as the effective phase voltage remains the same. Figure 8.24 shows the resulting center-aligned waveforms.

The ability to use regular center-aligned PWMs is significant because no special hardware is needed. The same PWM unit found on many motor control MCU devices can be used to implement the SVM techniques. In addition, the SWM techniques can be readily changed while the motor is running to compensate for different operating conditions. As an example, to minimize switching losses, the null = V_7, V_0 method may be implemented during motoring. However, for regeneration, the null = V_0, V_7 method should be implemented. This is achieved by changing the hightime calculations from Table 8.1. The state time calculations are the same.

TABLE 8.1 SVM High Time Calculations Using PWMs

	Sector 1	Sector 2	Sector 3	Sector 4	Sector 5	Sector 6
Alternating reversing	A = $T_1 + T_2 + 0.5T_0$ B = $T_2 + 0.5T_0$ C = $0.5T_0$	A = $T_1 + 0.5T_0$ B = $T_1 + T_2 + 0.5T_0$ C = $0.5T_0$	A = $0.5T_0$ B = $T_1 + T_2 + 0.5T_0$ C = $T_2 + 0.5T_0$	A = $0.5T_0$ B = $T_1 + 0.5T_0$ C = $T_1 + T_2 + 0.5T_0$	A = $T_2 + 0.5T_0$ B = $0.5T_0$ C = $T_1 + T_2 + 0.5T_0$	A = $T_1 + T_2 + 0.5T_0$ B = $0.5T_0$ C = $T_1 + 0.5T_0$
Null = V_0	A = $T_1 + T_2$ B = T_2 C = 0	A = T_1 B = $T_1 + T_2$ C = 0	A = 0 B = $T_1 + T_2$ C = T_2	A = 0 B = T_1 C = $T_1 + T_2$	A = T_2 B = 0 C = $T_1 + T_2$	A = $T_1 + T_2$ B = 0 C = T_1
Null = V_7	A = 100% B = $T_0 + T_2$ C = T_0	A = $T_0 + T_1$ B = 100% C = T_0	A = T_0 B = 100% C = $T_0 + T_2$	A = T_0 B = $T_0 + T_1$ C = 100%	A = $T_0 + T_2$ B = T_0 C = 100%	A = 100% B = T_0 C = $T_0 + T_1$
Null = V_7, V_0	A = 100% B = $T_0 + T_2$ C = T_0	A = T_1 B = $T_1 + T_2$ C = 0	A = T_0 B = 100% C = $T_0 + T_2$	A = 0 B = T_1 C = $T_1 + T_2$	A = $T_0 + T_2$ B = T_0 C = 100%	A = $T_1 + T_2$ B = 0 C = T_1
Null = V_0, V_7	A = $T_1 + T_2$ B = T_2 C = 0	A = $T_0 + T_1$ B = 100% C = T_0	A = 0 B = $T_1 + T_2$ C = T_2	A = T_0 B = $T_0 + T_1$ C = 100%	A = T_2 B = 0 C = $T_1 + T_2$	A = 100% B = T_0 C = $T_0 + T_1$

NOTE: All calculations referenced to top switch.

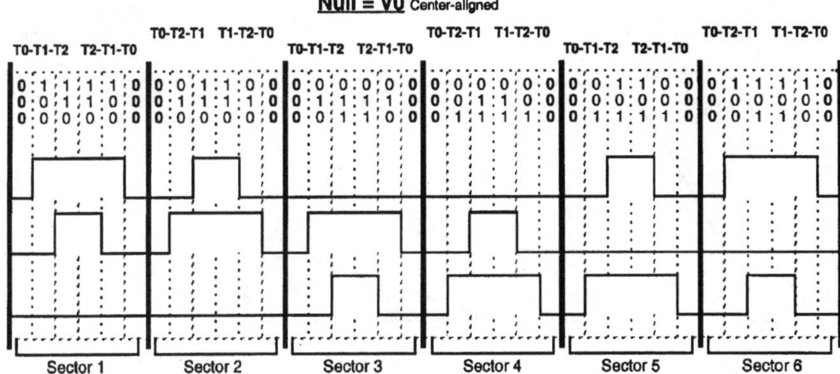

Figure 8.24 Null = V_0 center-aligned waveforms.

8.2.5 Overmodulation and six-step operation

Space vector modulation gives 15 percent more bus utilization than does sinusoidal PWM, but sinusoidal operation is limited by the boundaries of the hexagon (Fig. 8.16). The SVM methods can be adapted to run in an overmodulation mode. In overmodulation, the reference vector follows a circular trajectory that can extend the bounds of the hexagon. The portions of the circle inside the hexagon utilize the same SVM equations for determining the state times T_1, T_2, and T_0. However, the portions of the circle outside the hexagon are limited by the boundaries of the hexagon, and the calculations for T_1 and T_2 are given in Fig. 8.25. In this case, there is no null state time.

Although overmodulation allows more bus utilization than do standard SVM techniques, it results in nonsinusoidal phase winding operation. This yields a high degree of distortion, especially at low rpm. Overmodulation is normally used as a transitioning step from the SVM techniques into six-step operation. Six-step operation switches the inverter only to the six vectors (see Fig. 8.26), thereby minimizing switching. Before PWM techniques were developed, AC motor drives used six-step operations to accomplish variable-frequency AC. Six-step operation is problematic at low frequencies, where the staircase AC waveforms, as shown in Fig. 8.26, cause large current distortions, which result in unacceptable torque pulsations. However, at high frequencies (high rpm), this problem abates, and the drive can actually benefit from the reduced switching losses of six-step operation.

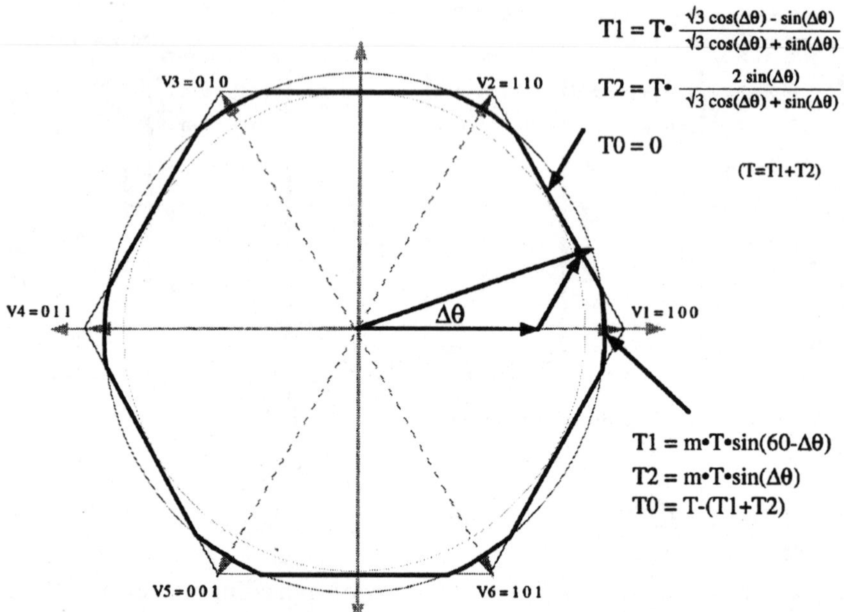

Figure 8.25 Overmodulation.

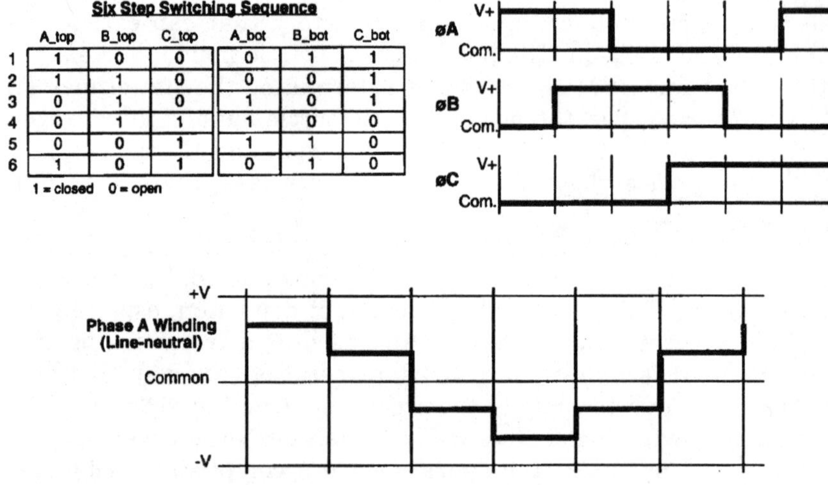

Six Step Switching Sequence

	A_top	B_top	C_top	A_bot	B_bot	C_bot
1	1	0	0	0	1	1
2	1	1	0	0	0	1
3	0	1	0	1	0	1
4	0	1	1	1	0	0
5	0	0	1	1	1	0
6	1	0	1	0	1	0

1 = closed 0 = open

Figure 8.26 Six-step operation.

TABLE 8.2 Modulation Summary

Modulation type	L-N voltage, V DC	L-L voltage, V DC	Phase operation	Bus clamping (switching loss savings)
Sine PWM	0.5	0.866	Sinusoidal	No
Third-harmonic PWM	0.57735	1	Sinusoidal	No
60° PWM	0.57735	1	Sinusoidal	Only at full modulation
SVM—Alt-Rev	0.57735	1	Sinusoidal	No
SVM—Null = V_0	0.57735	1	Sinusoidal	Yes (120° low)
SVM—Null = V_7	0.57735	1	Sinusoidal	Yes (120° high)
SVM—Null = V_7, V_0	0.57735	1	Sinusoidal	Yes (60° high, 60° low)
SVM—Null = V_0, V_7	0.57735	1	Sinusoidal	Yes (60° low, 60° high)
Overmodulation	*	*	Nonsinusoidal	Dependent on SVM type
Six-step	*	*	Nonsinusoidal	Yes (Lowest switching)

*These values are estimated to be higher than SVM but according to how much is undetermined.

8.2.6 Modulation techniques summary

Any modulation scheme can be used to create the variable-frequency, variable-voltage AC waveforms. Although sinusoidal PWM is the most obvious choice, third-harmonic PWM and SVM techniques have gained popularity because of their increased bus utilization. SVM techniques are not currently as popular as third-harmonic PWM because of the different timing requirements but are receiving more and more attention in the industry. SVM techniques can reduce current ripple and can offer lower switching losses than standard PWMs. The ability to use standard PWM units to implement SVM techniques should increase their viability. Table 8.2 gives a summary of the different types of modulation schemes.

8.3 AC Induction Motor Drive Algorithms

Once the AC waveforms are created via the modulation technique, the frequency and voltage are adjusted on the basis of the control algorithm. The algorithm can control torque or speed, it can be open-loop or closed-loop, and adaptive techniques can be implemented to compensate for variations in the motor due to manufacturing or temperature. As a result, the algorithms can take many forms. This section will outline a few of the basic algorithms, such as volts per hertz, slip control, and vector control (field orientation).

8.3.1 AC induction motor review

To fully understand the AC motor control algorithms, a basic knowledge of the AC induction motor operation is necessary. The three-phase AC induction motor was designed to be operated from the AC line, which has a fixed voltage and frequency. The synchronous speed (speed of the rotating field) of the AC motor is set by the applied frequency f and the number of poles in the motor p and is calculated as

$$\text{Synchronous speed} = 120 \cdot \frac{f}{p} \tag{8.8}$$

As an example, the synchronous speed of a four-pole motor running from a 60-Hz AC line is (120/4)*60 = 1800 rpm. The actual speed of the motor shaft is determined by the load, which sets a slip in the motor. "Slip" is defined as a percentage of the synchronous speed, or it can be defined as a slip frequency, which is the difference between synchronous speed and actual speed. The slip is significant in that it affects the torque and operation of the motor as seen in the torque versus speed (slip) curve in Fig. 8.27. In some control algorithms, slip frequency is the control variable.

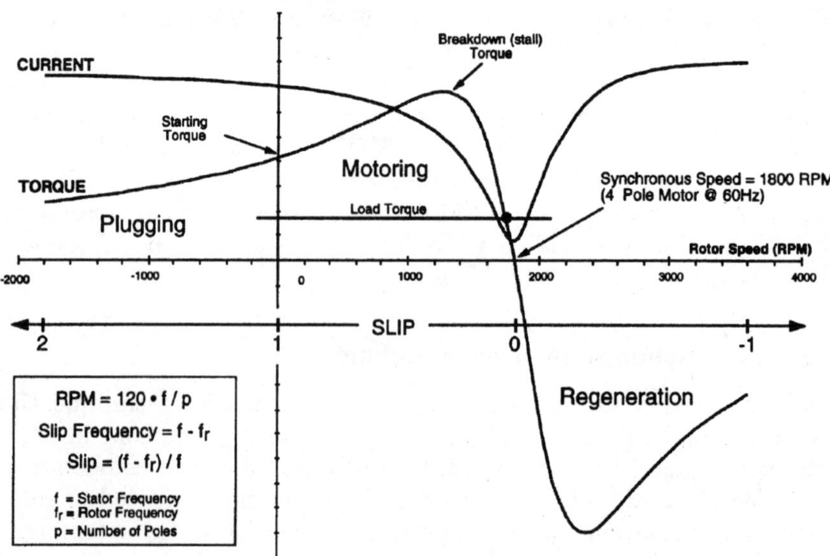

Figure 8.27 Torque and current versus motor speed (slip).

8.3.1.1 Motoring. During startup for normal motoring operation ($0<s<1$), the motor speed is zero, and the slip is equal to one. Very high currents occur during start-up because of the large slip. As the motor begins to turn, both the slip and the current decrease. The final speed of the motor is determined by the point in which the load torque equals the generated torque of the motor. This is at a point beyond the peak (breakdown) torque. With proper control, achievable torque can far exceed normal operating torque. This extends across a wide speed range. This high torque capability is one reason that AC motors have gained so much popularity.

8.3.1.2 Regeneration and braking. For many applications, the AC motor is chosen because it can also function as a generator. Generator action (regeneration) occurs when the slip is less than zero ($s<0$). This takes place when the load causes the rotor to turn faster than the synchronous speed of the motor, or the synchronous speed is lowered below the speed of the motor shaft. During regeneration, the motor generates a negative torque, which not only provides a braking action but also transfers energy back to the supply. Because of this, extreme caution is needed during regeneration to ensure that the supply can accept the regenerative energy, or the regenerative energy can be dissipated. Some industrial drives utilize a large power resistor (braking resistor) to dissipate the regenerative energy. Others use an active converter and transfer the energy back onto the AC mains. In an electric vehicle application, the regenerative energy is employed to recharge the battery pack and to brake the vehicle. Regeneration is free for the AC induction motor; no additional circuitry is needed. Therefore, implementing a drive that requires regeneration may actually be easier and more cost-effective with the AC induction motor than the DC motor.

8.3.1.3 Plugging. Braking also occurs in the induction motor when the slip is greater than one ($s>1$). This is known as "plugging" and is caused by either the direction of the field being changed while the motor is already running or the load causing the motor to turn in the opposite direction as the field. In this region, the torque is in the same direction as the field, which causes a braking action. However, after the motor stops, it will change directions. Although this type of braking does not provide energy back to the supply, the current is high, which can cause excessive heating in the motor and the drive. Therefore, the AC induction motor is not normally operated in this mode.

8.3.2 Open-loop volts per hertz (*V/F*)

With a basic knowledge of the AC motor, it is easy to see the drawbacks of typical fixed-frequency AC motor operation. This is where the control algorithm becomes beneficial. Open-loop volts per hertz (V/F) is the most commonly used AC induction motor control method. This scheme is typically employed in low-cost applications such as HVAC systems, fans, blowers, or pumps where the applied load is known. In many cases, V/F is used in fixed-speed applications to simply limit the inrush current during startup. In others, it is used to run the motor at variable speeds. The V/F algorithm is very simple and inexpensive to implement in comparison to more sophisticated AC induction motor control algorithms. No sensing elements are needed, the calculations are simpler, and a less powerful, less expensive MCU can be used. A block diagram of the V/F algorithm is shown in Fig. 8.28.

Motor speed varies with applied frequency. However, at low frequencies, the motor's inductive reactance ($X_L = 2\pi f$) drops, resulting in excessive current into the motor. The control algorithm must adjust the voltage as well as the frequency in order to reduce the current into the motor. The V/F algorithm controls both the frequency and the voltage in a proportional fashion up to the nominal operating frequency (base frequency). Beyond this frequency, the voltage is held at its maximum value while only the frequency is changed, as shown in Fig. 8.29. The voltage is a limit of the available supply. If more voltage is available, the voltage could be continuously increased even beyond the base frequency.

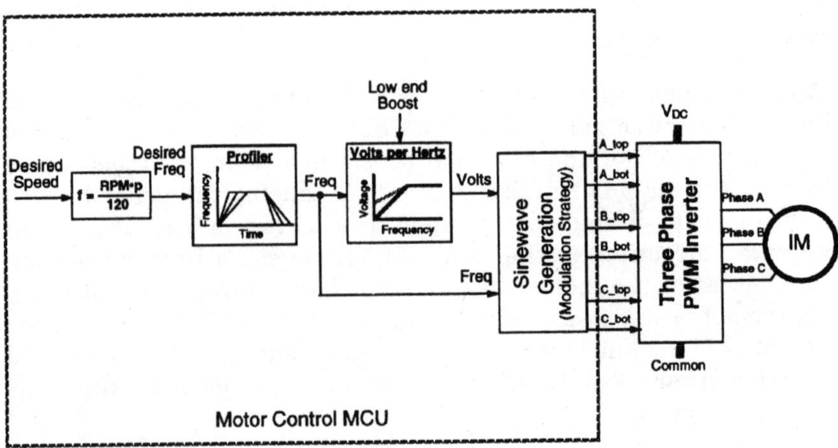

Figure 8.28 V/F block diagram.

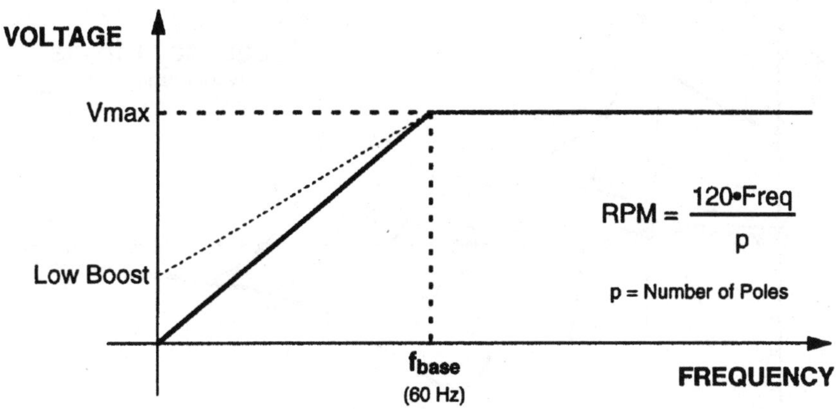

Figure 8.29 Volts/hertz operation.

8.3.2.1 Effects of voltage and frequency on torque curves. The effects of the voltage and frequency changes on the torque versus slip (speed) curves are shown in Fig. 8.30. Increasing frequency results in a torque decrease of $1/f^2$. Similarly, decreasing the applied frequency results in a torque increase of $1/f^2$. This is due to the inductive reactance of the motor. Increasing voltage causes the torque curves to increase by V^2 while still maintaining their shape.

The effect of the V/F algorithm on the torque curves is shown in Fig. 8.31. The torque curves retain their shape at all frequencies, as long as the voltage is adjusted proportionally.

8.3.2.2 Stator *IR* drop. The exception to the volts-per-hertz rule is at low frequencies where the increasing effect of the stator resistance causes a decrease in motor current and field flux and, therefore, a decrease in torque, as seen in Fig. 8.32.

The *IR* drop effect can be compensated for by increasing voltage at the affected frequencies. Figure 8.33 illustrates the inclusion of the low-end boost. The amount of low-end voltage boost is dependent on the motor and the *IR* drop. Determination of the proper voltage boost level is critical. Not enough boost results in a dropoff of current and field flux and, ultimately, poor torque. Too much boost results in high magnetizing currents with little usable torque, which can cause low-frequency instabilities.

8.3.2.3 *V/F* profiler (ramp rate). Because V/F is an open-loop algorithm, an important control variable is the ramp rate. How fast can the frequency change? If the acceleration rate is too fast, the slip will become too large and the motor will stall. Ramping down the motor is

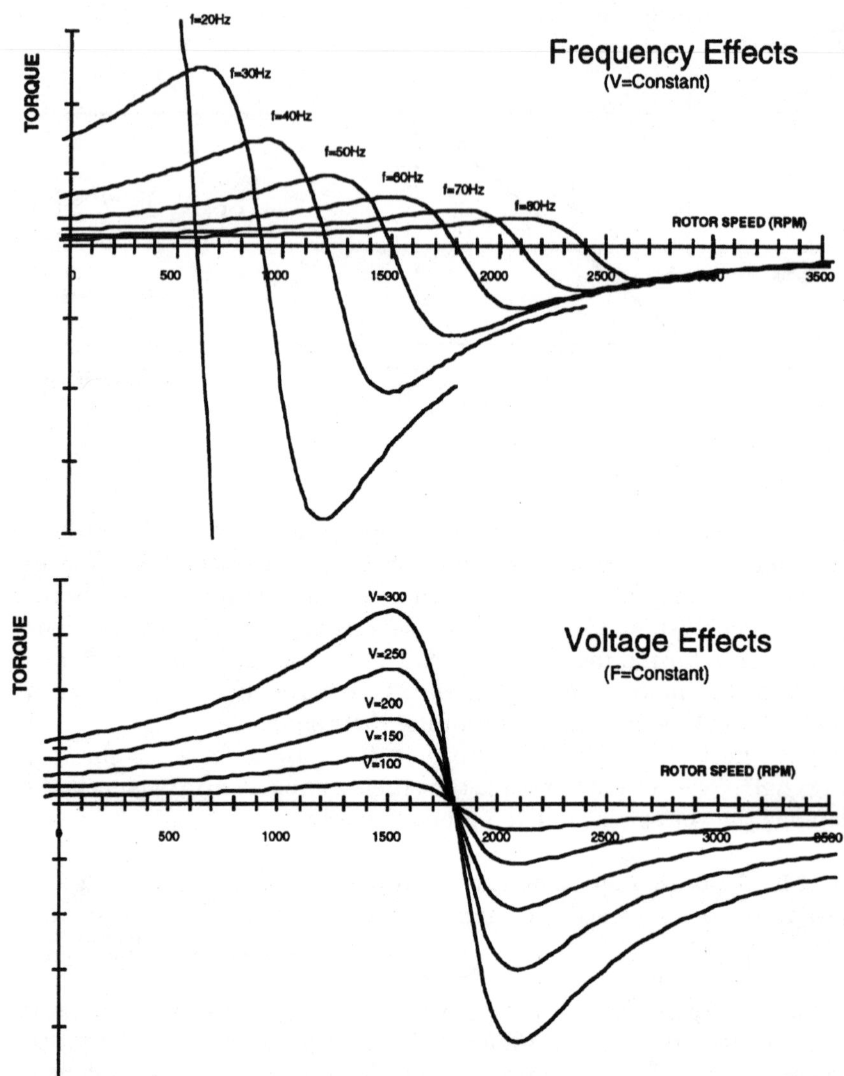

Figure 8.30 Voltage and frequency effects on torque curves.

of equal importance. If the motor is slowed too quickly, the slip will become too large and excessive regeneration will occur. If the drive cannot dissipate the regenerative energy, there is the potential to exceed the DC bus capacitor voltage ratings, which could damage the inverter. Because of this, a profiler is normally used to limit the acceleration and deceleration rate by only allowing predetermined incre-

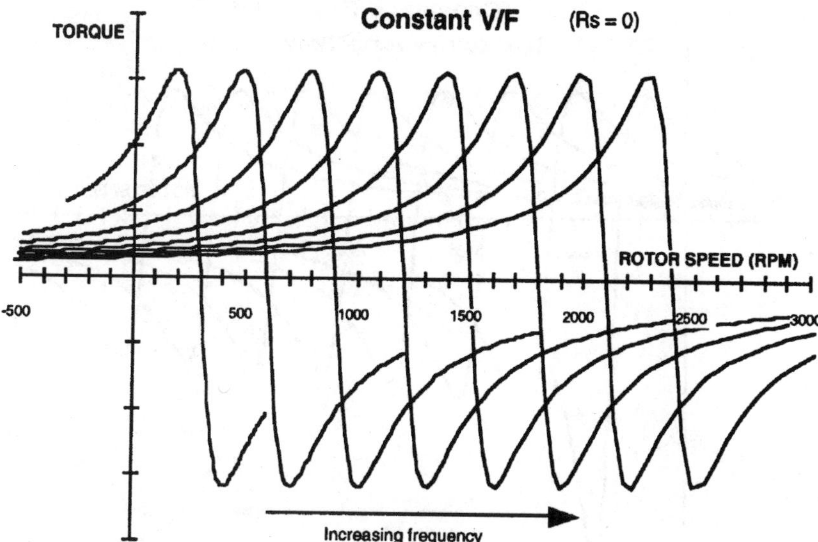

Figure 8.31 Torque curves with constant V/F.

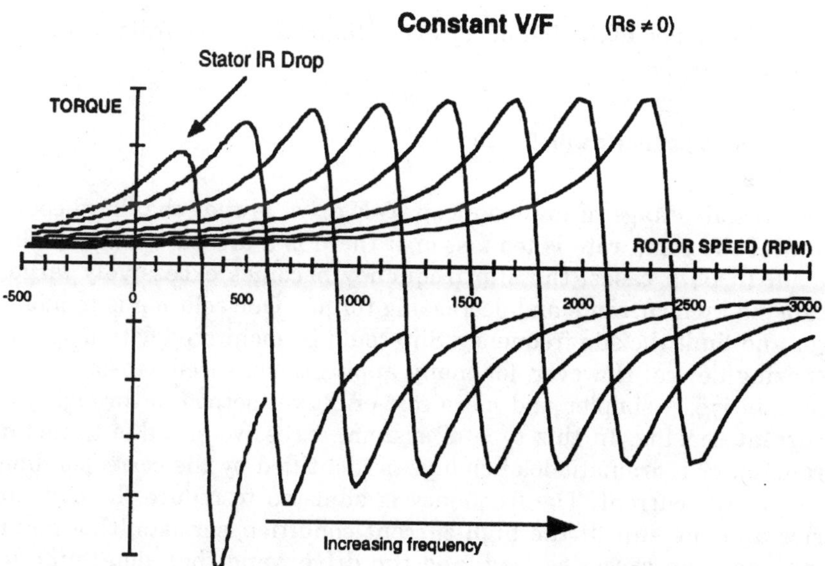

Figure 8.32 Effects of stator IR drop.

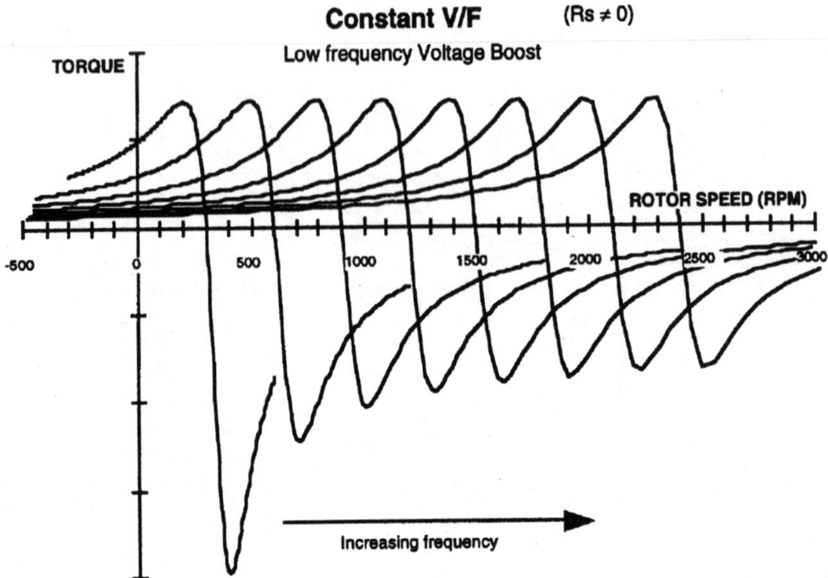

Figure 8.33 Torque curves with low-end boost.

mental changes to the frequency every time period, regardless of commanded frequency.

8.3.3 Volts per hertz with DC current sensing

One disadvantage of an open-loop V/F drive is that the drive could stall if the ramp rate is too fast or if there is a drastic change in the load. In both cases, the slip frequency becomes excessively large, causing high currents and decreasing torque. One solution is to monitor and limit the slip frequency. Slip could be monitored with a speed-sensing device. However, for many applications, speed sensor cost is prohibitive. A simpler and more cost-effective method is through DC current sensing. In this type of system, excessive slip due to motor rampup or a dramatic load change is identified by the corresponding rise in DC current. The frequency is adjusted to reduce the current rise and the slip. If the high-current condition persists, the motor may have an excessive load, and the drive could then determine to shut it down. Voltage sensing may also be incorporated to detect high currents due to a sag in DC voltage. Figure 8.34 shows a block diagram of using DC current sensing for slip limit.

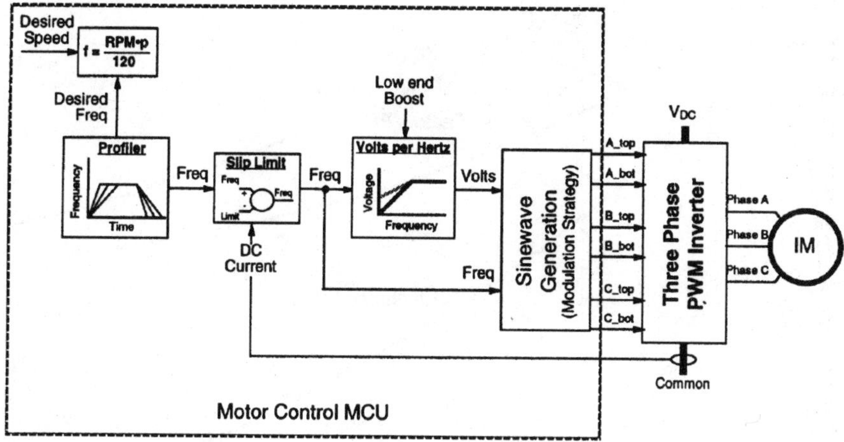

Figure 8.34 *V/F* with DC current sensing.

8.3.4 Slip optimization

Although limiting motor slip with DC current sensing is important in AC drives to prevent motor stalls or high-current conditions, control is limited. Better control can be gained by controlling the slip frequency directly. As seen in the torque/slip curves, there is a slip value for maximum torque (breakdown point). In addition, there is a slip value for maximum power factor and a slip value for maximum motor efficiency. These values will differ, as seen in Fig. 8.35. Typically, the rated slip (rated torque) of the motor falls somewhere between maximum power factor and maximum efficiency. The figure explains why a lightly loaded motor is less efficient than one operating at its rated load. Under lightly loaded conditions, the slip is very small.

The slip frequency can be controlled in a variety of ways. One method is shown in Fig. 8.36. In this system, a speed sensor is used to measure rotor frequency and the applied frequency is determined by maintaining a desired slip frequency. The desired slip frequency is determined by the parameter that is to be controlled. For example, an application concerned with system efficiency could ensure that the slip frequency is maintained to maximize efficiency. The voltage is controlled by a proportional-integral (PI) controller to maintain the desired speed. When the voltage reaches its maximum value, the drive would revert to classic *V/F* techniques in order to maintain speed regulation, and the slip frequency would vary, thereby losing the slip frequency control.

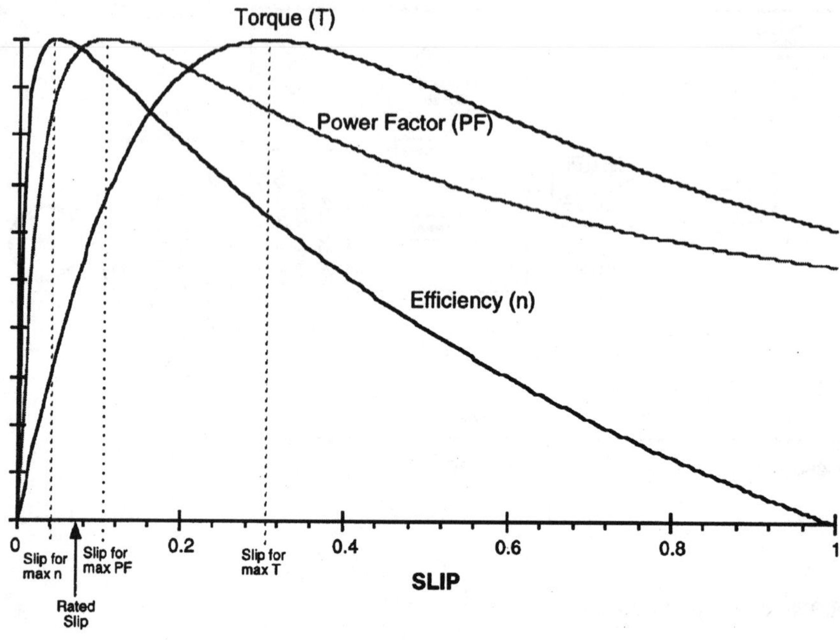

Figure 8.35 Torque, power factor, and efficiency versus slip.

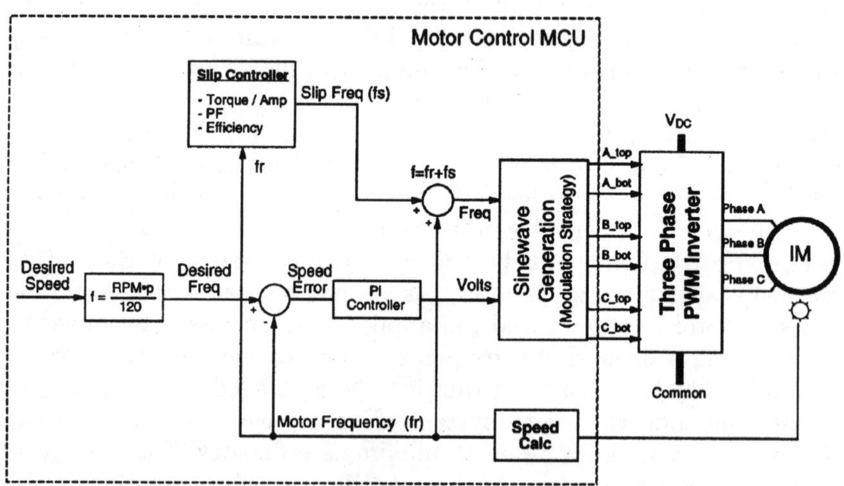

Figure 8.36 Slip optimization using speed sensing.

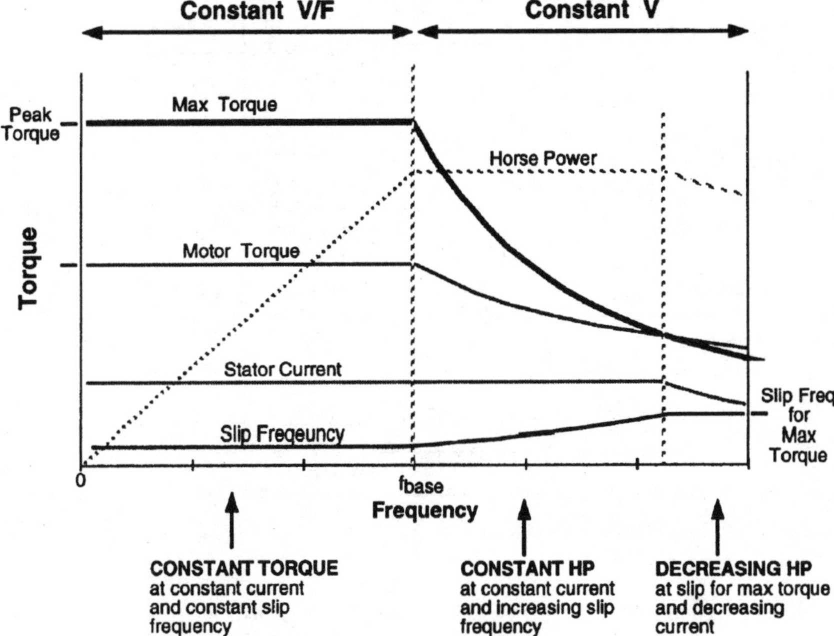

Figure 8.37 AC motor capability curves.

8.3.5 Slip frequency control

As seen in the section on volts per hertz, the torque/speed curve retains its shape if the voltage and frequency are adjusted proportionally. This can be visually represented by the capability curve shown in Fig. 8.37, which illustrates that the induction motor has the potential to achieve rated torque through a wide rpm range without exceeding the current ratings of the motor.

In the region at frequencies lower than the base speed, known as the *constant-torque region,* constant torque can be achieved as long as the stator current and slip frequency remain constant. Above the base frequency, the torque falls off as a result of the voltage limit. If more voltage were available, the constant torque region could be extended.

In the region above the base frequency, known as the *constant-horsepower region,* the torque falls off as $1/f^2$. Therefore, to maintain constant horsepower, the slip frequency must increase to reduce the effect of the decreasing torque. The horsepower can be maintained by increasing slip until the slip frequency reaches its breakdown limit, at which point the horsepower will decrease. The AC induction motor's capability curve is much like that of the DC motor, except that the AC machine's constant horsepower region is limited.

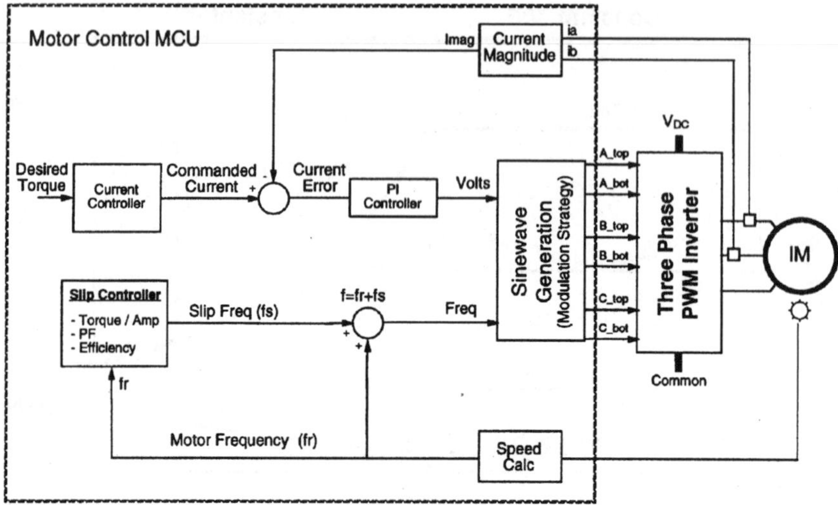

Figure 8.38 Slip control block diagram.

Given its capability curve, the induction motor can achieve good torque over a broad frequency range with slip frequency control. This performance is adequate for dynamic systems, such as an electric vehicle traction drive. In systems where the torque response is critical, such as servo drives, the slip control algorithm may not be suitable. In a slip control algorithm, the torque is controlled by the slip frequency and the commanded stator current, as seen in Fig. 8.38.

Frequency is controlled by the commanded slip frequency, and the voltage is adjusted by a PI controller to maintain the desired current level. Determining stator current can be accomplished by sensing two phase currents and converting the AC current measurements into a current magnitude. A transformation block is used for the conversion. This could consist of a 3–2 phase conversion process, as in a vector controller (described later in this chapter), with a rectangular-to-polar conversion as follows:

$$\text{mag} = (x^2 + y^2)^{-1/2} \tag{8.9}$$

Controlling slip frequency can result in excellent torque control. The drawback is that slip and current tables need to be created to generate the torque, since both slip frequency and current determine torque. In other words, a given torque can be achieved by different slip frequencies and different current levels. The primary question is how much slip frequency and how much current is needed for a given torque. One possibility would be to control the slip frequency by optimizing for motor efficiency or power factor and then adjusting current to generate the torque. Another would be to choose a slip frequency to achieve maximum torque per ampere so that power losses in the inverter could be minimized. The optimal system efficiency may be a point somewhere between maximum torque per ampere and maximum motor efficiency.

8.3.6 Vector control (field orientation)

Although slip frequency control offers performance which is satisfactory for many applications, generating the required slip and current tables can be a cumbersome task. In addition, slip frequency control does not separately control (decouple) the field-producing and torque-producing portions of the current. Because of this, instantaneous torque cannot be achieved. Field orientation (vector control) alleviates these problems. The majority of torque control drives implement vector control techniques. There are two types of field orientation: direct and indirect. In the *direct* method, flux sensing elements are used to determine the position of the flux vector. The *indirect* method, which is more widely used, measures stator current and rotor speed to determine the position of the flux vector. This is based on calculations derived from the motor model. Because of the sensing elements, indirect field orientation is used more often. In either case, vector control attempts to maintain a 90° spatial relationship between the flux and current vectors. This is intended to mimic what is achieved by the physical construction of a DC motor where the 90° relationship is maintained by the commutators and brushes. As a result, vector control allows the AC motor to be controlled in the same manner as a separately excited DC motor.

The vector controller utilizes speed sensing and phase current sensing to control the stator current vector. By mapping the measured phase currents as a vector onto a two-axis (d-q) coordinate system, the stator current is broken into two components, i_d and i_q, which are orthogonal to each other and are used to control the field flux and torque current, respectively. (See Fig. 8.39.) This enables independent control of both the flux and torque. If the d-q coordinate system is rotated synchronously with the rotor flux, i_d and i_q can be controlled as DC values just as if it were a DC machine. In fact, the torque equation is very similar to that of a DC machine.

From a practical point of view, the flux and torque currents are controlled through the stator current vector, which is controlled by the motor voltages and slip frequency. This is similar to slip control. However, in vector control the slip frequency is based on calculations derived from the AC motor model, as shown in Fig. 8.40. This is cru-

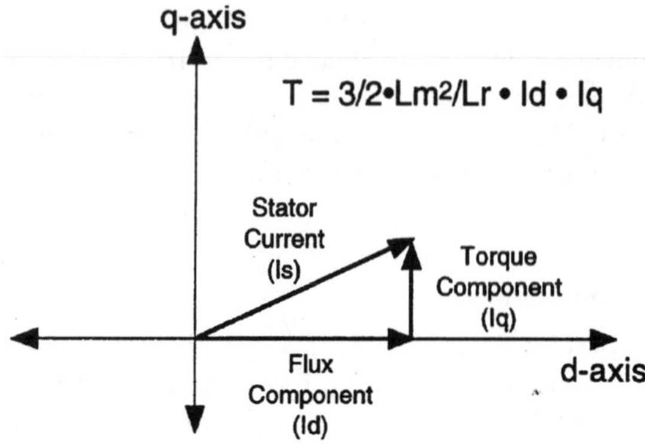

Figure 8.39 Current components.

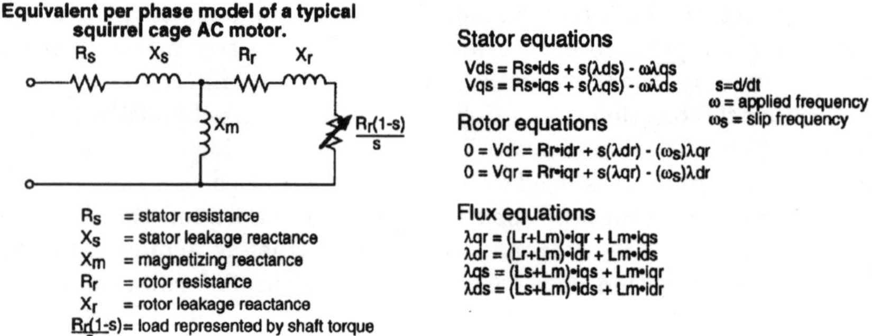

For vector control the rotor flux is oriented to the d-axis only ($\lambda qr=0$)
This is accomplished by maintaining a slip frequency defined by:

$$\omega_S = \frac{(1+Tr \cdot s)}{Tr} \frac{iqs}{ids} \qquad Tr = Lr/Rr = \text{rotor time constant}$$

Figure 8.40 Motor model and vector equations.

cial for correct field orientation. Therefore, a vector control algorithm must be tuned to the motor on which it is operating.

Several methods can be used to implement the vector control algorithm. Feedforward techniques, model estimators, and adaptive control techniques could all be employed to enhance response and stability. However, the principle to control the field flux and current vectors is the same. A block diagram of a simple vector controller is presented in Fig. 8.41.

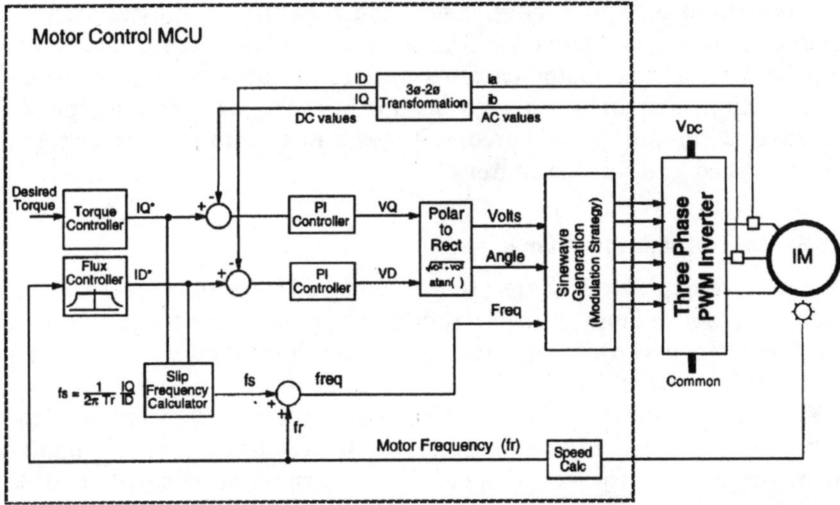

Figure 8.41 Vector control block diagram.

8.3.7 Sensorless vector control

The vector control drive's speed sensor often introduces cost, reliability, and noise susceptibility problems to the system. These issues have spurred the development of speed sensorless vector control techniques. In general, speed sensorless techniques utilize stator voltage sensing with stator current sensing to estimate rotor speed, as seen in Fig. 8.42.

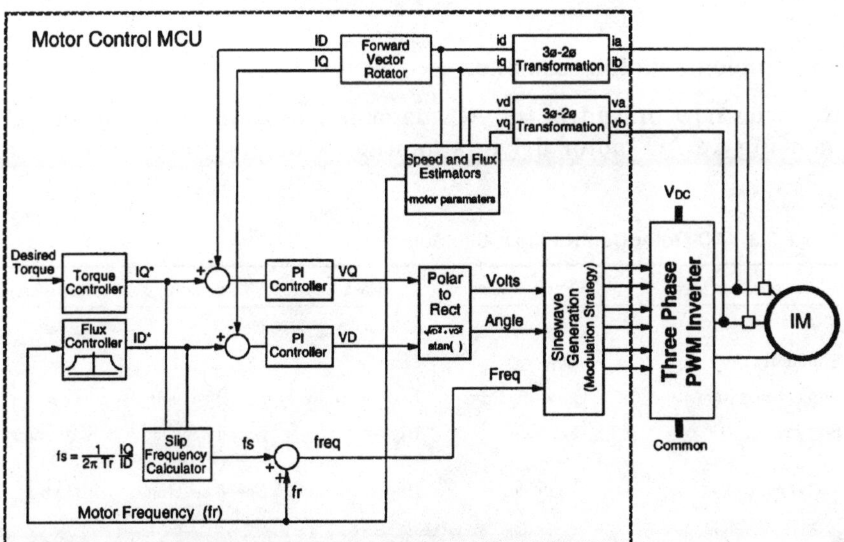

Figure 8.42 Sensorless vector control block diagram.

From these measurements, speed and rotor flux are estimated on the basis of the motor model. The accuracy of the measured quantities, as well as the motor parameters, greatly affects the algorithm. In addition, sensorless drives do not operate well at low speeds because of the small measured voltages and the integration process for the speed and flux estimators.

8.3.8 Control algorithm summary

As we have explained, there are a number of ways to control the AC induction motor. Some of these trade performance for simplicity. Some trade cost for performance. Table 8.3 shows the tradeoffs between the different algorithms.

For the most part, the application will determine the control algorithm. For example, high-horsepower (>10-kW) drives typically implement higher-performance, closed-loop algorithms because the cost of the control electronics is overwhelmed by the cost of the high-wattage motor and the large power devices. Similarly, since low-wattage drives are usually very cost-sensitive, they typically require low-cost, low-performance algorithms which utilize less expensive 8-bit MCUs without expensive sensing elements. The volts-per-hertz (V/F) algorithm is dominant in the AC motor control market today. However, as the cost per performance of the electronics decreases, vector control and sensorless vector control will continue to grow. The bottom line is that there are many acceptable implementations for a given application. The goal is to offer more performance or functionality at the lowest cost.

8.4 Implementation Examples

Knowing how to control the AC motor is one thing. Actually implementing an AC motor drive is another. Outlined below are some

TABLE 8.3 AC Motor Control Algorithm Summary

Algorithm type	Performance	Cost	Sensing	Processing
V/F	Low	Low	None	Low
Slip limit (V/F)	Low	Low	DC current	Low
Slip optimization	Low–medium	Low–medium	Speed-sensing	Low
Slip control	Medium	High	Speed-sensing 2-phase currents	Medium
Vector control	High	High	Speed-sensing 2-phase currents	High
Sensorless vector control	Medium–high	Medium	2-phase voltages 2-phase currents	High

application examples for implementing different types of motor control. The first is an MC68HC08MP16-based V/F drive. The second is an MC68332-based vector control drive. In these examples, various aspects of implementing an AC motor controller are outlined.

8.4.1 Volts-per-hertz system using the MC68HC08MP16

The processing requirements for a volts-per-hertz drive are very low and can be very easily implemented on low-cost 8-bit MCUs. Figure 8.43 illustrates an open-loop volts-per-hertz drive utilizing the MC68HC08MP16 (MP16). The MP16 is an 8-bit MCU that was specifically designed for open-loop V/F motor drives and incorporates all the necessary peripherals. For example, the on-chip motor control PWMs are used to develop the six center-aligned PWMs with deadtime, the analog-to-digital converter is used to sense the speed potentiometer and other miscellaneous analog signals, and the SCI is used for communications. (Chapter 10 gives a more detailed schematic of this MC68HC08MP16 system.)

8.4.1.1 Software implementation. The V/F drive is executed entirely in software. An internal interrupt generated from the on-chip PWM unit triggers an interrupt service routine whose only function is to implement the modulation strategy. The input variables to the interrupt

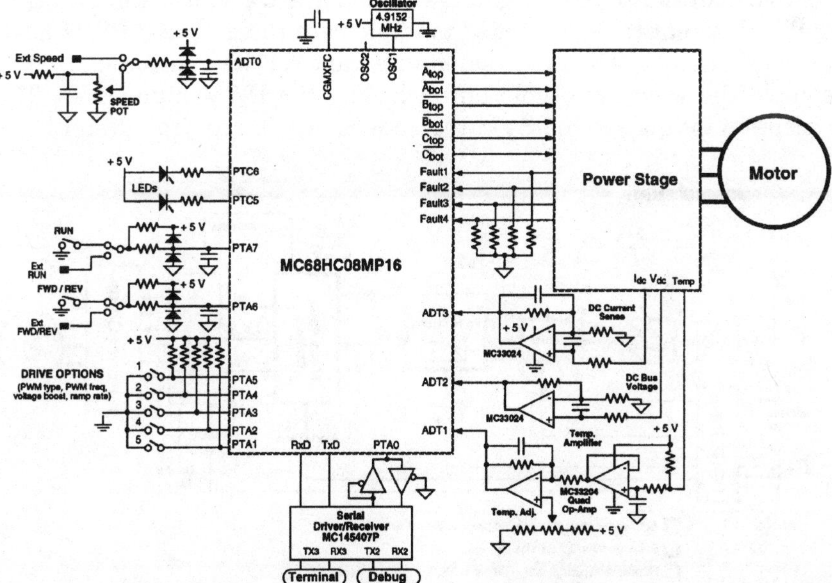

Figure 8.43 Volts/hertz example using the MC68HC08MP16.

service routine are frequency and voltage. The interrupt rate is determined by the PWM switching frequency and, in some sense, the interrupt service routine's execution time. For example, this system implementing 8-kHz PWMs has a periodic interrupt of 125 μs (1 ÷ 8 kHz)—plenty of time for the interrupt service routine's calculations to take place. However, when the switching frequency is changed to 16 kHz (interrupt_rate = 62.5 μs), the interrupt service routine's calculations cannot be processed in this time, and the interrupt rate is extended to two PWM periods (125 μs).

Because the interrupt service routine's only function is to implement the modulation strategy (create the AC waveforms), an outer loop is used to implement the V/F calculations, set the acceleration and deceleration rates (V/F profiler), and read the I/O pins to determine the motor drive options, such as modulation strategy, switching frequency, and low-end voltage boost. The outer loop is responsible for supplying the voltage and frequency values to the interrupt service routine. A block diagram of the control algorithm is shown in Fig. 8.44.

8.4.1.2 Interrupt service routine (PWM modulation). As mentioned in Sec. 8.2 (on modulation methods), the AC waveforms are created by modulating the PWM duty cycles in the shape of the desired modulation strategy (i.e., sine PWM, third-harmonic PWM, 60° PWM). This is done by utilizing a lookup table sampled as part of the interrupt service routine. A key advantage of using a waveform table is that only a different table is needed to implement the different PWM modulation schemes. The exact same interrupt service routine is used to sample the waveform table and calculate the PWM high times. The exception to this are the SVM techniques, which require different cal-

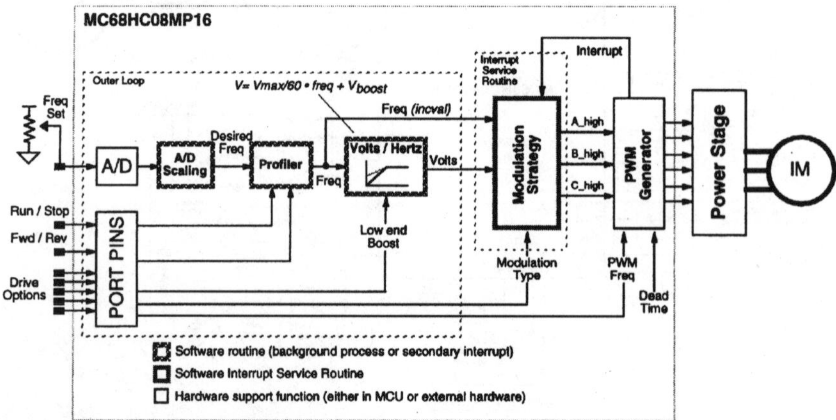

Figure 8.44 V/F software control loops.

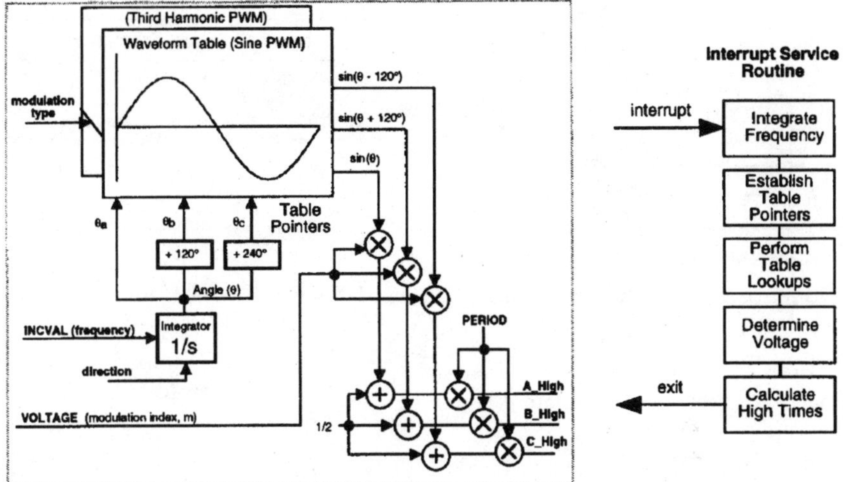

Figure 8.45 Interrupt service routine (ISR).

culations. The interrupt service routine's calculations and flowchart for implementing PWM techniques are shown in Fig. 8.45.

Integrator and table pointers. Sampling of the waveform table is accomplished by using an angle variable acting as a pointer into the waveform table. At every interrupt, the angle variable is incremented and a table lookup is performed based on the new angle:

$$\text{New angle} = \text{old\_angle} + \text{increment\_value} \tag{8.10}$$

This incrementing performs the integration function. The requested frequency, with the interrupt rate and angle variable's bit size, determine the amount the angle variable is incremented (increment_value) per the following equation:

$$\text{Increment value} = \text{frequency} \cdot \text{interrupt\_rate} \cdot 2^{\text{NUMBER\_OF\_BITS}} \tag{8.11}$$

A nice side effect of the MCU is that the circular binary number system corresponds well with the angular number system. For example, a 16-bit angle variable will roll over from \$FFFF to \$0000 corresponding with a rollover from 2π to 0 and a rollover to the beginning of the waveform table. In addition, the 120° offsets between phases can be accomplished by simply adding 120°(\$5555) to the angle variable. Therefore, angle_C = angle_A + \$5555 and angle_B = angle_A + \$AAAA.

Table lookup and interpolate. Since the angle variable is used as a pointer into the waveform table, the bit size of the angle variable in

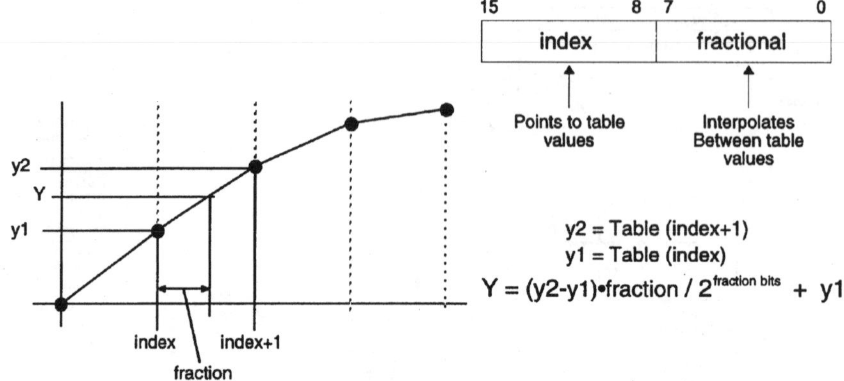

Figure 8.46 Table lookup and interpolate.

many ways determines the size of the table. For example, a 16-bit angle variable equates to 65536 (2^{16}) points in a cycle or 65536 points in the table. Generating a large waveform table, however, is not practical. To combat this, a smaller physical table is often used with linear interpolation. A 256-point physical table, for example, can be extended to an equivalent 65,536 points by using interpolation. This is done by segmenting the angle variable into an index portion (table index) and fractional portion (interpolator). The index portion is used to read the current table value and the subsequent table value, and the fractional portion is used to interpolate the value between these points. Figure 8.46 illustrates this.

In the MC68HC08MP16 implementation, 8 bits are used to index into the table and another 8 bits are used for the interpolation. This means that a 256-point physical table is required. It should be noted that linear interpolation can result in a degradation of accuracy, especially with a small physical table. However, much of the accuracy is dependent on the achievable voltage resolution and the error due to interpolation may be overshadowed by the voltage resolution. (This is reviewed in more detail later in this chapter.)

Table lookup without interpolation (90° table). Because voltage resolution plays such a big role in actual achievable accuracy, it may mean wasted calculation time to implement interpolation when these values cannot be realized anyway. With this in mind, interpolation can be avoided by simply utilizing the index portion of the angle variable to read the table directly. The fractional portion can either be discarded or rounded into the index portion. For an 8-bit table with 256 points and an 8-bit PWM, the error without interpolation is negligible.

Another method to reduce the size of the table and eliminate the need for interpolation is to implement only 90° of the table. The software would then need to check the angle variable to determine whether to step forward or backward through the table and whether to negate the table values. With this method, interpolation is avoided at the expense of more complicated software for angle checking.

Voltage determination and PWM high time calculations. After the table lookup and interpolation has been completed for all three phases, the voltage variable must be factored into the equations, and the resulting values must be converted to PWM high time values. These calculations are as follows:

$$\text{A\_HIGH} = \text{voltage} \cdot \text{va} \cdot \text{PWMperiod}/2 + \text{PWMperiod}/2$$

$$\text{B\_HIGH} = \text{voltage} \cdot \text{vb} \cdot \text{PWMperiod}/2 + \text{PWMperiod}/2$$

$$\text{C\_HIGH} = \text{voltage} \cdot \text{vc} \cdot \text{PWMperiod}/2 + \text{PWMperiod}/2 \qquad (8.12)$$

Two options can be employed to minimize calculation time. First, the table could incorporate waveform values that are already in terms of duty. This has the benefit of eliminating one multiply instruction per phase calculation, but a new table needs to be created for each different switching frequency. If the drive does not need variable switching frequencies, this is viable. The second option multiplies the requested voltage with 50 percent duty cycle only once (voltage · PWMperiod/2) and then multiplies this constant with the value read from the table, eliminating two unnecessary multiplications. Once the PWM high times are determined, the PWMs are updated and the interrupt service routine ends.

8.4.1.3 Outer-loop processing. The outer loop is used to read the drive options and A/D converters and to implement the *V/F* algorithm. The outer loop executes in an infinite manner. Figure 8.47 shows a flowchart of the outer-loop processing.

Setting the drive options is accomplished by reading the A/D and necessary port pins and updating the required variables: frequency, V_{boost}, modulation type, and switching frequency.

The V/F profiler takes the desired frequency and implements an increment of the current frequency each time through the outer loop if the new frequency and current frequency are different. The increment value determines the acceleration rate. A final comparison is used to ensure that the actual frequency does not exceed the desired set frequency. The final step in the outer loop is to execute the *V/F*

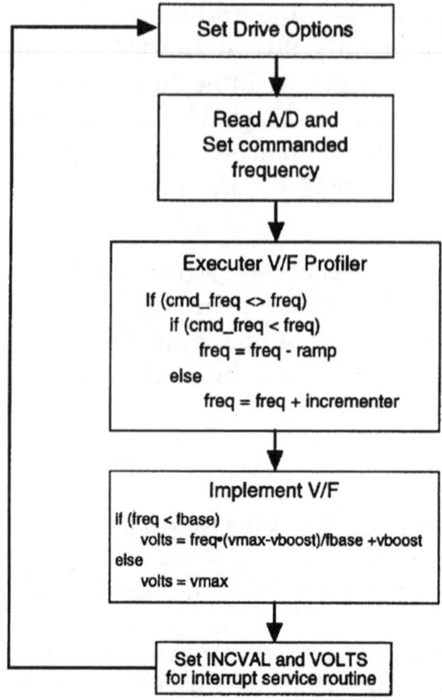

Figure 8.47 V/F outer-loop processing.

algorithm by the simple equation [Eq. (8.13)]. (The software program for a V/F system is shown in App. I.)

$$\text{Volts} = \text{v}_{\text{max}} \cdot \frac{\text{frequency}}{\text{f}_{\text{base}}} + \text{v}_{\text{boost}} \tag{8.13}$$

8.4.2 Example vector control system using the MC68332G

Vector control systems can vary greatly in implementation. A simple implementation of a vector control system utilizing the MC68332G is shown in Fig. 8.48. (Note that this design was also used in an electric vehicle; see Chap. 6.)

The time processing unit (TPU) on the MC68332G is used to generate the six center-aligned PWMs, as well as to perform the speed measurement. Although the TPU was not designed specifically for AC motor control, the TPU is a match/capture type of timer with a dedicated microengine to service the match/capture events. A time function known as *multichannel PWM* (MCPWM) is used to generate the six center-aligned PWMs with deadtime. This implementation requires external exclusive-OR gates to create the PWMs. Because no shutdown circuitry is incorporated, external hardware is also used for PWM enabling and protection.

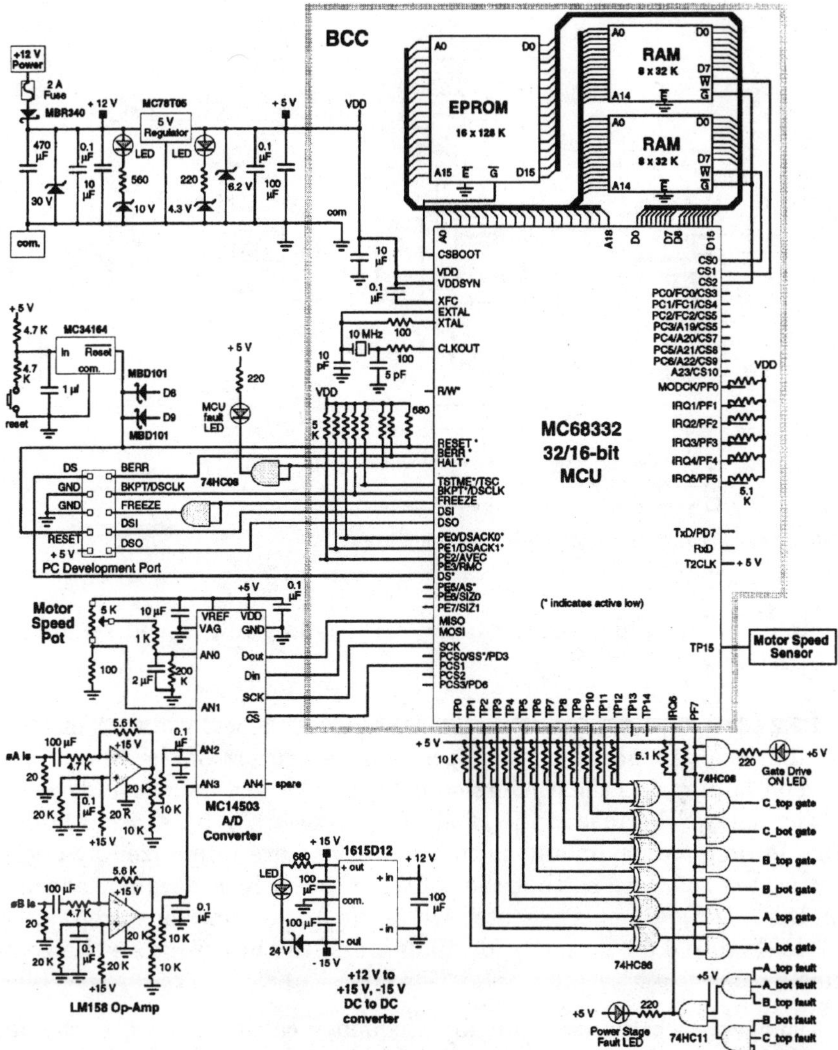

Figure 8.48 Example of vector controller using the MC68332.

8.4.2.1 Software implementation. The critical loop of the vector control system is implemented as an interrupt service routine. This loop performs the current measurements and transformations, the speed sensing and calculation, the PI controllers, slip calculator, modulation strategy, and any fault checking. The vector control loop is shown in Fig. 8.49. This system executes the interrupt service routine every 244 μs, which is two PWM periods (8.2-kHz switching). The outer loop is used to implement the torque and flux controllers, which are

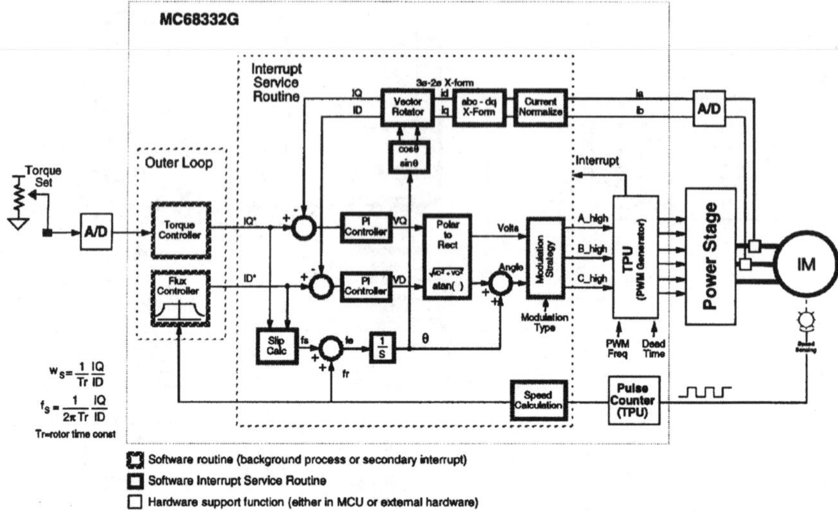

Figure 8.49 Vector control software processing loops.

simple lookup tables based on the A/D reading and speed readings, respectively.

8.4.2.2 Interrupt service routine. Stability and performance of this system are dependent on the speed and execution of the inner loop and on minimizing the conversion delays, transport delays, and calculation delays. Because of this, fast A/D converters ($<5\ \mu s$) are desired. In addition, transport and calculation delays are minimized by organizing the software so that the software code which does not depend on the A/D readings is executed while the A/D conversions are taking place. Figure 8.50 shows a flow diagram of the interrupt service routine.

Slip frequency calculator. The slip frequency calculator is the critical block for a vector control system. The slip frequency ensures correct field orientation between rotor flux and torque current. This is based on the I_Q and I_D currents, as well as the rotor open-circuit time constant (L_r/R_r). The equation for the slip calculator is given as

$$f_s = \left(\frac{1}{2\pi T_r}\right) \cdot I_Q/I_D \qquad (T_r = L_r/R_r = \text{rotor time constant}) \qquad (8.14)$$

If the correct value of the rotor time constant (L_r/R_r) is not known or is not correctly achieved, the vector control algorithm will operate in a "detuned" manner. This means that the slip frequency calculation will be incorrect and the flux and torque current are improperly

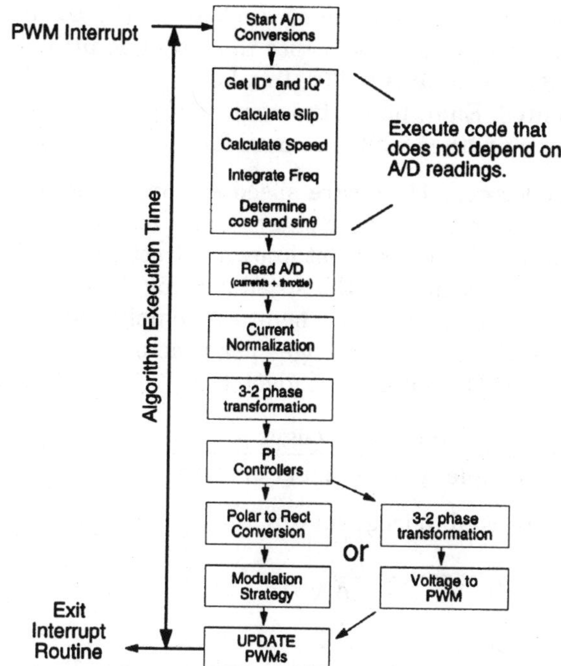

Figure 8.50 Interrupt service routine for vector controller.

aligned with the rotor flux; they are not truly decoupled. In other words, a change in torque current would cause a change in rotor flux—instantaneous torque control is not achieved. Depending on the importance of torque response, this may be acceptable. In the electric vehicle traction drive application in which this example is implemented, operating in a detuned manner is not critical.

Integrator. The integrator takes the frequency value and converts it to an angle for determining cos θ and sin θ for the vector rotators. The integration operation is done in the exact same way as it was done in the previous V/F example. Because this system has a 16-bit angle variable and a sample rate of 244 μs, the frequency resolution is 0.0625 Hz as calculated here:

$$\text{Frequency resolution} = \frac{1}{2^{\text{BITS}} \cdot \text{sample rate}} \tag{8.15}$$

cos θ and sin θ. The vector rotators in the 3–2 phase transformation and the 2–3 phase transformation require cos θ and sin θ. This is a lookup table and is implemented in exactly the same way as in the

V/F controller. Only one table needs to be created, since the cosine function is a 90° shift of the sine function. Since this example utilizes a 16-bit angle variable, $4000 equates to 90°. In addition, a sine table of 256 points is implemented. Eight bits of the angle variable are used to index into the table, and another 8 bits are used for interpolation.

Speed measurement and calculator. Hardware speed sensing is accomplished with a toothed disk on the rotor shaft. Only one signal is detected, so the direction of the motor is not known. Direction can be determined by having two signals from the speed sensing device and using a quadrature decoder. Speed measurement is accomplished by a pulse-type counter, where the number of edges per time are counted to give rotations per minute. The equation is given by

$$\text{rpm} = 60 \cdot \frac{\text{measured edges}}{\text{sample time} \cdot \text{number of teeth}} \tag{8.16}$$

The rpm value is then converted to frequency by

$$f_r = \text{rpm} \cdot \frac{p}{120} \tag{8.17}$$

where p is the number of poles.

The precision of the speed sensor in this system is 1 Hz. Better accuracy can be achieved by increasing the number of teeth on the disk or by slowing down the speed-sensing sample time. However, slowing down the sample time will introduce a delay in speed sensing and can cause stability problems. Another way to get better accuracy at low speeds is to measure the time between edges, which will be explained later in this chapter.

Current normalization. Current is sampled with the A/D converter. Because the A/Ds accept only a positive voltage level and the conversion result is unsigned, the A/D reading needs to be normalized. The software converts the unsigned A/D readings into positive and negative values by subtracting the zero-current reading from the present phase-current reading. The zero-current reading can be obtained from initial start-up before the drive is enabled. Normalization is attained by left-shifting the values so that full scale of the MCU data size is achieved. As a result, full current at the motor is equal to a signed fractional number whose number can be between -1 and 1.

3–2 phase transformation. The three- to two-phase transformation is used to convert the measured AC phase currents, i_a and i_b, into the two DC current components, I_Q and I_D. The conversion is composed of the transformation from a-b-c to d-q and a vector rotator, as defined in Fig. 8.51.

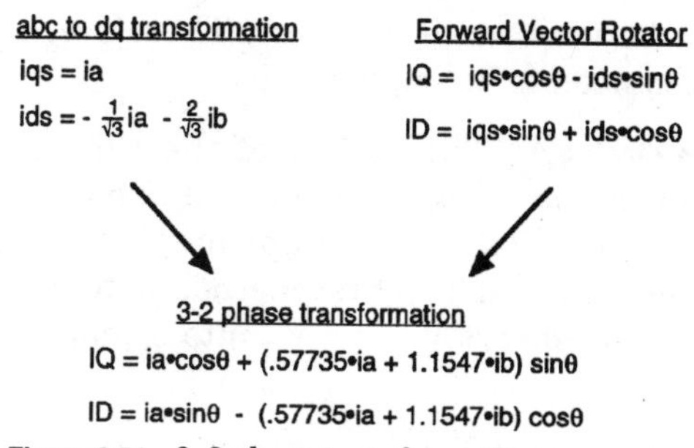

Figure 8.51 3–2 phase transformation.

PI current controllers. The two PI regulators control the current components, I_Q* and I_D*, defined by the torque and flux controllers. These are standard PI controllers of the form shown in Fig. 8.52.

The output voltage is the control variable which is adjusted to ensure zero error in actual and commanded current. In terms of the MCU, the PI controllers are defined as difference equations. Using $s = (1-z^{-1})$, the difference equations [Eqs. (8.18) and (8.19)] for the PI controllers become

$$V_Q = V_Q \cdot z^{-1} + K_p(\text{IQ\_err} - \text{IQ\_err} \cdot z^{-1}) + \text{Ki} \cdot \text{IQ\_err} \qquad (8.18)$$

$$V_D = V_D \cdot z^{-1} + K_p(\text{ID\_err} - \text{ID\_err} \cdot z^{-1}) + \text{Ki} \cdot \text{ID\_err} \qquad (8.19)$$

The z^{-1} is a delay operator. Therefore, $(V \cdot z^{-1})$ and $(\text{error} \cdot z^{-1})$ are previous values of the voltage and error, respectively. The PI controllers can be tricky to implement in an MCU because of the possibility for the numbers to roll over. A test needs to be put in place to ensure that the output voltages saturate to maximum limits. In addi-

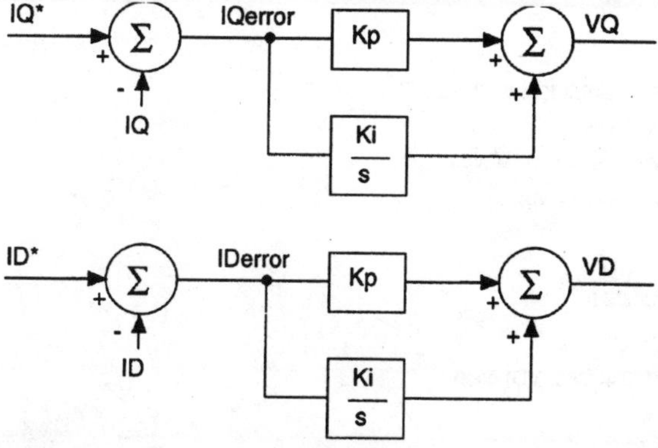

Figure 8.52 PI controllers.

tion, the Kp and Ki gains might be very small numbers. Therefore, a normalization procedure may also need to be implemented.

2–3 phase conversion (modulation strategy). The output of the PI controllers results in voltage. However, the voltage is split into two components: V_Q and V_D. There are two possibilities for implementing the modulation strategy based on these values. The first method is to use a rectangular-to-polar conversion to convert V_Q and V_D into a magnitude and an angle:

$$V_{mag} = (V_Q^{\,2} + V_D^{\,2})^{-1/2}$$

$$V_{angle} = a \tan \frac{V_Q}{V_D} \tag{8.20}$$

The magnitude would feed the modulation strategy block directly. The angle is added to the angle which was the integrator output. This resulting angle would be used to implement the modulation strategy in the standard sense, specifically, with a lookup table. In this way, any modulation strategy could be implemented, such as sine PWM, third-harmonic, or SVM.

If only sine PWM is desired, a 2–3 phase transformation can be implemented. This is similar to the 3–2 phase transformation that was used on the measured currents. The 2–3 phase transformation consists of the reverse vector rotator and a d-q → a-b-c transformation, as shown in Fig. 8.53.

The values v_a, v_b, and v_c would then need to be converted to PWM high times using the following relationships:

$$\text{A\_HIGH} = \text{va} \cdot \text{PERIOD/2} + \text{PERIOD/2}$$

$$\text{B\_HIGH} = \text{vb} \cdot \text{PERIOD/2} + \text{PERIOD/2}$$

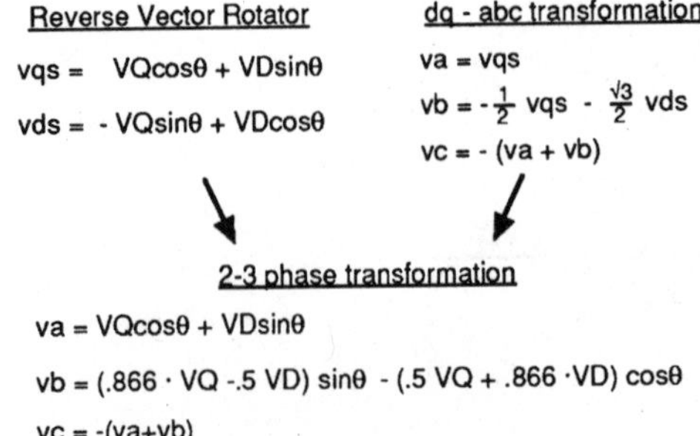

Reverse Vector Rotator
vqs = VQcosθ + VDsinθ
vds = - VQsinθ + VDcosθ
dq - abc transformation
va = vqs
vb = -1/2 vqs - √3/2 vds
vc = - (va + vb)
2-3 phase transformation
va = VQcosθ + VDsinθ
vb = (.866 · VQ -.5 VD) sinθ - (.5 VQ + .866 ·VD) cosθ
vc = -(va+vb)

Figure 8.53 2–3 phase transformation.

$$C\_HIGH = vc \cdot PERIOD/2 + PERIOD/2 \tag{8.21}$$

8.4.2.3 Outer-loop processing (torque and flux controllers). Where the interrupt service routine implements the current regulators and vector control strategy, the outer loop reads the torque input and supplies the commanded control currents, $I_Q{}^*$ and $I_D{}^*$. This is done through the torque and flux controllers. The torque and flux controllers can be implemented in a number of ways. In this simple implementation, lookup tables are used. The flux value is held constant until the point of field weakening. At this point, the flux current is decreased inversely proportional to the speed. If flux weakening is not required, the flux controller simplifies to a constant value. In this system, the torque controller is a simple lookup table. In a speed control system, the torque controller may be a proportional-integral (PI) controller closing the loop on speed. Current limiting is accomplished through these tables so that the current magnitude of the resulting vector does not exceed the maximum stator current:

$$I_s = (I_Q{}^2 + I_D{}^2)^{-1/2} \tag{8.22}$$

8.4.3 Implementing SVM techniques

In both the vector control system and the V/F system, standard PWM modulation strategies were employed. However, SVM techniques can be substituted for the PWM modulation strategies. As discussed before (Sec. 8.2), SVM techniques can be attained with PWMs and the MCU. Implementing SVM techniques is very similar to PWM techniques in that the interrupt routine must integrate the frequency and calculate the PWM high times. One difference is that the sine table requires only 60° of the sine function. (*However, it may be easier to implement 90° of the sine function.*)

The high time calculations are based from sector information of V_{ref}. This requires five comparison instructions. Once the sector information is defined, the state times (T_1, T_2, T_0) are calculated per the calculations in Fig. 8.54, and the high times are determined on the basis of Table 8.1. The calculations for T_1, T_2, and T_0 can be done in a common subroutine. The flowchart in Fig. 8.54 shows the SVM implementation. In addition to utilizing only a 90° sine table, the SVM calculations require one less table lookup (or lookup table) and one less multiply instruction than in conventional PWM methods. (A software routine for implementing space vector modulation is shown in App. F.)

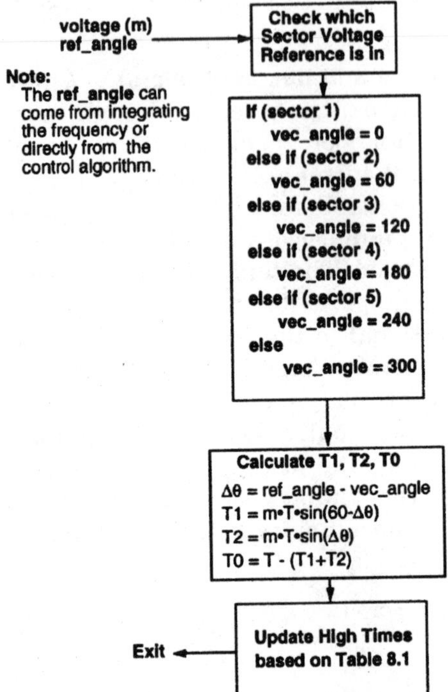
voltage (m)
ref_angle
Note:
The ref_angle can come from integrating the frequency or directly from the control algorithm.
Check which Sector Voltage Reference Is In
If (sector 1)
vec_angle = 0
else if (sector 2)
vec_angle = 60
else if (sector 3)
vec_angle = 120
else if (sector 4)
vec_angle = 180
else if (sector 5)
vec_angle = 240
else
vec_angle = 300
Calculate T1, T2, T0
Δθ = ref_angle - vec_angle
T1 = m•T•sin(60-Δθ)
T2 = m•T•sin(Δθ)
T0 = T - (T1+T2)
Update High Times based on Table 8.1
Exit

Figure 8.54 SVM interrupt service routine.

8.5 AC Motor Drive Implementation Considerations

There are many factors involved in implementing AC motor drives. In this section we outline some things that need to be taken into consideration.

8.5.1 PWM bit resolution (voltage resolution)

PWM resolution has a dramatic effect on output waveform creation. For example, 8 bits of PWM resolution will have 256 different voltages levels for creating the sine wave. For a 300-V bus, this is 1.17 V of resolution. For full-voltage operation, this may not be a big issue, but for lower voltages this causes very crude steps in the creation of the sinusoid. Sixteen bits of resolution results in a voltage resolution of 0.004 V for the same 300-V DC bus. Although PWM timers are categorized by their bit size (i.e., 12-bit PWM timer), the usable resolution of the PWM timer is a function of the timer resolution (timer clock) and the switching frequency. Table 8.4 outlines timer resolution, switching frequency, and usable PWM bit resolu-

TABLE 8.4 PWM Bit Resolutions versus Timer Resolutions and Switching Frequencies

PWM* resolution, bits	Timer clock resolutions				
	240.0 ns (4.2 MHz)	120.0 ns (8.3 MHz)	60.0 ns (16.7 MHz)	40.0 ns (25.0 MHz)	30.0 ns (33.3 MHz)
8	16.276 kHz	32.552 kHz	65.104 kHz	97.656 kHz	130.21 kHz
9	8.138 kHz	16.276 kHz	32.552 kHz	48.828 kHz	65.104 kHz
10	4.069 kHz	8.138 kHz	16.276 kHz	24.414 kHz	32.552 kHz
11	2.035 kHz	4.069 kHz	8.138 kHz	12.207 kHz	16.276 kHz
12	1.017 kHz	2.035 kHz	4.069 kHz	6.104 kHz	8.138 kHz
13	0.509 kHz	1.017 kHz	2.035 kHz	3.052 kHz	4.069 kHz
14	0.254 kHz	0.509 kHz	1.017 kHz	1.526 kHz	2.035 kHz
15	0.127 kHz	0.254 kHz	0.509 kHz	0.763 kHz	1.017 kHz
16	0.064 kHz	0.127 kHz	0.254 kHz	0.381 kHz	0.509 kHz

*PWM resolutions depend on switching frequencies and timer clock resolutions.

$$\text{Switching frequency} = \frac{1}{2^{\text{bits}} \cdot \text{timer resolution}}$$

tion. Better resolutions are achieved with a lower switching frequency or a faster PWM timer.

It should be understood that a 16-bit PWM which is driven from a slow clock may not utilize all the available bits for a given switching frequency. Because of this, timer clock frequency is a very important parameter when selecting a PWM timer. The number of bits necessary is dependent on the application, but fewer bits will affect control stability in closed-loop systems. As a reference, the trend is to have bit resolutions in excess of 10 bits.

8.5.2 Effects of the angle variable's bit size on frequency resolution

The *frequency resolution* is defined as the fineness of the step at which it can be adjusted. Generally, smoother control can be achieved with finer frequency resolution. Frequency resolution is established by the number of bits in the angle variable and the interrupt rate:

$$\text{Frequency resolution} = \frac{1}{2^{\text{number of bits}} \cdot \text{interrupt\_rate}} \tag{8.23}$$

Even though the interrupt rate affects frequency resolution, it is usually set by other factors. Therefore, the size of the angle variable is the only practical way to adjust frequency resolution. Table 8.5

TABLE 8.5 Angle Variable Bit Size and Frequency Resolution

Bits	Table size	Frequency resolution* $\Delta T = 125\ \mu s$	$\Delta T = 250\ \mu s$	$\Delta T = 500\ \mu s$
8	256	31.25 Hz	15.625 Hz	7.8125 Hz
10	1024	7.8125 Hz	3.90625 Hz	1.953125 Hz
12	4096	1.953125 Hz	0.9765625 Hz	0.48828125 Hz
14	16384	0.48828125 Hz	0.24414063 Hz	0.12207031 Hz
16	65536	0.12207031 Hz	0.06103516 Hz	0.03051758 Hz
18	262144	0.03051758 Hz	0.01525879 Hz	0.00762939 Hz
20	1048576	0.00762939 Hz	0.0038147 Hz	0.00190735 Hz
22	4194304	0.00190735 Hz	0.00095367 Hz	0.00047684 Hz
24	16777216	0.00047684 Hz	0.00023842 Hz	0.00011921 Hz

*Frequency resolution $= \dfrac{1}{(2^{\text{bits}} \cdot \Delta T)}$.

illustrates that a larger angle variable will result in better frequency resolution. However, as mentioned previously in this section, this may be at the expense of a larger table.

8.5.3 Effects of sample rate (interrupt rate) on waveform creation

As indicated in the previous section, the interrupt rate affects frequency resolution. In addition, the interrupt rate affects the number of samples used to create a high-frequency waveform. As the interrupt rate is decreased, the number of samples in producing a high-frequency waveform is decreased. (See Table 8.6.)

TABLE 8.6 Update Rate and Number of Samples

Output frequency,* Hz	Number of samples† $\Delta T = 63\ \mu s$	$\Delta T = 125\ \mu s$	$\Delta T = 250\ \mu s$	$\Delta T = 500\ \mu s$
500	32	16	8	4
400	40	20	10	5
300	53	27	13	7
200	80	40	20	10
100	160	80	40	20
50	320	160	80	40

*Frequency = 1/(samples · ΔT).
†Samples = 1/(frequency · ΔT).

Speeding up the interrupt rate (higher PWM frequency) will produce better high-frequency waveforms. In addition, control loop stability can be improved by speeding up the interrupt rate. However, the MCU's processing capability may limit this. The desired output PWM frequency is typically dependent on other factors such as audible noise and inverter switching losses.

8.5.4 Speed sensing and calculating considerations

In the slip control and vector control algorithms, speed sensing is a requirement. Normally, the speed-sensing element consists of a toothed or slotted disk that feeds a pulse train or series of pulse trains (quadrature encoding) back to the speed calculator. (*Quadrature decoding is necessary to determine the direction of the motor.*) The speed can be calculated from these signals in two ways: (1) the time between edges can be measured; and (2) the number of edges can be counted in a given time period. Both methods are useful and may be used for different situations.

Counting the time between edges is the most common technique and works well, especially at low rpm, but is very sensitive to errors such as shaft jitter or encoder tolerances at higher rpm levels. With this method, the calculation is dependent on the number of edges (teeth) in a revolution, the number of measured timer ticks between edges, and the timer resolution. The resolution is given by

$$\text{rpm} = \frac{60}{\text{measured ticks} \cdot \text{timer resolution} \cdot \text{number of teeth}} \tag{8.24}$$

From rpm, rotor frequency is calculated by

$$f_r = \text{rpm} \cdot \frac{p}{120} \tag{8.25}$$

The results in rotor frequency are given by

$$f_r = \frac{p/2}{\text{measured ticks} \cdot \text{timer resolution} \cdot \text{number of teeth}} \tag{8.26}$$

A curve of rpm versus timer counts is shown in Fig. 8.55. Low-speed resolution is determined by the timer's bit size. For example, a 16-bit timer with a 4-MHz clock (250-ns timer resolution) and a 256-tooth speed-sensing element could resolve down to an rpm of 60/(65535 · 250 ns · 256) = 14.3 rpm. For a four-pole motor, this is equal to 14.3 rpm/30 = 0.48 Hz. Figure 8.55 also illustrates that the relationship between rpm and timer counts is not a linear function. Therefore, at increasing frequencies (high rpm), the count values

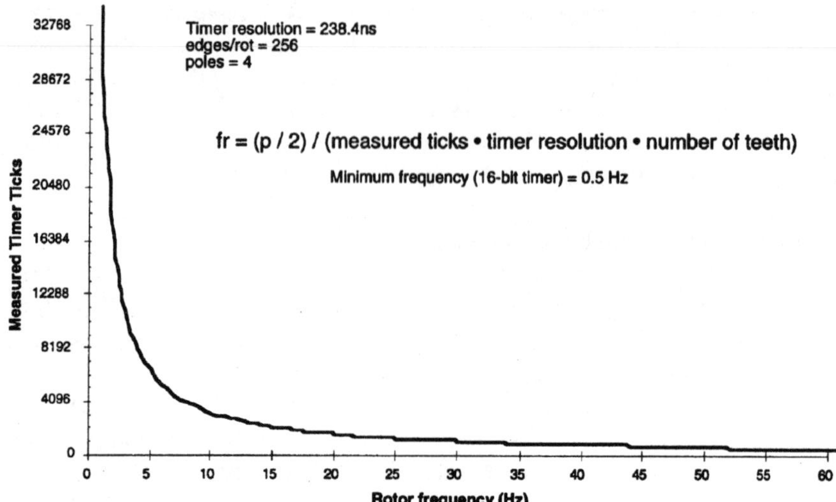

Figure 8.55 Speed sensing by measuring time between edges.

become increasingly small, and a one timer tick error could result in a large discrepancy in measured rotor frequency.

8.5.4.1 Measuring edges. Counting the number of edges in a given time period is the other speed calculation method. The speed is determined by the number of teeth (pulses) in a revolution, the sample time, and the number of measured pulses. The relationship is given by

$$\text{rpm} = \frac{60 \cdot \text{measured edges}}{\text{sample time} \cdot \text{number of teeth}}$$

$$f_r = \left(\frac{p}{2}\right) \cdot \frac{\text{measured edges}}{\text{sample time} \cdot \text{number of teeth}} \tag{8.27}$$

The key benefit of this method is that the function is linear, as shown in Fig. 8.56. Therefore, the same resolution applies to all speeds. The resolution is determined by one measured edge and the sample time. As an example, a 256-tooth speed-sensing element sampled every 7.8 ms would have a speed resolution of $60 \cdot 1/(0.0078 \cdot 256) = 30.05$ rpm. For a four-pole motor, this would equate to a frequency resolution of $30.05/30 = 1.002$ Hz. Better speed resolution can be achieved by increasing the sample time. However, since speed sampling time can affect the control system, there is a tradeoff between speed resolution and sample time. Many systems implement both types of speed calculations. At low frequencies (rpm), the first method is used to achieve better frequency resolution at the low end. At high frequencies, the

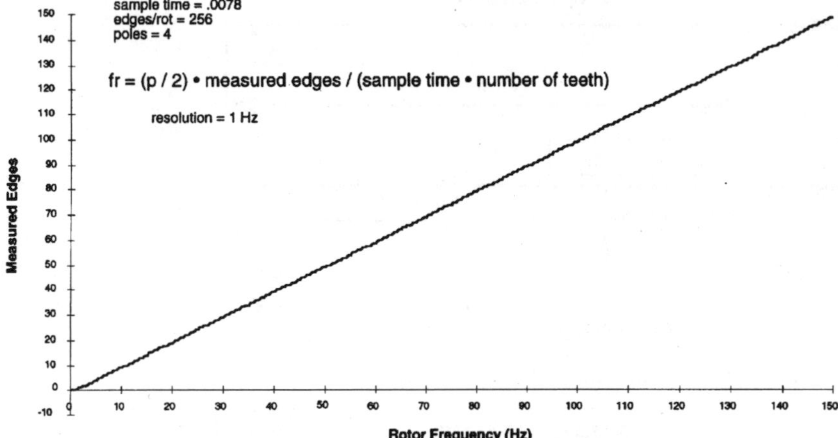

Figure 8.56 Speed sensing by counting edges in a sample period.

second method is used, since the time between edges may be too small to accurately measure.

8.5.5 Event timing

In a closed-loop system, it is desired to minimize any system delays that can result in system instabilities. In the vector controller, the inner loop must sample the A/D converters, implement the calculations, and update the high times to the PWMs. For maximum performance of the system, all these should be synchronized and minimized. Therefore, an event timer would be beneficial to trigger the A/D converters and the interrupt service routine; the conversions are complete when the interrupt service routine expects them, and the interrupt service routine is complete when the PWM unit is ready to update the high times for the next cycle. (See Fig. 8.57.)

8.5.6 PWM switching frequency and deadtime distortion

AC motor drives utilize switching frequencies that range from 2 to 20 kHz. Determining the PWM switching frequency is dependent on a number of factors. Reducing audible noise is one factor that leads to high switching frequencies. Reducing inverter switching losses leads to lower switching frequencies. In addition, switching frequency has a direct impact on waveform distortion caused by deadtime. Deadtime distortion causes torque pulsations, additional harmonics, and loss of voltage utilization. The degree of distortion is dependent on the

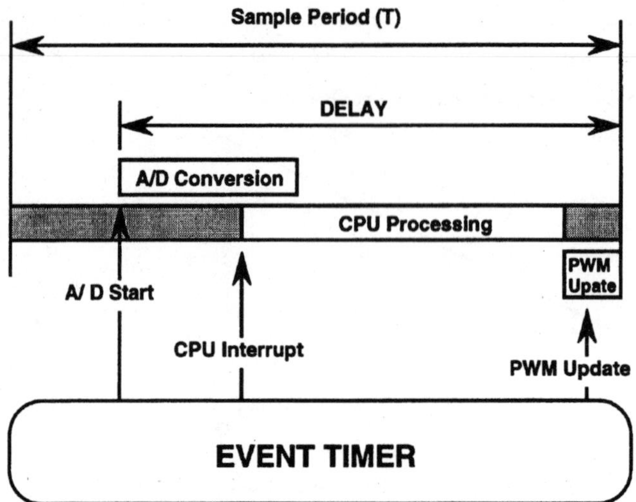

Figure 8.57 Event scheduling.

switching frequency and the amount of deadtime. The percentage of deadtime for a PWM period determines the distortion. A smaller deadtime region and a lower switching frequency result in less deadtime distortion. Unfortunately, the deadtime requirement is set by the hardware. Therefore, to minimize deadtime distortion, a lower switching frequency is needed. (For methods to correct for the distortion caused by deadtime, see Chap. 9.)

Summary

The implementation of digital controls for AC induction motors requires an understanding of microcontrollers, software design, and AC induction motor operation. In this chapter we have discussed the basic principles needed for MCU-based AC motor drives. The AC induction motor drive is still evolving, in addition to higher-performance MCU and power modules. Key points in this chapter are listed below.

- The inverter stage develops the variable-frequency and variable-voltage AC from the DC voltage. The DC voltage is obtained from the Ac power line with a converter stage.
- Typical motor drive inverters use power IGBT and FWD modules, filter capacitors, gate drivers, optoisolators, current sensors, and EMI filter networks.
- Digital control techniques simplify the implementation of control algorithms and make more sophisticated algorithms feasible. In

addition, digital control techniques allow the control algorithm to be easily modified through software without the need for component changes or costly board redesign.

- Center alignment not only reduces the likelihood that more than one power device will switch at any given time, thereby minimizing noise in the system, but also causes a frequency doubling effect in the motor, which reduces current ripple and acoustic noise.
- Sinusoidal PWM is a common AC motor drive PWM method, while third-harmonic PWM and SVM techniques have gained popularity because of their increased bus utilization.
- SVM techniques can reduce current ripple and can offer lower switching losses than standard PWMs. The ability to use standard PWM units to implement SVM techniques also helps to increase their viability.
- There are two basic parts in the AC motor control software: (1) variable AC waveform generation and (2) control of the voltage and frequency of these AC signals.
- The basic AC induction motor algorithms are volts per hertz, slip control, and vector control (field orientation).
- The application will determine the control algorithm. Large motor drives (>10 kW) typically implement higher-performance, closed-loop algorithms. Low-wattage motor drives are usually very cost-sensitive and use low-performance algorithms that utilize less expensive 8-bit MCUs without expensive sensing elements. Volts per hertz is dominant in the AC motor control market today. However, as the cost per performance of the electronics decreases, vector control and sensorless vector control will increase.
- The PWM element's bit size and time resolution, interrupt rate, rpm speed sensing, event timing, and deadtime distortion effects are all considerations when implementing an AC motor drive.

Further Reading

Bose, B. K., *Power Electronics and Variable Frequency Drives,* IEEE Press, 1997.

Koyama, M., Yano, M., Kamiyama, I., and Yano, S., *Microprocessor-Based Vector Control System for Induction Motor Drives with Rotor Time Constant Identification Function,* IEEE, 1985.

Pinewski, P., *Understanding Space Vector Modulation,* EDN Products ed., March 7, 1996.

Plunkett, B. K., "Direct Flux and Torque Regulation in a PWM Inverter-Induction Motor Drive," *IEEE Transactions on Industry Applications,* March/April 1977.

Rajashekara, K., Kawamura A., and Matsuse K., *Sensorless Control of AC Motor Drives,* IEEE Press, 1996.

Rashid, M. H., *Power Electronics—Circuits, Devices and Applications,* Prentice-Hall, 1988.

University of Wisconsin, Madison, *Dynamics and Control of AC Drives,* Course Notes, May 16–19, 1994.

Valentine, R., and Pinewski, P., *Electronics for Electric Vehicle Motor Systems,* SAE 1995 FTTC Conference 951888.

Valentine, R., Pinewski, P., and Huettl, T., *Advanced Motor Control Electronics,* SAE 1996 FTTC Conference 961693.

Chapter

9

Deadtime Distortion Correction

David Wilson

Motorola Semiconductor Products

Introduction

As an aspiring musician, I sometimes add a little distortion to my guitar amp to make the sound more interesting, or to cover up my mistakes. However, when dealing with a drive connected to a motor, the best distortion is *no* distortion. As Dr. Ned Mohan put it, the goal is harmony, *not* harmonics! In the case of AC induction motors, you want to achieve the cleanest possible sine waves coming out of the drive.

But first, the bad news. The six-transistor inverter topology commonly used in most voltage-sourced inverters requires that a "deadtime" must be inserted between the turnoff of one transistor in a half-bridge and the turnon of its complementary device. Otherwise, both transistors in a half-bridge may be on momentarily at the same time, which can do some nasty things to a drive. As a result of inserting this deadtime, a distortion is introduced in the output voltage and current waveforms when the inverter is driving an inductive load (such as a motor).

The good news is that this distortion can be satisfactorily corrected in most situations. With the use of a current-sensing device, correction waveforms can be generated which are synchronous to the motor phase currents and applied to the PWM signals. The great news is that Motorola has developed a patent-pending technique to accomplish this without the need for current sensors and has integrated this feature into the MC68HC708MP16 microcontroller! The hope is that, for the first time, the benefits of distortion correction can be brought to the arena of low-cost motor control applications, which cannot afford expensive current-sensing techniques.

9.1 The Problem

To better understand the effects of the distortion, it is necessary to review the equations governing the modulation process. Referring to Fig. 9.1, if we assume that the deadtime is zero and sinusoidal modulation is employed, the averaged or filtered output for unity power supply voltage can be calculated:

$$\overline{V_{o,1}(t)} = \frac{t_h(t)}{T} = \frac{1}{2} + \frac{M}{2} \sin(\omega_o t + \theta) \tag{9.1}$$

where $\overline{V_{o,1}(t)}$ = averaged phase 1 output voltage
$t_h(t)$ = high time of the PWM signal

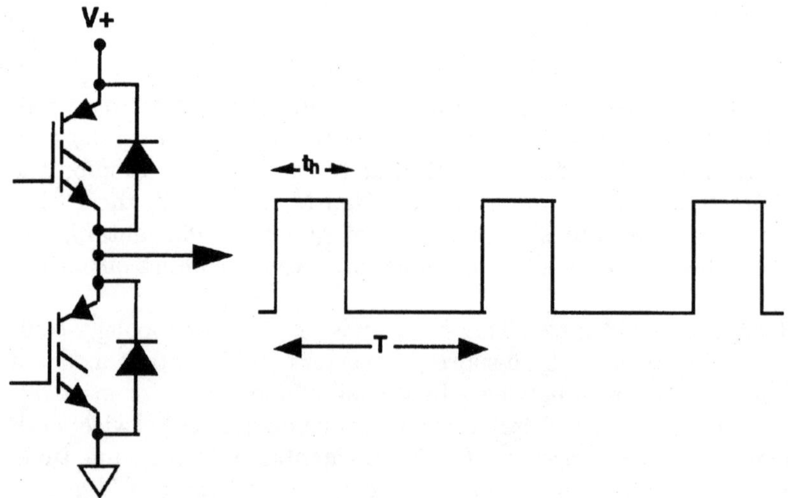

Figure 9.1 PWM output of a typical half H-bridge.

T = PWM period
M = modulation index (from 0 to 1)

It should be pointed out that the analysis presented here can be applied to other modulation waveforms as well. Solving for the high time [Eq. (9.2)], we obtain

$$t_h(t) = \frac{T}{2} + \frac{TM}{2} \sin(\omega_o t + \theta) \tag{9.2}$$

If we could just find a way to keep deadtime equal to zero, I could conclude with the high time equation [Eq. (9.2)] and finish this chapter now (which would make for a pretty boring entry in this book). However, as I already stated, deadtime is a necessary evil in order to prevent shoot-through current. Figure 9.2 illustrates a half-bridge circuit composed of IGBTs which are attempting to generate a desired

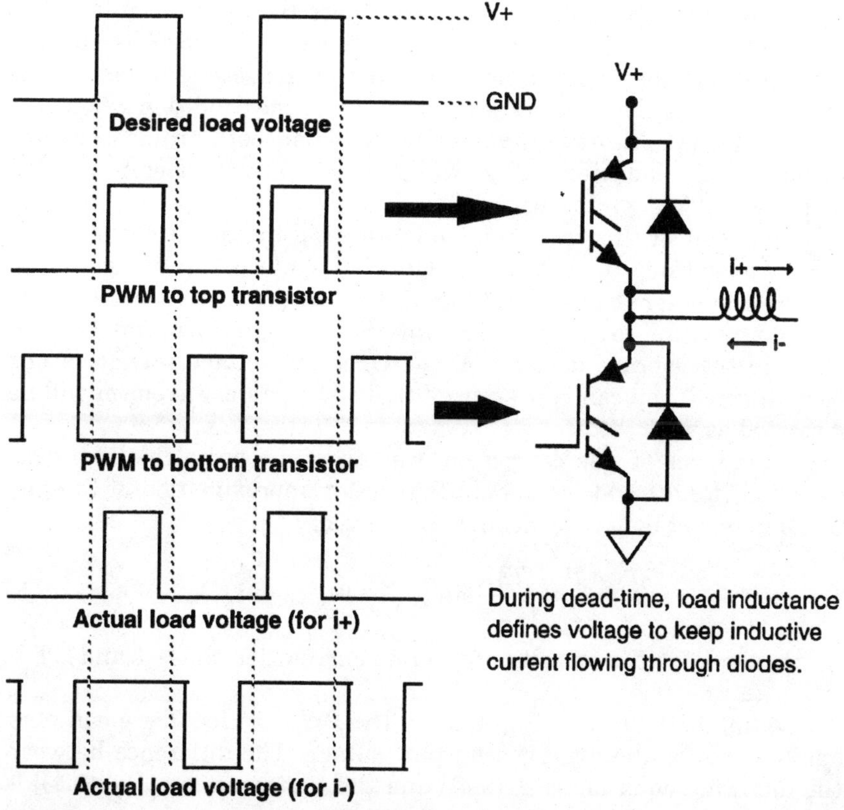

Figure 9.2 Distortion created from load inductance.

PWM output waveform with 50 percent duty cycle. With no deadtime, this could easily be achieved by turning on the top transistor for half of the cycle and then turning the bottom transistor on for the remainder of the cycle. However, with deadtime inserted, the ON time of both the top and bottom transistors are shortened evenly, so that neither the top nor bottom PWM signal corresponds to a 50 percent duty cycle.

What actually happens to the output voltage during deadtime, when both transistors are off, and the inverter is in a high impedance state? It depends on the load characteristics. What if the load has inductive properties, such as a motor winding? Rule 1 about inductors: If current is flowing in an inductor, and you try to interrupt that current, the inductor will fight you. It immediately turns into a voltage source and will achieve whatever voltage is necessary to protect its current flow. (Now you know why the commutation diodes are there; otherwise, the inductor would jack up its voltage so high that they would blow out the transistors.) Therefore, if current is flowing out of the inverter, and both transistors are turned off, the inductor voltage seen by the half-bridge will go negative to keep its current flowing out of the inverter, as illustrated in the second to last waveform of Fig. 9.2. The bottom waveform shows the condition when current is flowing into the inverter. Now, when both transistors are turned off, the inductor voltage will go positive in an attempt to keep its current flowing into the inverter.

As a result of this inductive action during the deadtime interval, the pulse width of the output voltage will be affected. It will either be larger or smaller than desired (depending on the current polarity) by an amount equal to one deadtime interval. This, in turn, produces an offset in the average output voltage. Once the voltage waveform has been affected, you can rest assured that the current waveform will be distorted as well.

On the basis of this discussion, we can now rewrite the high time equation [Eq. (9.2)] to include a first-order approximation of the distortion created by having deadtime inserted:

$$t_h(t) = \frac{T}{2} + \frac{TM}{2}\sin(\omega_o t + \theta) - \mathrm{sgn}(i_1)\mathrm{DT} \tag{9.3}$$

where $\mathrm{sgn}(i_1)$ is the polarity of the line current for phase 1 and DT is the deadtime.

Dividing both sides of Eq. (9.3) by the PWM period T, we can once again solve for the averaged output voltage. The difference between the following equation [Eq. (9.4)] and the first equation [Eq. (9.1)] is that the effects of the deadtime distortion are now included.

$$\overline{V_{o,1}(t)} = \frac{1}{2} + \frac{M}{2}\sin(\omega_o t + \theta) - \frac{\text{sgn}(i_1)\text{DT}}{T} \tag{9.4}$$

This equation suggests that the distortion of the phase voltage can be approximated as a bipolar square wave, which is synchronized to the phase current signal. Figure 9.3 shows an actual phase voltage waveform obtained from a hybrid power module which confirms this. A pure sine wave is also shown for reference to illustrate the effects of the distortion. To avoid any potential confusion, it should be mentioned that the two waveforms in Fig. 9.3 were taken consecutively and are not aligned in time.

To understand the shape of the current waveform, it is necessary to consider the distortions and interactions of all three phases. If we assume steady-state load conditions, it is safe to assume that the angular separation of each line current is 120°. Since the distortion voltage is 180° out of phase with the current for each phase, then each of the distortion waveforms is also separated by 120°, as illustrated in Fig. 9.4.

Assuming that the motor load is linear, we can use superposition and analyze the system response to the distortion alone (i.e., modulation signal equals zero). Under this condition, if we assume that the motor load is balanced, we can average the three distortion voltages to obtain the motor neutral voltage n. Unlike three phase sine waves, the distortion waveforms do not result in a neutral voltage of zero over time. Instead, it is a square wave which transitions every 60°.

For the sake of our analysis, let's suppose that the motor is a three-phase Y-connected load (although the results apply equally to Δ-connected loads). By subtracting n from d_1, we obtain the distortion voltage, which is actually impressed across the phase 1 leg of the load, shown as the bottom waveform of Fig. 9.4. This voltage sets up a current in that phase which is a function of the load impedance. When the original sine-wave current is added back in by superposition, the result is a current waveform whose peaks are clipped, as seen in Fig. 9.3. In fact, under certain conditions, when the distortion waveform is large with respect to the modulation waveform, the distortion can actually cause the current to dip at its peaks.

Inquiring minds will also wonder about the mysterious flat spots of the current waveform at the zero crossings in Fig. 9.3. Since the source of this distortion is intimately linked with the proposed correction technique, the flat spots will be covered in detail in the next section.

From a review of Eq. (9.4) and Figs. 9.3 and 9.4, there are several characteristics about deadtime distortion worth remembering:

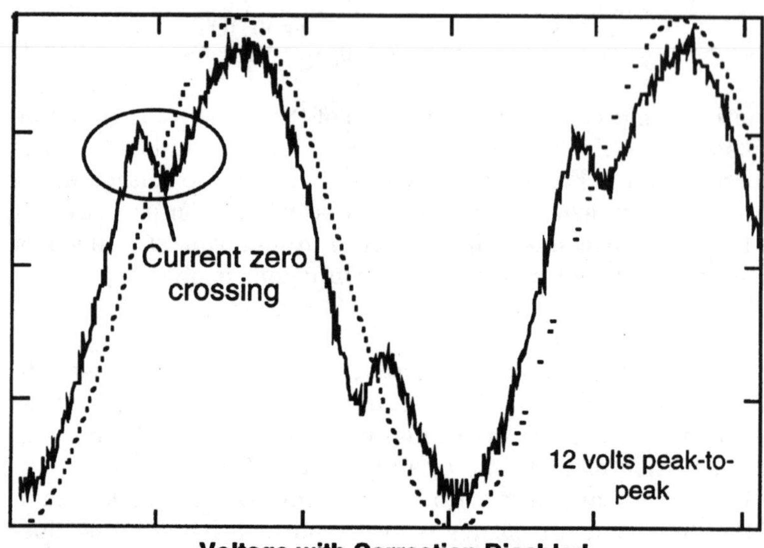

Voltage with Correction Disabled

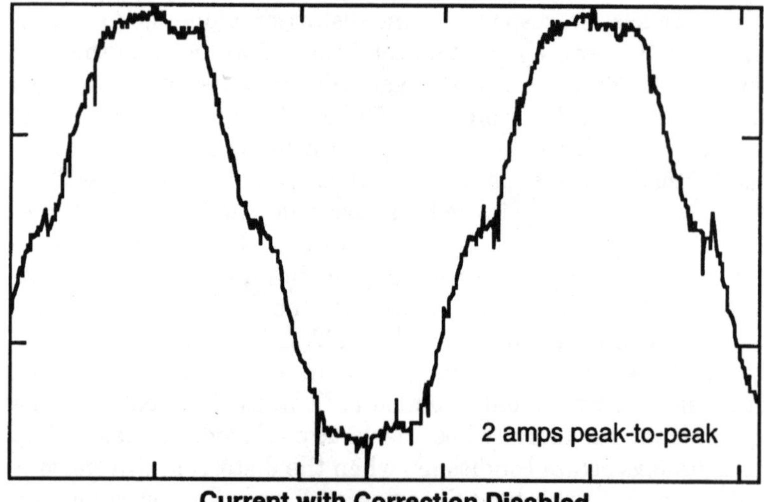

Current with Correction Disabled

(Test Conditions: 372 watt 3ø motor, PWM freq. = 7.3 kHz, dead-time = 3 μs, Output ω = 1.7 Hz)

Figure 9.3 Voltage and current distortion due to deadtime.

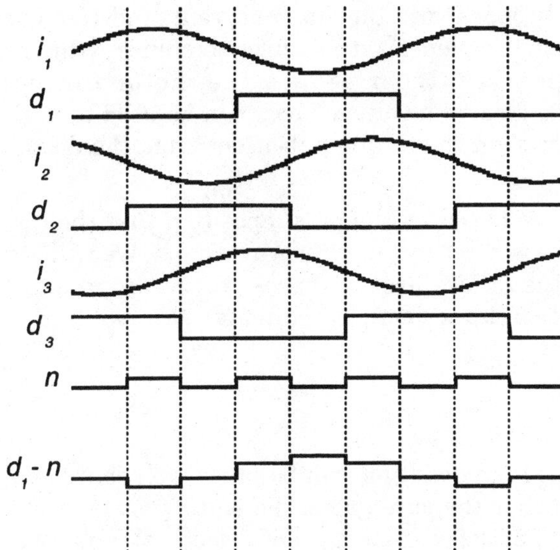

Figure 9.4 Calculated distortion waveforms.

1. The amplitude of the distortion per unit of bus voltage is equal to the ratio of the deadtime to the PWM period. Since deadtime is a system parameter that is usually fixed in accordance with the switching characteristics of the power devices, the problem is usually associated with higher PWM frequencies. In some cases, an engineer will specify a higher PWM frequency to minimize total harmonic distortion (THD) without realizing that the distortion from deadtime actually gets worse.
2. The voltage distortion causes a current distortion, and the net result is torque pulsations felt on the motor shaft. Some engineers have reported that this also translates into stability problems between the motor and drive under certain conditions.
3. Unlike the modulation signal, which is purposefully impressed on the motor windings, the amplitude of the distortion is *not* affected by the modulation index. This means that the problem, as viewed from a signal-to-noise vantage point, is minimized when the applied voltage to the motor is large. The other way to interpret this is that the distortion is most severe when the voltage is small. On an AC induction motor, this occurs when the motor rpm is also small, and the momentum of the rotor cannot smooth out the torque pulsations, making them even more apparent.

4. The distortion is synchronous with the motor current, and 180° out of phase with it. At low frequencies, this effect combines with the stator resistor losses to further reduce the motor torque. Overmodulation in the form of a voltage boost can be used to mitigate this problem. However, the torque pulsations from the distortion are still present.
5. The previous analysis is based on the presupposition that the distortion waveform has a perfect rectangular waveshape. We will see in the next section that there are variances from this premise, which have an impact on the correction technique.

9.2 The Solution

Since the distortion effect from deadtime can be fairly well characterized, it stands to reason that the cure should be equally straightforward. With the exception of a few second-order effects, this is true. Since the output waveform has a distortion, and the characteristics of that distortion can be closely approximated, the answer is to "countermodulate" the original PWM signal to essentially provide noise cancellation. In other words, superimpose a correction signal on top of the sine-wave signal in the processor to exactly cancel the output distortion. Equation (9.5) is obtained by modifying Eq. (9.3) (which defines the high time of the output waveform in the presence of distortion) to include a correction term to counteract the distortion. Since the distortion signal resembles a square wave, the correction term is also a square wave. Since the distortion signal is synchronized to the current waveform for that phase, the correction term must also be synchronized to the same current waveform.

$$t_h(t) = \underbrace{\left[\frac{T}{2} + \frac{TM}{2}\sin(\omega_o t + \theta) + \text{sgn}(i_1)\text{DT}\right]}_{\text{modulation}} \underbrace{-\text{sgn}(i_1)\text{DT}}_{\text{distortion}} \tag{9.5}$$

Another way to view the correction process is illustrated in Fig. 9.5. Recall from Fig. 9.2 that the deadtime was balanced between the top and bottom PWM signals. For that particular example, 50 percent duty cycle was desired. With the insertion of deadtime, the top and bottom PWM signals had their ON times reduced by an equal amount to something less than 50 percent. What would happen if we did not choose to split the deadtime up evenly? Figure 9.5 shows the case in which all the deadtime is completely relegated to the "recessive" PWM signal. In other words, because the load inductor drives the

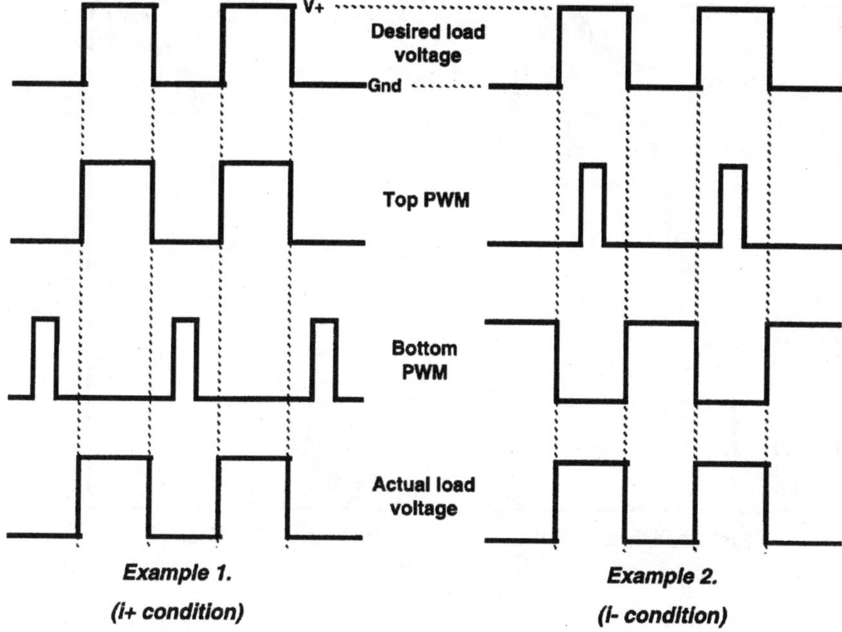

Figure 9.5 Distortion correction accomplished by redistribution of deadtime.

phase voltage according to its current polarity, the turnon of one of the transistors is redundant, since the inductor drives the voltage in that direction, anyway. The problem is identifying which PWM signal (top or bottom) is recessive, since it changes depending on the output current polarity. It is therefore necessary to know the output current polarity for each phase.

With no distortion correction applied, the PWM module on the MC68HC708MP16 uses one PWM register to derive the top and bottom PWM signals for each motor phase and automatically inserts a programmable deadtime into these signals. When distortion correction is enabled, the PWM module uses two PWM registers for each motor phase—one when the current polarity for that phase is positive, and the other when it is negative. The waveforms of Fig. 9.5 are obtained by programming one register with the desired pulse width *plus* the deadtime and the other register with the desired pulse width *minus* the deadtime. (Actually, when generating center-aligned PWMs on the MC68HC708MP16, to obtain an output pulse-width delta of plus or minus one deadtime, the correction value should be plus or minus *one half* of the value in the dead-time register.)

One economical way to measure the current polarity for each phase is to use the current polarity sense inputs on the MC68HC708MP16.

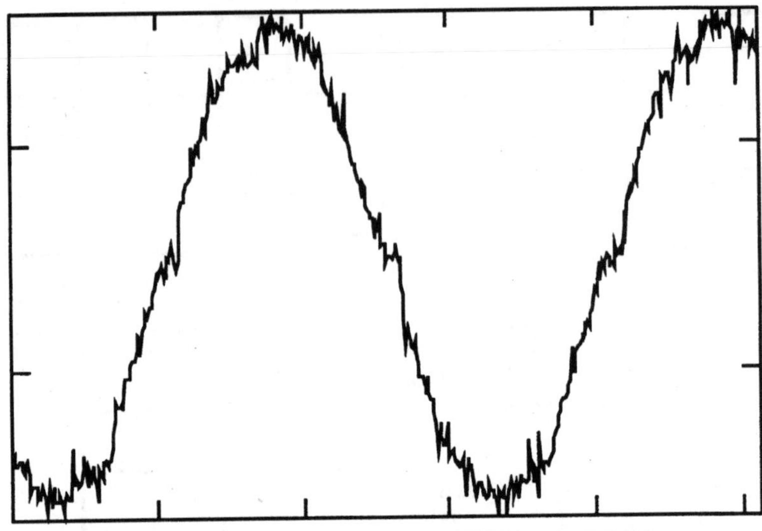

(Test Conditions: 372 watt 3ø motor, PWM freq. = 7.3 kHz, dead-time = 3 μs, Output ω = 1.7 Hz)

Figure 9.6 Partially corrected current waveform.

These pins are used to monitor the three PWM voltage waveforms supplied to the motor. Each input is sampled during the deadtime for that particular waveform. If the input is high, the current polarity is determined to be negative, and vice versa. This information is then used to automatically toggle between one of two PWM registers, as discussed above.

When the distortion is corrected in this manner, the current waveform of Fig. 9.6 is obtained. The modulation index is scaled so that the current peak amplitude matches that of Fig. 9.3. Otherwise, the same modulation index results in a 50 percent increase in the peak amplitude, which demonstrates the severity of the distortion. Although most of the distortion is gone, some still exists at the zero crossings, causing torque pulsations that can be detected on the motor shaft. To eliminate this distortion, we need to go beyond our first-order explanation of the distortion source to understand what is happening to the system during the current zero crossings.

Up to now, we have assumed that the distortion waveform is a perfect rectangular waveshape, and therefore has "snappy" rising and falling edges. Related to this premise, we have also implied that the inverter output voltages are either high or low at any given time, including the deadtime intervals. (At this point, you may be getting

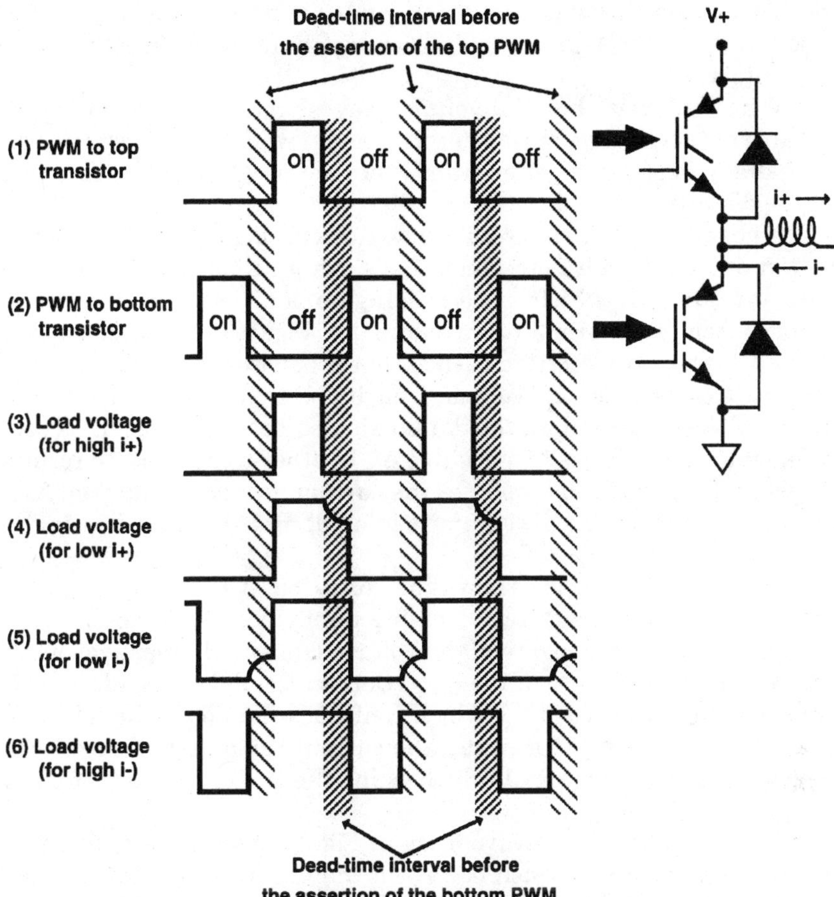

Figure 9.7 Output voltage waveforms under various current conditions.

suspicious that I'm about to modify these assumptions; if so, your insight serves you well.)

Figure 9.7 shows the voltage waveforms out of one half-bridge of the inverter under various current conditions. As can be seen, our assumptions are valid when the current amplitude is high. However, under low-current conditions, which occur at a zero crossing, the inductor is less aggressive in snapping the voltage high or low, presumably due to parasitic capacitance in the motor and drive that can support the inductor's current flow. Prior to the deadtime interval, if the output voltage were already driven in the direction that the inductor would drive it anyway on the basis of the current polarity, this phenomenon would not be observable. The problem occurs when the other transistor was already on, and the inductor must drive the output voltage through the full power supply range during the deadtime interval.

The net effect is that the voltage waveshape does not transition instantaneously between waveforms 3 and 6 when the current polarity changes. Instead, a softer transition occurs, which takes the edge off the distortion waveform.

The next question to address is how this effect causes the current to flatten out at the zero crossings. Let's examine the case of positive inductor current, which is decreasing so that the output voltage changes from waveform 3 to waveform 4 of Fig. 9.7. As stated earlier, the polarity of the distortion is such that it opposes the polarity of the current. However, as the voltage transitions from waveforms 3 to 4, the Δv of the distortion (in the darker shaded deadtime region) is positive, which contributes to a positive Δi. In other words, the Δv results in negative feedback that opposes the current decreasing further. As a result, the current magnitude is regulated, and the zero crossing is delayed.

Waveform 5 in Fig. 9.7 shows the condition of the output voltage shortly after the current zero crossing finally occurs. As the current continues to go more negative, the voltage transitions from waveform 5 to waveform 6. The Δv of the distortion (in the lighter shaded deadtime region) is still positive, which continues to oppose the negative trend of the current. As a result, the current remains flattened until waveform 6 is achieved, at which point the Δv in the distortion is zero.

The progression of the waveforms in Fig. 9.7 correlates with negative trending current, which occurs on the negative slope of the current sine wave. After the current reaches its negative peak and starts positive again, the waveforms of Fig. 9.7 occur in reverse order from 6 to 3. The current flattening occurs once again, this time with the negative Δv of the distortion opposing the positive trend of the current.

Applying the previous discussion to the current waveform of Fig. 9.6, we realize that even with correction enabled, if we wait for the current polarity to change before toggling the correction value, the current has already flattened out at the zero crossings, and a minimal amount of distortion is still present. However, what would happen if we didn't wait for the polarity to change? Figure 9.8 shows an uncorrected current waveform with proposed thresholds indicating where the correction value should be toggled. If this can be accomplished, the output voltage waveform will change before the undesirable effects of the Δv of the distortion waveform can be felt and immediately drive the current through the zero crossings. The flat spots at the

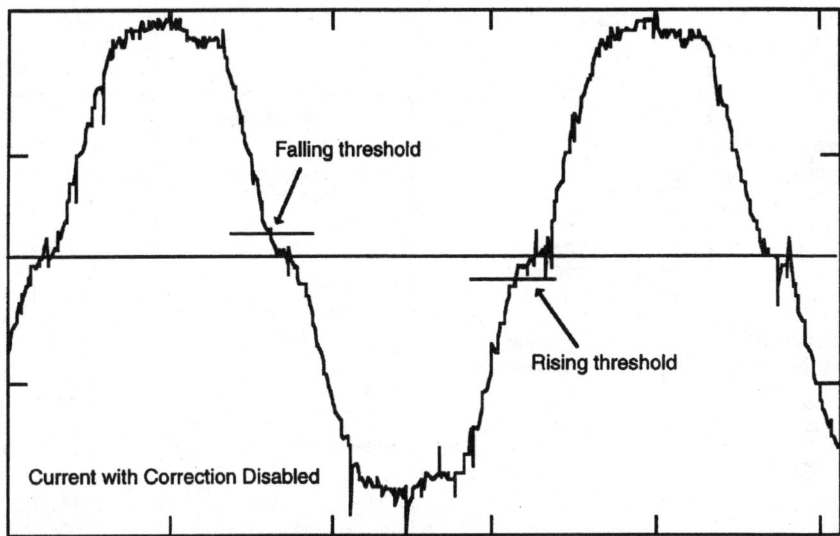

Figure 9.8 Proposed current thresholds for correction toggling.

zero crossings will thus be eliminated at the expense of small vertical distortions in the waveform.

If a current sensor is used, the current waveform could be directly measured and the correction values toggled at the appropriate points on the waveform. However, current sensors (especially with isolated outputs) are expensive. Is there a way to obtain this information without a current sensor?

Let's return to Fig. 9.7 and examine the output voltage waveforms under different current conditions. If we compare the (slash-marked) deadtime region with the (shaded) region, we notice that when the current magnitude is large, the voltages during the deadtimes are the same, regardless of polarity. However, when the current magnitude is small, the voltage waveforms are different in the deadtime intervals. This suggests a strategy for detecting when the current waveform is approaching a zero crossing, allowing us to act before it actually occurs. What we need is a "waveform discriminator" that can tell the difference between these voltage waveforms.

Figure 9.9 illustrates a distortion correction system, which incorporates such a sensor. By using either hysteresis or a simple low-pass filter, the sensor detects whether the load inductor aggressively snaps the voltage waveform during the deadtime intervals. At the end of each deadtime region, the comparator output is sampled by the D-type flip-flops, and the results are stored for both deadtime intervals.

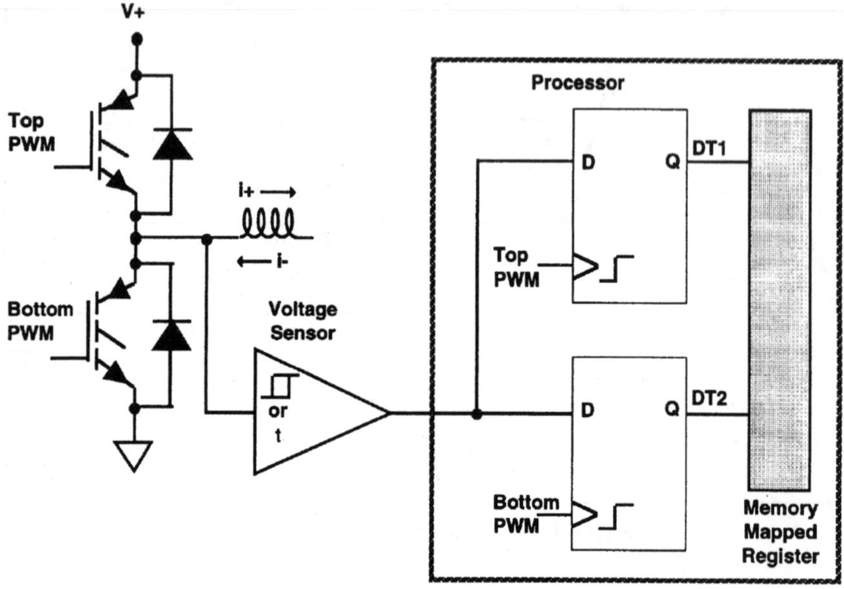

By sampling the output of the voltage sensor as shown (inside the processor), the following information can be deduced:

DT1	*DT2*	*Load Current Condition*
0	0	High amplitude i+
1	1	High amplitude i-
0	1	Low amplitude, either polarity

Figure 9.9 Sense scheme for optimized deadtime distortion correction.

As seen in the chart of Fig. 9.9, if the voltage waveform is cleanly snapped by the inductor, both results will agree with each other. For example, if both outputs are low, the current is large and flowing out of the inverter. If both outputs are high, the current is large and flowing into the inverter. However, under low-current conditions, regardless of polarity, the sampled results will be different, indicating to the control algorithm that a current zero crossing is looming in the near future. The distortion correction value can thus be toggled before the current begins to flatten out.

Figure 9.10 shows an example schematic of a voltage sensor using hysteresis. In this case, the resistors are sized to work with a bus voltage of about 150 V. The thresholds on this circuit correspond to input voltages of approximately 15 and 125 V. Also shown as part of

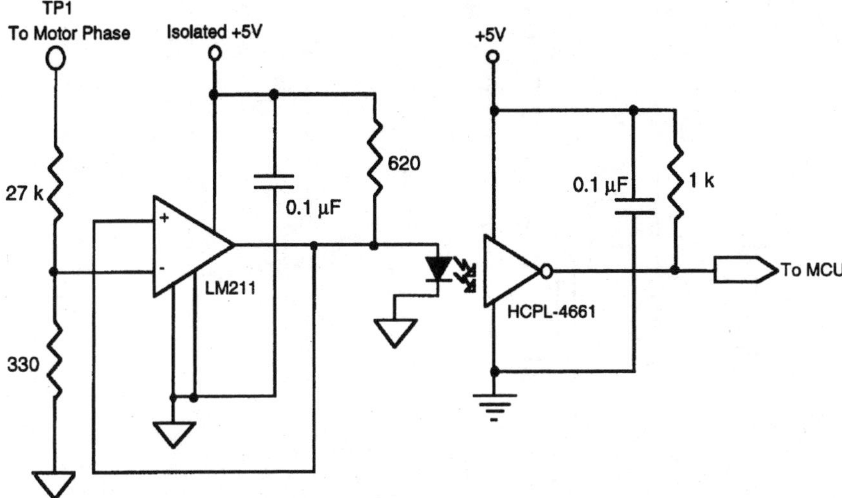

Figure 9.10 Voltage sensor based on hysteresis.

the interface is an optoisolator, which may not be required if the processor is referenced to the same ground as the bus voltage.

If the bus voltage is constant, the circuit in Fig. 9.10 works just fine. However, if the load must be decelerated rapidly, or the motor is connected to a load that can also supply energy, regeneration can occur, which can "pump up" the bus voltage. The need thus exists for a sense circuit whose threshold is scaled from the bus voltage, as shown in Fig. 9.11. In this case, the sensed motor voltage waveform is filtered in such a way that only a sharp transition of the waveform at the start of the deadtime interval will result in the comparator threshold being reached by the end of the deadtime, which is when the comparator output is sampled. With the circuit shown, optimum results were achieved when the filter time constant was about 60 percent of the deadtime value, plus or minus 20 percent. Again, if the MC68HC708MP16 is referenced to the bus voltage ground, the optoisolator may not be required.

The software can access the outputs of the double-action samplers (2 bits per phase for a total of 6 bits) via location $0024 in the memory map (the $ symbol indicates hexadecimal value). The bit pattern progression per phase should be %00 at the positive peak of the current waveform, %01 around the current zero crossing, %11 at the negative peak of the waveform, %01 again near the rising zero crossing, and back to %00 at the positive peak (the % symbol indicates binary value). The software is responsible for sensing when the current is near the

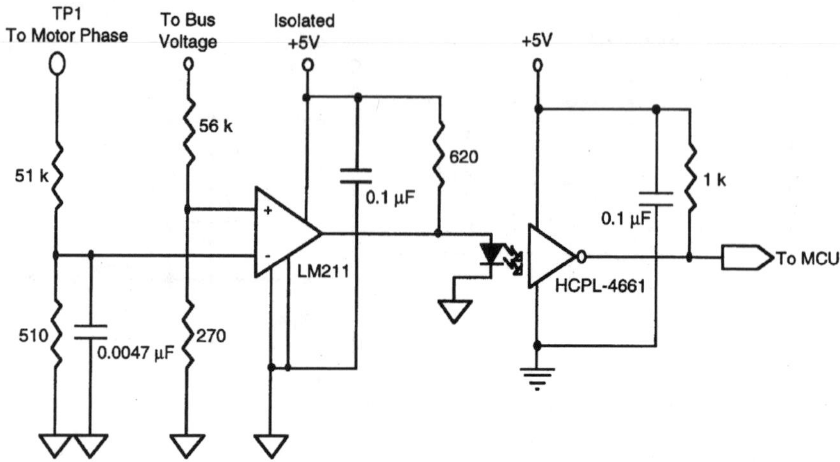

Figure 9.11 Voltage sensing scaled off of the bus voltage.

zero crossing (or has changed polarity in the case of high modulation frequency) and changing the correction value appropriately. Figure 9.12 shows a state diagram which illustrates one proposed method of accomplishing this on a per-phase basis. DT_1 and DT_2 are taken from Fig. 9.9 and correspond to the two outputs of the double-action sampler. The output from the algorithm is the $IPOL_n$ bit for that particular phase. States 5 and 10 correspond to the condition of a near-zero current being detected. During these states, the waveform pointer value (θ_c) is recorded, and further polling of the DT_n bits is disabled until the waveform angle changes by more than 60°, allowing for a strong %00 or %11 reading to be reestablished. Also notice that multiple readings of the DT_n bits are taken to eliminate potential noise.

Code based on Fig. 9.12 has been written using a C compiler. The algorithm performs correction on all three phases simultaneously by driving the $IPOL_n$ bits in the PCTL2 register that in turn control which PWM VALUE REGISTER is used to drive the output's pulse width. The execution of the algorithm is synchronized to run once each time the PWMs are updated. The state information for each phase is preserved as static variables so that execution may resume in the proper states the next time the routine is run.

Assuming an 8-MHz MC68HC708MP16 clock frequency, the algorithm processes all three phases in about 45 µs (microsiemens). States 6, 7, 11, and 12 are included to synchronize the states with the current waveform every half-cycle. To speed up execution, states 6 and 7 could probably be eliminated, which would result in synchronization only once per cycle on the positive peak of the current wave-

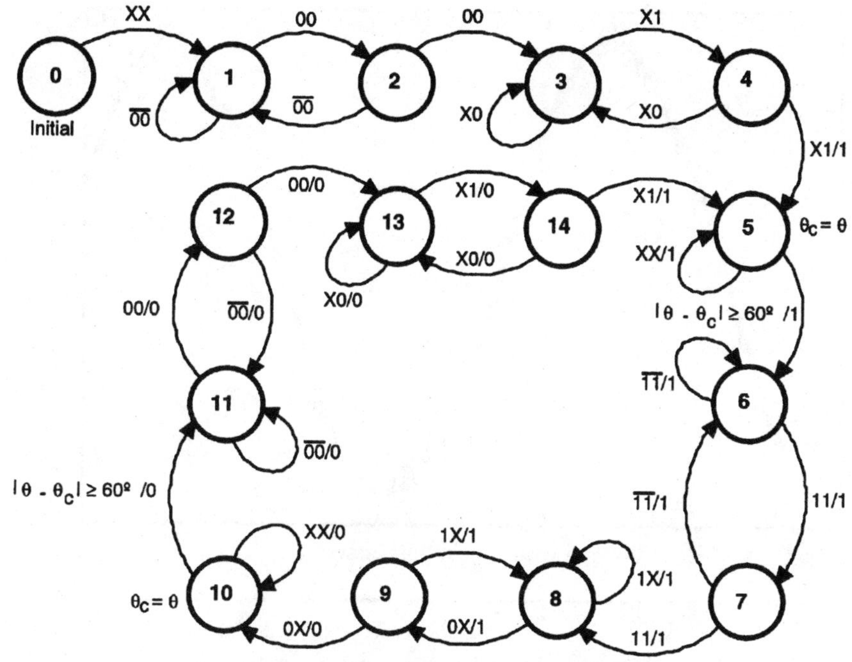

Figure 9.12 State diagram for distortion correction.

form. Also, even though the C compiler generates remarkably tight code, faster execution speed may be possible with code that takes full advantage of this particular MCU's instruction set.

9.3 The Results

If you have made it this far through the chapter, congratulations! Having read through the section covering the source of the distortion, and lasted through the discussion on how to correct it, you're now ready for dessert! After all, technology is neat, and cool, and all that stuff, but it's the results that count.

By using the techniques described in the previous section, we can re-create the conditions of Fig. 9.3. Once again, the modulation index is reduced to match the amplitude of the waveforms in Fig. 9.3, since removing the distortion creates larger sine waves for a given modulation index (and thus more torque).

The resulting waveforms shown in Fig. 9.13 were obtained consecu-

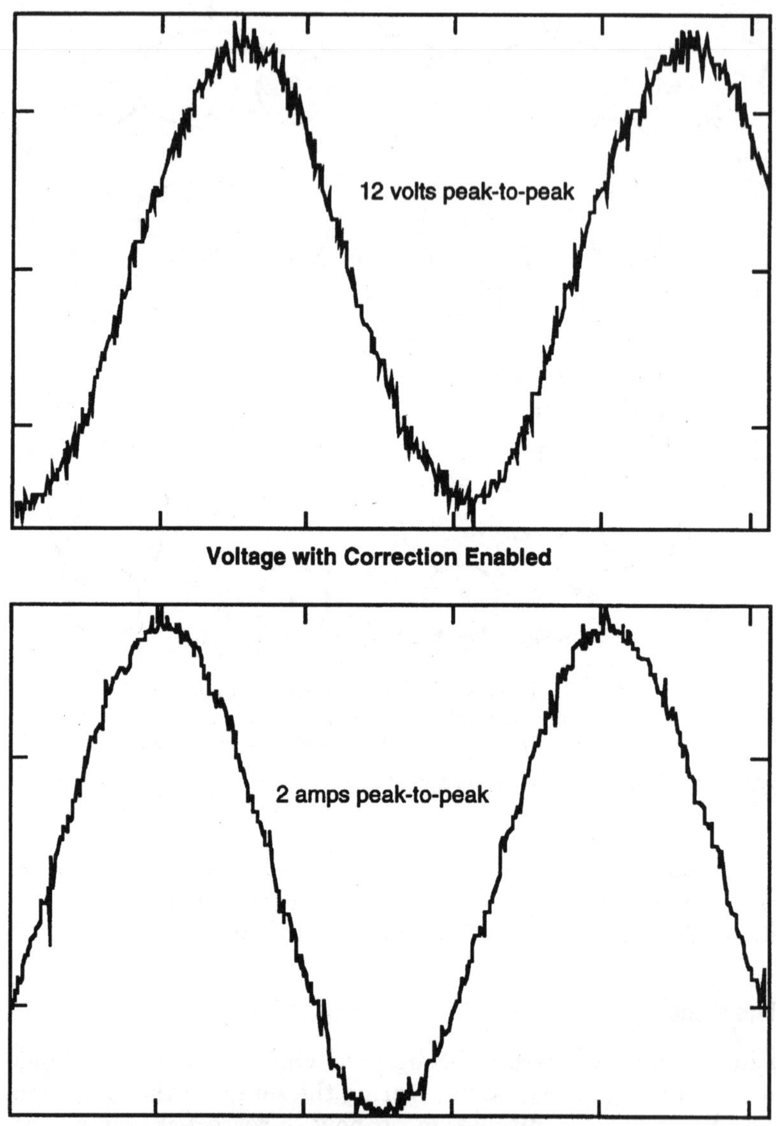

Figure 9.13 Results of distortion correction using the MC68HC708MP16.

tively, and their time axes are not aligned. Careful examination of the current waveform reveals the early transitions before the zero crossings. However, this distortion is minimal compared to allowing the current to flatten out at the zero crossing. But, best of all, little or no torque disturbance can be detected on the motor shaft, even when loaded. *With the use of the voltage information obtained during the deadtime, the software can "countermodulate" the PWM waveforms to cancel the distortion.*

Conclusion and Summary

In this chapter, we have identified a source of distortion that is common to power stages employing totem-pole transistor configurations requiring deadtime. On AC induction motors running open-loop, the problem typically manifests itself as poor low-speed performance, such as torque ripple and rough operation.

We presented a solution that involves sensing the motor phase voltages during the deadtime intervals and has been implemented on the MC68HC708MP16 microcontroller. Software running on the microcontroller can poll the data from the voltage-sensing circuitry and modify the modulation waveform to cancel the effects of the distortion. This results in quieter, smoother-running, happier motors! Key points about this chapter are listed below.

- Deadtime applies to half H-bridge power stages and is a specific time interval in which both the top and bottom power transistors are switched off to minimize shoot-through currents.
- Deadtime intervals may be set for 1 to 5 μs and are affected by the power stage's switching performance.
- Distortion from deadtime occurs because the motor's current flow (which is controlled by the PWM of the power stages) is nonlinear during the deadtime interval. The net result is torque pulsations on the motor's shaft and possible instability in the motor drive electronics.
- The PWM frequency directly affects deadtime distortion—higher frequencies raise the distortion. Because the amplitude of the distortion is unaffected by the PWM modulation index, the AC induction motor is most affected by the deadtime distortion at low rpms.
- Deadtime distortion correction can be fixed by "countermodulating" the original PWM signal to essentially provide noise cancellation. In other words, a correction signal is superimposed on top of the sinewave signal in the processor to exactly cancel the output distortion.

- The motor's voltage remains fairly stable at high current levels. At low motor current levels, the motor's voltages change during the deadtime intervals.
- Software running on the microcontroller can poll the data from the voltage-sensing circuitry and modify the modulation waveform to cancel the effects of the distortion.
- A motor voltage sensor can consist of a comparator that (with hysteresis or a simple low-pass filter) detects whether the load inductor aggressively snaps the voltage waveform during the deadtime intervals. At the end of each deadtime region, the comparator output is sampled by the D-type flip-flops, and the results are stored for both deadtime intervals.
- Motor current polarity sense inputs (as in the MC68HC708MP16) feed in vital data that enables the MCU's program to compute a deadtime correction strategy.

Chapter

10

Motor Control MCUs and Development Tools

Jim Gray, Bill Lucas, Warren Schultz, and Richard Valentine
Motorola Semiconductor Products

Introduction

In this chapter we will review general MCU architectures and examine two MCU motor control development tools. The use of MCUs has been shown in several circuits throughout this book; in order to successfully implement an MCU-based motor control, it is essential to understand a few specific items about MCU devices and their supporting development tools. The advantage of an MCU for a motor control design is that the MCU's program (software) can be changed or revised by changing a few lines of text in a data file. This data file (usually referred to as *source code*) is then converted into machine code by a software compiler or assembler and is programmed into the MCU. Such program flexibility requires the use of development tools designed specifically for each type of MCU. The MCU type is determined mostly by its architecture, peripheral elements, and special functions.

Considering the trends in electronic control systems, you don't need to wonder whether MCUs will be used in motor controls; rather, you need to determine how many and which types of MCUs will be utilized. Learning how to apply and test MCUs will be crucial for control design engineers and technicians. The MCU manufacturers are continuing to introduce higher-performance MCUs every few months. The intricacy of programming and debugging these complex MCU devices can be overwhelming without easy-to-use development tools. In some instances, the success of an MCU-based application depends more on the development tool's performance than on the MCU itself. User-friendly, but powerful, development tools allow fast program debugging, enabling programmers to focus their efforts on the software design rather than on the nuances of the development system.

Another challenge when using an MCU is designing the software. This task can sometimes be expedited by searching software libraries that contain control application-type algorithms. Keyboard scan routines and display drivers, as well as math equations, are only a few examples. The software can be developed with high-level instructions such as C or C++. High-level languages are more MCU-independent, but usually increase memory size and may exhibit longer throughput. Assembly language is used by some programmers who are intimately familiar with one particular MCU family and want to maximize the code efficiency, throughput, and density.

10.1 MCU Types for Motor Controls

Single-chip microcontroller (MCU) devices are ideal for many types of control applications, including motor systems. The MCU can measure the motor's current, voltage, speed, temperature, and even magnetic flux and can then compute the best operating strategy for the motor. The MCU can allow the motor's speed or torque to be managed with high precision and can protect against or at least detect motor fault conditions. Before motor-specific MCUs are discussed, a review of the different types of MCU architectures may be useful.

10.1.1 MCU basic architecture

The microcontroller is a single-chip computer. It operates as a stored-program machine; that is, it must read its program code and data values from its memory in order to operate. Two common methods are used to accomplish this. One is called *von Neumann architecture* and has been employed in many MCUs and microprocessors (e.g., the MC68HC05MC4 8-bit MCU from Motorola Inc.). This method uses one data bus and memory space for both program code and data values, saving cost but slowing down the code execution, since the program code and data values cannot be loaded simultaneously. The other approach, called *Harvard architecture,* separates the program code and data values into two memory structures, allowing parallel loading of both at the same time. This technique speeds up execution time but requires more data pins. There are some modified MCU versions of the Harvard architecture that use only one external memory bus but use both program and data buses internally (e.g., the TMS320C240 16-bit DSP MCU from Texas Instruments Inc.). It is important to remember that the design of the program code (software program) allows the MCU to perform a long set of tasks or solve complex mathematical equations. The speed at which the MCU can execute these instructions becomes important in high-speed, closed-loop control applications, but from a motor control design perspective, it is the program code design that is more crucial. Executing poorly designed or deficient code with either a low- or high-performance CPU still results in an inadequate and unstable control system.

Figure 10.1 shows the general architecture commonly used for many MCU types. Each of these basic structures in the MCU plays a critical role in the MCU's operation.

10.1.1.1 Central processor unit (CPU). The central processor unit is the brains of the MCU. This structure controls most of the MCU's functions and receives instructions from the software program. There are different methods for devising a CPU, but, essentially, a typical CPU will contain an arithmetic logic unit (ALU) and several registers for loading the program code, data values, and computed results. Some CPUs implement instructions such as multiply and accumulate

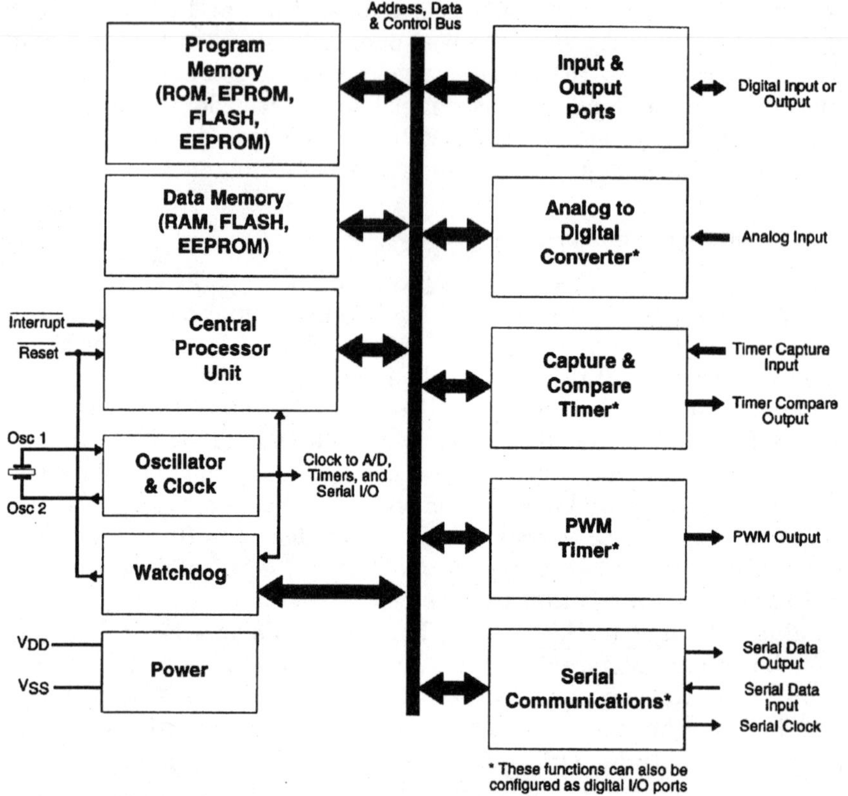

Figure 10.1 Basic motor control MCU architecture.

(MAC instruction) using special logic to allow one-cycle execution times. These are called *digital signal processors* (DSPs) and are useful for math-intensive, high-speed controller applications. Some confusion arises in CPU terminology when a complex instruction set computer (CISC) also includes a MAC instruction that can execute at high speed. There are other CPU variations, such as the reduced instruction set computer (RISC), that use less instructions and operate faster. RISC-type MCUs normally execute most of their instructions in one clock cycle. Coprocessors are also added to improve floating-point math operations. No matter which CPU design is selected, the bottom line is how fast and how precise the CPU can execute its program for a given cost. The execution time of one instruction by itself is not a reliable benchmark on which to judge an MCU's performance. The actual throughput—or the time it takes to read an input line, compute a value, and then output an appropriate signal—is a more important measurement for MCU applications. This requires program code to be written, debugged, and measured for each type of MCU to determine its actual throughput. Selecting only the most difficult and time-dependent, closed-loop subroutines for benchmarking can expedite MCU comparison testing.

10.1.1.2 Memory elements. The memory elements store the program code and its data values as well as computed values generated during its operation. Normally, the program code and its data values are stored in read-only memory (ROM) or in one-time programmable (OTP) memory. While both permanent, the difference between ROM and OTP memory is that the ROM memory patterns are fabricated into the MCU, while an OTP memory is electrically programmed. When using a ROM MCU, you must keep in mind that the code changes must be done by the MCU maker, and a large inventory of ROM coded devices could be scrapped when a latent program bug is discovered. The OTP approach allows a minimal number of MCUs to be programmed—perhaps, for instance, one day's worth of production, reducing the cost of program revisions. ROM-coded MCU devices are usually lowest in cost but should not be considered until the program code has been totally proved and only if the motor control application requirements do not vary every few weeks.

Another approach to storing the software program to memory is to use electrically erasable programmable memory (EEPROM) or erasable programmable read-only memory (EPROM), which can be erased by a special ultraviolet light source. EPROM-type MCUs usually have a clear window, which is covered with an opaque sticker to prevent accidental erasure from weak ultraviolet sources such as sunlight or other light sources. EEPROM MCU devices allow fast program patches and are suitable for both prototyping and low-volume production. A variation on the EEPROM design is a flash erasable memory. This type of memory is erased in one operation and then electrically reprogrammed. MCUs with either flash erasable memory or EEPROM will cost more than EPROM because of the size of the memory logic elements. The extra initial cost of the flash-memory- or EEPROM-type MCU may not be that significant in the long term if several program revisions are encountered. Some MCU devices incorporate both EEPROM and flash memory (e.g., the 68HC908AT32 8-bit MCU or the 68HC912B32 16-bit MCU from Motorola Inc.). This type of device allows fast reprogramming of the MCU with the flash memory and can store operational values during power-OFF conditions in its EEPROM.

The other memory structure used in an MCU, called *random-access memory* (RAM), stores its operational values and usually some user control settings. If the application must resume its operation during power interruptions, its RAM must retain its values. This can be done by using EEPROM-type elements or by placing the MCU in a stop mode with an independent power source that keeps the RAM powered up.

10.1.1.3 Input/output and peripheral elements. MCUs differ from microprocessors (MPUs) in several ways. The MCU contains internal memory, has special peripheral functions built in, as well as input/output (I/O) ports. The input and output ports are usually designed in groups of eight lines (or pins). The input and output lines are normal 0 or 1 logic-type signals. In many MCUs an I/O pin can be programmed to be either an input or an output. Generally, the program initializes the I/O lines to be either input or output, with some programs actually changing an input line to an output line and back again to perform certain tasks.

Many types of peripheral elements have been used in MCUs. A few of the more popular ones include a wide variety of timers, communications, analog-to-digital converters, digital-to-analog converters, phase-locked loops, high-current outputs, and special display drivers. Motor control MCUs should contain a timer element that can supply at least six channels of PWM signals that need minimal intervention from the CPU. Analog-to-digital inputs are also necessary to measure the motor's current, temperature, and other analog signals.

10.1.1.4 Clock and bus control structures. An MCU requires a clock oscillator to trigger the sequence of steps in the MCU's logic. The clock's oscillator can be generated with a crystal, ceramic resonator, or an *RC* network, or can be supplied from an external generator. The frequency of the MCU's oscillator can be misleading in determining its effect on the MCU's performance. This is because most MCUs divide the oscillator by a factor of 2 or more. For example, a typical motor control MCU (e.g., an MC68HC05MC4 8-bit MCU from Motorola Inc.) uses a 6-MHz crystal frequency that is divided by 2, resulting in a 3-MHz internal bus clock frequency or a 333 ns cycle time. It is the internal bus frequency that sets how fast the program code can be sequenced. Some MCU designs use low-frequency oscillators, for example, a 32-kHz oscillator, and an internal frequency multiplier to minimize EMI problems from the oscillator.

It is important to note that the MCU's oscillator directly affects the operation of the MCU. If the oscillator fails or becomes erratic, the MCU will cease to function. Some MCUs contain internal logic that checks for a clock oscillator failure and can revert to an internal *RC*-type oscillator mode of operation to ensure that the program can enter a controlled shutdown or "limp-along state." This is important for a control system in which the output lines are controlling a motor and could remain stuck in one state if the oscillator failed.

The oscillator also provides the clock signal to other MCU's structures, such as its timing units and A/D converters. The timing units may use the fundamental oscillator's frequency as their input and provide programmable dividers that may range from 1 to more than 128. Timers are critical in most control applications. For example, the MCU can compute actions on the basis of its input signals and toggle the appropriate output, but a timer is required to set the length of time that the output can remain active. Timers are needed to generate the pulse-width-modulation (PWM) signals that form the basis of most motor speed controls.

Another important MCU element is its internal bus structures. Some MCUs are now designed with an internal bus structure that allows modularity. This means that different peripheral elements, such as liquid-crystal display (LCD) drivers, can easily be added into the MCU. A high-volume design, for example, may require a combination of functions that are not standard but can be accommodated without totally redesigning the MCU. This concept is similar to the plug-and-play idea in personal computers, except that it is implemented at the submicrometer silicon level.

10.2 Motor Control MCUs

There are three general motor control MCU groups. The first group includes basic brush-type DC motor speed (and/or directional) controls and requires only a general-purpose low-cost 8-bit MCU with internal A/D and timer. The second motor control group consists of brush motor control systems that require more capability than a simple speed regulation design. For example, the MCU may need to control more than one motor and interface with a large control panel. Electronic commutated motors require an MCU that can drive six or more power stages with PWM signals. The second group of motor controls requires medium-performance 8-bit MCUs that contain specific motor control structures such as six PWM outputs, A/D conversion, and input timer capture. The third motor control group includes precision motor systems that are found in machine tools and industrial automation. This group uses 16-bit or 32-bit MCU with high-speed CPUs and complex timer structures. New MCUs are being introduced for motor and power controls every few months. (See App. C for Internet Web Site listings of some MCU manufacturers.)

When comparing various MCU products for motor controls, attention should be given to several items: instruction cycle time, memory size, memory type, PWM bit size, PWM time resolution, A/D bit size, A/D conversion time, fault inputs, and development tool support. The MCU's cost is mostly determined by its silicon die size and package type. The challenge for the MCU manufacturer is to reduce the cost of the MCU while increasing its performance. This is being accomplished by scaling down the MCU's die size through the use of more precise designs and manufacturing equipment, and by utilizing CPU architectures that are optimal for real-time controller applications. One interesting aspect of MCU designs is that their internal operating voltages are being reduced to accommodate submicrometer line spacing. The MCU's external lines, however, still have to be able to operate in an electrically noisy environment, which is quite different from the gentle environment of MPUs (microprocessor units).

10.2.1 Low-cost motor control MCU examples

A *low-cost MCU* is defined as one that sells for under $2 and is usually a 4-bit or 8-bit device. Two examples of this type of MCU are a 68HC705JJ7 8-bit 20-pin MCU (Motorola Inc.) and a PIC12C672 8-bit 8-pin MCU (Microchip Technology Inc.). Each of these devices has some unique features: The 68HC705JJ7 has a 16-bit programmable timer that is combined with a single-slope A/D converter. The PIC12C672 is in a small 8-pin package and yet still offers A/D.

The 68HC705JJ7 8-bit MCU, shown in Fig. 10.2, utilizes a CISC-type architecture based on the HC05 core. This MCU uses the von Neumann architecture and loads both instruction code and data values from the same memory bus. Its instruction set consists of 62 instructions and eight addressing modes. The eight addressing modes provide flexibility in accessing data that is needed to execute an instruction. Instructions require 2 to 10 clock cycles with most running between 3 and 5 cycles. The fastest cycle time is 476 ns with a 4.2-MHz oscillator. This device is available in 20-pin package and is rated to operate over a −40 to +85°C ambient temperature. The MC68HC705JJ7 has 6144 bytes of user EPROM, 16 bytes for user vectors, and 224 bytes of RAM. The MCU also has a special-personality EPROM memory consisting of 64 bits. This special 64-bit memory allows each MCU to be programmed to retain some unique numbers or values (a serial number, for example), and cannot be erased except by a UV light source.

The MC68HC705JJ7's general-purpose I/O pins are rated to sink 1.6 mA and source 0.8 mA. There are six high-current I/O lines that

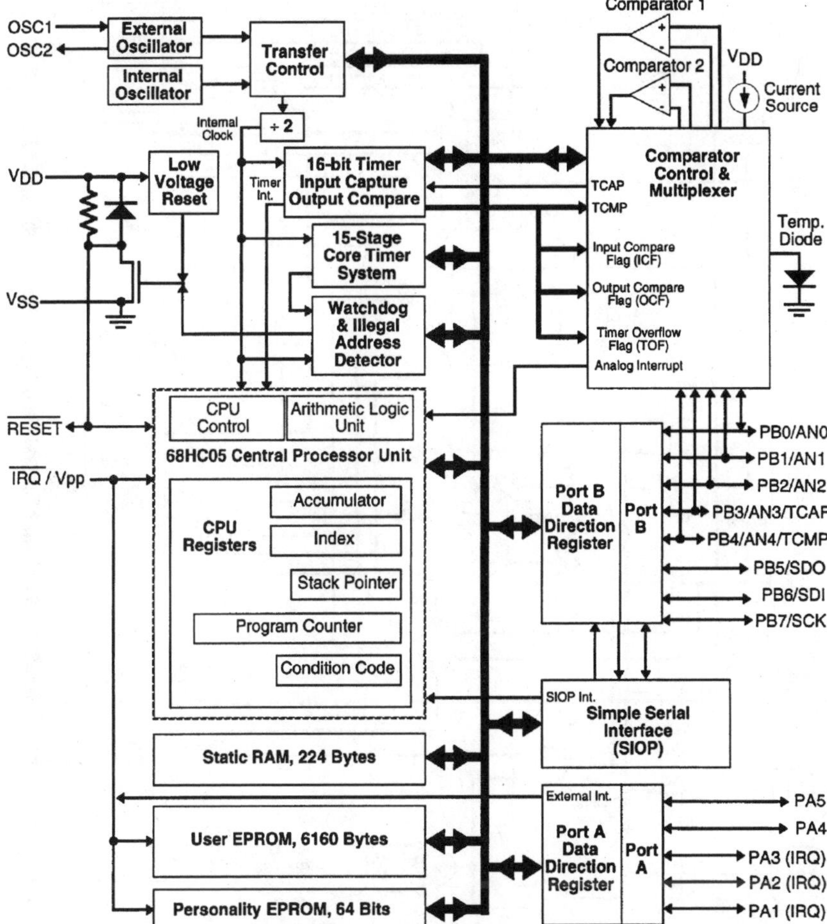

Figure 10.2 MC68HC705JJ7 8-bit MCU block diagram.

are rated to sink 15 mA and source 5 mA each. The total current rating of all six high current I/O pins is limited to sink 40 mA and source 15 mA. The 15-mA sink rating is sufficient to control LEDs, some optocouplers, and sensitive-gate triacs. The designer must take care not to exceed the 40-mA total sink current specification.

One unique feature of the MC68HC705JJ7 is its A/D subsystem. This A/D converter is based on internal comparators and a selectable current charge/discharge function, as shown in Fig. 10.3. An internal temperature-sensing diode can be selected as an A/D input, which is useful for tracking the local ambient temperature. This type of A/D converter requires more intervention from the program but is low-

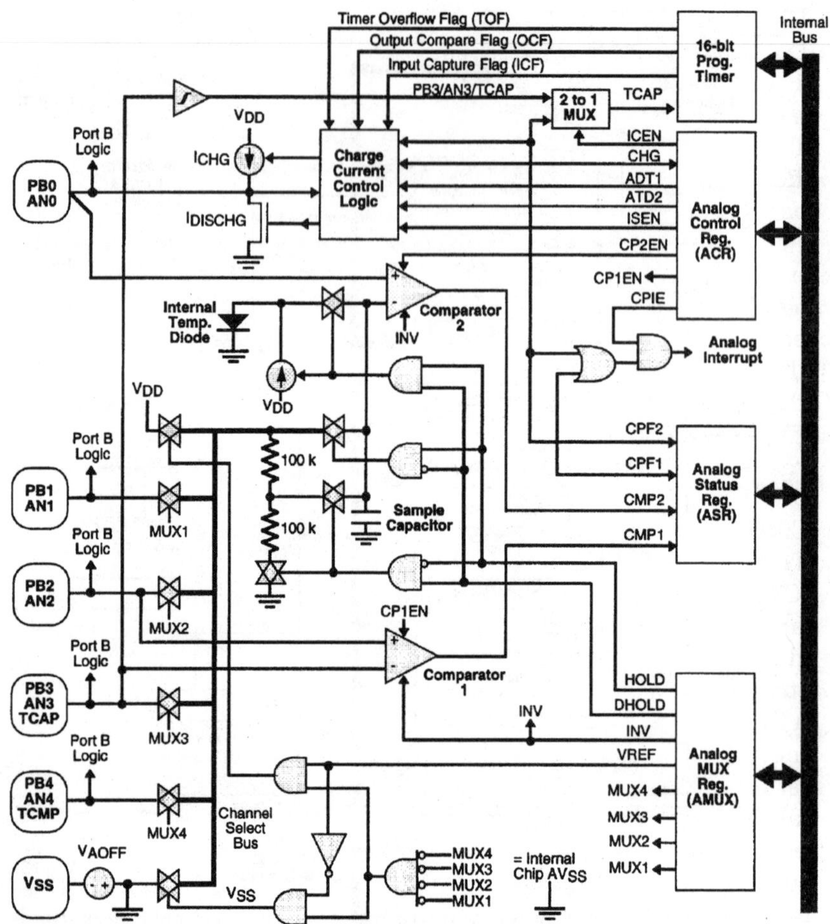

Figure 10.3 MC68HC705JJ7 A/D subsystem block diagram.

cost and can resolve to 12 or more bits using a ratiometric method. The ratiometric method employs a reference reading that is obtained from a known source, such as the MCU's supply voltage, and then compares that value to the unknown value. This requires software instructions.

The A/D conversion time is determined mostly by the external ramp capacitor's value, the unknown voltage, and the MCU's internal charge current (100 μA). For example, if an external 0.234-μF ramp capacitor is used, the charge time will be 3.5 ms for an unknown voltage of 3.5 V. Setting the oscillator clock frequency to 4.2 MHz and the timer prescaler to divide by 8 results in an A/D's resolution of 12 bits or 1 part in 4096 for a charge time of 3.5 ms. [*Note:* When using this

A/D's maximum resolution (16 bits), the electrical noise levels in the A/D control logic may cause errors. This is because a 16-bit resolution is 1 part in 65,536 which means that a 3.5-V analog signal, in theory, can be resolved to 5.3 μV.]

The charge time is measured by the MCU's internal 16-bit timer, indicating that there is a definite relationship between the A/D ramp capacitor charge time and the timer count rate. Increasing the capacitor charge time will allow more timer counts to accrue, allowing a more precise A/D resolution. The external ramp capacitor's value can range from 0.01 to 2 μF. Capacitors larger than 2 μF should not be used since they will create high discharge energy in the MCU and will also be subject to higher internal leakage. The fastest 8-bit conversion A/D time is in the 500-μs range, which is too long for detecting short circuits and shutting down a power stage, but for the measurement of slow-moving variables, such as heatsink temperatures or DC supply voltages, this type of A/D is sufficient. (See MC68HC705JJ7 data sheet for more detailed A/D description.)

The PIC12C671 MCU (Microchip Technology Inc.) uses a RISC type of CPU. The CPU employs separate buses for instruction and data (Harvard architecture). This MCU uses 35 single-word instructions and can execute most instructions in four oscillator cycles or one instruction cycle, which is specified at 1 μs. One interesting feature of the PIC12C671 MCU is its 8-pin package. This low pin count requires some functions, such as the external oscillator, to be optional. An internal oscillator can be used, which frees up two pins and allows a total of five I/O pins and one input pin. The remaining two pins are used for power and common. This device is rated to operate over a −40 to +125°C ambient temperature and can sink or source 25 mA on any I/O pin. The maximum combined sink or source current for all its general-purpose I/O pins should not exceed 100 mA. Combining two of its pins in parallel allows 50 mA of current capability, which is sufficient to directly drive standard non-sensitive-gate triacs. The 25-mA rating is adequate to directly drive LEDs, optocouplers, and sensitive-gate triacs.

Four lines can be configured for A/D operation. The PIC12C671's A/D converter is an 8-bit unit, which gives a resolution of 1 part in 256 or 0.01953 out of 5 V. The maximum A/D conversion speed is 15.2 μs (at V_{ref} = 3 V). This conversion speed is adequate for control measurements such as temperature and average currents. It would not be fast enough to protect an IGBT power stage against a dead short, since most IGBTs are rated to sustain a shorted circuit for up to only 10 μs. The A/D's source impedance should not be higher than 10 kΩ, or errors may occur due to the A/D internal leakage currents.

The program code uses 14-bit wide instructions and is stored in a 1024 × 14 or 2048 × 14 EPROM element. The RAM element size is 128 × 8. This EPROM and RAM memory size is adequate for small control applications that need to perform only a few simple tasks or one medium-complex task. The RAM's contents can be kept alive with less than 10 μA of supply current when the MCU is operated in a standby mode. Under certain conditions, this can be further reduced to less than 1 μA.

10.2.2 Medium-cost motor control MCU examples

A *medium-cost MCU* is defined as one that usually sells between $2 and $5 in high volumes and in a ROM version. These midperformance MCUs are normally 8-bit, but, depending on their die size and complexity, can be 16-bit units. (Note that the price will also vary per the MCU's memory type—ROM, OTP, EPROM, flash, and EEPROM.) There are hundreds of MCU types that can be considered midrange, but only a few are designed specifically for motor controls. Two of these motor control MCUs—the MC68HC705MC4 and the MC68HC708MP16—have been shown in application examples in this book and will now be covered in more detail. The MC68HC705P9A is also a very popular 8-bit MCU with A/D for control applications and can be employed in uncomplicated motor controls. The MC68HC05MC4 is designed for basic brushless DC motor control applications, while the MC68HC08MP16 is more powerful and can be used for some AC induction motor drive applications. These two MCUs are CISC types and utilize von Neumann architecture.

The 28-pin MC68HC705MC4 8-bit MCU has a two-channel PWM timer module with a commutation multiplexer for brushless motor control. It also has a six-input 8-bit A/D, serial communications interface (SCI), 176 bytes of RAM, and 3584 bytes of EPROM. Its high-speed PWM subsystem is the key element that makes this MCU especially effective for motor control applications, generating the six PWM signals for controlling a brushless motor. The basic elements of the MC68HC705MC4 are shown in Fig. 10.4.

The 64-pin MC68HC708MP16 uses a higher-performance 8-bit CPU (Motorola designers call this the *CPU08 core*) than the MC68HC705MC4. The architecture of this '08 family is based on a module concept which allows new modules such as PWM units or MUX ports to be easily added to meet different application requirements. Such a modular design approach, as shown in Fig. 10.5, allows each module type to be further optimized or revised without redesign-

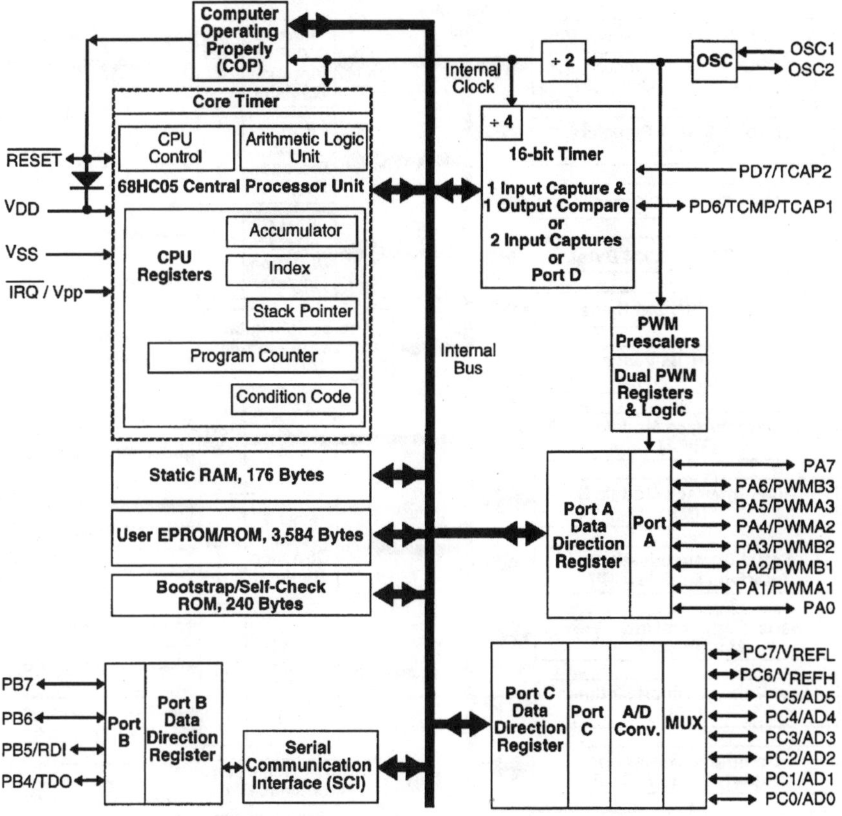

Figure 10.4 MC68HC705MC4 8-bit motor control MCU block diagram.

ing the entire MCU. Note how each module interfaces with the internal bus and is designed to perform a specific task.

This MCU has several features: 16,384 bytes of EPROM, 512 bytes of RAM, 12-bit 6-channel PWM unit, serial peripheral unit (SPI), serial communication unit (SCI), a 2-channel 16-bit timer, a 4-channel 16-bit timer, a 4-line fault input that can selectively disable certain PWM outputs, and a 8-bit 10-channel A/D converter. This device uses an 8-MHz internal bus frequency. Its software is upward-compatible from the MC68HC705MC4. Again, it is the PWM subsystem in this MCU that makes it a motor control device. The MC68HC708MP16 is designed primarily to drive three-phase AC induction motors with a V/F-type control strategy. It can support 30° PWM and certain types of space vector modulation. When used to control brushless DC motors, the MC68HC708MP16 allows fast shaft position computation with its input timer capture lines. Cycle-by-cycle current limiting and

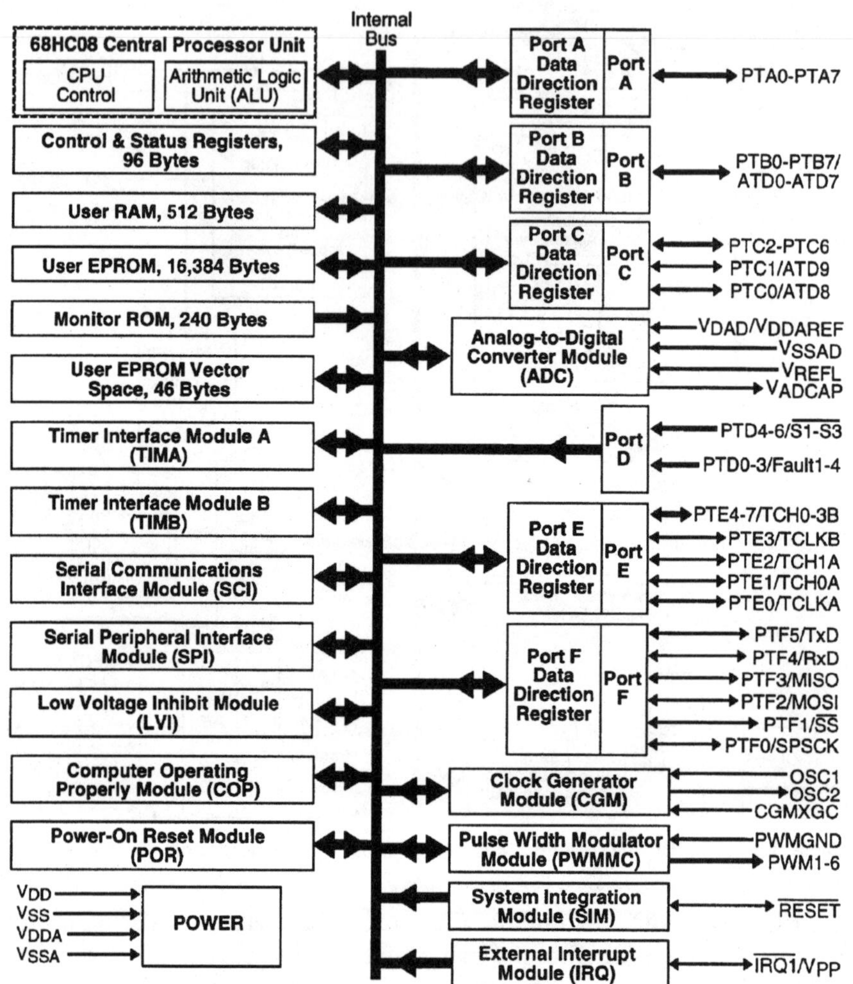

Figure 10.5 MC68HC708MP16 8-bit motor control MCU block diagram.

phase advance can be accomplished for BLM controls. The phase advance is useful for increasing the BLM's performance at high rpm.

The MC68HC08MP16 PWM unit also includes deadtime regulation, a feature that ensures minimal shoot-through currents in a half H-bridge but does introduce distortion in the motor's current. (A method to correct this problem is described in Chap. 9.)

10.2.3 High-cost motor control MCU examples

A *high-cost motor control MCU* is defined as one that usually sells for more than $5 with some units costing up to $30. These units are employed in precision motor control systems and in applications that require complex motor control strategies. Most high-cost motor control MCUs use 16-bit or 32-bit cores and employ a complex timer system. Two examples of high-performance MCUs that are designed for motor control are the TMS320C240 (Texas Instruments Inc.) and the 86XC196MH (Intel Corp.). MCUs such as the MC68HC16Y1 (Motorola Inc.), MC68332 (Motorola Inc.), and SAB-167 (Siemens Components Inc.) are just a few of the other high-performance MCU products that can be employed in motor drives because of their powerful timer units and overall fast throughput.

The 132-pin TMS320C240 is a 16-bit fixed-point DSP controller device with several features: 288 words of data RAM, 256 words of data/program RAM, 16,000 words of ROM or flash EEPROM, dual-channel 10-bit A/D with 16 inputs, three 16-bit general purpose timers, several timer compare units, a timer capture unit with quadrature encoding capability, SCI, and SPI, as well as other various features. Each of the three timers includes several subtimer elements that can be configured to generate up to 12 PWM channels. Note the special block labeled "test/emulation." Most high-performance MCUs require special lines to facilitate their testing and to simplify development tool interfacing.

The TMS320C240 has some special features for motor control. Its dual-channel A/D allows two simultaneous analog inputs to be sampled and converted, which is very handy for measuring the exact motor's current relationships of two phases. This MCU has one external interrupt (power drive interrupt) to disable the PWM output lines, which means that external logic may be required to combine several motor fault conditions to drive this one interrupt pin.

The 84-pin 86C196MH MCU is a 16-bit CISC-type device that is claimed to offer increased performance through a special register-to-register architecture. This device has several special features for three-phase motor control: a waveform generator unit that generates three complementary nonoverlapping PWM pulses in either an edge-triggered or center-aligned mode, deadtime generator, and an output disable that shuts off the PWM outputs.

10.3 Selecting an MCU for a Motor Control System

Besides the motor type and control strategy requirements, the physical layout of the motor control application can also affect the selection of the MCU. It may make sense to use two or more MCUs if the user control panel is fairly complex and located away from the motor control module. For example, you might choose one MCU for the user panel control, one for the overall system control, and one for just controlling the motor. A single large MCU that can accomplish everything is usually technically feasible but may not be cost-effective. In the manufacturing of MCU products, the cost is not linear for building an MCU whose die is twice as large. The cost will be somewhat exponential because of die defects, testability, and higher fault rates. Therefore, a large complex MCU die will cost more than twice of one that is half its size. Another issue of concern focuses on packages and printed-circuit-board (PCB) layout. High-pin-count MCUs (over 48 pins) require more expensive packages and are more difficult for routing the PCB traces. Multilayer PCBs are sometimes needed to accommodate these high-pin-count MCUs.

Other factors need to be considered when deciding to use more than one MCU. One is that a second MCU can also act as a safety device. In a critical motor control system that may cause damage if the motor control becomes unstable, a second MCU is necessary. Its prime function is to monitor the activity of the main motor control MCU and shut down the motor's power if necessary. A second MCU can also perform other tasks, such as control panel interfacing, communications, and diagnostics.

The bandwidth of the MCU that is consumed by the motor control program is an important consideration when other complex motor system tasks must be accomplished. Different motor types, control strategies, and system requirements affect the capacity of the MCU. A device such as the MC68HC705MC4 is designed to operate a BLM or simple motor speed controls and will have little bandwidth or program capacity remaining to accomplish other complex tasks. The MC68HC708MP16 can be used in a BLM control and will have about 60 percent of its bandwidth intact to accomplish other tasks. High-performance 16-bit or 32-bit MCUs can accomplish the BLM control function and may have over 80 percent of their bandwidth available to run other complex tasks. *If only one MCU is to be used in the motor control system, it is important to allow extra capacity or bandwidth in this single MCU for performing indirect or non-motor-control tasks.* These tasks may include keyboard scan routines, display drivers, overall system control, external diagnostics through a serial port, special test routines, predication of possible failures, and so forth.

Choosing the minimal MCU to accomplish the fundamental motor control makes sense if no other tasks are required. Sometimes one may question the wisdom of using a MCU just to do a simple motor control. As mentioned previously, it is the programmability of the MCU that allows easy changes to the motor control strategy, PWM modulation type, and other nuances that make it attractive for basic motor control applications. The downside is that the motor control designer must become familiar with MCUs, programming languages, and development tools.

Other considerations for selecting an MCU motor control device focus on its motor control attributes. It should have more than one fault input line to shut off its PWM output lines. This allows more than one fault source to be detected and minimizes external logic. Its PWM unit should be very flexible, easy to program, and require minimal software intervention. The PWM timing should allow implementation of several types of modulation methods and be able to change the modulation method without disturbing the motor's operation. This can be important for motor drives that operate from zero speed to very high speeds. Other important MCU criteria are fast math ability for vector control strategy—the MCU will probably need to perform a multiply, add, and/or accumulate instruction in one clock cycle.

10.4 Motor MCU Development Tools

The implementation of an MCU-based motor control system can be simplified by using development tools built specifically for motor applications. This will save time in designing hardware and in programming because you are not starting from scratch. The circuit-board layout itself can be a challenge for a novice. Writing program code can also be a formidable learning experience for the beginner. A motor control development system will be examined for two 8-bit MCUs—the MC68HC05MC4 and MC68HC08MP16—and two power stage boards.

A complete MCU development system would include several elements, as shown in Fig. 10.6. A desktop or laptop personal computer

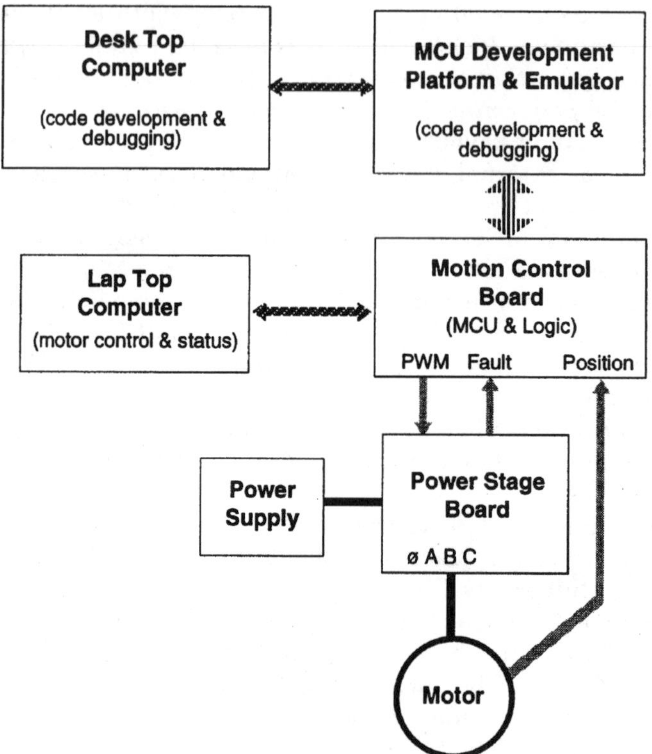

Figure 10.6 Motor control MCU development system.

is used for creating the program code. After the code has been written, it is complied or assembled and tested in a code development station and then downloaded into the MCU's target memory. MCU manufacturers and instrument manufacturers may offer two or three levels of development tools. The lowest-cost versions usually do not allow real-time debugging and are best suited for low-speed, open-loop-type applications. The medium-cost development tools allow real-time debugging, dual-port memory, and a bus state analyzer. The dual-ported memory enables the designer to fine-tune filter circuits or monitor changes to program variables while the program is running at full speed. The bus state analyzer's full range of sequenced or logical event triggering and data-capture modes can verify nested or complex program flows. High-cost MCU development instruments usually operate with their own internal computer system rather than a PC. These high-cost MCU development stations offer more debugging features and higher performance.

The MCU development station should include real-time debugging and should be compatible with most desktop computers. One such system is the Motorola Modular Development System (MMDS). A complete MC68HC05MC4 motor control development system consists of a platform (M68MMDS05), an emulator (M68EM05MC4), a target head adapter (M68TA05MC4P28), a low-noise cable (M68CBL05A), and a motion control board (KITITC127/D), plus a power stage board [KITITC122/D 48-V 3ø (three-phase) BLM or KITITC132/D 380-V 3ø AC motors]. This MMDS system has the following specifications and capabilities:

- Real-time, nonintrusive, in-circuit emulation at the operating frequency of the target device
- Control of code execution in run or step mode
- Real-time bus state analyzer
- Four complex breakpoints (any combination of address, data, control, and/or I/O)
- 64 instruction breakpoints
- 32 real-time variables
- 32 bytes of real-time memory
- 64 kbytes of fast emulation memory (SRAM)
- Current-limited target input/output connections
- Six software-selectable oscillator clock sources
- Command and response logging to save session history and/or to generate script files
- SCRIPT command for automatic execution of a sequence of commands (such as system configuration setup and automated test runs)
- Assembly source-level debugging
- Host/emulator communications speeds as high as 57,600 baud
- Extensive on-line device information via CHIPINFO command
- Integrated development environment (IDE) including editor, assembler, and source-level debugger
- Support for full suite of separately purchased device-specific emulator modules (EMs) and flexible target system cables
- Station module with built-in power supply, 85 to 264 V AC
- Compact size: 15.38 in deep × 10.19 in wide × 2.75 in high (39 cm deep × 25.9 cm wide × 7 cm high)

- Lightweight: 6 lb (2.7 kg)
- 9- to 25-pin serial cable adapter and 9-lead RS-232 serial cable
- IDE including editor, assembler, and assembly source-level debugger
- MMDS user manuals

The MMDS requires a device specific emulation module (EM) that contains circuitry that emulates the specific features of the target 68HC(7)05 or 68HC(7)08 microcontroller. The EM attaches to the MMDS through two expansion connectors located on the bottom side of the EM board. Connection to your target system is then made through a separately purchased target cable and target head adapter that attaches to a target connector located on the top side of the EM board. In addition to the target connector, each EM includes a connector that provides the ability to interface the EM to a logic analyzer for external hardware trace of code execution. The target head adapter attaches to the motion control board. The motion control board is connected to the power stage board which is wired to the motor.

10.5 Low-Voltage Motor Development Boards

A low-voltage brush or brushless DC motor system can be implemented by hooking up power and a motor to the MC68HC705MC4 motion control board (KITITC127/D) and low-voltage power stage board (KITITC122/D). The application example shown in Fig. 10.7 illustrates system connections to a low-voltage power stage and a brushless DC motor. This arrangement can stand alone, or the MC68HC705MC4 motion control board can be connected to a software development sys-

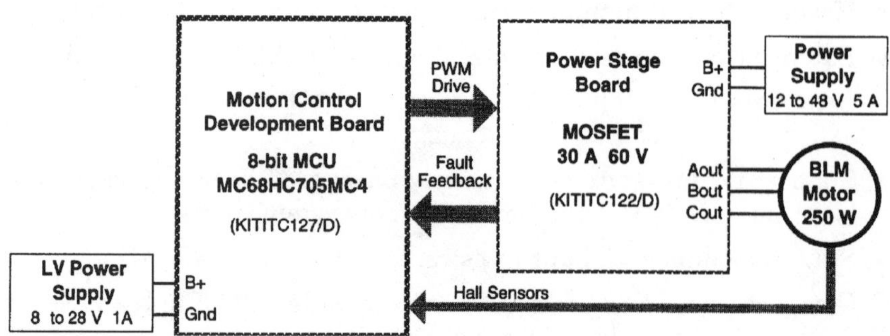

Figure 10.7 Brushless DC motor MCU development system.

tem (MMDS05). The two boards are designed in such a way that the drive and feedback ribbon connectors line up and are connected with the supplied ribbon cables. Once they are plugged in, it is only a matter of connecting power supply, motor, and Hall sensor leads to get a system up and running. The MC68HC705MC4 is preprogrammed to operate a three-phase brushless DC motor. The source code is also provided to enable you to modify it to meet specific motor application requirements. A review of both the MCU logic and power stage board provides insight into their actual circuit design and is applicable to any low-voltage MCU-based motor control system.

10.5.1 Brushless motor MCU development system

The MC68HC705MC4 MCU motion control board is designed with most of the necessary components that are required for use in a motor control system with the exception of the power stage. The power stage will be reviewed later. The MC68HC705MC4 evaluation board provides control signals for either brush or brushless DC motors. Input signals from switches and a potentiometer are used to provide RUN/STOP, FORWARD/REVERSE, and SPEED commands. These commands can also be provided from an external source. Three Hall inputs are provided for connection to the brushless DC motor. The control and Hall signals are logic level, except for the potentiometer input, which is a 0- to 5-V analog signal.

10.5.1.1 Using the MC68HC705MC4 motor control development board. This board can run in two configurations. It will operate independently with the preprogrammed MCU or with the MCU removed, under control of the development system (MMDS) described previously, and is connected to an M68HC05MC4 emulator via M68CBL05A cable and M68TA05MC4P28 target head adapter. The output side of this board connects to a low-voltage or high-voltage (KITITC132/D) power stage with a 14-pin ribbon cable. Six outputs provide power device control signals for controlling a three-phase brushless DC motor. Brush DC motors can be controlled by using either one or two of the three available phases. All six outputs are buffered and will sink 25 mA, making them suitable for directly driving optocouplers in isolated gate drives. A switched +5 V is also provided to serve as the B+ power source for optocoupler input diodes. It is turned off at reset to facilitate orderly powerup and powerdown of the gate drives.

The MC68HC705MC4 MCU logic board schematic (Fig. 10.8) shows several additional elements that have been added to facilitate motor development and to minimize possible catastrophic failures. A special

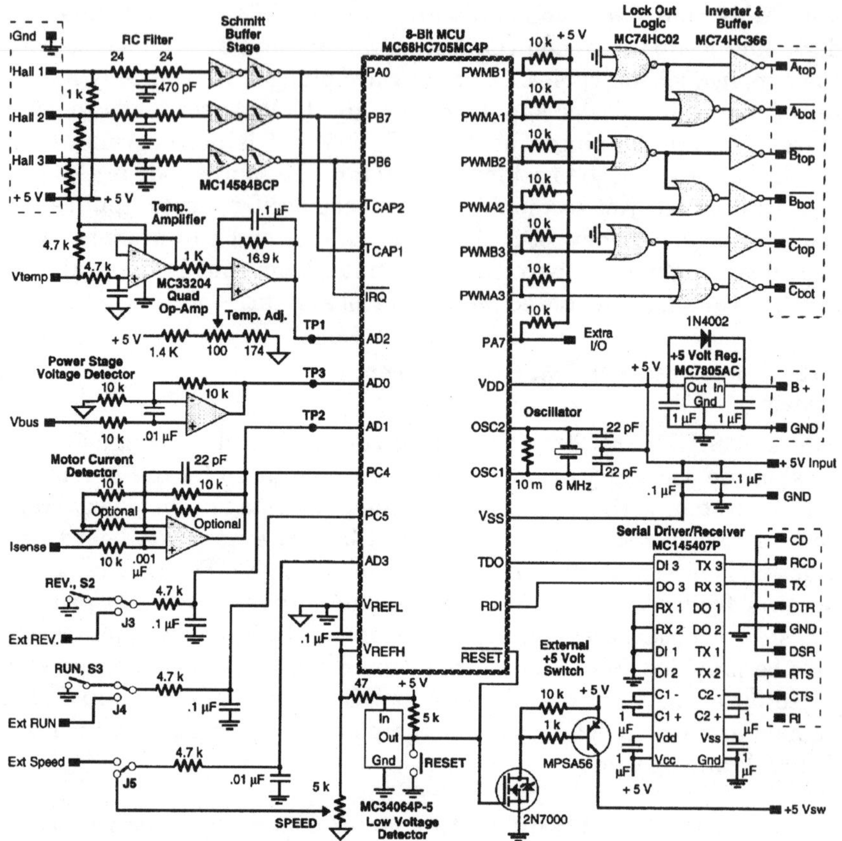

Figure 10.8 MC68HC705MC4 motor control development circuit.

logic stage ensures that the PWM output lines do not operate the half H-bridge power stage in a destructive manner. This is a very cost-effective method to protect several dollars' worth of power electronics.

The following descriptions of the board's input and output lines or pins reveals the type of signals that should be employed. The inputs and outputs are grouped into six connectors. One connector contains three Hall sensor inputs, a + 5-V connection for the Hall sensors, and a ground. Input connection B+, if used instead of the + 5-V input, is supplied through a two-pin connector. It will accept power supply voltages from 7.5 to 28 V when driving a low-voltage power stage, and 7.5 to 15 V when driving a high-voltage power stage. The lower voltage limit for driving the high-voltage power stage comes from the need to supply more current from the 5-V bus to drive optocoupled inputs. A 14-pin ribbon connector contains six outputs for driving a power stage and a switched 5-V power line. Feedback signal inputs are located on another 16-pin ribbon connector, where there is provision for temperature, bus voltage, and current-sensing feedback signals. There is also a DB-9 connector with the standard configuration for RS-232.

The board is powered in one of two possible methods. Either an unregulated DC voltage—a 7.5- to 15-V-DC supply power supply, for example—can be fed into the 2-pin connector, or a + 5-V regulated supply can be connected to the + 5-V input terminal. One or the other, not both, are required. An advantage of using the unregulated input is that an external + 5-V regulated supply is not necessary, and the on-board + 5-V regulator is utilized. If the motor is powered from a 12- to 15-V supply, this same voltage bus can be used to power the MCU logic board. Watch out for excessive power dissipation in the on-board + 5-V regulator, a MC7805AC. It is not heatsunk, so its input voltage should be limited to + 15 V, and its case temperature should be checked if extra circuitry is added that is powered from this regulator.

The other inputs include a speed control input and an analog ground for the speed control input. The speed input can be used to control motor speed with an external 0- to 5-V analog signal. Zero volts corresponds to zero speed and 5 V, to full speed. To use it, a jumper (J5) needs to be moved to the external position, thus disabling the on-board speed control pot.

A pin is provided to externally reverse the motor when it is grounded. A jumper (J3) needs to be moved to the external position, which disables the on-board FORWARD/REVERSE switch. A motor run input pin enables the motor when it is grounded. To use it, a jumper (J4) needs to be moved to the external position, which disables the on-board RUN/STOP switch.

A 5-pin connector is employed for the brushless DC motor's Hall shaft position sensors. Three pins are intended to interface with open collector Hall sensors and are buffered with Schmitt triggers and filtered for noise immunity. The other 2 pins connect to the Hall effect sensors + 5 V and ground. It is important not to mix this ground with the motor's power ground. Doing so will result in motor instability, since the Hall-effect signals will be corrupted with ground noise.

A 16-pin ribbon cable connector is used for feedback signals from the power stage board. These motor feedback signals include motor current, diode temperature sensor, and motor bus voltage. The current sense input is connected to A/D channel AD1 through a gain-of-2 noninverting amplifier. The gain of this amplifier has been made easy to adjust with optional resistors. Motor bus voltage is connected to A/D channel AD0 through a gain-of-2 noninverting amplifier.

Temperature sensing is accomplished by translating the forward voltage of a diode into a usable A/D voltage with an amplifier. The output of this amplifier is connected to A/D channel AD2. Two pins on the feedback ribbon cable serve as a ground for the analog circuits. This ground is routed in such a way that all the analog returns connect with digital ground at just one point. Again, it is important that the AGND not be looped into or mixed into the motor's power ground.

The six PWM outputs are brought out with a 14-pin ribbon connector. Two of its pins are connected to the 5-V bus through a switch that is open at reset. The resulting switched 5 V can power optocouplers in isolated off-line motor drives. The off-at-reset feature keeps output transistors off during reset. Six other pins are tied to ground. They provide a return for the switched + 5 and noise isolation between output lines in the ribbon cable.

The PWM outputs drive three phases of half-bridge configured output transistors and are set up in an active low configuration. Processor outputs are routed through cross-coupled NOR gates to provide protection from inadvertent simultaneous turnon of a top and bottom transistor in one phase. They are also buffered to allow direct drive of optocouplers in isolated off-line motor drives.

A DB-9 connector is set up for RS-232 communication with the HC705MC4. It has standard RS-232 pinouts. Test points (TP1, TP2, and TP3) provide access to buffered feedback signals for temperature, motor bus current, and motor bus voltage. These voltages are seen by A/D converter inputs AD2, AD1, and AD0. The temperature feedback voltage can be calibrated with a potentiometer. Test instruments can be grounded with the GND and AGND (digital and analog ground) test points. All six outputs and a ground are also available as test points.

10.5.1.2 BLM application. The PWM signal is generated by a dedicated PWM generator in the MC68HC705MC4 and can pulse-width-modulate the output transistors at a 23-kHz rate with the processor clocked by a 6.00-MHz crystal. Speed is controlled by the duty cycle of this PWM signal, while direction is determined by commutation sequence. Hall sensor inputs provide commutation feedback from the motor. They are buffered with Schmitt triggers and filtered, to provide noise immunity.

An important application consideration is loss of power. If the MCU and power stage boards are powered from separate power supplies, loss of power to the MCU board will pull all the PWM inputs to a low state. With all its inputs low, the power stage board will turn on all three upper half-bridge output transistors and abruptly brake the motor. During powerdown, removal of the MC68HC705MC4 board's power before removal of the motor's power supply produces this condition and brakes the motor.

A number of design considerations are important in brushless DC motor drive systems. Sensor inputs, simultaneous conduction lockout, powerup, grounding, and optocoupler drives are discussed below.

10.5.1.3 Sensor inputs. For brushless motors that use sensor inputs for commutation, noise immunity of the sensor inputs is a key design consideration. Noise on these inputs can be particularly troublesome, since commutating to the wrong state can jerk the motor and increase power dissipation. To facilitate noise robust sensor inputs, Schmitt triggers have been placed between the Hall sensor input connector and ports PA0, PB7, and PB6. Hysteresis makes the Schmitt trigger significantly more robust than using processor ports directly. In addition, these signals are filtered with 100-ns single-pole filters. Using relatively low value pull-up resistors, on the order of 1 kΩ, provides an additional measure of noise immunity.

The way that the code is written also has a significant influence on noise robustness. Since the sequence of commutation is known, on the basis of the state of the forward/reverse input, it is relatively easy to detect an out-of-sequence Hall sensor input. Generally speaking, it is desirable, when this occurs, to turn off all the power transistors until a valid Hall code is received.

10.5.1.4 Lockout. Especially on a machine that will be used for code development, it is important to prevent simultaneous conduction of upper and lower power transistors in the same phase. This is easily accomplished with the cross-coupled NOR gates shown in Fig. 10.8. These logic gates lock out the bottom-transistor drive signals whenever the top transistor is on in the same phase. Code errors will, therefore, not destroy power stage output transistors by turning on the top and bottom of one half-bridge simultaneously. This arrangement also prevents simultaneous conduction in the event of a noise-induced software runaway.

10.5.1.5 Powerup/powerdown. When power is applied or removed, it is equally important that top and bottom output transistors in the same phase are not turned on simultaneously. Since the outputs are low when unpowered, sequencing is important in optocoupled drives where the output buffers drive optocouplers.

10.5.1.6 Grounding. Finally, board layout is an important design consideration. In particular, the way in which grounds are tied together influences noise immunity. To maximize noise immunity, two grounds are used. Digital ground (GND) is common to the power supply return and serves as a general-purpose ground. An analog ground (AGND) ties the speed control input return, op amp signal grounds, and the A/D converter's V_{refl} together before connecting with digital ground at only one point. AGND also runs as a separate trace to pins of the feedback connector.

10.5.2 Brushless motor MCU development board software

Four software modules and a batch file for assembling them are contained on a 3.5-in floppy diskette included with the MC68HC705MC4 motion control development board. Modules A and B contain source code for very basic brush and brushless motor control. Modules C and D contain source code for a somewhat "smarter" version of the same software. The brushless modules are written for motors with 60° Hall sensors. These files can be assembled with the Motorola "MASM" assembler. Some syntax changes may be necessary for assembly with assemblers from other vendors. (See Apps. G and H for MC68HC705MC4 software code listing.)

To assemble the files, consult the WHATSUP.DOC file on the diskette. Note the handling of the 68HC705MC4's watchdog timer bit in the MOR register when emulating the processor with a MMDS05/MMEVS05 or when using the EPROM device on the ITC127 board. The watchdog must be disabled when the processor is used in a standalone mode on the ITC127 board. The MOR register code must be commented out for loading the S-records into the MMDS05/MMEVS05, as location $F00 is considered to be ROM in the emulator's memory map. The 68HC705MC4 residing in the emulator must have its watchdog timer bit in the MOR register disabled.

This software shows basic methods to commutate and provide basic control for brushless and brush-type DC motors using the 68HC705MC4. Complexity was intentionally kept to a minimum so that the user could spend a small amount of time studying the software and still understand the steps required to spin a motor. All the software modules provide open-loop motor control. Closed-loop algorithms, such as PID, can be added to provide better speed control under varying load conditions.

The modules have been developed using ITC127 control boards, connected to ITC122 and ITC132 power stages. Brushless DC motor software modules were verified with fractional horsepower motors from several companies (Pittman, Hurst, Astromec, and Kollmorgen). Brush DC motor software modules were tested using an automotive radiator fan motor. A brief description of each module follows.

10.5.2.1 SMPLBRLS.ASM. This software module provides very basic commutation of a brushless DC motor. There is no acceleration software control provided by the module. The software is very straightforward and relatively well documented. For a first-time motor control user, an explanation of this software may be helpful.

There are four sources of interrupts to the processor. Each Hall-effect rotor position sensor transition provides one of three interrupts. The fourth interrupt is provided by the processor's timer overflow interrupt. The RUN/STOP toggle switch is monitored by the program. When the switch is switched to the RUN position, the potentiometer is read by the A/D converter and that value is placed into the PWM data register. Because the motor is not rotating, no Hall sensor interrupts are received. Nothing will happen until the timer overflow interrupt has incremented a variable named TIMEOUT to 3. When this occurs, the program looks at the position of the rotor and communicates it to its next position in the rotation sequence. As the motor rotates, the Hall effect sensors will then interrupt the processor and commutation will continue. At this point, unless the motor is rotating very slowly, the timer overflow interrupt that helped the motor to start will no longer have an effect on the program, since rotation will keep the values of TIMEOUT to less than 3. For simplicity, this software rotates the motor only in a clockwise direction as viewed from the shaft end of the motor. Speed control is provided by reading the value of the speed pot. on the board and placing the pot. value in the PWM data register.

10.5.2.2 BRUSHLES.ASM. This software module provides somewhat more intelligent commutation of a brushless DC motor than SMPLBRLS.ASM. To protect power transistors, PWM limits are checked at the lower and upper ends of the PWM duty cycle. Motor direction control is provided by sensing the position of the FORWARD/REVERSE toggle switch. To help eliminate jerking at initial start-up and on direction changes, a simple form of acceleration control is provided. Hall-effect sensor error code checking/recovery is included to enhance noise immunity. In addition, the speed control A/D value is integrated to eliminate jitter at speed extremes. Motor commutation is the same as that provided in the SMPLBRLS.ASM software.

10.5.2.3 SMPLBRSH.ASM. Controlling a brush-type DC motor is much easier than driving a brushless DC motor. This software sets up the PWM generator in such a way that A_{out} will switch B+ voltage to the motor and C_{out} will be the ground output to the motor. The A_{out} output will receive the PWM signal. This software module simply reads the RUN/STOP toggle switch and, if it is in the OFF position, the program will force the PWM duty cycle to zero, thus stopping the motor. If the switch is in the ON position, the pot. value is read by the A/D converter and then that value is placed into the PWM data register, controlling the speed of the motor.

10.5.2.4 BRUSH.ASM. This software module provides somewhat more intelligent control of a brush DC motor than SMPLBRSH.ASM. Similar to BRUSHLESS.ASM, it provides PWM limits, forward/reverse, and controlled acceleration. Speed control is the same as in SMPLBRSH.ASM.

10.5.3 Low-voltage motor power stage

In this section, we are introducing a MOSFET power stage that is designed to interface with microprocessor development tools for three motor control MCUs: the MC68HC705MC4, MC68HC708MP16, and MC68HC16Y1. This power stage is intended to run brush or brushless DC motors at up to 4 A of continuous current from DC bus voltages up to 48 V. A summary of the information required to use an ITC122 low-voltage micro-to-motor interface board follows.

10.5.3.1 Power stage board function. The low-voltage power stage board is designed to provide a direct interface between microcomputers and fractional horsepower motors. It accepts six logic inputs, which control three complementary half-bridge outputs and is arranged in such a way that a logic ZERO at the input turns on the corresponding power transistor. This type of configuration is applicable to pulse-width-modulated (PWM) systems in which the PWM signal is generated in a microcomputer, digital signal processor, or other digital system. It is suitable for driving fractional horsepower brush and brushless DC motors. In addition to controlling the motor, it provides current sense, temperature sense, and bus voltage feedback and is designed to interface directly with Motorola HC05MC4, HC08MP16, and HC16Y1 motor control development tools.

Coupled with logic control boards for the HC05MC4, HC08MP16, and HC16Y1, it completes the link between microcomputer development tools and a motor. Its use allows code to be written before hardware design is completed in an environment where mechanical outputs can be seen while the code is written. The design, as shown in Fig. 10.9, can also be used as a reference for expediting hardware development.

10.5.3.2 Electrical characteristics. The low-voltage power stage can operate from 12 to 48 V and up to 5-A current levels per phase. The maximum current ratings are based on the application of PWM sig-

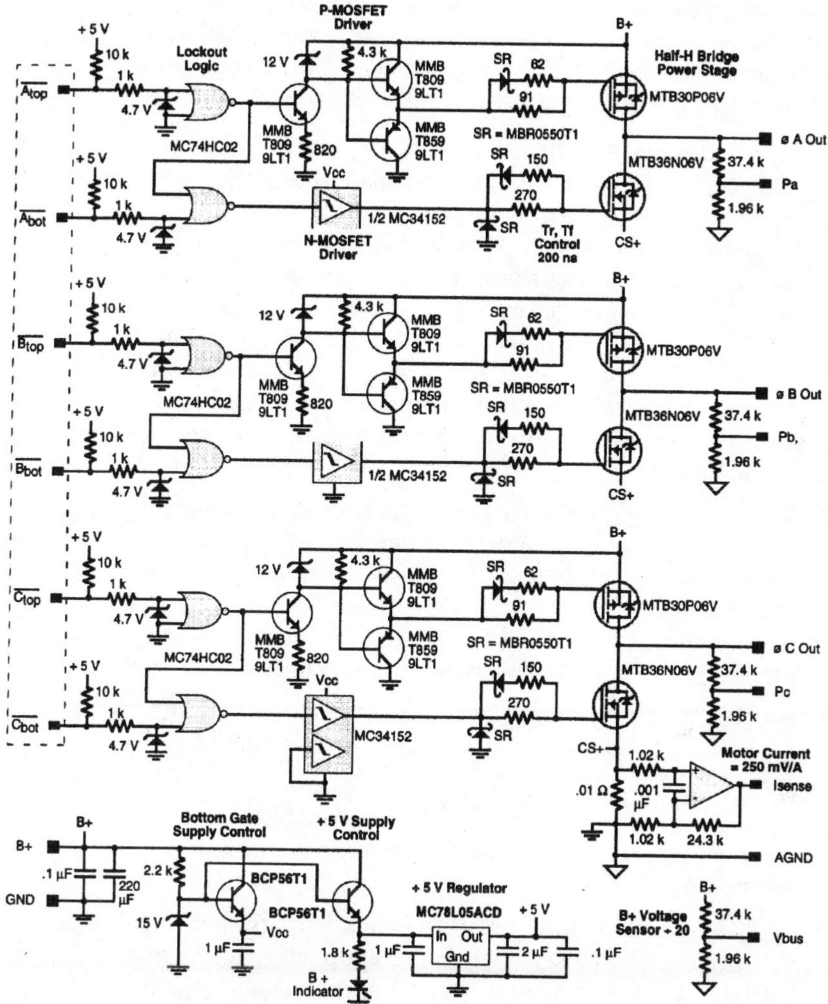

Figure 10.9 Low-voltage motor power stage development circuit.

nals only to the top inputs, a PWM frequency of 23 kHz, and a supply voltage of 40 V. Bonding heatsinks to the back of the board and/or providing airflow will significantly increase both power dissipation and output current ratings.

10.5.3.3 Pin-by-pin description. Inputs and outputs are grouped into four connectors. Two connectors are provided for inputs, one with screw terminals and the other for a ribbon cable. Either can be used; they are wired in parallel. Outputs to the motor, B+, and ground are also supplied on a screw connector. Feedback signals are grouped together on a separate ribbon cable connector. In addition, through-hole pads have been placed adjacent to the current sense feedback resistor and the temperature sensing diode for easy access.

1. *Inputs:* Inputs A_{top}, A_{bot}, B_{top}, B_{bot}, C_{top}, and C_{bot} are logic inputs. A logic 0 turns on the input's corresponding output transistor; that is, a logic 0 on A_{top} turns on output transistor A_{top}, and so on. Logic levels are standard 5-V CMOS. Input current is higher, typically 500 μA, since each of these inputs is pulled up with a 10-kΩ resistor. In the absence of any inputs, all output transistors are turned off. If a logic 0 is inadvertently applied to both top and bottom inputs for one phase (i.e., A_{top} and A_{bot}), the bottom input is locked out and only the top output transistor is turned on. This feature helps protect the bridge from errors that may occur during code development.
2. *Motor outputs:* Motor output terminals are labeled A_{out}, B_{out}, and C_{out}. This output configuration can be used to drive a fractional-horsepower three-phase brushless DC motor, a reversible brush DC motor, or three brush DC motors unidirectionally. When driving a single brush DC motor, thermal performance is optimized by using A_{out} and C_{out} for the motor connections.
3. *B+:* B+ is the motor power input connection. It is the only supply required. Acceptable input voltage range is 12.0 to 48 V DC. B+ is located on the output connector.
4. GND: There are multiple ground connections. One of the two grounds on the output connector should be used as the power supply return.
5. I_{sense}: I_{sense} is a current sense feedback voltage that appears on one pin of the feedback ribbon cable connector. It is derived from a 0.01-Ω low-inductance surface mount sense resistor that is in series with the ground return. The voltage across this resistor is amplified with a gain of 25. I_{sense}, therefore, represents motor current with a scale factor of 250 mV/A. Since only return current is measured, this output will not detect shorts from the motor outputs to ground or B+.
6. V_{bus}: V_{bus} is a bus voltage feedback signal that appears on one pin of the feedback connector. It is derived from a 37.4-kΩ/1.96-kΩ divider which scales B+ at a ratio of approximately 50 mV per volt. This is an unfiltered and unbuffered signal.
7. V_{temp}: This is a temperature output signal derived from a forward-biased diode's ON voltage and appears on one pin of the feedback

connector. The temperature diode is mounted in such a way that it measures board temperature adjacent to a power transistor.

8. *Phase voltage feedback:* Phase voltage feedback signals P_a, P_b, and P_c are also included on the ribbon cable connector. Three pins provide motor phase voltages divided with the same 50-mV/V ratio as V_{bus}. They are also unfiltered and unbuffered signals.

10.5.3.4 Power stage application example. An important application consideration is PWM topology. The MCU controller is programmed to pulse-width-modulate the A_{top}, B_{top}, and C_{top} inputs, and just commutate A_{bot}, B_{bot}, and C_{bot}. This configuration performs better from a power dissipation standpoint than does its more commonly used alternative, namely, lower half-bridge PWM with upper half-bridge commutation. When the upper half-bridge is pulse-width-modulated, circulating currents flow through the lower $r_{ds,on}$ and lower forward diode voltages of the N-channel transistors. Since transition times for both P-channel and N-channel transistors are approximately the same, high-side PWM is considerably more efficient. With more efficient operation, available output power to the motor is maximized.

10.5.3.5 Low-voltage power stage design considerations. A block diagram that provides an overview of the design is illustrated in Fig. 10.10. Top and bottom inputs for each phase are coupled to gate-drive circuits through cross-coupled NOR gates. This arrangement locks out the bottom input when both inputs for one phase are low, thereby adding robustness for lab use. If all six inputs are low, transistors A_{top}, B_{top}, and C_{top} are turned on, braking the motor. This condition can occur when the MCU logic board is being powered down.

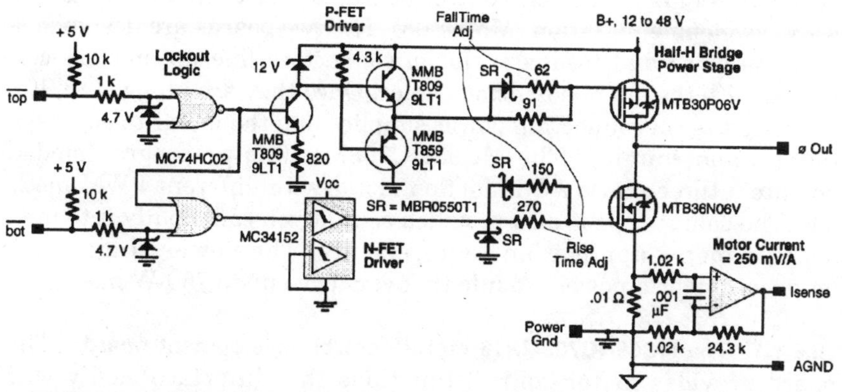

Figure 10.10 Low-voltage motor gate drive circuit.

The output is a three-phase bridge that is made from complementary 30-A surface mount MOSFETs. Gate drive for the N-channel transistors is provided by MC33152 MOSFET drivers and for the P-channels by a discrete circuit. Both the P-channel and N-channel transistors have transition times targeted for approximately 200 ns. This target allows sufficiently high-value-gate-drive resistors to somewhat soften diode snap, yet it produces switching losses that are less than static losses at 23 kHz.

10.6 High-Voltage Motor Development Boards

The design of high-voltage motor electronics is less forgiving than low-voltage motor systems. A high-voltage drive will usually fail (sometimes in a spectacular fashion) when problems occur. The same problems may cause erratic operation only in a low-voltage motor drive. Both the power stage and logic board must be designed to operate in an electrically hostile environment. In this section we will discuss the design and application of a high-voltage motor development system.

10.6.1 AC induction motor MCU development system

An AC induction motor system can be implemented by hooking up power and a motor to an MC68HC708MP16 motion control board (KITITC137/D) and a high-voltage power stage board (KITITC132/D). An application example, shown in Fig. 10.11, illustrates system connections to a high-voltage power stage board and a three-phase AC induction motor. This arrangement can run standalone, or the MC68HC708MP16 motion control board can be connected to a software development station (MMDS08). The two boards are designed so that the drive and feedback ribbon connectors line up and are connected with the supplied ribbon cables. Once they are plugged in, it is only a matter of connecting power supplies and the motor to get a system up and running. The MC68HC708MP16 is preprogrammed to operate a three-phase AC induction motor with different PWM methods. The source code is also provided so that you can modify it to meet specific motor application requirements. The power stage board employs a hybrid power module that can drive up to 750-W motors.

10.6.1.1 The MC68HC708MP16 motor control development board. This board provides motor control functions that interface easily with power stages and emulators. Its preset configuration is applicable to

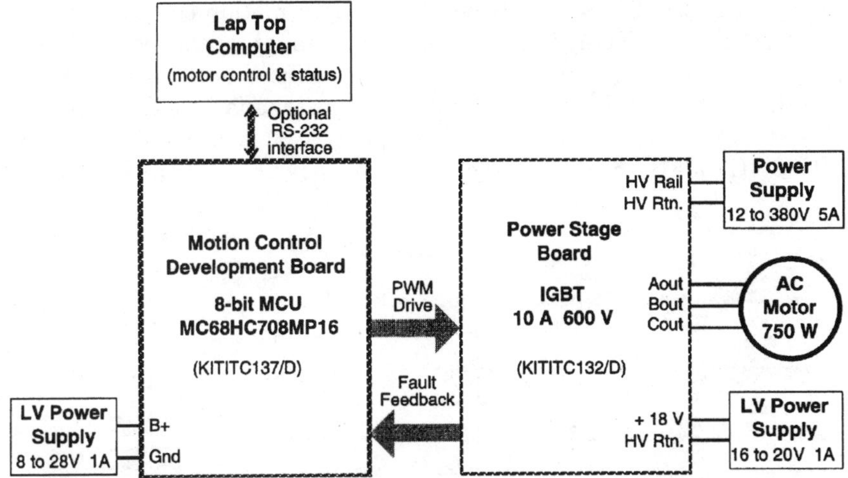

Figure 10.11 AC induction motor development system.

AC induction motors but can be adapted for brush and brushless DC motors. A summary of the information required for use of the MC68HC708MP16 motion control development board follows. Discussions of hardware design and software are included under separate headings.

Inputs are accepted from switches and a potentiometer on the board or external RUN/STOP, FORWARD/REVERSE, and SPEED signals. The speed input is a 0- to + 5-V signal with 0 V corresponding to zero speed and 5 V producing full speed. RUN/STOP and FORWARD/REVERSE are logic inputs, with logic lows producing run and reverse outputs. Hall 1, Hall 2, and Hall 3 inputs are also provided for connection to brushless DC motors.

The MC68HC708MP16 motion control development board is designed to run in two configurations. It will operate on its own with the processor supplied, and, with the processor removed, it will connect to an M68HC08MP16 emulator with an M68CBL08A cable and M68TA08MP16P64 target head adapter. The output side of this board connects to a high-voltage or a low-voltage (KITITC122/D) power stage via ribbon cable. Six outputs provide power device control signals for three-phase induction or brushless DC motors. Brush DC motors can be controlled by using either one or two of the three available phases. All six outputs will sink 25 mA, making them suitable for directly driving optocouplers in isolated gate drives. A switched +5 V is also provided to serve as the B+ power source for optocoupler input diodes. It is turned off at reset to facilitate orderly powerup and powerdown of the gate drives.

Figure 10.12 shows the MC68HC708MP16 motion control development schematic. Note that there is no special external logic stage to prevent the upper and lower PWM lines from both switching on simultaneously, which may invoke a power stage failure. To minimize this risk, it is important to ensure that the FAULT input lines on the MC68HC708MP16 are connected to the power stage and that the

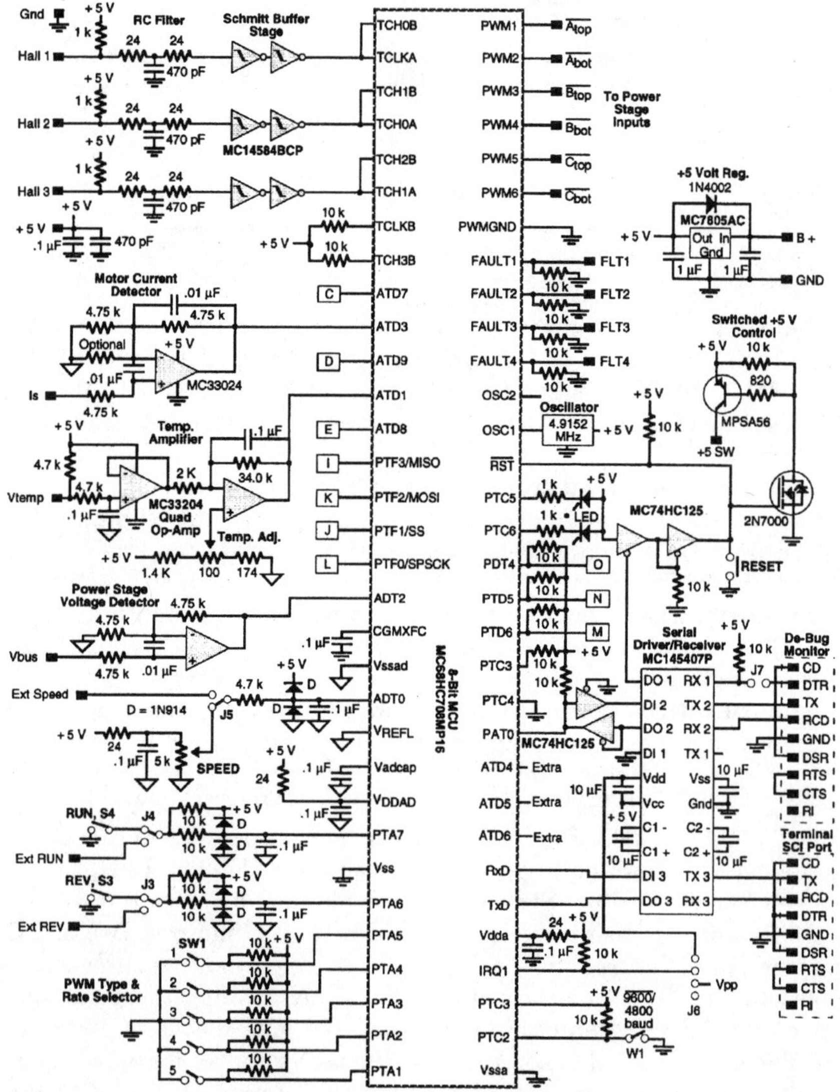

Figure 10.12 MC68HC708MP16 motor control development circuit.

MCU's software is written to allow a fault input shutdown of the PWM output lines.

As we describe the board's input and output lines or pins, you can see the type of signals that should be employed. The inputs and outputs are grouped into six connectors. Control signal inputs are located on a 9-pin connector. They are optional external interfaces that include a provision to power the board with + 5 V if the B + input on the 2-pin connector is not used. A 5-pin screw connector contains 3 Hall sensor inputs, a + 5-V connection for the Hall sensors, and a ground. The B + input, if used instead of the + 5-V input, can accept power supply voltages from 7.5 to 28 V when driving a low-voltage power stage board, and 7.5 to 15 V when driving a high-voltage power stage board. The lower voltage limit for driving the high-voltage power stage comes from the need to supply more current from the 5-V bus to drive optocoupled inputs. The 14-pin ribbon connector contains 6 outputs for driving a power stage and a switched 5-V power line. Feedback signal inputs are located on another 16-pin ribbon connector, where there is provision for temperature, bus voltage, and current sense feedback signals. There is also one DB-9 connector with the standard configuration for RS-232 for communication with personal computers. When connected to a serial port on a personal computer, it can be used to allow keyboard control of motor drive functions. Another DB-9 connector is used for a monitor mode—it supports background debugging for the MCU.

The board is powered in one of two possible methods. Either an unregulated DC voltage—for example, a 7.5- to 15-V-DC supply power supply—can be fed into the B + connector, or a + 5-V regulated supply can be connected to the + 5-V input. One or the other, not both, is required. An advantage of using the unregulated B + input is that an external + 5-V regulated supply is not required and the on-board + 5-V regulator is utilized. You should watch out for excessive power dissipation in the on-board + 5-V regulator, a MC7805AC. Since it is not heatsunk, its input voltage should be limited to + 15 V, and its case temperature should be checked if extra circuitry is added that is powered from this regulator.

The other inputs include a speed control input and an analog ground for the speed control input. The speed input can control motor speed with an external 0- to 5-V analog signal. Zero volts corresponds to zero speed, and 5 V to full speed. To use it, a jumper (J5) needs to be moved to the external position, which disables the on-board speed control potentiometer.

A pin is provided to externally reverse the motor when it is grounded. To use it, a jumper (J3) needs to be moved to the external position, which disables the on-board FORWARD/REVERSE switch. A motor run input pin enables the motor when it is grounded. To use it, a jumper (J4) needs to be moved to the external position, which disables the on-board RUN/STOP switch.

If a brushless DC motor is to be controlled, the brushless DC motor's Hall shaft position sensors interface with Schmitt trigger buffers and filtered for noise immunity. The other two pins connect to the Hall effect sensors + 5 V and ground. It is important not to mix this ground with the motor's power ground. Doing so will cause motor instability, since the Hall-effect signals will be corrupted with ground noise.

A 16-pin ribbon cable connector is used for feedback signals from the power stage board. These motor feedback signals include motor current, diode temperature sensor, and motor bus voltage. One pin of the feedback connector is intended to be a current sense input. It is connected to A/D channel ATD3 through a gain-of-2 noninverting amplifier. The motor bus voltage input is connected to A/D channel ATD2 through a gain-of-2 noninverting amplifier. Temperature sensing is accomplished by translating the forward voltage of a diode into a usable A/D voltage with an amplifier. The output of this amplifier is connected to A/D channel ATD1. Two pins on the feedback ribbon cable connector serve as a ground for the analog circuits. This ground is routed in such a way that all the analog returns connect with digital ground at just one point. Again, it is important that the AGND not be looped into or mixed in with the motor's power ground.

Phase feedback signals phase A, phase B, and phase C are also included on the feedback connector. When used with the high-voltage power stage, a divided down phase voltage appears at these pins. With a low voltage power stage, no signals appear at these pins unless they are supplied by the user.

The six PWM outputs are brought out with a 14-pin ribbon connector. Two pins are connected to the 5-V bus through a switch that is open at reset. The resulting switched 5 V can be used to power optocouplers in isolated off-line motor drives. The off-at-reset feature keeps output transistors off during reset. Several pins provide a return for the switched + 5 and noise isolation between output lines in the ribbon cable.

The PWM outputs provide control signals for three phases of half-bridge-configured output transistors and are set up in an active low configuration. They have the current sinking capability to drive opto-coupled power stages as in the high-voltage power stage.

Test points (TP1, TP2, TP3) provide access to buffered feedback signals for temperature, motor bus current, and motor bus voltage.

These voltages are seen by A/D converter inputs ATD1, ATD3, and ATD2. The temperature feedback voltage can be calibrated with a potentiometer. Test points (TP4, TP5, TP6) allow access to buffered phase voltage feedback signals from the feedback connector. Test instruments can be grounded with the GND and AGND test points. All six outputs and a ground are also available as test points.

A five-position DIP switch (SW1) enables modulation parameters to be changed while a motor is running. Position 1 sets full modulation for either 60 or 120 Hz. Position 2 selects either sine-wave or third-harmonic modulation (using PWM). Positions 3, 4, and 5 select one of eight PWM frequencies that range from 2000 Hz to 22,000 Hz in 2-kHz increments.

A push-button switch (SW2) resets the processor and turns off the switched 5 V supplied on the output ribbon cable connector.

The speed control potentiometer (R2) controls motor speed unless a jumper (J5) is set to the external position, or control is taken over by the SCI port. A small pot. adjusts the analog signal that represents temperature. Setting the voltage at a test point (TP1) to 1 V at 25°C with this pot. is the recommended default calibration.

10.6.1.2 AC induction motor application. A three-phase, center-aligned, sine-wave PWM signal is generated by the MC68HC708MP16. Speed is controlled by the frequency and amplitude of this signal, while direction is determined by phase-to-phase sequence. Systems parameters are easily changed while a motor is running. Switch 1 on the MC68HC708MP16 board will change PWM rate, modulation type, and full modulation frequency. If the RS-232 communications interface (or motor SCI port) is used, the personal computer keyboard can issue commands for changing the boost voltage and selecting different types of space vector modulation.

A number of design considerations are important in AC induction motor drive systems. AC induction motor drives are relatively complex in terms of software design, processing power, and the PWM timer's hardware. Brushless DC motor drives, reviewed in the MC68HC705MC4 BLM section, tend to be complex in regard to the noise management of the sensor inputs.

On a machine that will be used for code development, it is desirable to prevent simultaneous conduction of upper and lower power transistors in the same phase. This feature is built into the HC708MP16's PWM timer. Once the timer has been initialized correctly, simultaneous conduction of a top and bottom output transistors in the same phase is locked out. Code errors, therefore, that occur after initialization is completed will not destroy power stage output transistors by turning on the top and bottom of one half-bridge simultaneously. This arrangement also prevents simultaneous conduction in the event of a noise-induced software runaway.

10.6.1.3 Deadtime. In AC induction motor drives, providing deadtime between turnoff of one output transistor and turnon of the other output transistor in the same phase is an important design consideration. Deadtime is also a feature that is built into the MC68HC708MP16's PWM timer. It is programmable, to accommodate a variety of gate drives and output transistors. In the software, 2 μs of deadtime has been selected for operation with the high-voltage power stage board.

To ensure proper sequencing, a switched + 5 is provided for sourcing drive current to the optoisolators. This supply is held off until RESET occurs and input voltage is high enough for safe operation. Connection to an opto input is illustrated in Fig. 10.13. It applies to operation with the high-voltage power stage.

10.6.2 AC induction motor MCU development software

Software included with the MC68HC708MP16 motion control development board provides basic AC induction motor control. Firmware for this application is programmed into the MC68HC708MP16 for immediate use. Source code is also provided on diskette. Open-loop volts-per-hertz (V/F) drive from 0 to 120 Hz, and PWM rates of 1800 to 28,800 Hz, are supported. Other options include sine, third-har-

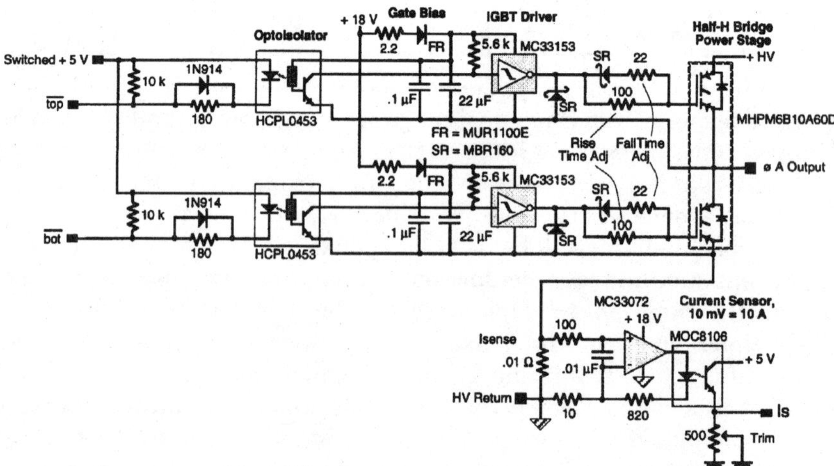

Figure 10.13 Optoisolator high-voltage gate drive circuit.

monic injection, or space vector modulation waveforms; full modulation at 60 or 120 Hz; and RUN/STOP and direction control. Two different operating modes are possible with the supplied software: STANDALONE MODE and TERMINAL MODE. (See App. I for MC68HC708MP16 software listing.)

10.6.2.1 Standalone mode. When the MC68HC708MP16 is initialized after reset, it is operating in STANDALONE MODE. In this mode, all options (speed, direction, etc.) are read from controls on the board. Since the software ensures coordination of the actual changes in PWM, voltage, and other parameters, changes may safely be made in real time while driving a motor. One exception to this is if the motor load has a large amount of inertia. In this case, the rate of speed change allowed by the software may not be slow enough to prevent regeneration of excessive DC bus voltage.

SPEED, or drive frequency, is determined by a speed potentiometer. Frequency may be set from 0 to 120 Hz in 1-Hz increments. Large changes are not instantly applied; instead, a slow ramp to the new setting is implemented.

A FORWARD/REVERSE switch (SW3) sets the drive direction. When direction is reversed, speed is ramped down to zero and then ramped up to the current speed setting in the new direction. A RUN/STOP switch (SW4) allows speed to be forced to zero. SPEED is ramped to zero when stop is selected and then ramped up to the current speed setting when the switch is returned to RUN.

10.6.2.2 Terminal mode (control of motor with external computer terminal). The MC68HC708MP16's serial port (SCI) is also enabled and monitored for activity. A terminal, or terminal emulation software running on a personal computer, communicates with this port. Any basic serial communications software that is set for 9600 baud, 8 data bits, no parity, and 1 stop bit, will work.

When commanded to do so via the terminal, the MC68HC708MP16 development board can be switched to a TERMINAL MODE, where all control is by PC keyboard entries. This can be done in real time without disturbing motor drive parameters. When TERMINAL MODE is activated, it uses the board's switches and pot. settings as defaults. When TERMINAL MODE is deactivated, settings revert back to the board's switches and pot.

10.6.2.3 Software functional overview. The core function of the demonstration code is to synthesize three-phase waveforms for variable-frequency drive of AC induction motors. This task is simplified greatly by the six-channel motor control PWM unit on the MC68HC708MP16. In general, waveforms are synthesized by looking up values in a table for each point along a curve and then converting these values to PWM duty cycles. The repetition rate, or carrier frequency, determines how many data points define the curve. Drive frequency, typically 0 to 120 Hz, is determined by how rapidly the microprocessor steps through the table values.

The time base for this process is the rate at which the PWM unit interrupts the HC08 CPU. If the PWM unit is configured to interrupt every cycle, this rate is identical to the carrier frequency. It is common practice to service the PWM unit less often than this at higher carrier frequencies. The entire process is performed 3 times on each interrupt, to create three waveforms that are each offset 120°.

Only about 1800 bytes of code are needed for this basic motor operation. The user terminal interface, additional demonstration features, and factory test routines use about 11,000 bytes. The code is modular and written in C language (except for the SINESCALE routine), in order to encourage experimentation and reuse. A brief summary of each module is listed below. (Consult the C source code for complete details in App. I.)

MAIN:

Initializes PWM and SCI units

Resets communication, A/D data, and waveform data table pointers

Enables interrupts for PWM and communication

Enters SCAN loop

PWM:

PWM interrupt handler

Services COP (computer operating properly)

Passes data table pointer to QUADZ for each phase

Loads PVALX registers from global RAM value pwmmod

Exchanges two phases if reverse direction is set

Maintains waveform data table pointers

Sets PWM unit LDOK bit

QUADZ:

Accepts waveform data table pointer

Translates full waveform pointer into quadrant pointer

Selects sine or third-harmonic injection according to settings

Calls SINSCALE to scale table value with global RAM value vscale

Modifies global RAM value pwmmod

SINSCALE:

Accepts waveform data and scaling value

Scales with 24-bit accuracy

Returns integer formatted for use in PVALX registers

SVM:

Accepts waveform data table pointer

Calculates SVM (space vector modulation) time segments via CALCULATE function

Modifies PWM PVALX registers

CALCULATE:

Calculates times for SVM modulation

SCAN:

Scans hardware for speed, PWM, etc. settings

Scans serial communication buffer for commands via GETCH

Parses commands, sets control flags

Calls RECALC to execute setting changes

Calls MENU to reflect changes back to terminal user

RECALC:

Recalculates correct PWM modulus, load frequency, and interrupt frequency

Recalculates correct data table pointer increment value

Updates PWM registers and RAM variables coherently with interrupt mask

MENU:

Transmits command menus and current settings to terminal user

RECEPT:

SCI interrupt handler—writes to buffer

Maintains pointer

PUTCHAR:

SCI transmit

GETCH:

Parses input string in buffer

10.6.2.4 Software development. The MC68HC708MP16 motion control development board may be used in an emulation environment with Motorola MMDS08 or MMEVS08 development tools. Executable code in S-record format is included on the source code diskette. It is very important to note that emulator operations, such as stopping emulation or setting breakpoints, can cause possible harm to power stages. If, for example, the emulator is stopped in a state that energizes the motor, the relatively low winding resistance in most AC motors will allow excessive current to flow. Under these circumstances, it is relatively easy to stress power devices. Any software change, no matter how minor, should be checked out with the procedure described above before applying power to the motor.

10.6.3 High-voltage motor power

An IGBT power stage that complements microprocessor development tools for the HC05MC4, HC08MP16, and HC16Y1 is presented here. This high-voltage power stage is intended to run three-phase motors up to 750 W from DC bus voltages up to 380 V.

Caution: Before using this high-voltage motor development board or any high-voltage power electronics, a few words of warning are in order. Since the motor control power stage is capable of operating at dangerous (lethal) voltages and of supplying dangerous amounts of power to rotating machines, it is critically important that safety rules are followed. To facilitate safe operation, input power for the high-voltage rail should come from a current-limited DC laboratory power supply. Before moving scope probes, making connections, and so on, it is generally advisable to power down the high-voltage supply. When high voltage is applied, using only one hand for operating the test setup minimizes the possibility of electrical shock. Operation in lab setups that have grounded tables and/or chairs should be avoided. Wearing safety glasses, avoiding ties and jewelry, and using shields are also advisable.

10.6.3.1 Evaluation board description. The high-voltage power stage board shown in Fig. 10.14 is designed to provide an optically isolated interface between microcomputers and induction motors up to 750 W.

It accepts six logic inputs that control three IGBT half-bridge outputs and is arranged so that a logic ZERO at the input turns on the corresponding power transistor. The inputs are directly tied to optoisolator input diodes and require an ability to sink 25 mA. This type of configuration is applicable to PWM systems in which the PWM signal is generated in a microcomputer, digital signal processor, or other digital system. It is suitable for driving induction motors up to 1 hp off DC bus voltages up to 380 V. In addition to controlling the motor, current sense, bus voltage, and temperature feedback, signals are provided. This board is designed to interface directly with Motorola KITITC127/D, KITITC137/D, and M68MCD16Y1 motion control development tools.

Coupled with logic control boards for the HC05MC4, HC08MP16, and HC16Y1, it completes the link between microcomputer development tools and a motor. It allows code to be written before hardware design is completed in an environment where mechanical outputs can be seen while the code is written. The design, as shown in Fig. 10.14, can also be used as a reference for speeding hardware development.

10.6.3.2 Electrical characteristics. The high-voltage power stage can operate from 12 to 380 V and up to 10 A peak phase current. The nominal operating high voltage is 320 V, 18 V of which is used as the gate drive supply. It is important to ensure that the IGBT power module's power dissipation is limited to under 50 W with normal fan operation.

10.6.3.3 Pin-by-pin description. Inputs and outputs are grouped into four connectors. Two connectors are provided for inputs, + 5 V, and ground. One consists of screw terminals, and the other is for ribbon cable. Either can be used; they are wired in parallel. Outputs to the motor, + 18 V, and the high-voltage rail are supplied on a large screw connector. Returns for the high-voltage rail and + 18 V are also on this connector. Feedback signals are grouped together on a separate ribbon cable connector. In addition, two through-hole pads have been placed immediately adjacent to the current feedback resistor.

Inputs: Inputs A_{top}, A_{bot}, B_{top}, B_{bot}, C_{top}, and C_{bot} are logic inputs. A logic 0 turns on the input's corresponding output transistor; that is, a logic 0 applied to input A_{top} turns on output transistor A_{top}, and so on. Logic levels are standard 5 V, with sink currents that are typically 20 mA. They are pulled up to + 5 with 10-kΩ resistors. Therefore, in the absence of any inputs, all output transistors are turned off.

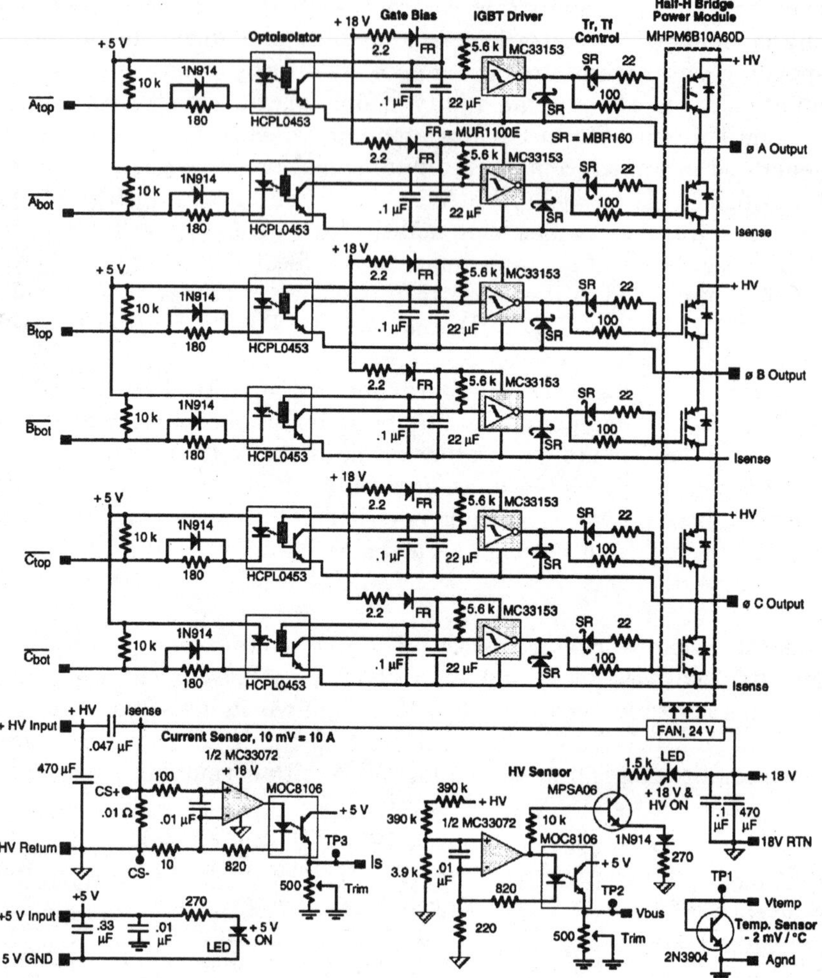

Figure 10.14 High-voltage motor power stage development circuit.

HV RAIL: HV RAIL is the motor power connection. It is intended for use with current limited, line-isolated, laboratory power supplies. Acceptable input voltage range is + 12 to + 380 V DC. It is located at the top of the output connector.

HV RETURN: HV RETURN is the power supply return for the motor power supply HV RAIL.

+18: +18 is the gate drive supply. Its tolerance is 16.2 to 19.8 V DC.

RTN: Two terminals on the output connector are labeled RTN. They are connected together; either one may be used as the + 18-V gate drive supply's return. The other can be used for instrument grounds. These terminals are common to the motor power supply return, HV RETURN, but do not have its current-carrying capability. They should not be used for connecting the motor power supply.

Motor outputs: Motor output terminals are labeled A_{out}, B_{out}, and C_{out}. They can be used to drive a three-phase AC induction motor, three-phase brushless DC motor, a reversible brush DC motor, or three separate brush DC motors unidirectionally.

I_{sense}: I_{sense} is a current sense feedback voltage that appears on one of the feedback connectors. It is derived from a 0.01-Ω low-inductance surface mount sense resistor that is in series with the ground return. The voltage cross this resistor is optoisolated and amplified with a nominal gain of 25. I_{sense}, therefore, represents return current with a scale factor of approximately 250 mV/A. This signal is isolated with a simple open-loop optocoupler and is therefore not particularly accurate. A spot on the board is provided to trim this output. Software trim techniques can also be used to improve accuracy.

V_{bus}: V_{bus} is a bus voltage feedback signal that appears on one of the feedback connectors. It is derived from the high-voltage rail and, like I_{sense}, is isolated with an open-loop optocoupler. It also has provisions for a trim resistor.

V_{temp}: V_{temp} is a temperature output signal derived from a forward-biased diode's ON voltage and appears on one pin of the feedback connector. The diode, a small signal 2N3904 transistor, is mounted so that it measures ambient temperature. Two test points immediately adjacent to the temperature sensor transistor are intended to make it easy to remove and reconnect with a twisted pair of wires. Connected to a twisted pair, it can be placed on the heatsink or motor.

AGND: A separate ground is provided for the analog signals. It is labeled AGND. AGND and GND are tied together on control boards ITC127 and ITC137. If a different input source is used, it may be necessary to tie GND and AGND together.

Test points TP1–TP3: Test points TP1, TP2, and TP3 provide access to feedback signals for temperature, motor bus voltage, and motor bus current, respectively. TP1 is connected to the forward-biased diode that is used for measuring temperature. Its ground is the analog ground AGND. TP2 is connected to the optocoupled output signal for motor bus voltage, and TP3 is connected to the optocoupled output signal for motor bus current. Both are also referenced to AGND.

An important application consideration occurs at powerup and powerdown. When the controller's power is off, all of its outputs look like logic lows. To avoid turnon of all six IGBTs simultaneously, it is necessary to take precautions. The MCU controller board must be powered up first before the power stage. This is accomplished by supplying the + 5 V that powers the opto inputs from the controller board. On the MC68HC705MC4 and MC68HC708MP16 logic boards, this + 5 V is switched off at reset in order to disable high-voltage power stage inputs when power to the controller is not present.

10.6.3.4 Power stage design considerations. A simplified schematic of one phase is illustrated in Fig. 10.13. Top and bottom inputs are optocoupled to inverting gate drivers. This arrangement isolates the inputs from the high-voltage power stage, making it suitable for use with microcomputer development tools. It also facilitates board layout and improves noise immunity, since each gate drive can be returned directly to the emitter of its corresponding IGBT. To simplify things, the IGBTs have a 6-V gate threshold, making negative gate bias unnecessary. The MC33153 gate drivers have an undervoltage lockout that is designed for the 6-V threshold.

From a systems point of view, using optocouplers for both top and bottom gate drives is a very effective strategy for minimizing design time. Although optocouplers are not necessary for the lower half-bridge gate drives in many applications, they effectively minimize conducted noise from microcontrollers. Since it is noise management that typically takes the most design time, the improved noise robustness of an optocoupled topology can get products to market more quickly.

To further improve noise robustness, several components in addition to optocouplers are used. Referring again to Fig. 10.13, diodes (1N914 types) hold the optocouplers off with a lower impedance when the inputs are high. Resistors in series (2.2-Ω values) with the 18-V gate drive supply protect the gate drive bus from di/dt-induced voltage transients. Between the driver output and IGBT gates, two resistors and a diode are used instead of a single-gate drive resistor. The additional components allow the IGBTs to be turned on more slowly than they are turned off. Because of the characteristics of the freewheeling diodes, this arrangement produces less noise than a single resistor.

The upper gate drive bootstrapped power supply has a 22-μF surface mount storage capacitor. This value has been chosen for induction motor drives that use sine-wave or third-harmonic pulse-width modulation. Space vector modulation and brushless DC motor drives will work better with larger storage capacitors. Pads are provided on the board for adding larger capacitors if they are needed.

Summary

This chapter has described the use of microcontrollers for motor controls. As the price-to-performance ratio of MCUs continues to improve, it is very likely that the vast majority of future motor control systems will employ at least one MCU. Key facts that should have been learned from this chapter are listed below.

- Microcontrollers will become a standard component in motor control systems because they allow the operation of the motor control system to perform more effectively.
- The programmability of the MCU allows many changes to the motor control system, by just changing a few lines of instructions in a text file, rather than by replacing or adding fixed components.
- If the MCU's oscillator fails, its output lines will remain locked in whatever state they were at the time of the oscillator's clock failure. This event can cause an electronic commutated motor to lock up since the power stage is no longer switching, allowing the motor's windings to conduct DC current.
- There are a number of features in a motor control MCU:

  An A/D converter unit for measuring the motor's current and voltages.
  A programmable timer unit that can generate at least six channels of PWM or space vector modulation signals
  Fault inputs to shut down the PWM outputs
  Quadrature-type input timer capture
  Fast math computations
  Event timer

- A complex instruction set computer (CISC) is a general-purpose type MCU with a large variety of instructions. Digital signal processor (DSP) and reduced instruction set computer (RISC) MCUs use less instructions and usually execute those instructions very quickly. For motor control applications, the choice of the MCU type is dependent on the specific performance requirements of the motor control design.

- The reliability of a motor control design will be determined mainly by the MCU's software design and power stage protection methods, and the stability of the feedback elements.
- Read-only memory (ROM) is the most economical type of memory to fabricate, but it cannot be changed. If several software program iterations occur, the ROM may turn out to be more costly than EPROM or OTP types. The ROM is a good choice for high-volume, proven, and slowly evolving applications.
- An electrically programmable read-only memory (EPROM) is a good choice for minimizing the risk of lost inventory. One-time programmable (OTP) devices are EPROM types in a lower-cost package. The OTP devices cannot be reprogrammed since they lack the ultraviolet light window.
- Electrically erasable read-only memory (EEPROM) and flash memory MCU types are good for prototyping and for small-volume production. Some MCUs include just a few bytes of EEPROM, which is useful for storing critical data values during power-OFF conditions.
- The MCU uses its timer units for performing motor control tasks in several ways. Internal timers in the MCU can be programmed to begin timing when an input line is switched. This is very useful for measuring motor speed or shaft position. A timer can also be used to track an overcurrent condition in the motor's power stage.
- The MCU's oscillator is normally used as a clock source—its accuracy will directly affect all time calculations.
- MCU devices will continue to offer higher performance while maintaining cost parity because the MCU's die size is shrinking by the use of more precise designs and manufacturing equipment. The MCUs internal operating voltages are also reduced, allowing submicrometer line spacing.
- The most difficult aspect of employing an MCU for a motor control application is usually the initialization software. Any mistake or omission in the MCU's start-up software routines will directly affect its operation. Some MCUs have dozens of registers that must be configured to meet certain operating requirements of an application.
- The ease of implementation in motor control design is the greatest advantage of using an MCU and power stage development boards.

The motor control hardware and software are already working and only need to be customized for the user's application.

Acknowledgments

The authors would like to thank Peter Pinewski and John Deatherage for their assistance with this chapter on motor control MCUs and development systems.

Further Reading

Bennett, W., Evert, C., and Lander, L., *What Every Engineer Should Know about Microcomputers,* 2d ed., Marcel Dekker, 1991.
Gray, J., Lucas, B., and Schultz, W., ITC137 68HC708MP16 Motion Control Development Board, Motorola Application Note AN1624, 1997.
Intel Corp., 86XC196MH data sheet.
Lucas, B., and Schultz, W., *ITC132 High Voltage Micro to Motor Interface,* Motorola Application Note AN1606, 1997.
Lucas, B., and Schultz, W., *ITC122 Low Voltage Micro to Motor Interface,* Motorola Application Note AN1607, 1997.
Lucas, B., and Schultz, W., *ITC127 68HC705MC4 Motion Control Development Board,* Motorola Application Note AN1717, 1997.
Microchip Technology Inc., PIC12C672 data sheet.
Motorola Inc., MC68HC705JJ7, MC68HC05MC4, MC68HC08MP16, General Release Specifications.
Sibigtroth, J., *Understanding Small Microcontrollers,* Motorola Inc., M68HC05TB/D, rev. 1.
Texas Instruments Inc., TMS320C240 DSP controller product specification.

Chapter

11

Networking for Motor Control Systems

Chuck Powers and Thomas Huettl
Motorola Semiconductor Products

Introduction

In this chapter we will review various methods that motor control systems may use to communicate with internal subsystems and external devices. Electronic communication in control systems can be compared to human communication. For example, when two people are talking to each other, they first need to hear each other above any background noise. If they are too far apart, their voices cannot be distinguished. Once the other person is heard, one must decide if the speech is intelligible—is it a recognizable language? If so, the thought process is enabled and a response is given. If several people talk to the same person at the same time, communication will once again become confused. Verbalization can also be disrupted if the background noise level drowns out the message. Electronic communications suffer from the same problems—noise, distance, languages, priority of talkers and listeners, and so forth. A communication network for motor controls must consider these same kind of problems.

Today's motor control applications are transforming from simple on/off-type controls into precise control systems that must operate in highly integrated control environments. As a result, the type of electronics integrated into a typical motor control system has become more sophisticated. This, in turn, has greatly increased the need for improving the flow of control and status information to and from these systems. To achieve the full potential that a distributed control environment can offer, a natural evolution is taking place, linking motor control electronics to distributed network I/O components via serial communication networks. (*Note:* In this chapter the term *component* refers to a node, subsystem, or any device capable of serial data communications.)

As will become obvious in this chapter, one challenge when designing a communications network for a motor control is determining which type of bus and protocol to use. Several books have been written on digital networking; this chapter will cover only the basic serial communication methods that are applicable to motor controls. (*Editor's note:* A more difficult aspect to grasp when reviewing digital communication methods is the manner in which the protocol layers are described. The material presented here focuses mainly on motor control drives and their subsystems, including those for automotive use, but does not include factorywide automation networks.)

11.1 Why Motor Control Communication?

The high cost of system installation, the need to improve information sharing between system components, and the desire to improve overall system reliability are some of the factors that are driving the current trend toward serial communication networks. Serial communication technology solutions offer the designers of complex motor control systems a variety of benefits. Connecting components to a single serial communication, or multiplex, network can improve reliability over traditional wiring harnesses because of reduced wiring complexity, including the number of physical connections required. The improved data sharing and communication capabilities can also make it much easier to identify and replace faulty components.

These networks can ease the addition or enhancement of system features, which in turn allows system designers to differentiate their products from systems with more traditional, and fixed, feature sets. Components can be added to a network at a reduced cost, if all the information required by that component is already being shared across the network. Overall, the number of I/O points in a network may be reduced through more efficient collection and distribution of I/O and status information. Systems in which motor controls play a big part can certainly benefit from these cost and reliability improvements. For example, motor drives and conveyors can be driven more precisely, stopping in a safe manner when a problem develops, which can prevent possible system damage. The benefits of implementing a serial communication network in a distributed control system can be considerable.

11.2 What Is Multiplexing?

Serial multiplex communication is the process of merging multiple communication paths into a single serial signal path. A serial multiplex communication network is created by connecting each module or system which must collect, calculate, exchange, or display data to a single serial communication bus. All necessary data can then be collected and manipulated by the module or system which is closest to the source of the data. A data source can be anything from a temperature sensor for a motor to a conveyor-belt speed sensor. Each module which collects or calculates data then transmits it onto the serial multiplex network where all other modules or systems which need that data can retrieve it simultaneously. For example, a motor controller that requires conveyor-belt speed to correct motor torque can also transmit this data onto the serial multiplex network, where a display console can receive and display the information.

11.2.1 ISO OSI protocol reference model

To provide a basic reference for developing and implementing serial communication protocols, the International Standards Organization (ISO) has published International Standard ISO 7498, *Information Processing Systems—Open Systems Interconnection (OSI)—Basic Reference Model.* This standard is a reference model for specifying the properties of telecommunications and data communications protocols. The OSI model is subdivided into seven layers, with different protocol attributes and functionality assigned to each layer.

11.2.2 OSI protocol layers

Figure 11.1 illustrates the basic structure of the ISO OSI reference model for communication protocols. Services provided by the *physical layer* (layer 1) typically include the physical signaling and communication media used. The *data-link layer* (layer 2) provides the message structure and bit timing. This layer is typically divided into three sublayers, the lower media access control (LMAC), upper media access control (UMAC), and the logical link control (LLC) sublayers. The *network layer* (layer 3) allows for transparency of addressing, routing, and the quality of service employed over the network.

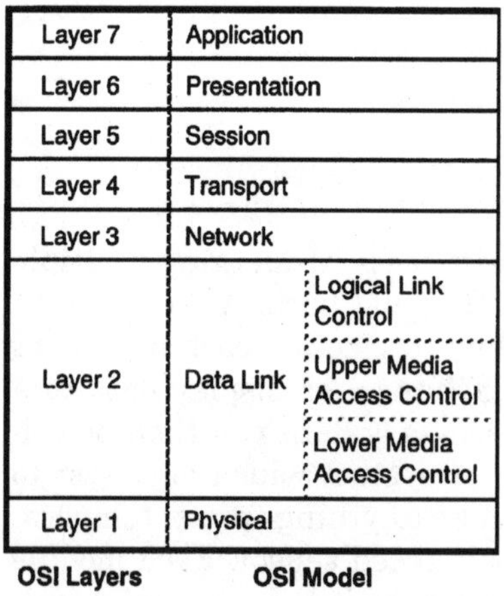

OSI Layers	OSI Model	
Layer 7	Application	
Layer 6	Presentation	
Layer 5	Session	
Layer 4	Transport	
Layer 3	Network	
Layer 2	Data Link	Logical Link Control
		Upper Media Access Control
		Lower Media Access Control
Layer 1	Physical	

Figure 11.1 ISO OSI reference model.

Transport layer (layer 4) services include responsibility for making and maintaining requested connections, controlling flow, and performing sequence control between nodes on the network.

When multiple types of connections are possible (e.g., simplex, half-duplex, or full-duplex), the *session layer* (layer 5) is responsible for establishing, configuring, and releasing these connections. The *presentation layer* (layer 6) transforms the data received from the *application layer* (layer 7) into a format suitable for the remaining protocol layers, and vice versa. The application layer, which acts as the interface between the protocol stack and the application requiring communication services, defines the application messages used for network communication.

Many protocols, particularly control network protocols, do not specify services for every layer outlined in the OSI reference model. Instead, only the layers deemed appropriate for the targeted applications are specified. In these cases, while some services normally attributed to other layers are usually needed, they are typically merged into other layers in order to simplify the protocol definition. For example, many protocols used in automotive networking applications, which can also apply to motor drives, only formally specify the application, data-link, and physical layers. If other layer services are needed, they are included as part of the application or data-link layer.

11.3 Physical Media

A basic communications system consists of a transmitter, a receiver, and an information channel. Selection of the medium for this information channel, which physically carries the data signals from one component to the next, must be made from a wide variety of materials and network topologies. The choice of possible network media includes single or dual wires (twisted or untwisted), coaxial cable, plastic or glass optical fiber, and high- or low-frequency radio waves, among others. The different network topologies which can be utilized include a traditional drop bus, a hub-tree structure, a daisy chain, passive or active stars, or a wide variety of hybrid combinations. The selection of physical media for a multiplex communication system is a complex one, with significant compromises to be considered, and is heavily dependent on the requirements of the particular application. Factors in this decision include cost, ease of installation and maintenance, communication bandwidth and protocol requirements, and the physical and electrical environment in which the system must operate.

11.3.1 Media selection

When selecting media for a communication network, it is best to start with the application requirements, rather than raw cost or environmental capabilities. What is the minimum data throughput required for the control system network to achieve the desired operation? The data throughput calculation should include enough spare bandwidth to account for the issues typically encountered in network systems, such as message retries due to collisions, or lost messages due to electrical noise appearing on the network. Also factored into the equation must be the overhead associated with the protocol selected. The data throughput of some protocols can be 50 percent or less, and the bandwidth requirements must take this into account. For example, some studies have shown that the effective throughput on heavily loaded 10-Mbit/s Ethernet networks can fall to 35 percent or less. In some cases this can be offset by other protocol features, such as nondestructive arbitration, thus eliminating the need for transmit retries due to message collisions.

Another protocol-related factor to be considered when selecting a network medium is the maximum propagation delay which can be tolerated by a system. One element directly affecting propagation delay limits is whether the chosen protocol utilizes nondestructive arbitration or is collision-based. Typically, collision-based protocols can tolerate much greater network propagation delays than can protocols requiring communicating nodes to synchronize to each other during a transmission. Propagation delay limits are directly related to the maximum distance between two components, the communication data rate used, the bit encoding techniques, and the electrical characteristics of the physical medium itself (copper vs. fiber). Many systems today specify a matrix of network configurations and data rates. This allows systems implemented in more compact areas to benefit from higher data rates, while still permitting more extended systems to be implemented, but at the price of lower network bandwidth. Once the bandwidth and protocol requirements have been determined, the next set of factors to consider pertain to the environment in which the network must operate.

The physical environment variables in which the system must work include the operating temperature range, the humidity and corrosiveness of the ambient environment, and physical shock and vibration which the network components must endure. The physical media and required network connectors must be specified to survive the physical stress that may be encountered in the most extreme operating conditions which can be anticipated. Failure to consider the physical environment can result in repeated component or network failure. Motor drive systems in general will always require a more robust communication network than will office computer networks. An electrical noise burst in the office network may produce an inconvenience such as a printer error, but this same noise burst could cause a large motor drive to operate erratically, with possible physical damage.

The electrical environment is also a major factor in the choice of a physical medium. Electromagnetic compatibility (EMC) requirements, for both susceptibility and emissions, are becoming more stringent. The use of media which is more susceptible to external electromagnetic interference (EMI) or is more prone to electromagnetic emissions can result in the imposition of strict limitations on the placement of modules and the routing of wiring harnesses. Factors which have a direct impact on EMC include the media materials, media construction, the communication data rates, and the electrical characteristics of the physical signaling. For lower data rates, typically <50 kbits/s, low-cost single wire networks are feasible. As data rates increase toward 1 Mbits/s, the use of advanced wave-shaping techniques, twisted-pair wiring, and shielding become necessary, but at increased component and media cost. For EMC sensitive networking applications, data rates in excess of 1 Mbits/s are typically considered the domain of fiberoptic media.

11.3.1.1 Fiberoptics. The use of fiberoptic networks can eliminate the physical media as a source or sink of EMI, although the network connectors can sometimes cause EMC problems at extreme data rates. This capability, however, comes at a higher price than wire-based media. While the cost of plastic optical fiber has dropped dramatically in the last few years, the use of fiberoptic media still requires connectors which are more expensive than more traditional interconnects. The physical network topologies are more limited as well, due to current connector technology and the physical restrictions of dealing with light-based systems. Extreme physical environments can also limit the use of fiberoptic solutions. One distinct advantage of an opto cable is that it is inherently nonconductive. This fiberoptic cable attribute may be required in motor control systems where external components, such as remote sensors, may be exposed to dangerous voltages. If the fiber cable were dropped onto a high-voltage source, this voltage would not produce catastrophic results.

When all the protocol bandwidth and environmental factors have been evaluated, the media cost tradeoffs can be determined. For systems which can operate using lower data rates, the more cost-effective wired media (single wire, unshielded twisted pair) can be utilized. When it is determined that higher data rates may be necessary, more costly media, such as shielded twisted pair, coaxial cable, or optical fiber need to be selected, depending on the data rate and application requirements.

11.4 Bit Encoding

When designing multiplex systems for motor controls, it is important to choose a bit encoding technique that meets the noise immunity, decoding, and data rate of the system. Criteria for choosing the encoding technique include minimal transitions per bit, sufficient time between transitions, and bit synchronization (to allow time for bit testing).

Noise immunity in the multiplex system requires a validation test performed at the message level. Bit error algorithms such as checksum, parity bit, and a cyclical redundancy check (CRC) are among the most common testing used. Most protocols can perform message length by the message type or length in the data. These and other message level tests are independent of the bit encoding method and should not influence data integrity of the bit encoding technique.

Many types of encoding techniques are available for designing a multiplex communication system. In this chapter we discuss non-return-to-zero (NRZ), pulse-width modulation (PWM), variable pulse-width modulation (VPWM or VPW), and Manchester encoding.

11.4.1 Non-return-to-zero encoding

One of the most basic bit encoding techniques is NRZ. NRZ encoding represents each binary data bit as a static value during the entire bit time, with a high (passive) level representing a logic 1 and a low (active) level representing a logic 0. The Universal Asynchronous Receiver-Transmitter (UART) communication port, widely used in computer, industrial, and automotive networks, utilizes a 10-bit NRZ byte-level structure. Figure 11.2 illustrates typical NRZ bit encoding.

For extended data streams, the NRZ encoding technique has some distinct disadvantages. Unlike other, more complex, bit encoding techniques, it is not inherently self-clocking. Each node on the network must provide an internal sampling clock for its communication

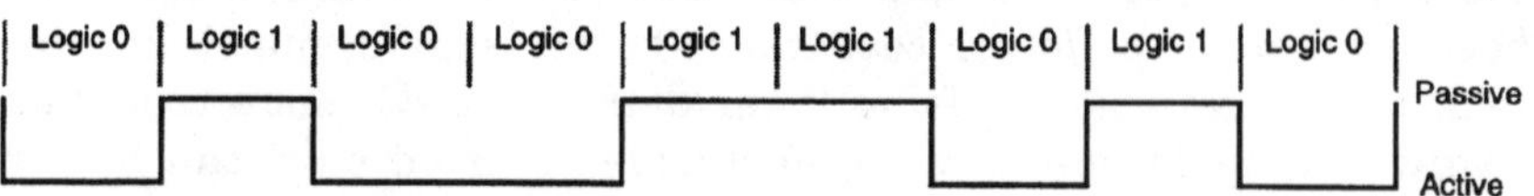

Figure 11.2 Non-return-to-zero (NRZ) bit encoding.

port. Also, because there are no guaranteed edges in each bit time, long data streams of the same logic level can produce synchronization problems among receive data with slight oscillator variations. For this reason, many protocols that use NRZ also perform bit stuffing. "Bit stuffing" is a technique in which a bit of the opposite logic level is inserted, or stuffed, into a data stream that consists of more than a certain number of bits at the same logic level—typically 5. This ensures that the bit stream will go no more than 5 bit times before an edge is detected, thus allowing the communicating nodes to resynchronize their receivers. Special hardware is necessary for inserting and removing these stuffed bits.

11.4.2 Pulse-width modulation bit encoding

Traditional PWM bit encoding, as illustrated in Fig. 11.3, is an encoding method in which each bit period is a constant length and is composed of two segments, an active segment T_A and a passive segment T_P. The binary data bits are encoded using different lengths for T_A and T_P. Many protocols use PWM bit encoding, with a variety of definitions for the active and passive bit segments. To achieve the greatest tolerance to oscillator variations and network EMI, most PWM encoding methods divide the total bit time into thirds, with T_A equal to 50 percent the length of T_P, or T_P equal to 50 percent the length of T_A, depending on which logical bit is being encoded. This $\frac{1}{3}$–$\frac{2}{3}$ or $\frac{2}{3}$–$\frac{1}{3}$ encoding always ensures that at least the final one-third of the total bit time will be at the passive level, allowing the receiver to easily differentiate between data bits.

There are some disadvantages to the PWM bit encoding method. The two edges per bit time inherent in PWM encoding can increase possible EMI problems. Also, encoding and decoding of PWM pulses is more complex, because at least three samples per bit time are required to determine whether a data bit is a logic 1 or logic 0. NRZ, on the other hand, usually samples the level of the bus once per bit time. (*Note:* There are instances in which NRZ elements sample the bus level several times per bit period.)

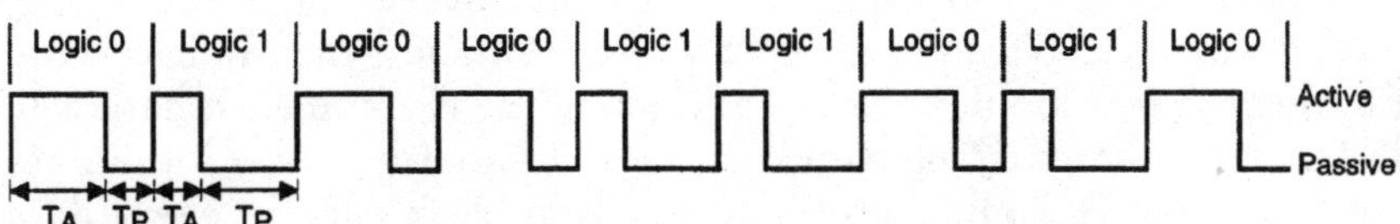

Figure 11.3 Pulse-width-modulation bit encoding.

11.4.3 Variable-pulse-width modulation bit encoding

Variable-pulse-width (VPW) modulation is a variation of PWM and has been used in harsh environments such as automotive data networks. Also known as *Huntzicker encoding,* variable PWM uses a combination of active and passive logic levels and two different bit lengths to encode the data bits. For example, a logic 1 can be encoded either as an active short bit or as a passive long bit. Conversely, a logic 0 is encoded as either an active long or a passive short. The logic level of each bit in a byte is the same, so the length of each byte varies depending on the data being encoded.

The single edge per bit time of VPW encoding can provide an improvement in EMI performance over traditional PWM encoding. But the data-dependent length of each data byte can be a disadvantage in some systems, particularly in respect to the communication software and its handling of each byte in a timely manner. For example, in current VPW systems, the length of a single byte can range from 512 to 1024 μs, depending on the data. This variability requires the software to be able to manage communication based on the shortest byte possible, even though the average byte length will be 50 percent greater.

11.4.4 Manchester bit encoding

One popular bit encoding scheme for data communications is known as *Manchester encoding*. The name "Manchester" code has become the default name for the more correct term of Manchester II (also known as *biphase-L*) encoding. Manchester encoding represents the data value with a level transition in the middle of each bit time (or data cell). In its most common usage, a low-to-high transition represents a logic 1 and a high-to-low transition represents a logic 0. Figure 11.4 illustrates Manchester II bit encoding, in contrast to NRZ; note the biphase technique, in which one transition always occurs in the middle of the bit time, and one transition sometimes occurring at the start of a bit time.

Manchester is a self-clocking encoding scheme, with the mid-bit-time transition representing both data and clock. This allows a single serial bit stream to provide both, creating several advantages over more traditional NRZ encoding that receives and synchronizes data. First, the transition defines the position of the clock in each data cell, so there is inherently no skew or phase difference between the data and the clock. Second, there is an inherent clock edge in every data cell, allowing frequent resynchronization of the receiver or data

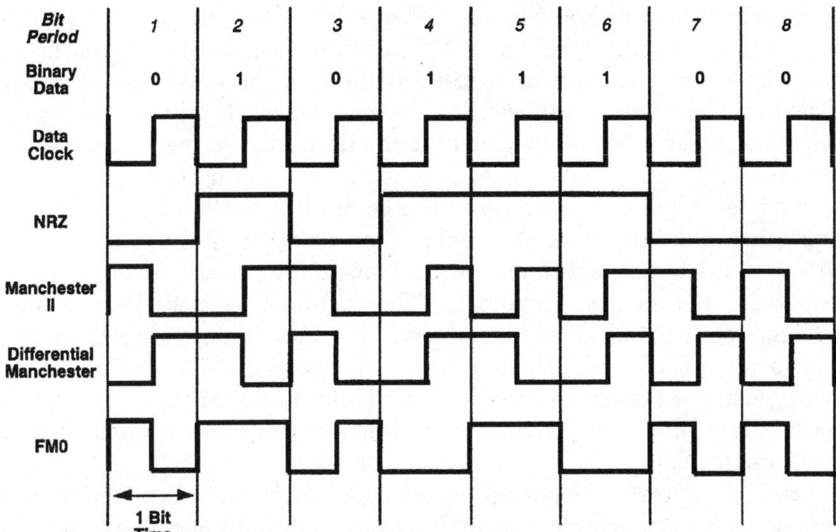

Figure 11.4 Manchester II, differential Manchester, and FM0 bit encoding.

decoder (as often as once per bit time, if desired). This eliminates cumulative clock drift or jitter over successive data cells, which occur in NRZ bit streams. Also, with NRZ encoding there is no guarantee of any minimum spacing between transitions (consider a long string of 1s or 0s). This can contribute a DC component in NRZ coding, which can prevent the use of AC coupling to the communication media. In contrast, Manchester II can easily be either capacitor- or transformer-coupled to the media.

Of course, one disadvantage associated with Manchester II encoding is the increased EMI potential due to a data stream that averages one and a half transitions per bit time, with a worst-case scenario of two transitions per bit time. Also, the more complex waveform requires more complex, and more costly, physical layer devices to perform the symbol encoding and decoding, as well as clock recovery, if needed.

11.4.5 Differential Manchester bit encoding

Even with its advantages over NRZ data, Manchester II encoding has one significant disadvantage for some communications media. It is sensitive to the data polarity (the direction of the data transitions). For example, if a receiver uses transformer coupling to connect to a network that uses Manchester bit encoding, it can be very easy to invert the signal polarity, which would result in inverted data.

A second form of Manchester, called *differential Manchester encoding,* eliminates this problem. Differential Manchester is similar to Manchester in that a mid-bit-time transition must always occur for clocking. However, in differential Manchester, each data bit is encoded with either a transition, or no transition, at the beginning of the bit time. A logic 0 has a transition at the beginning of the bit time, and a logic 1 has no transition at the beginning of the bit time.

Figure 11.4 illustrates the comparison between Manchester II and differential Manchester encoding. Note that the direction of each transition is no longer important. The mid-bit-time transition is now present only for clock recovery, and the data is determined by the presence of the bit time boundary transitions. Differential Manchester therefore retains all the advantages of Manchester II and adds the additional advantage of insensitivity to communication media polarity.

Like Manchester II encoding, differential Manchester also poses potential EMI concerns, although the average number of transitions is lower. The required physical layer would also require a similar complexity, and cost, to a Manchester II device, although it would have the advantage of being designed insensitive to signal polarity, allowing easier component installation.

11.4.6 FM0 bit encoding

Another variation of differential Manchester encoding is sometimes referred to as "FM0," as shown in Fig. 11.4. FM0 always has a transition at each bit time boundary for clock recovery, with the data encoded as a transition, or no transition, in the middle of the bit time. A logic 0 has a transition in the middle of the bit time, and a logic 1 has no transition in the middle of the bit time. FM0 encoding is very similar to differential Manchester, except that it is shifted in time. Although both encoding schemes are virtually the same, with most of the same attributes, advantages, and disadvantages, FM0 has one possible advantage for some types of data decoders (or receivers). Because the clock edge occurs at the start, rather than in the middle, of a bit time, the clock recovery can be easily aligned to the bit time boundary.

11.5 Protocols

Computer, consumer, communication, automotive, and industrial manufacturers use a variety of protocols for their serial data communication needs. There are over 25 protocols and bus types currently on the market for serial data communications, many that are proprietary. A few, however, are becoming somewhat standardized for high-volume applications such as those used in the automotive industry. These automotive protocols and bus types, such as J1850, CAN, MI-bus, and UART, can also apply to motor drive systems. (Note that some automobiles employ over fifty motors, as discussed in Chap. 4.)

11.5.1 Automotive protocols

Standardization of protocols for use in the automotive market have concentrated on both J1850 and CAN. To distinguish between system requirements, the Society of Automotive Engineers (SAE) J2057 specifies three classes of data communication networks—Class A, B, and C, each for a particular type of network. Class A is used for low-speed data communication, generally with speeds of 1 to 10 kbits per second (kbits/s). Typical applications for a Class A network would be operator-interface systems or other similar (non-safety-related) systems that do not require a fast response. A Class B network can meet all Class A requirements but generally uses more complex error detection and checking methods and operates at faster speeds (10 to 125 kbits/s). A Class B system might be used for status and diagnostic data sharing between the engine control unit, the speed control system, the transmission control system, and the instrumentation console, thereby eliminating some duplicated sensing systems. For high-speed, real-time data communication, a Class C network would be used. Class C networks have the highest speed capability (125 kbits/s to >1 Mbits/s), allowing them to be employed for real-time control functions. Communication of control functions between systems such as engine control, traction control, active suspension, and antilock breaking will require high speed and low error rates. Because of the needs of these applications, Class C networks typically combine their high data rates with more advanced data integrity features, which also make them the most costly system to implement in Class B and C communications. The automotive manufacturers using Class A communications typically utilize proprietary protocols as a result of the cost sensitivity of their applications. For Classes B and C networks, standardization has focused on two protocols: SAE Standard J1850 and the Controller Area Network (CAN). SAE Standard J1850 has become a U.S. standard for medium-speed automotive networking applications. CAN is already seeing use in some automotive high-speed networks, while other manufacturers are evaluating it for their real-time control applications. For Class A networks, one protocol that is the basis for several systems currently in use is the Universal

TABLE 11.1 Bit Encoding

Protocol	Transmission rate	Bit encoding
CAN (controller area network)	≤1 Mbits/s	NRZ + bit stuffing
J1850 (SCP)	41.6 kbits/s	PWM
J1850 (Class 2)	10.4 kbits/s	VPW
UART	<115 kbits/s	NRZ

KEY: NRZ = non-return to zero, PWM = pulse-width modulation, VPW = variable pulse width.

Asynchronous Receiver-Transmitter (UART). Table 11.1 shows the transmission rate and bit encoding for these automotive protocols.

11.5.2 SAE Standard J1850

This protocol, SAE J1850, is currently being used by several automotive manufacturers for various body electronics applications as well as for meeting the CARB OBD II requirements. J1850 protocol encompasses the lowest two layers of the International Standards Organization (ISO) open system interconnect (OSI) model, the data-link layer, and the physical layer. It is a multimaster system that uses the concept of carrier sense multiple access with collision resolution (CSMA/CR), in which any node can transmit if it has determined the bus to be free. Nondestructive arbitration is performed on a bit-by-bit basis whenever multiple nodes begin to transmit simultaneously. SAE J1850 allows for the use of a single- or dual-wire bus, two data rates (10.4 or 41.7 kbits/s), and two bit encoding techniques [pulse-width modulation (PWM) or variable pulse-width modulation (VPW)]. The transmission wire length set by SAE is 40 m, with 35 m in the vehicle and 5 m for diagnostic hookup.

An SAE J1850 message, or frame, consists of a start of frame (SOF) delimiter, a 1- or 3-byte header, 0 to 8 data bytes, a cyclical redundancy check (CRC) byte, an end of data (EOD) delimiter and an optional inframe response byte, followed by an end-of-frame (EOF) delimiter. (See Fig. 11.5.) Frames using a single-byte header are transmitted at 10.4 kbits/s using VPW modulation. Frames with a 1-byte consolidated header or a 3-byte consolidated header can be transmitted at either 41.7 or 10.4 kbits/s, utilizing either the PWM or VPW modulation technique. All SAE J1850 frames, regardless of the header style, employ a CRC byte for error detection.

Each frame can contain up to 12 bytes (VPW) or 101 bit times (PWM), with each byte transmitted MSB first. The optional in-frame

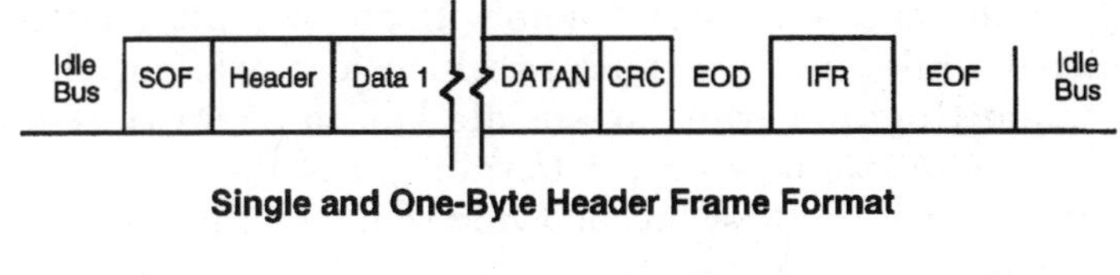

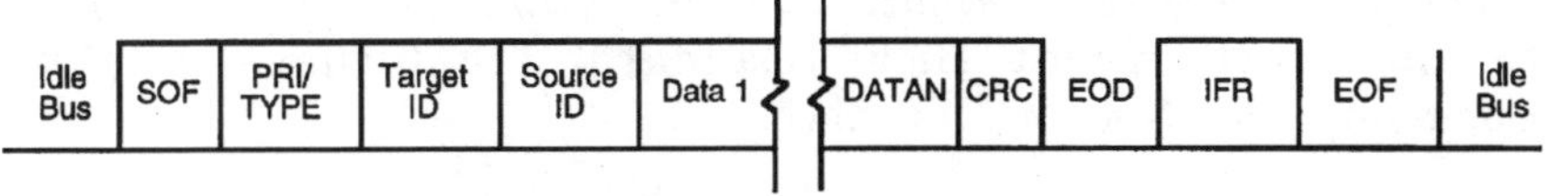

Three-Byte Header Frame Format

Figure 11.5 SAE J1850 frame formats.

TABLE 11.2 SAE J1850 Protocol Features

Feature	Headers		
	Single byte	1 and 3 bytes	1 and 3 bytes
Bit encoding	VPW	VPW	PWM
Bus medium	Single wire	Single wire	Twisted-wire pair
Data rate	10.4 kbits/s	10.4 kbits/s	41.7 kbits/s
Data integrity	CRC	CRC	CRC

KEY: VPW = variable pulse width, PWM = pulse-width modulation, CRC = cyclic redundancy check.

response can contain either a single byte or multiple bytes, with or without a CRC byte. Table 11.2 summarizes the allowable features of the SAE J1850 protocol. The requirements of each individual network determine which features are used.

Frames transmitted on a J1850 network can be either physically or functionally addressed. Since every node on a J1850 network must be assigned a unique physical address, a frame can be addressed directly to any particular node by placing that node's physical address in the header of the frame (peer-to-peer addressing). This is useful in applications such as diagnostic requests, where a specific node's identification may be important. With functional addressing, a frame containing data is transmitted with the function of that data encoded in the header of the frame. All nodes which require the data of that function can then receive it at the same time. This is of particular importance to networks in which the physical address of the intended receivers is not known, or could change, while their function remains the same.

If multiple nodes on the network begin transmitting message frames at the same time, nondestructive "bitwise" arbitration takes place. This allows the message frame with the highest priority to be transmitted while any transmitters which lose arbitration simply stop transmitting and wait for an idle bus to begin transmitting again. When multiple nodes begin to transmit at the same time arbitration starts with the first bit following the SOF delimiter and continues with each bit thereafter.

For more detailed information on the features of SAE J1850, refer to the most recent SAE Standard J1850-Class B Data Communication Network Interface.

11.5.3 Controller Area Network (CAN)

The Controller Area Network protocol, or CAN, is a communication protocol originally developed by Robert Bosch GmbH for use in high speed (Class C) automotive networking applications. However, with the increase in the variety of CAN devices available, some auto manufacturers have even begun using CAN for Class B and Class A networking applications.

Like SAE J1850, CAN is a multimaster, CSMA/CR protocol which uses nondestructive bitwise arbitration to ensure that the highest-priority message gets transmitted successfully if multiple nodes begin transmitting simultaneously. However, there are significant differences between CAN and SAE J1850 which make CAN more suited for high-speed networks. The transmission wire length set by SAE is not a specified length but is dictated by physical media and data rate.

Unlike the two versions of SAE J1850, the CAN protocol specification allows for a wide range of transmission rates, ranging from 5 kbits/s to 1 Mbits/s. The physical medium itself is not addressed by the CAN protocol, but is left to the user to define. This allows the user to select the physical medium that is most appropriate for the application, provided the bit timing requirements are met. Acceptable media include (but are not limited to) twisted-pair (shielded or unshielded), single wire; fiberoptic cable; or transformer coupled to power lines. RF transmitters are also being developed by some users for CAN systems. The most prevalent implementation to date is a twisted-pair bus using NRZ bit encoding. A variety of twisted-pair CAN transceivers from different semiconductor manufacturers are currently available that can simplify the CAN network design.

The message format CAN uses is an NRZ-encoded fixed frame with a variable number of data bytes. The minimum length of a

CAN data frame with zero data bytes is 44 bit times. The number of data bytes in a CAN data frame can range from 0 to 8, with each data byte transmitted MSB first. Four message types are defined: data frame, remote frame, error frame, and overload frame. Data and remote frames are used to transmit and request data, respectively. The error frame is transmitted by all nodes whenever an error is detected during a data or remote frame transmission. The overload frame is used by a node to delay the transmission of further message frames until it can begin transmission. CAN data and remote frames are routed to receiving nodes through user-defined message identifiers which are contained in the identifier field of each data or remote frame. These message identifiers can be either 11 bits (standard ID) or 29 bits (extended ID), giving the user much flexibility when developing a message addressing strategy. The CAN protocol does specify the use of functional addressing to allow multiple nodes to act on a single message, but physical addressing can be employed if the application requires node-to-node communication.

A variety of features are defined in the CAN protocol to help maintain the high level of network integrity required by Class C systems. One feature is the use of a 15-bit CRC to ensure the integrity of data and remote frames. In addition, each message transmitted onto the network must be acknowledged by all other nodes to guarantee that it appeared on the network correctly. This acknowledgment (ACK) is transmitted by all nontransmitting nodes on the network, no matter which node(s) are the intended receivers. When errors are detected on the network, CAN has built-in transmit and receive error counters that help contain errors and prevent nodes that have catastrophic failures from restricting communication on the bus. A faulty transmitting node always increases its error counters more quickly than other nodes, therefore becoming the first node to go "bus off." This allows the other nodes to resume normal message traffic. Automatic retransmission of corrupted messages also minimizes the possibility of a message being lost due to an error. These features give CAN nodes the ability to distinguish between temporary errors and permanent node failures, making CAN an ideal protocol for noisy environments such as those encountered in automotive applications. (For more information on the CAN protocol, refer to the CAN specification, version 2.0, published in 1991 by Robert Bosch GmbH, or refer to ISO 11519 and ISO 11898 for the two existing International Standards Organization publications which describe different implementations of the CAN protocol.)

11.5.4 Universal Asynchronous Receiver-Transmitter (UART)

The Universal Asynchronous Receiver-Transmitter (UART) has been widely used for years as a means for asynchronously communicating between computers or other microprocessor-based systems. Desktop computer serial I/O ports, for example, typically used to connect PCs to modems and other serial I/O devices, are UART-based communication systems. UARTs have been employed for many years to implement low-cost simple serial communication networks. The ISO diagnostic protocol ISO 9141 is based on the UART, as is the SAE J1708 protocol for communication networks in trucks and buses.

The UART is basically a simple asynchronous communication port that is byte-based: the host computer transmits each message on a byte-by-byte basis and also receives each message a byte at a time. This simple method allows UART communication to be added to a system with very low-cost hardware, while the CPU burden required is dependent primarily on the messaging strategy used. The transmission wire length set by SAE is not a specified length but is dictated by physical media and data rate.

The byte format of the UART is one start bit, 7 or 8 data bits, an optional address or data indicator bit, and 1 or 2 stop bits. UART systems generally use an active low, NRZ bit encoding with the bus idling high, corresponding to a logic 1 level. Besides the data byte itself, the only other communication features defined for the UART are the BREAK signal, consisting of 10 or 11 consecutive active bit times (depending on the byte format) and the idle bus indication, which occurs following 10 consecutive idle bit periods. All other features of any UART-based network—how messages are addressed, how collisions are handled, the data rate used, and how data integrity is ensured—are defined by the user on an application-by-application basis.

UART-type communication is employed in low-cost systems where it is impractical to provide a clock signal between devices to ensure synchronized communication. Each transmitter/receiver provides its own internal clock, with the receiver typically sampling the bus multiple times per bit, helping to protect against noise that may appear on the bus during a transmission. To work within the limitations of this basic communication system, data rates on UART networks are low, usually less than 120 kbits/s, and more typically in the 9600-bit/s to 19.2-kbit/s range. This helps improve the noise immunity of the system and also reduces potential EMC problems. The physical layer in a UART network is also defined by the user, although some existing standards, such as EIA RS232-D and RS485, have been designed for UART-type networks. While these physical layer standards can provide fairly robust performance, in many cases a simple one- or two-transistor driver/receiver network is all that is needed to meet the system's performance and EMI requirements.

11.5.5 Interconnect bus, MI-bus

Another serial communication bus, called *MI-bus,* is supported by a bipolar IC device (MC33192). The MI-bus was developed by Motorola Inc. and is designed for stepper motor control applications. The MI-bus serial push/pull communications protocol efficiently supports distributed real-time control while exhibiting a high level of noise immunity. Under the SAE vehicle network categories, the MI-bus is a Class A bus with a data stream transfer bit rate in excess of 20 kHz and thus inaudible to the human ear. It requires a single wire to carry the control data between the master MCU and its slave devices. The bus can be operated at lengths of up to 15 m.

At 20 kHz the time slot used to construct the message (25 μs) can be handled by software using many MCUs available. The MI-bus is suitable for medium-speed networks requiring very low-cost multiplex wiring. Except for the ground, it requires only one signal wire connecting the MCU to multiple slave MC33192 devices with individual control. A single MI-bus can accomplish simultaneous system control of multiple stepper motors.

The MI-bus interface shown in Fig. 11.6 is composed of a single NPN transistor, Q1. The two main functions of this transistor is first to drive the push field with approximately 20 mA of current while also exhibiting low saturation characteristics, $V_{ce,sat}$, and then to protect the input/output pins of the MCU against any electromagnetic interference captured on the bus wire. The MCU input pin (P_{in}) is used to read the pull field of the bus and is protected by two diodes, D2 and D3, and two resistors, R5 and R6. Any transient EMI-generated voltage present on the bus is clamped by the two diodes to a voltage value not to be greater than the V_{DD} or less than the V_{ss} supply voltages of the MCU.

An 18-V zener diode, Z1, protects the MCU output pin, P_{out}, from overvoltages commonly encountered in automotive applications as a result of load dump and inductive transients generated on the battery line.

The MI-bus is resistor-loaded according to the number of MC33192 devices installed on the bus. Each device has an internal 10 kΩ pull-up resistor to 5.0 V. The external pull-up resistor, R7, is recommended to optimally adjust termination of the bus for a load resistance of 600 Ω.

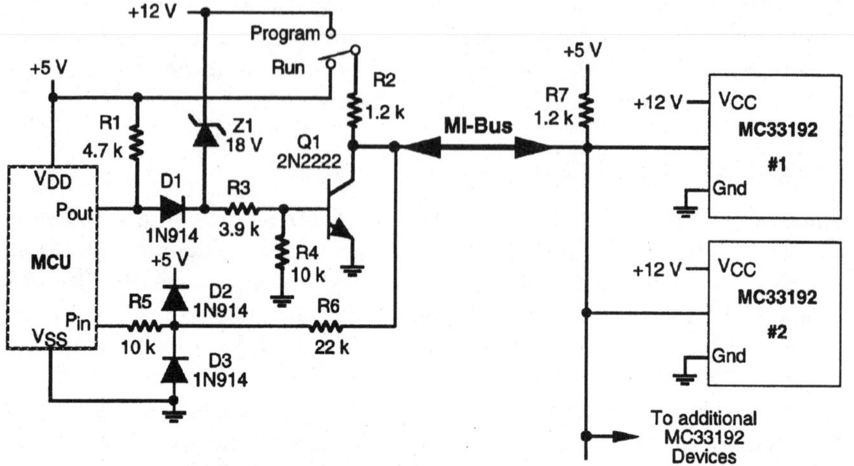

Figure 11.6 MI-bus interface to MCU.

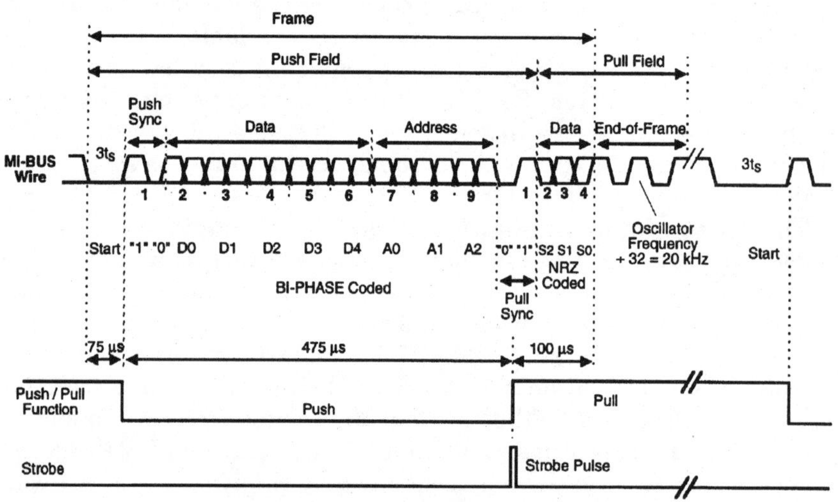

Figure 11.7 MI-bus timing diagram.

The MI-bus can have one of two valid logic states, recessive or dominant. The recessive state corresponds to a logic 1 and is obtained through use of the internal 10-kΩ pull-up resistor. The dominate state corresponds to a logic 0 which represents a voltage less than 0.3 V and is created by the $V_{ce,sat}$ of Q1.

The information on the MI-bus is sent in a fixed message frame format, as shown in Fig. 11.7. The system MCU can take control of the bus at any time with a start bit which violates the law of Manchester biphase code by having three consecutive time slots (3 ts) held constantly at a logic 0 state.

Communication between the system MCU and slave MC33192 devices always use the same message frame organization. The MCU first sends 8 serial data bits consisting of 5 control bits followed by 3 address bits. This communication sequence is called a "push field," since it represents command information sent from the MCU. Five control data bits (D0–D4) with three address bits (A0–A2) are sent in sequential order. The condition of the bus during any of the control bit time windows defines a specific control function. A "pull sync" bit is sent at the end of the push field, the positive edge of which causes all data sent to the selected device to be latched into the output circuit.

After the pull sync bit is sent, the MCU will listen for serial data bits on the bus sent back from the previously addressed MC33192 device. This now starts the pull field data sequence, representing information pulled from the addressed MC33192 and received by the MCU. The communication between the MCU and the selected device is valid only when the MCU reads (receives) the pull field data having the correct codes (from the S2, S1, S0 data) followed by an end-of-frame (EOF) signal. The frequency of the EOF signal may be a submultiple of the selected device's local oscillator or related to an internal or external analog parameter using a voltage-to-frequency converter.

The *MI-bus interface stepper motor controller* is intended to control loads in harsh environments. This communication device offers a noise immune system solution for difficult control applications involving relay drivers, solenoids, and motor controllers. When used with an appropriate MCU, the MI-bus provides an economical solution for applications requiring a minimum amount of wiring and optimized system versatility.

11.6 Network Systems Considerations

Most computer networks are configured as a star or ring configuration, as illustrated in Fig. 11.8, and are quite efficient for office systems. Systems in factory floors or systems that deal with critical function generally use the approach of a distributed operating system, shown in Fig. 11.9. The choice of the control ring is defined by the type of network required to meet the overall system needs. In this section we will discuss and show some examples of both automotive and industrial networks that are applicable to motor drives.

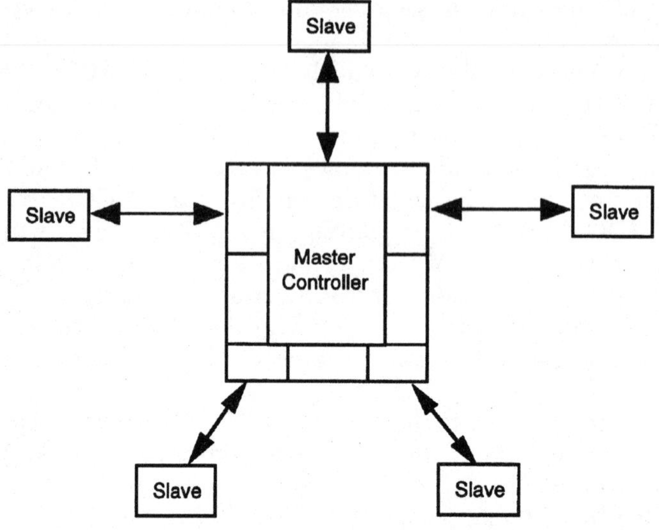

Star Configuration

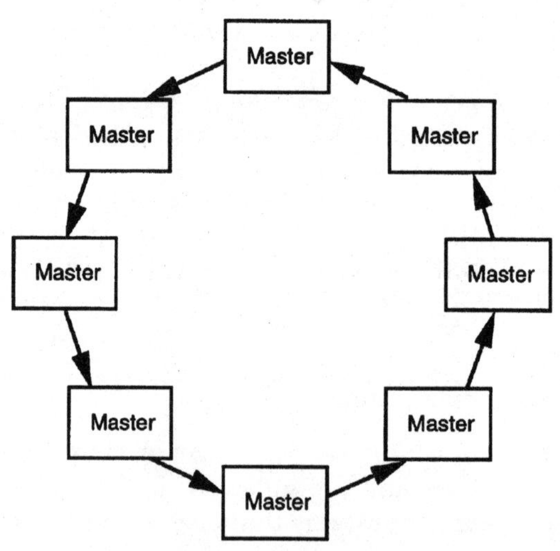

Ring Configuration

Figure 11.8 Star and ring bus control networks.

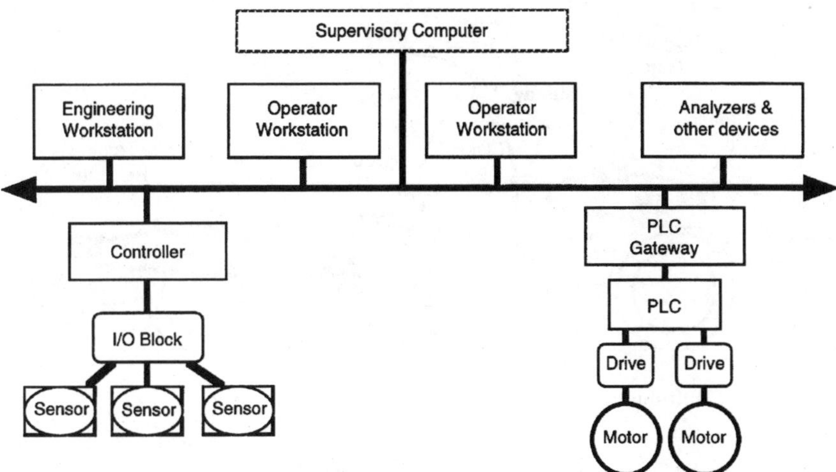

Figure 11.9 Distributed bus control networks.

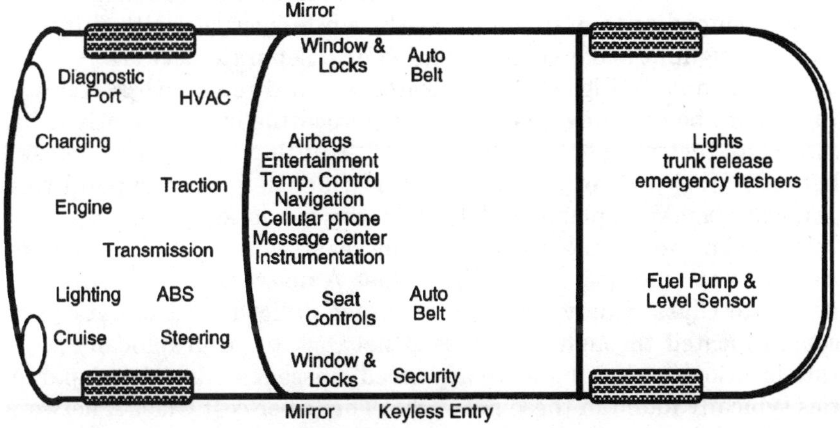

Figure 11.10 Single-bus control network.

11.6.1 Automotive and motor control networks

When designing a communication system for motors in vehicles or other similar applications, consideration must be given to whether a single- or multiple-network architecture is preferable. Interconnecting all the systems onto a single network in a vehicle (Fig. 11.10) is the most cost-effective solution, but it poses significant problems. Multiple nodes attempting to transmit simultaneously for an extend-

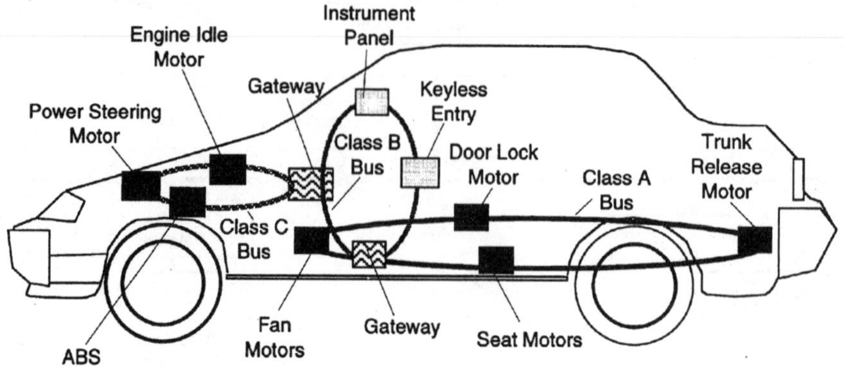

Figure 11.11 Multibus control network.

ed period of time could result in important information being delayed. One reason why three classes of networks were specified by the SAE was to allow functions and systems with similar data exchange needs to be grouped onto a single network, while systems with different data throughput needs could be placed on a separate network.

The example in Fig. 11.11 illustrates how a three-network architecture might be implemented, each communicating independently using different techniques and data rates. To allow data to be passed between networks or to allow systemwide diagnostic capabilities, gateways can be implemented between the networks.

However, the complexity and cost increase with multiple networks. An alternate method is to utilize Class A networks as subnets connected to Class B networks. This approach allows control data to be communicated through the Class B network to geographically separate locations, using the more advanced message addressing capabilities typically found in these networks. The lower-cost Class A network provides the interconnection between limited intelligence nodes, such as simple switches, sensors, motors, or actuators.

11.6.2 Industrial networks

With factory automation evolving, the network design is becoming a challenge. Industrial controls have undergone a change that is similar to the one that occurred when desktop computers replaced central processing systems. Sensors and actuators have evolved from change of state devices to intelligent mechatronic transducers that are able to make decisions on the basis of their environment and operator inputs. Such devices have opened the door to distributed machine control and a host of ancillary benefits. These benefits include diagnostics, adaptive response, and software that can be configured to manufacturing specifications.

Since many factories are now switching to multiplex networks, system designers are finding benefits which include easier design and installation, higher reliability, and lower maintenance costs. In the past designers had to plan the entire factory floor before electricians could connect wires; any last-minute changes would drastically disrupt the process. The system accuracy thus depended on the wiring integrity. With multiplex network communications, floorplans do not need to be as precise because modifications can be adapted as system needs occur. Also, maintenance and reliability depend less on wiring and more on software.

Figure 11.12 compares point-to-point wiring and a distributed system. Note how the distributed network system uses smart sensors

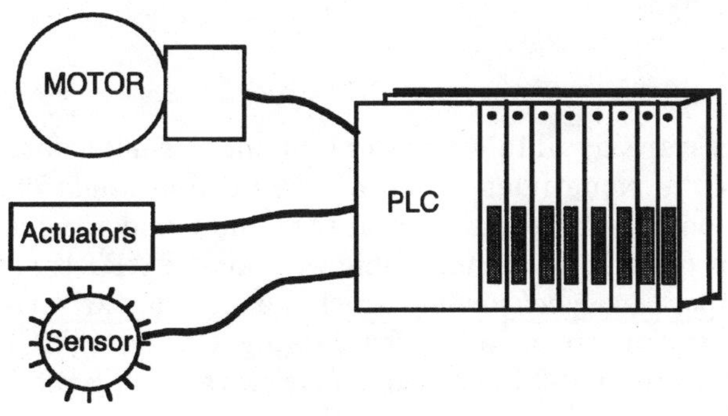

Point to point wiring

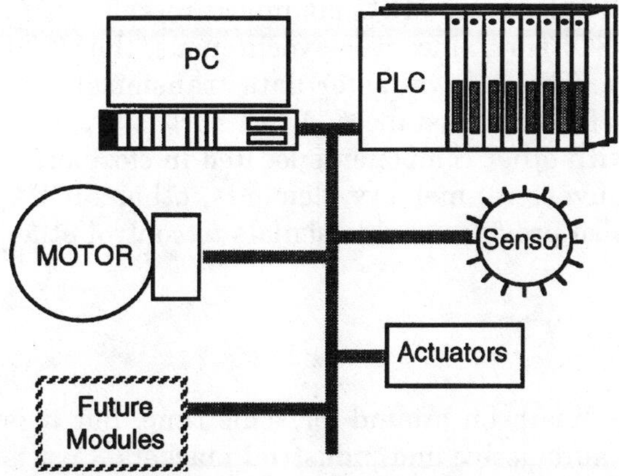

Figure 11.12 Point-to-point versus distributed network.

and actuators that share information over the bus and can make some decisions independent of the central computer or controllers.

11.7 Network Implementation

One way to simplify the implementation of a serial data network is to use microcontrollers that have a built-in MUX port. Most MCUs have one or two serial data communication ports, and several specialized MCUs are available or under development that provide either CAN or J1850 ports. Two examples are the MC68HC705V8 and MC68HC912B32 (Motorola Inc.). These MCUs have built-in logic for dealing with the J1850 bus and require minimal external components. Other MCUs are available with CAN ports. (Refer to the MCU manufacturer's data sheet or application notes for their data network design recommendations.)

11.7.1 SCI and SPI

Most microcontrollers are available with one or more serial data ports. The most basic is sometimes called a *serial communication interface* (SCI). This port usually has separate transmit and receive lines, uses NRZ bit encoding, and is compatible with most UART communication. The SCI port is *asynchronous,* which means that the serial data bit stream must contain extra bits for clocking. Generally, SCI ports are limited to less than 100 kbit/s transmission rates. A typical use of the SCI port is for communication with an external device, such as a display logic unit or another SCI port in other equipment. Usually some type of interface device is required between the MCU's SCI port and the external network. (See Chap. 10 for examples.)

The other serial data port commonly found in MCUs is called a serial peripheral interface (SPI) by some MCU manufacturers. The SPI port is different from the SCI port, since it is synchronous, that is, it uses a separate clock line. This allows faster data transmission, 1 Mbits/s or faster, but only for short distances. A SPI port is typically used for communicating with other components located in close proximity such as fast A/D converters, memory elements, other MCUs, and so on. The SPI port also provides enable signals to control other SPI devices.

Summary

Multiplex communications has been around for some time, but it is just now coming of age in automotive and industrial markets. Cost is still an issue and, as systems become more integrated, system manufacturers will be required to look for methods that reduce overall systems cost. Research and development for new communication standards, as well as modifying existing standards for specific applications, are continuing. The future of motor control applications will require serial multiplex communication, and we will see wider use of multiplex as a solution to meeting the demands of the industry. Key points about this chapter are listed below.

- A serial data network can offer increased reliability and easier troubleshooting over normal point-to-point wiring for each remote sensor or other modules. This is due to reduced connections and complexity in the wiring harness. With a serial data network, it becomes easier to add additional modules.
- Speed, distance, operating environment, fault tolerance, number of nodes, and reparability are a few of the issues that need to be addressed when designing a serial data network.
- A basic communications system consists of a transmitter, a receiver, and an information channel (medium).
- The seven layers used in the ISO OSI reference model are
  Layer 1—physical
  Layer 2—data link
  Layer 3—network
  Layer 4—transport
  Layer 5—session
  Layer 6—presentation
  Layer 7—application
- The main problem associated with high-voltage motors that employ a serial data network is the EMI that corrupts the serial data. The data can become unreadable. Also, if the data wires come in contact with the motor's high-voltage lines, catastrophic results may occur.
- A single wire for serial data transmission, while inexpensive, is subject to electrical noise problems, and is generally limited to short distances and low-speed transmission rates. A twisted-wire pair that uses differential signals allows fairly high transmission rates and is used in many local area computer networks. Shielded twisted-pair or coaxial cable offers high transmission rates but must be properly terminated. Fiberoptic cable permits very high EMI tolerance but requires costly opto-to-electronic interfacing.
- The SAE J1850 recommended standard lists three network classifications: A, B, and C, for automotive usage.
- The controller area network (CAN) protocol specification has a speed range of 5 kbits/s to 1 Mbits/s.

- Non-return-to-zero (NRZ) encoding represents each binary data bit as a static value during the entire bit time and is commonly used with Universal Asynchronous Receiver-Transmitter (UART) communication ports.
- A serial data network can be implemented with minimal components by the use of a microcontroller with a built-in J1850 or CAN port.
- Three common tests for message validation are checksum, parity bit, and cyclic redundancy check (CRC).

Acknowledgments

The authors would like to thank Tom Newenhouse and Tom Balph for their assistance with this chapter.

Further Reading

Emaus, B. D., *Elements, Definitions, and Timing of Network Conversion,* SAE Technical Paper Series 941660, 1994.

Motorola Data Sheet MC33192, *MI-Bus Interface Stepper Motor Controller,* Motorola, Inc. 1995.

Pinewski, P., Valentine, R., and Huettl, T., *Electronics for Electric Vehicles Motor Systems,* SAE Technical Paper Series 95188, 1995.

Powers, C., *Example Software Routines for the Message Data Link Controller Module on the MC68HC705V8,* Motorola Semiconductor Technical Data AN1224/D.

Powers, C., and Huettl, T., *Using Existing Multiplex Communication Technology to Implement an Electric Vehicle Communication Network,* SAE Technical Paper Series 951887, FTTC, Aug. 7, 1995.

Roden, M. S., *Digital and Data Communications Systems,* Prentice-Hall, 1982.

SAE Recommended Practice J1708, *Serial Data Communications between Microcomputer Systems in Heavy Duty Vehicle Applications,* Nov. 1989.

SAE Recommended Practice J2178/1, *Class B Data Communication Network Messages, Part 2, Detailed Header Formats and Physical Address Assignments,* Sept. 1993.

SAE Recommended Practice J2178/2, *Class B Data Communication Network Messages, Part 2, Data Parameter Definitions,* Jan. 1993.

SAE Recommended Practice J2178/3, *Class B Data Communication Network Messages, Part 3, Frame IDs for Single Byte Forms of Headers,* April 1993.

SAE Standard J1850, *Class B Data Communication Network Interface,* March 1994.

SAE Standard J551/5 (draft), *Performance Levels and Methods of Measurement of Magnetic and Electric Field Strength from Electric Vehicles, Broadband, 9 kHz to 30 MHz,* Feb. 1995.

Chapter

12

MCU Motor Control Reliability

Richard J. Valentine
Motorola Semiconductor Products

Introduction

In the preceding chapters we have discussed motor control circuits, some electronic principals, software, and how electronic motor controls can be developed. This chapter examines the dark side of motor control electronics, namely, the problem areas that occur when mixing sensitive microcontroller circuits with high-energy power stages.

12.1 Motor Control Design Principles

Electronics technology continues to have an impact on all types of motor-powered products that are in use around the world. In many cases, some of these products have critical control elements that are managed by some type of microcontroller (MCU) device. Exercise equipment, home appliances, and automotive systems are just a few examples. During the design of any product, safety and reliability factors must be considered. Designing and manufacturing a completely risk-free product that is reliable over the course of many years is a challenge, one that is not always met. Parts fail from abuse, defects, and poor implementation. In some products, such as large motor control systems, serious injury to the users or bystanders may result when a critical component, such as its microcontroller, fails to perform properly. An automobile that leaves its occupants stranded because the fuel pump motor control has failed can be more than an inconvenience in some situations. Designing safety elements into electronic controls requires a thorough understanding of the product's application and possible mishaps. Understanding how to protect the MCU will lead to more reliable controls, since the MCU usually manages most, if not all, of the control design's functions. (Protecting the MCU alone, however, will not ensure total fail-safe operation or eliminate catastrophes, since other critical electronic components can also fail.)

12.1.1 Circuit reliability issues: understanding the odds

The infamous "Murphy's Law" reveals that engineers have developed a sense of humor and disdain about design failures. It is important to realize that parts which can cause the most damage will eventually fail, as a result of either some undetected minuscule factory defect or an application-related stress. In the long term, thermal cycling and moisture contamination are especially harmful to active electronic components, while electrostatic discharge (ESD) and inductance-generated voltage spikes can also cause damage. Providing fail-safe protection to any product that could directly or indirectly cause harm should be part of its design methodology. This, however, can be a difficult task when the protective elements represent a significant portion of the overall cost. If the added cost of incorporating fail-safe designs can be turned into a selling feature, its implementation is easy to justify; the challenge then is to design the most cost-effective protective element or elements. Microcontroller devices will continue to be incorporated into more types of products. When you recognize areas that can present problems, designing a reliable fault-tolerant MCU-based product is not difficult.

12.2 Motor Control Problem Areas

Figure 12.1 shows a poor motor control design, while Fig. 12.2 shows an improved version. An electronics power and logic system manages the motor per the user or application inputs. There are several potential failure modes that may occur; a few are listed below. Some of these are caused by an MCU failure or could be detected and handled by the MCU. This is why many mission-critical designs use more than one MCU.

- Intermittent power or sensor connection (loose wires or connectors).
- Reversal of power connections (user hookup error).
- Power stage transistor short (induced by poor thermal interface or short).
- MCU computational error (caused by undiscovered software bug).
- Latent ESD damage to active components (static charge during assembly).
- Motor-bearing failure (poor lubrication or excessive rpm).

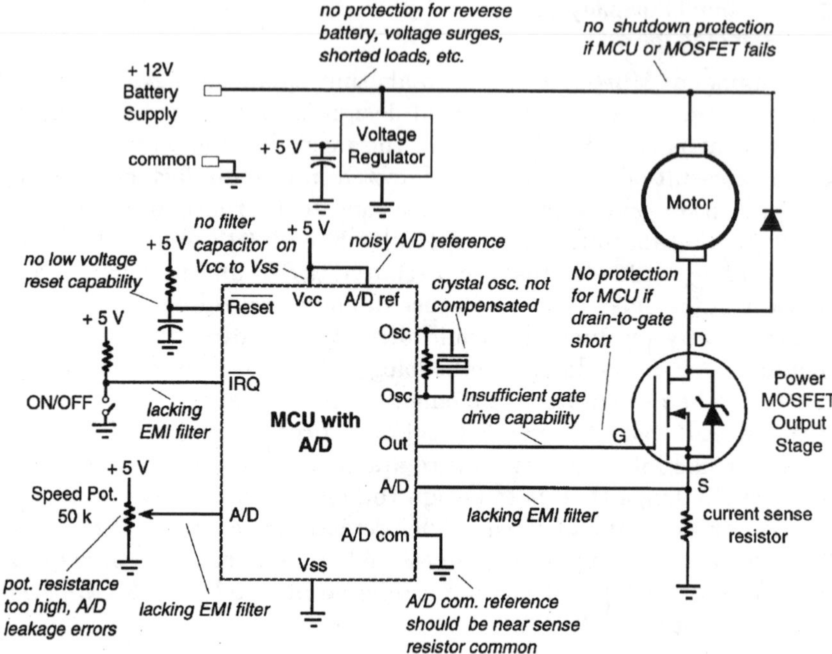

Figure 12.1 Poor motor control design.

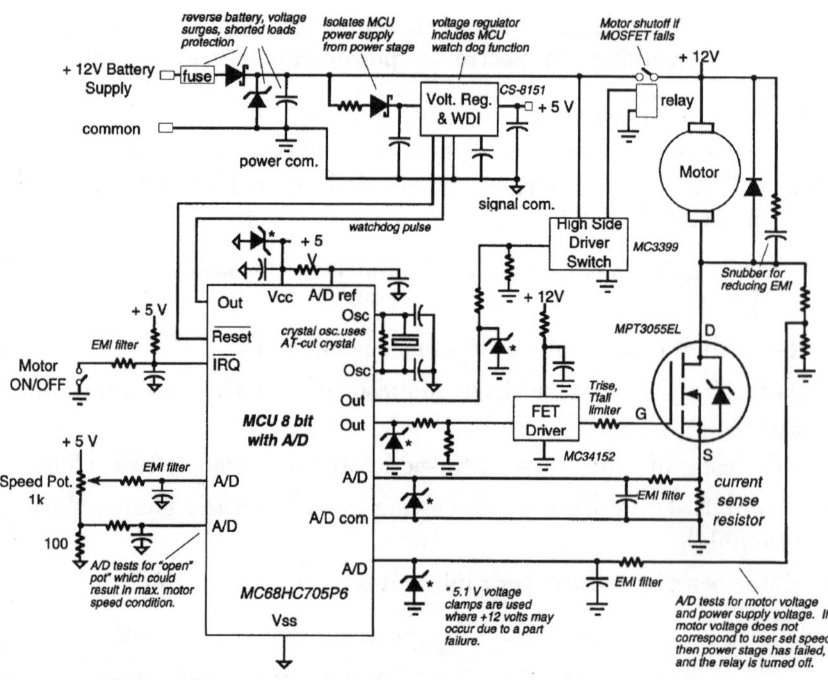

Figure 12.2 Improved motor control design.

- Motor shaft lockup due to external mechanical mishap (user misuse).
- *Unknown failure modes; these cannot be imagined but will happen.*

Unknown failure modes can be the most disconcerting category. You need to understand which is the most potentially dangerous scenario of any electronics system and then decide how to deal with it. In the example of a motor control, the motor itself might be considered the most hazardous element. In the case of an open sensor or software glitch, turning off the power stage, which turns off the motor, would seem to be the appropriate action, since it disables the motor and stops any mechanical movement managed by the motor that could be harmful. Such thinking is a good start, but it assumes that the power stage or the MCU control device is not one of the parts that can fail.

12.2.1 Indirect problem areas

MCU-based designs that control motors are also more prone to damage than, for instance, an MCU that is used in office computer equipment. Office computers typically utilize well-regulated power supplies, usually operate in a mild operating environment, and have probably been subjected to EMI design rules. An MCU that drives an inductive load directly or indirectly is subject to voltage transients from the kickback voltage of that inductive load. An indicator lamp or LED driven by an MCU in a motor control system would seem a safe design. A problem arises, however, when the lamp or LED is installed in a place where a person could touch the lamp or LED and induce sufficient ESD into the connecting wires that lead back to the MCU. Battery-powered MCU equipment is susceptible to several problems: ESD from the user, EMI from nearby high-powered electrical/electronic equipment, reverse battery hookups, and extreme temperature/humidity environments.

12.3 MCU Vulnerabilities in Motor Controls

The typical MCU data sheet, MC68HC05 (or similar MCU device), for example, lists maximum voltage and current levels that the MCU can survive. You may not realize, however, that the MCU may still succumb to an operation failure, even though it does not suffer permanent electrical damage when subjected to voltage or current levels that are within its maximum data sheet's specifications.

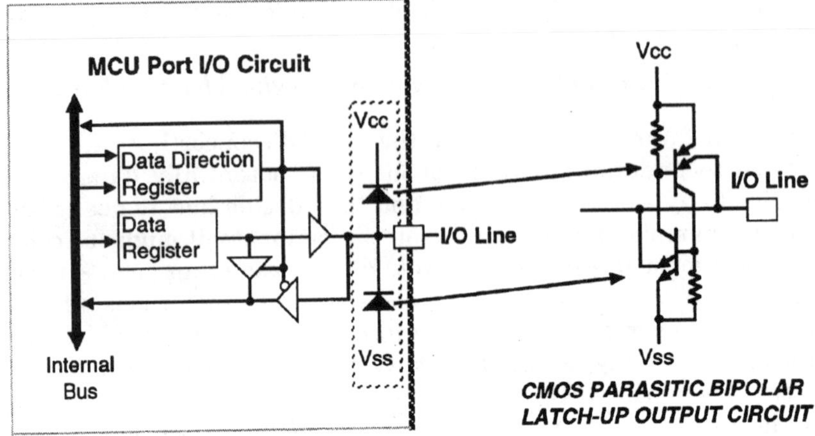

Figure 12.3 Internal clamp diodes are not that simple.

12.3.1 Excessive MCU input voltages

Figure 12.3 illustrates a typical MCU input/output (I/O) port circuit design. Note that the internal diodes are connected from the port line to both supply voltage (V_{cc}) and common (V_{ss}). These diodes will conduct if the port line voltage level exceeds the V_{cc} by about + 0.5 V or the V_{ss} by about − 0.5 V. In effect, the internal diodes clamp the input voltage to about 0.5 V above the V_{cc} supply or below V_{ss} common. Using the internal diodes to contain excessive voltage transients on the MCU I/O lines is not recommended, especially if the clamp current levels are above a few microamperes. When internal diodes are used as clamps, the current must flow through micrometer-size pathways or metalization in the MCU and out to the common or V_{cc} bus. If several port lines are in a clamp mode, the total clamp current can reach levels that create significant voltage drops in the internal V_{cc} or V_{ss} pathways. These voltage drops can affect the MCU logic states or, even worse, can generate enough heat to cause permanent damage to the MCU's structure.

Using the MCU port lines to conduct clamp current can also lead to more trouble than just excessive MCU power dissipation or internal voltage drops if the external voltage transients are of an RF nature. The MCU internal clamp diodes are likely to be more complex bipolar elements that are subject to latchup under certain conditions. These conditions include voltage transients of an RF nature that are found in many types of applications, such as cellular or cordless phones, wireless modems, and mobile radios.

12.3.2 Effects of noise spikes on MCU lines

Externally generated noise spikes on any MCU line can induce erratic MCU operation: The program may crash or lock up, data direction registers can change state, or timers may be reset or corrupted. Noise spikes should be contemplated as a normal design consideration in any motor control MCU design. One quick test recommended by EMI experts is to operate a high-powered cellular or mobile transceiver in close proximity to the MCU prototype design. If the unit MCU locks up or operates erratically, serious EMI lab testing must be undertaken. Operating the MCU prototype design near an automotive ignition system will give a good indication of EMI tolerance. ESD test generators are also available that can apply ESD to the external wiring, switches, control panel, and so on, to verify ESD endurance.

12.3.3 Program problems

There are some op-codes that can invoke some MCUs into a self-test mode. If, by chance, an out-of-control program or a "hole" in the program manages to allow the MCU program register to load and execute a special self-test code, the MCU ignores its user program and performs a built-in self-test routine. Usually, only a complete power-down and normal start-up will reset the MCU out of a self-test mode. Some programmers actually search all their assembly object listings for any "self-test" data codes, and, if found, the program is rewritten to eliminate these particular codes. MCU self-test modes can also be invoked by raising a certain pin to an abnormal voltage level during a reset. It is wise to check with the MCU manufacturer to verify how the self-test mode operates. This will ensure that it is not inadvertently invoked during normal operation in the actual application. Some initialization routines can be written to perform a limited circuit test of the application, and, if a failure is detected, the controller is put in a "service required" mode. Testing both critical input sensor lines and power stage control lines is a recommended practice, and, in some applications, an MCU may be dedicated to perform just this function at all times.

A technique to catch an out-of-control program is to add several "no-op" instructions followed by a software interrupt at the end of each program sequence or subroutine. (The software interrupt halts other program activity and is used instead of a jump instruction.) The software interrupt routine is written to force a reset of the MCU and can toggle an I/O line to alert the user or programmer that a major program error has occurred.

Another method to help ensure reliable MCU operation is to use two identical MCUs whose output lines are fed to an AND gate. This means that both MCUs must compute the same results and output the identical logic 1 signals before the AND gate will allow the logic 1 signals to pass through to a power control stage (or other critical output element).

The normal watchdog routine in which an internal or external timer must be reset periodically from the program will also detect a program lockup. The "no-op" and SWI test may detect a program fault faster than the watchdog, depending on the watchdog timer value. Other factors that can cause mysterious program faults are a poor reset control circuit, floating interrupt lines, or power supply brownouts. Often the program runs fine in the development system environment but fails after the MCU is electrically programmed and runs in a standalone mode. In many cases, this is due to an incorrect reset or interrupt design in the user's circuit. The development system usually has some type of built-in reset and interrupt control circuits that will allow the MCU to operate. All unused MCU pins should be terminated with pull-up or pull-down resistors. Directly connecting unused lines to $V_{ss(common)}$ or $V_{cc(+\text{ MCU supply})}$ can spell trouble if the port line's data direction registers are inadvertently set as outputs, which could happen with an out-of-control program or simple program error. Another reason to use termination resistors rather than direct connections is that during reset all of the MCU port lines will go to a high impedance, allowing output lines to float. This can raise havoc if the output lines are directly driving a high-impedance power device, such as a power MOSFET. Any slight leakage or EMI can turn on the MOSFET during the reset period, which lasts until the MCU starts executing the port initialization code. (*Note:* Some MCUs are designed with internal I/O port termination resistors to minimize this problem.)

12.4 Understanding the Root Causes

We need to review some traditional engineering fundamentals to ascertain what causes the EMI problems and other impediments that can affect an MCU or other sensitive semiconductor devices in a motor control design. You should also be familiar with the limitations of CAE (computer-aided engineering) tools and MCU development systems.

12.4.1 Electromagnetic theory

Controlling an inductive load is normally accomplished by switching the current through the inductor with a controlling device, such as a relay/contactor or power transistor. An MCU output port line can also directly control low-power loads. But when an inductive load is switched off, a collapsing magnetic field builds up a countervoltage (CEMF) whose amplitude is related to the inductance, coil current, coil resistance, and the rate at which the current was switched off. The formula used to calculate the inductive kickback voltage is

$$E_{\text{peak}} = L * \frac{Di}{Dt} \tag{12.1}$$

where E_{peak} = kickback voltage
L = inductance, H
Di = current amplitude delta, A
Dt = time period of switching event, s

12.4.2 Inductance effects with relays and motors

Putting Eq. (12.1) into practice reveals that driving miniature 5-V relays or any inductive load, such as a small motor, can generate a significant kickback voltage. The MCU's fast switching speed (typically 50 ns) is responsible. Relays are sometimes used to switch the power in a motor drive. Miniature 5-V relays usually have coil inductances in the tens of millihenries. Generally, lower coil currents mean more windings are used, resulting in higher inductance values. A test of three generic 5-V miniature relays indicated coil values of 55 Ω 26 mH, 70 Ω 65 mH, and 250 Ω 77 mH. Theoretical peak voltages of over 20,000 V are possible if these miniature relays are switched off at fast (50 ns) switching speeds. Adding a simple clamp diode across the coil is the usual fix but introduces another set of problems, discussed later. First, there is more to understand about relays.

When the relay coil is initially energized, its current will rise at a slope controlled primarily by the coil's inductance value, its DC resistance value, and the applied voltage. The inductor's current will continue to ramp up, with the DC resistance generally limiting the maximum level. If the coil's pulse width is too narrow, the coil current may not reach its maximum value, which reduces its kickback voltage, and may not operate the relay mechanism correctly.

It is important to keep track of the maximum inductance values of any component that is being switched by electronic devices. This holds true for so-called noninductive devices, such as lamps or resistors. All components have a certain amount of inductance, which will become critical when the switching edges are too fast.

When any inductor is switched with an open-collector (bipolar-type) or open-drain (MOSFET-type) MCU output line, the inductor's kickback voltage can easily exceed the transistor's breakdown voltage rating and may cause a second breakdown failure. Once the transistor enters second breakdown, its blocking voltage diminishes, and the remaining inductive energy will literally burn a hole into the silicon chip. Some power transistors, such as avalanche rated power MOSFETs, will act like a zener diode if their nominal blocking voltage is exceeded and will not fail until their power dissipation rating is surpassed.

12.4.3 PCB layouts and autorouters

Printed-circuit boards are usually designed with PCB layout software tools. There are several PCB layout programs in use that incorporate an autorouter that selects how the PCB traces should be connected. This method works fine for low-speed and low-current circuits. Big problems usually occur when an autorouter is allowed to place high-speed and high-current lines without some guidance from an EMI-knowledgeable engineer. (Some companies now actually require the PCB layout tools to be operated by experienced EMI engineers.) A few general rules are listed below that can help to minimize EMI difficulties in PCB layouts:

- The overlying principle for PCB high-current switching lines is short and fat; thus, make the traces short but as wide or as fat as possible.
- A "3-to-1" layout rule has been suggested by EMI experts. The trace width should not be less than one-third of its length.
- Route clock lines or high-speed lines near V_{cc} or V_{ss}.
- Slow down switching edges if possible.
- A/D lines should be routed carefully; the A/D reference pins should not be mixed into the V_{cc} and V_{ss} current-carrying traces.
- Follow the MCU data sheet or application-note-recommended PCB layout rules, if available.

When analyzing EMI problems, remember the basics: maximum frequency and amplitude of the signal that can either be generated or received, the time and place that the EMI occurs (failures occur every time a high-power radar antenna swings into sight or in close proximity to high-voltage power distribution lines, etc.), the impedance of the suspicious network, and the physical aspects of the traces or wiring.

MCU clock or oscillator circuits can be a potential EMI problem. Most MCU manufacturers usually show a recommended oscillator

PCB layout in their specification sheet or product-related application note. Recent MCUs are using lower-power oscillator designs to minimize radiated emissions. Placing a small metal shield around the entire MCU and associated components can also be effective in minimizing EMI problems.

The MCU V_{cc} line should be adequately decoupled, since it is conducting high-frequency currents. The placement and values of the MCU's V_{cc} decoupling capacitors are somewhat critical. The capacitors should be located in close proximity to the V_{cc} and V_{ss} pins. (See Chap. 7 for further information.)

12.4.4 Program bugs

As we said before, an out-of-control program can be caused by an external voltage spike upsetting the MCU's internal logic. Undetected errors in the program can also cause a failure. Programs in MCUs tend to be hand-packed to minimize ROM size and increase throughput. Unfortunately, this is where errors can creep in. Even programs written in high-level languages can fail, if there is a bug in the compiler, for example. It is important to note that most MCU control designs deal with external events and generally use timers as part of some calculation. When a timer generates an interrupt, or an external event needs servicing, the normal program flow is halted, and the interrupt is processed. Because the user's program can be interrupted at any time, the main routine or its subroutines may be halted at any time (depending on preset control logic), which may allow the main routine or a subroutine to get "lost" during one of these numerous interruptions. (This can happen if too many subroutines are nested together, allowing the program stack RAM to read or write into normal user RAM.) Because of the random nature of external events, every possible combination of events, in relation to the MCU's timer values and program flow, may not be accounted for.

If a random glitch does occur during the course of debugging an MCU program or testing the MCU system, it should not be dismissed without recording the conditions that were present. If another random glitch occurs again, it can be compared to the first, to determine if it is all that random, that is, do the same conditions exist for both glitches? The MCU development tools can help if they are capable of real-time analysis. Most low-cost MCU development systems are adequate for preliminary program development but lack the sophistication to show exactly what is happening inside the MCU's CPU and control registers during reset, interrupts, timer rollovers, and so on.

12.4.5 Low-voltage reset

Some MCUs may not recover well if the V_{cc} sags momentarily. This is different from a normal powerdown or powerup condition when the V_{cc} drops to zero and then goes back to normal value. The MCU's internal power-ON reset (POR) usually can deal with a normal zero to full V_{cc} voltage swing. If the V_{cc} slowly sags by 50 percent and then slowly rises to normal, the MCU may not recover. External MCU supervisory devices (MC34160, MC34164, MAX814, CS-8151, etc.) are used to monitor the V_{cc} bus and will generate a reset when the V_{cc} is sagging, as during a brownout. A good test to determine if the MCU is vulnerable to brownout is to run the MCU at normal voltage, then slowly decrease the V_{cc} until MCU errors occur, and then raise the V_{cc} slowly back to normal. If the program does not recover, a power supply supervisory design is required. It has been this writer's experience that MCU supervisory devices are frequently worth the extra cost in MCU power control designs and should be designed in at the beginning of the project. If rigorous testing shows that the supervisory circuit is not required, the parts are simply not assembled onto the PCB. This trades a little PCB space for redesigning the PCB later when testing shows that a supervisory circuit was required for stable MCU operation.

One final note about MCU operating voltages: If a low-voltage detection and supervisory device is used, it should match the MCU's minimum V_{cc} specifications. For example, if the MCU is rated to operate at 5 V plus or minus 10 percent, the supervisory device should force the MCU into a reset mode if the V_{cc} sags below 4.5 V.

12.4.6 Operating temperatures and environment

Long-term reliability of most active electronic components in a motor control, including MCUs, is affected by temperature, thermal shocks, and moisture. The MCU package and internal die mounting methods determine, to a high degree, how well the MCU will hold up long-term in a stressful environment where many power control systems are employed. Using a low-cost plastic package may suffice for applications in an office environment, but for automotive under-the-hood or motor controls in a factory setting, superior MCU packages should be specified. There are some ways, in addition to specifying high-reliability packages and die attach methods, to further increase the MCU chances of survival in a hostile environment. The power dissipation of the MCU should be kept low by using the lowest clock frequency possible. This is determined by the application's throughput time requirements. Using a heatsink or other cooling device on the MCU can also help reduce the MCU's die temperature. Potting or encapsulating the MCU can give added protection against moisture or external chemical contamination.

It is also wise to conduct extreme temperature testing of the MCU-based product to determine where the failure points occur. If the failure points are close—within 20 percent, for example—to the expected worst-case operating temperatures, a redesign is probably required. High temperatures increase leakage currents, while cold temperatures decrease switching and internal propagation delay times in CMOS semiconductor devices. Always assume that the product will be operated in a worst-case temperature condition and design accordingly. (An operating temperature range of −40 to +85°C is a reasonable prospect for many products.)

12.5 MCU Circuit Design Techniques

We have reviewed the causes of many problems in motor control MCU designs. Design methods that can minimize these problems start with a basic design principle. Each MCU line should be connected through a low-pass filter that sets the upper frequency limit for that line per the needs of the application. Unless this is done, the MCU lines will act as antennas, and the MCU's internal logic circuits and program stability will suffer.

12.5.1 Simulation limitations

Circuit simulation programs can help evaluate the results of excessive voltage spikes or other possible abnormalities if the design engineer knows where these voltage spikes or events will occur. The simulation program's accuracy is limited by the model's specifications and, in general, should not be totally trusted to find all the potential problem areas. Many semiconductor models are not 100 percent complete or up-to-date and may have not been thoroughly tested in unusual modes of operation. The simulation program is a powerful tool for developing circuits and even system-level designs but cannot guarantee that the actual product is totally perfect. In some cases, the simulation results that do show problem areas are not immediately obvious. Generally, trouble spots occur on or near signal transitions. These areas should be examined on at least a 1-μs scale with a 10-ns minimum simulation resolution.

A few comments about hardware debugging—digital oscilloscopes can hide or filter out fast, narrow voltage spikes. Always search for voltage spikes with a time base setting that can reliably show 10- to 50-ns signals. Use of a slow (100 μs) time base will probably fail to show the trouble spots. It is not always easy to catch a fast single event that is mixed in with a lower-frequency repetitious signal when using an analog oscilloscope. Digital scopes are very good at capturing fast single-event signals, provided they are correctly set up to do so. It is easy to miss fast nonrepetitive pulses when the digital scope is set to display a much slower repetitive-type signal. (The writer uses three types of oscilloscopes for debugging MCU control designs—a 120-MHz analog storage scope, a general-purpose 150-MHz digital scope, and a 2-GHz digital scope. A spectrum analyzer is also used for EMI testing.)

12.5.2 Protecting the input lines

Noninductive generated voltage spikes can also be coupled by parasitic capacitance into the motor control MCU lines by various means: poor PCB layouts, insufficient isolation between MCU lines to the power control stage, or mounting the MCU too close to a high-voltage field source. One starting point in protecting the MCU input lines would be to assume that all MCU lines can be exposed to voltage spikes of a magnitude similar to that of the worst-case voltage levels that are used in the MCU's application or system. For example, in motor control applications, any line that leads to the outside of the electronic module can be exposed to either vehicle ground or full AC line voltage. The degree of protection that the MCU's lines require against excessive voltage depends on the application. Figure 12.4 shows protection methods for MCU input lines. The basic method is to use zener diodes for voltage clamping and series resistors for current limiting. Capacitors are also used to form a *RC* filter that minimizes RF voltages. These methods work well for MCU input lines. The *RC* filter time constant is set for the lowest frequency possible that will still allow the application to work as planned.

Magnetic coupling can also be a problem if the MCU's lines or the MCU itself are located near an intense magnetic field source, such as a high-power transformer or high-current power conductors. Metal shields can be utilized to reduce magnetic coupling.

MCU input ports, such as timer inputs, are sometimes set to toggle on switching edges and require a minimum Dv/Dt to operate correctly. For this type of input the *RC* filter method may not suffice and a buffer element may be required, as shown in Fig. 12.5. The buffer consists of a comparator that has high gain and whose trip point is set about midway between V_{cc} and V_{ss}. The comparator design method

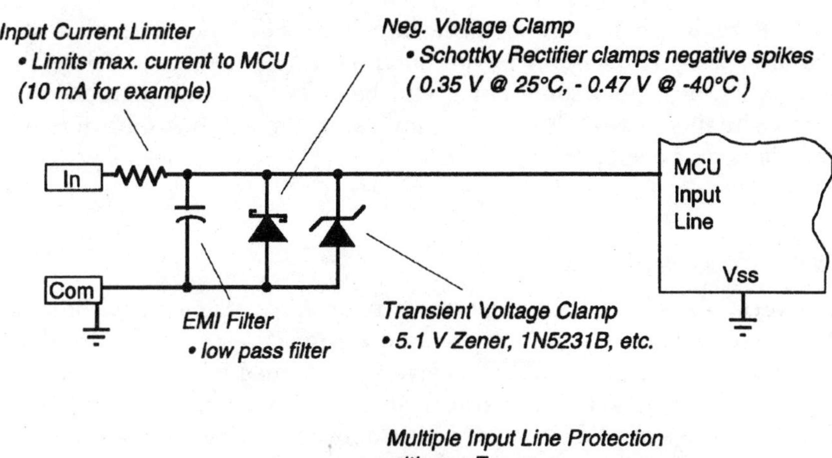

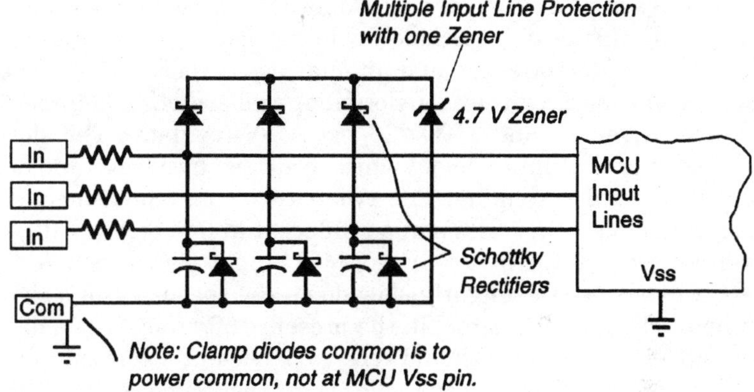

Figure 12.4 Input line protection methods.

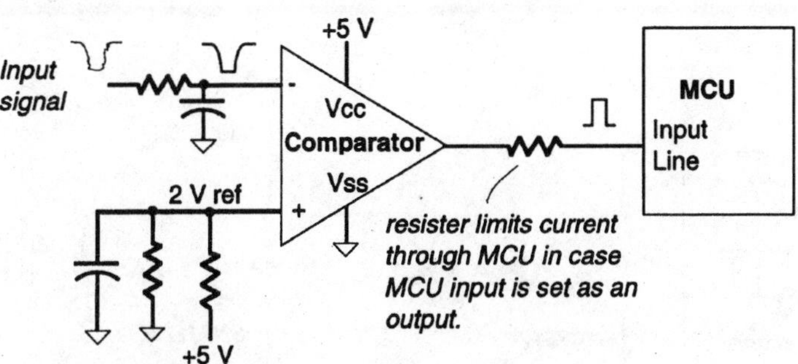

Figure 12.5 Buffer comparator element.

can be modified to allow hysteresis, which will have the effect of a Schmitt trigger; an input signal such as a sine wave with slow rise and fall times, plus some noise will be converted into a pulse with fast-switching edges. (Refer to comparator applications data for hysteresis circuit design information.)

12.5.3 Serial data line buffer

For serial data lines that must work in a electrical noisy environment a twisted-wire-pair (TWP) driver and receiver is sometimes used, as shown in Fig. 12.6. The TWP receiver is designed with high common-mode noise rejection and operates at a low impedance. The differential line drivers and receivers will operate −7 to +12 V common mode, making this design suitable for many electrical noisy environments, such as a motor speed and angular position sensor. The speed sensor transmitter circuit uses a slotted optical switch, a high-performance line driver IC, and a TO-220-size +5-V regulator. The slotted opto detector uses Schmitt logic, which provides hysteresis for noise immunity and pulse shaping. The twisted-wire pair is connected to the receiving circuit that uses a quad differential line receiver IC. The receiver connects to the motor drive's MCU input port. A simple low-pass *RC* filter is used to slightly slow down the speed sensor's signal, which helps clean up the signal in the presence of strong RF fields.

Interrupt and reset lines should receive particular attention to minimize the chances of extraneous voltage spikes. Some MCUs can be programmed to respond to either edge-triggered or level-sensitive external interrupts. Using the level-sensitive option allows for more noise immunity at the expense of reduced response time to an external event that forces an interrupt.

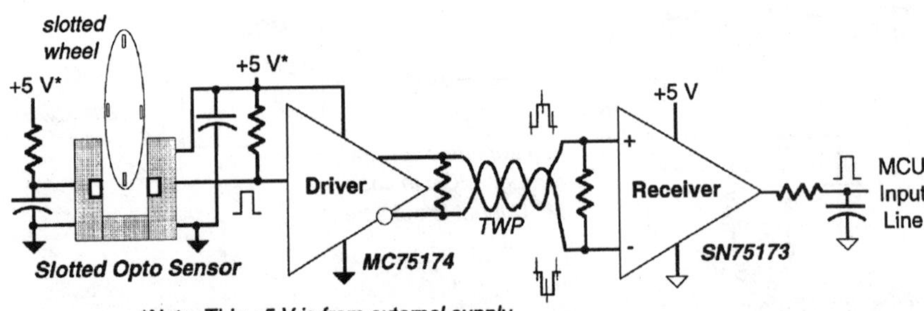

Figure 12.6 Serial data wire transceiver.

12.5.4 Program debugging methods

One debugging method to verify that an external event or any particular routine is functioning correctly is to toggle an unused output port when a particular event happens. For example, if an external pulse-type signal is supposed to drive an internal MCU counter, whose value is used to perform some calculation, it is important to know that the input pulse increments the counter only once. A noisy input pulse may advance the timer incorrectly. If an output pulse is generated every time the MCU increments its counter, an external oscilloscope could verify the operation. Sometimes the timer may have been toggled from both an errant program routine and a poor pulse input. It is important to realize that the MCU can be programmed to perform some simple debugging steps and can help locate problem areas if a real-time MCU development station is unavailable.

12.5.5 General output line considerations

As explained previously, connecting a microcontroller directly to control inductive loads usually requires voltage clamping networks to protect the MCU's internal silicon structures. Leaving out some form of voltage transient protection may result in minor erratic program operation or, in the worst case, failure of the MCU or other devices. Even an MCU indirectly driving an inductive load is subject to some degree of voltage transient energy that propagates through the load control power transistor and back into the MCU's output port line.

Sometimes the actuator control power device fails first because of its inadequate protection against excessive CEMF or other load faults. If the power device design does not limit the actuator's CEMF voltage or protect against power supply voltage spikes, the power device may self-destruct and allow high energy levels to propagate through the MCU device, which is more expensive and more difficult to replace.

Several CEMF protection methods are possible. One solution involves selecting a semiconductor switching device with sufficient voltage breakdown rating so that it will "never" enter a secondary breakdown condition. Unfortunately, the word "never" is covered by Murphy's law, and the transistor will probably be subjected to some degree of overvoltage stress during its lifetime. If the power MOSFET shorts out with a drain-to-gate short circuit (as happens in many cases), the MCU output line will be subjected to the MOSFET's full drain voltage, which could be several hundred volts.

Figure 12.7 shows protection for the MCU output line against excessive positive and negative voltage transients. A Schottky barrier

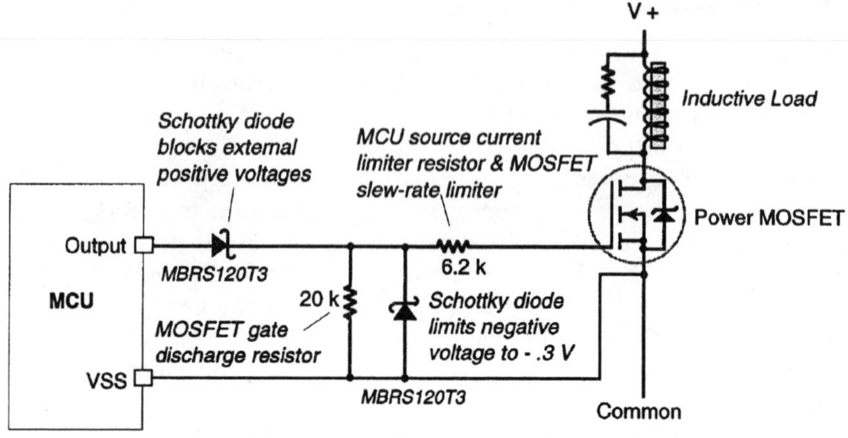

Figure 12.7 Output line protection methods.

diode in series with the output line to the MOSFET provides isolation against positive going voltages from the MOSFET's drain-to-gate path. This does compromise the circuit operation by lowering the available gate voltage by one Schottky diode drop and does not allow the gate to discharge through the MCU output. A resistor pull-down discharges the gate. This resistor's value cannot be sized too low, or the MCU output will be loaded down too far to ensure a sufficient gate bias level. A typical 5-V-type MCU "1" output can supply 0.8 mA at a level of 4.2 V. This calculates to a load value of 5250 Ω, which leaves only about 3.9 V gate bias with the Schottky diode drop subtracted out. A resistor in series with the gate limits the maximum current from the MOSFET and establishes a passive element that mostly controls the MOSFET's turnon switching time.

A variation on this method is to add an active turnoff element, as shown in Fig. 12.8. One additional note about directly driving a MOSFET—the MOSFET's gate input capacitance can draw a high peak current from the MCU. A series resistor value (≥500 Ω) should be selected to limit the MCU peak current to under 10 mA. The MOSFET's turnon switching time is affected by how fast its gate input capacitance is charged up, which may require a significant current peak for a large MOSFET. The MCU output will thus require a buffer stage to drive large MOSFETs. Several MOSFET and IGBT drivers (MC33153, MPIC2113, etc.) are available for interfacing the MCU output line to drive MOSFETs or similar power devices. There are many other IC devices available for driving lamps, relays, LEDs, or displays that easily interface to the MCU.

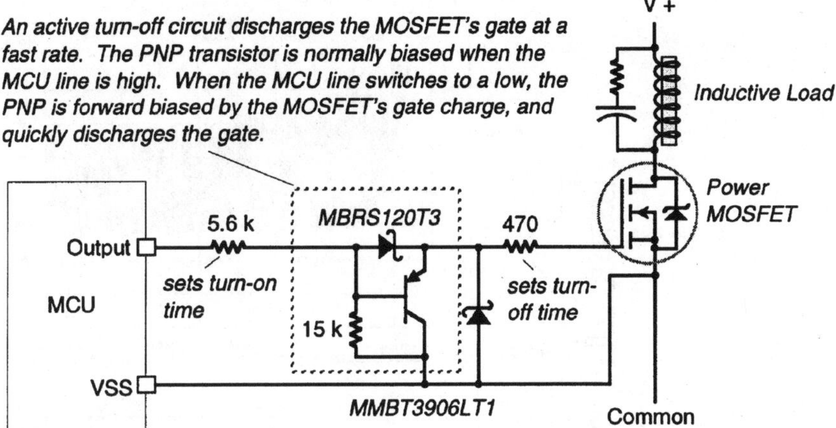
An active turn-off circuit discharges the MOSFET's gate at a fast rate. The PNP transistor is normally biased when the MCU line is high. When the MCU line switches to a low, the PNP is forward biased by the MOSFET's gate charge, and quickly discharges the gate.
V +
Inductive Load
Power MOSFET
MCU
Output
VSS
5.6 k
sets turn-on time
MBRS120T3
15 k
470
sets turn-off time
MMBT3906LT1
Common

Figure 12.8 Using an active turnoff element.

12.5.6 EMI reduction in the MCU power control stage

Figure 12.9 shows a switching circuit in which an MCU drives a simple power control stage and the resultant EMI. The power stage's transistor is selected to satisfy a few essential requirements: maximum voltage rating above supply voltage, current rating above the solenoid operating current level, a low ON voltage to keep power dissipation down (no heatsink), and the ability to be driven from the MCU's 5-V logic. The relay's inductive kickback voltage is contained with an avalanche-rated power MOSFET. Unfortunately, this design will also generate significant EMI that can affect the MCU or other nearby sensitive devices. There are two conditions in this switching circuit that contribute to EMI: the MCU output high-speed transitions which drive the power MOSFET and the inductive kickback voltage causing the MOSFET to avalanche (i.e., the MOSFET acts like a zener diode).

Adding a diode [this is sometimes called a "free-wheeling diode" (FWD)] across the relay stops the power MOSFET from avalanching but does not eliminate EMI, as shown in Fig. 12.10.

Slowing the switching edges of the MCU output signal that is driving the power MOSFET's gate will help minimize EMI because the power MOSFET's switching times are no longer in the RF range. The MCU output signal can be less than 100 ns, and, if the power MOSFET is switched at anywhere near this speed, serious EMI will result. One simple method to slow down the driving signal edges of the MCU is to add a series resistor and a gate-to-source capacitor. This is simi-

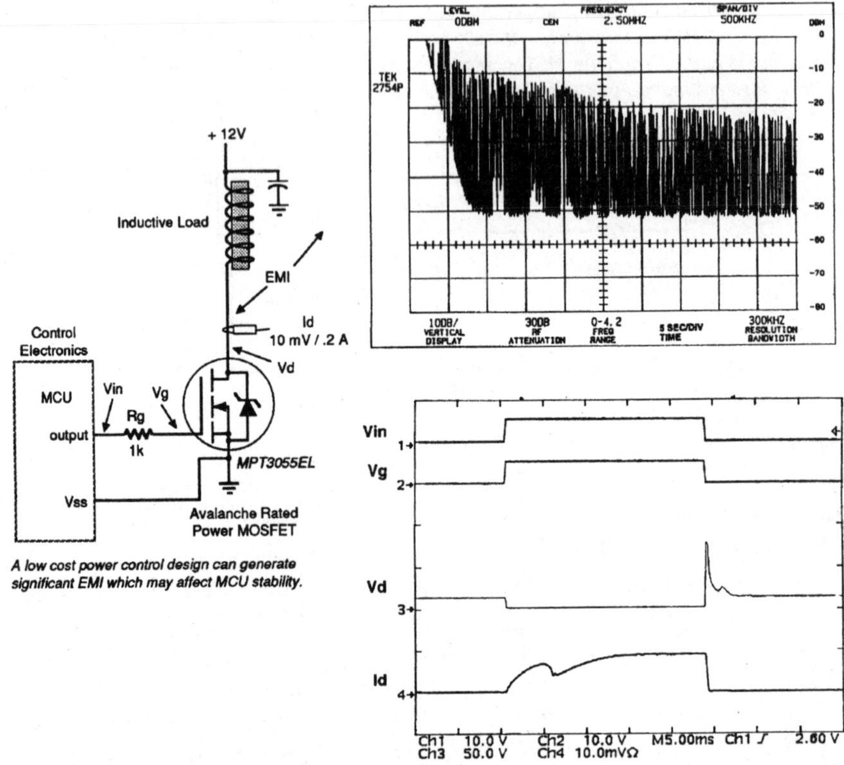

Figure 12.9 Simple switching circuit and EMI problems.

lar to the *RC* filter scheme for the MCU's input lines. To minimize the chance of parasitic RF oscillations during switching, the series resistor value should be 1000 Ω or less, and the gate-to-source capacitor value should be fairly large, 0.01 to 1.0 μF, in order to slow the gate drive signal transition times. The gate drive source lines must be connected close to the source lead and placed out of the source lead's load current path.

Figure 12.11 shows the effect of just adding a 0.1-μF gate capacitor, which reduces EMI by slowing down the power MOSFET's switching speed. The gate *RC* network is a low-cost fix but does introduce a timing delay and extra power dissipation in the MOSFET. The extra power dissipation occurs because of the power MOSFET's slow turnoff and may require increased heatsinking for the MOSFET.

Figure 12.12 shows the delay and switching time effect for different gate capacitor values. Generally, a 50 to 100 μs delay time should not affect the application, since the operating frequency for relays or sole-

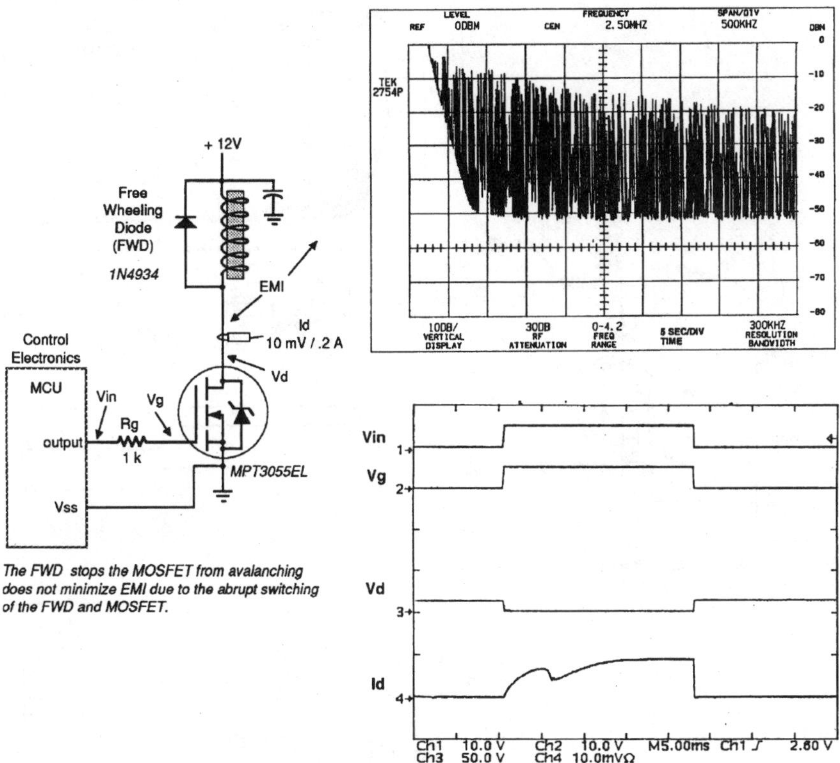

Figure 12.10 Adding a FWD diode does not eliminate EMI.

noids is usually less than 10 Hz. Obviously, this much delay time would be a problem for high-frequency motor control PWM signals, but this *RC* network design is satisfactory for a power stage that simply turns the motor on or off.

One other point about using an *RC* filter in the MOSFET's gate drive: A typical power MOSFET exhibits high transconductance, which means that it only takes a small gate voltage variation to turn the load on or off. The gate voltage variation can be determined by dividing the load current by the MOSFET's forward transconductance. For example, a MTP3055EL's forward transconductance (g_{fs}) is 5.0 mhos. If the load current is 0.2 A, then the ΔV_{gs} is only 0.04 V ($\Delta V_{gs} = 0.2 \div 5$). Therefore, only a small portion of the gate's voltage transition time affects the switching times of the load. This is why a large gate-to-source capacitor must be used to slow down the gate voltage transition time.

Slowing down the power MOSFET may not eliminate EMI, since it is usually not possible to switch relay or similar type loads with very

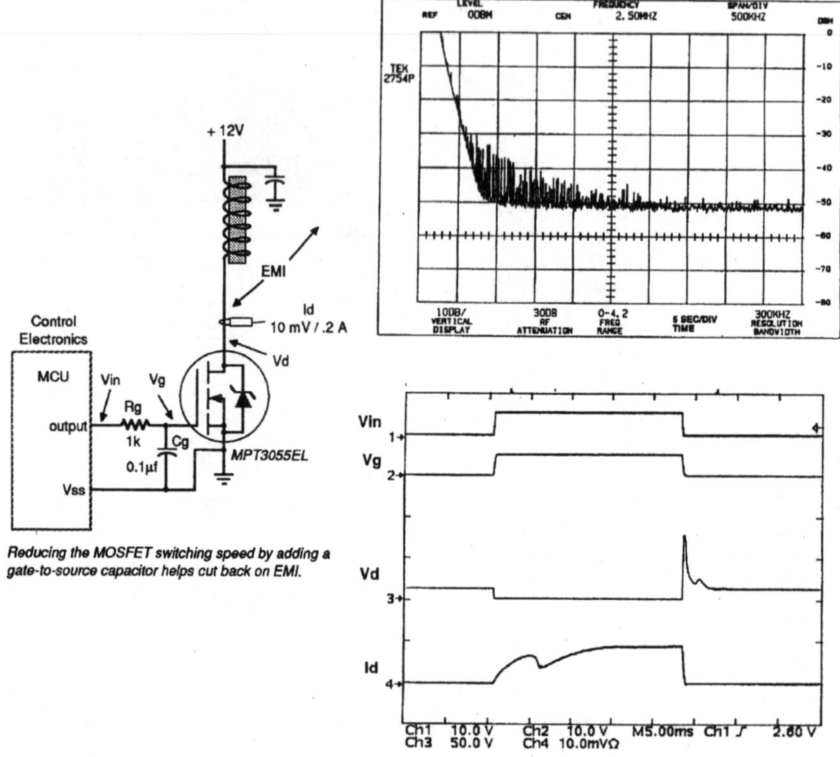

Figure 12.11 The effect of a 0.1-μF gate capacitor with FWD left out.

slow switching times. The other EMI source is from the inductive kickback voltage clamp. When this device, either a free-wheeling diode or zener, is activated, it generates EMI, as shown in Fig. 12.13. The free-wheeling diode also slows down the solenoid's or relay's mechanical turnoff performance.

Replacing the free-wheeling diode with an *RC* snubber, as seen in Fig. 12.14, can reduce EMI and still allow good turnoff mechanical response. The snubber capacitor value should be large enough to prevent the power MOSFET from avalanching. The exact value of the *RC* snubber is determined by the specific application. A 1-μF metalized film capacitor and 10-Ω resistor are good starting values for the snubber.

In summary, by adding an *RC* filter to the gate drive to slow down the power transistor's switching times and an *RC* snubber across the load to contain the inductive kickback voltage, the MCU can drive a power MOSFET with minimal chance for EMI.

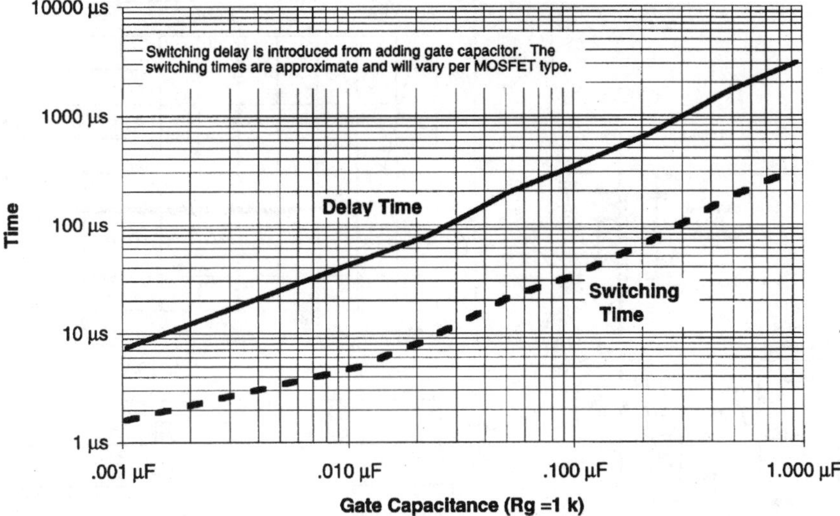

Figure 12.12 Gate capacitor value versus switching time.

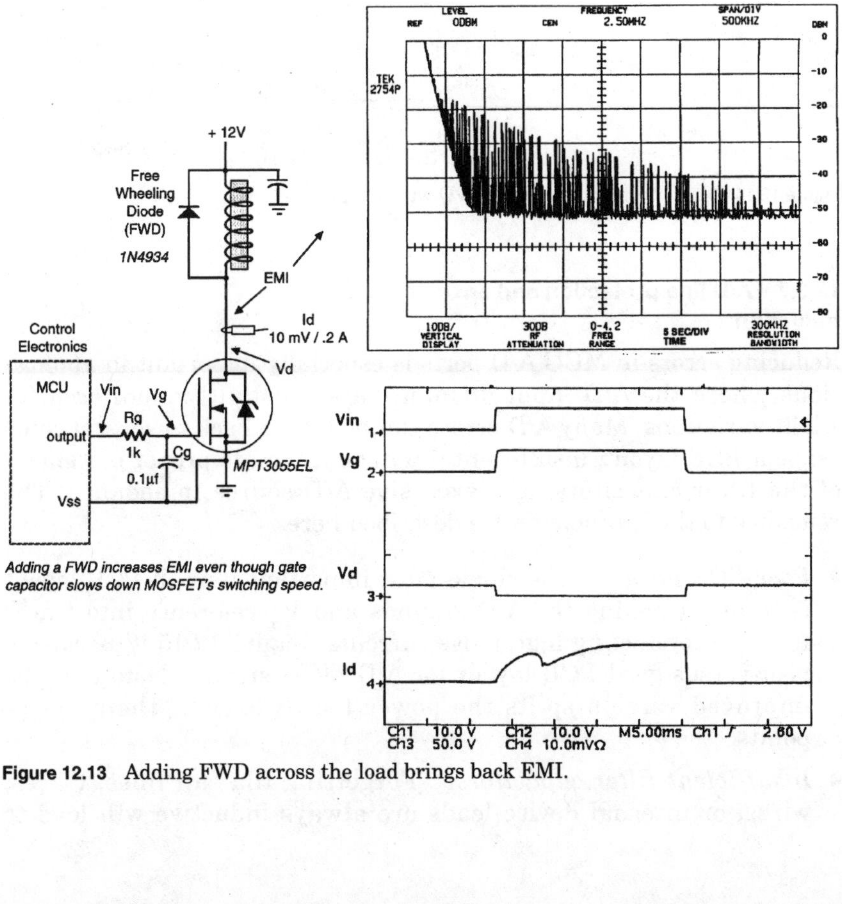

Adding a FWD increases EMI even though gate capacitor slows down MOSFET's switching speed.

Figure 12.13 Adding FWD across the load brings back EMI.

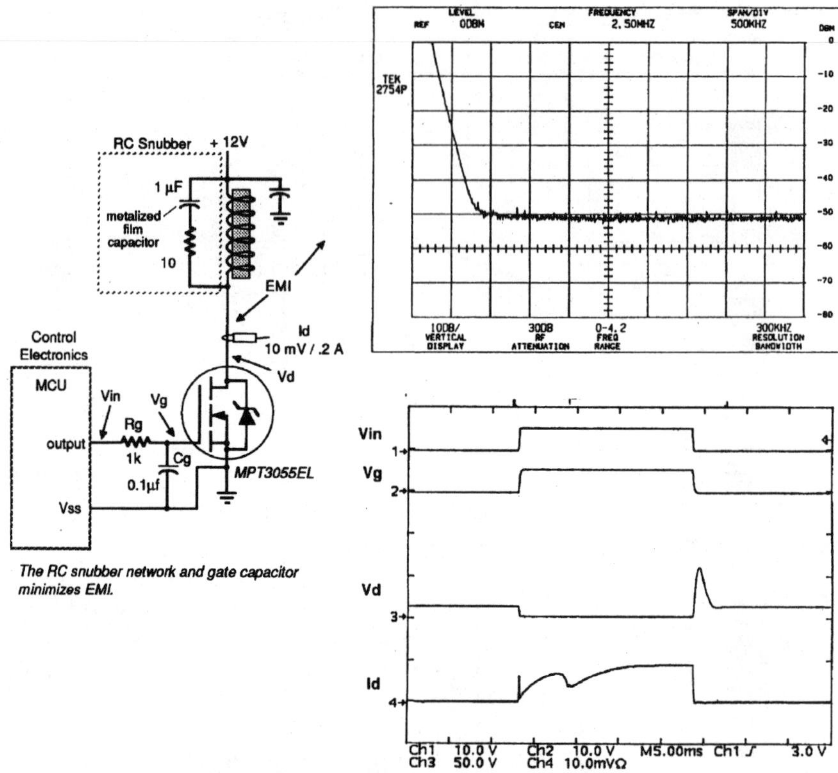

Figure 12.14 The effects of replacing FWD with *RC* snubber.

12.5.7 A/D line protection and error reduction

Reducing errors in MCU A/D ports is especially important in applications where the A/D input or inputs are constantly monitored for slight variations. Many A/D errors can be traced back to several causes: poor PCB layout, insufficient filter capacitors, improper placement of the filter capacitors, and excessive A/D source impedance. The remedies to these problems are described here:

- *Poor PCB layout:* The single most important item in PCB layout is to avoid mixing the A/D grounds and V_{cc} reference into traces that are conducting high pulse currents. Figure 12.15 illustrates a poor-versus-good PCB layout for A/D MCU signals. Note how the improved version splits the power traces back to their source points.
- *Insufficient filter capacitors:* Forgetting that all lines, traces, wires, or internal device leads are always inductive will lead to

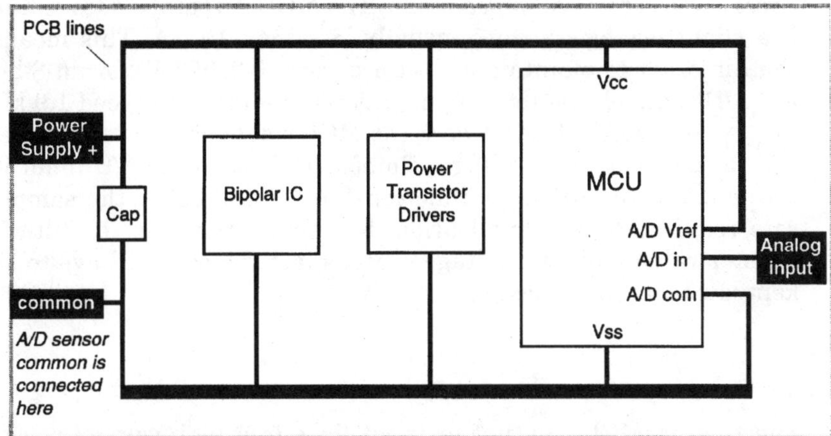

This PCB layout can lead to disaster if implemented for motor control

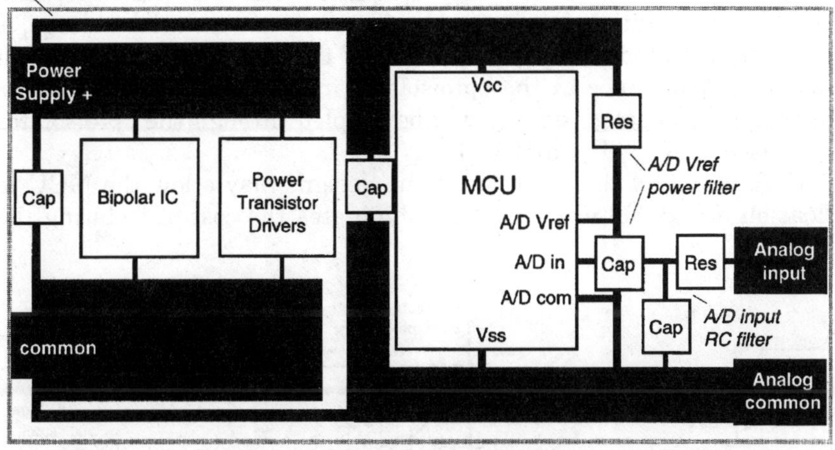

These improvements will help minimize A/D noise

Figure 12.15 Poor versus good PCB (printed-circuit-board) layout for MCU and A/D lines.

noisy V_{cc} and commons if filter capacitors of sufficient size and type are not used. In some cases, a separate regulator or power supply diode isolator may be required to supply the MCU's A/D V_{ref}^{+}.

- *Improper placement of filter capacitors:* Bypass or decoupling capacitors need to be placed as close as possible to the source of the high-frequency noise. For an MCU design, this means locating bypass capacitors near the MCU's V_{cc} and V_{ss} pins, as well as near the voltage regulator output and input pins.

- *Excessive A/D source impedance:* The MCU A/D stage does exhibit a slight leakage current, usually less than 10 μA. This means that, in order to maintain a 1 bit accuracy (0.01953 V) for an 8-bit 5-V A/D channel, the total input resistance must not exceed 1.9 kΩ (R_{input} = 0.0195 V/10 μA). Check the MCU's data sheet for the exact A/D leakage current. Adding a simple *RC* filter to the A/D input is acceptable if the *RC* filter time constant is well below the sample rate required by the application. Remember that the *RC* filter's resistor value, plus the voltage source's impedance, will have to be kept low to minimize errors.

12.5.8 Isolating MCU lines

Connecting an MCU's output or input lines to a high-voltage power stage requires the use of excellent isolation devices. The use of optoisolators, as shown in Fig. 12.16, provides a high degree of isolation from high-voltage waveforms that are encountered in large motor drives or power line conditioning equipment. The important specification for any isolation device, besides its maximum isolation voltage, is the common-mode voltage rejection. If there is a large amount of internal capacitance in the optoisolator input-to-output, the chances are high that enough energy will be coupled through the optoisolator to cause trouble with an MCU line.

In the interest to save costs, some designs may allow the MCU to float on a high-voltage bus. This eliminates the cost of isolating the

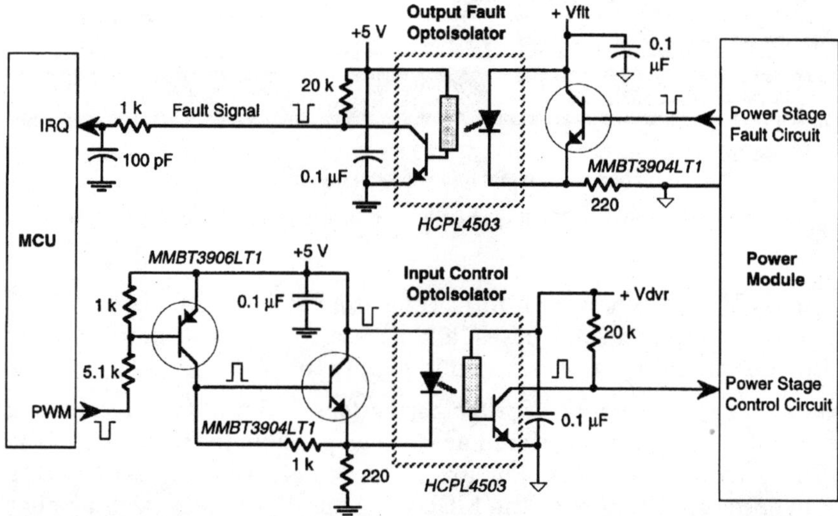

Figure 12.16 Using optoisolators.

MCU from the high-voltage bus but requires some form of isolation to the user-operated inputs. The additional difficulty with this method is in field servicing. The technician may accidentally ground out the MCU's common with catastrophic results. As a matter of principle, if the MCU is accessible for debugging, connecting test probes to it should not represent a lethal hazard for the service personnel.

12.5.9 Power supply reversal and fluctuations

The MCU's stability will only be as good as its power supply. In motor control circuits it is possible to have intermittent power connections or users that are "playing" with the power switch. Another common problem is reversed power connections caused by battery reversal or improper hookups. Reversing the V_{cc} and V_{ss} on a MCU is usually catastrophic to the MCU die. Figure 12.17 shows some simple methods to protect against fluctuating power and reversed power voltages. Adding a zener diode across the MCU's V_{cc} and V_{ss} pins that is rated just below the maximum MCU operating voltage is good insurance against a shorted regulator or other mishaps that would allow excessive MCU supply voltage.

A good principle to follow when connecting or designing an MCU's power supply is to establish its maximum output current level. This can be done with small fuses (which are available in surface mount styles). In some cases, two different-size fuses are required—one large size for the power stage and a small size for the control logic.

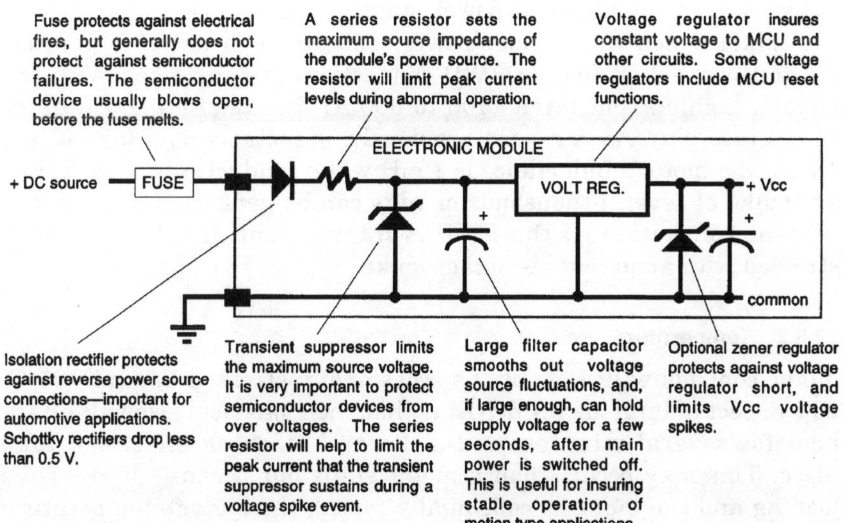
Fuse protects against electrical fires, but generally does not protect against semiconductor failures. The semiconductor device usually blows open, before the fuse melts.
A series resistor sets the maximum source impedance of the module's power source. The resistor will limit peak current levels during abnormal operation.
Voltage regulator insures constant voltage to MCU and other circuits. Some voltage regulators include MCU reset functions.
ELECTRONIC MODULE
+ DC source
FUSE
VOLT REG.
+ Vcc
+
+
common
Isolation rectifier protects against reverse power source connections—important for automotive applications. Schottky rectifiers drop less than 0.5 V.
Transient suppressor limits the maximum source voltage. It is very important to protect semiconductor devices from over voltages. The series resistor will help to limit the peak current that the transient suppressor sustains during a voltage spike event.
Large filter capacitor smooths out voltage source fluctuations, and, if large enough, can hold supply voltage for a few seconds, after main power is switched off. This is useful for insuring stable operation of motion type applications.
Optional zener regulator protects against voltage regulator short, and limits Vcc voltage spikes.

Figure 12.17 Protecting against fluctuating and reversed power voltages.

Adding a series resistor to the power supply regulator input will also limit the maximum peak current levels. For example, an MCU design that requires only 0.01 A and is powered from a 12-V supply could use a 5-V regulator with a 100-Ω 2-W series resistor in its 12-V input side. The 100-Ω resistor would normally drop 1 V but would limit the maximum current to 0.12 A in a short-circuit condition. More importantly, with the addition of a zener transient suppressor to the regulator's input as well, the resistor will limit the zener's maximum peak current when the 12-V supply is subjected to serious voltage spikes.

12.6 Other Motor Control Protection

Other electronic circuits besides the MCU and its related components require protection in a motor drive. Failure to protect these other circuits can lead to failures or erratic MCU operation.

12.6.1 Power stage voltage spike protection

Power semiconductor devices are usually fairly robust when operated within their rated parameters. Unfortunately, the environment in which most large motor controls operate can produce electrical voltage levels that will induce catastrophic power stage failures unless precautions are taken. The use of overvoltage protective devices, such as MOVs (metal oxide varistors) or zener transient suppressors in the motor's power stage, is highly recommended. Intermittent connections, in or at the motor or power stage, can generate very high voltage spikes which will damage power semiconductor devices. An intermittent connection (e.g., a FWD connection) is dangerous: With the motor's high current levels (100 to 600 A) the equation $E = L\,di/dt$ comes into play. As reviewed previously, inductance is a fact of life, and, if the motor's inductance is 1 mH while conducting 300 A, a voltage spike of several thousands of volts can be generated by an intermittent connection. If the FWD is intermittent, the IGBT will be stressed with an excessive voltage spike.

12.6.2 Temperature sensing

Motor controls will probably be occasionally stressed beyond +85°C. One aspect of large motor drives is that their physical size will retain heat for several minutes, and a thermal stacking effect can take place. This may occur when a motor stalls out because of excessive loading and the motor is continually cycled; the motor's temperature rises every time the motor is cycled on again. Temperature monitoring of the power stage is important because of the possible hundreds or even thousands of watts of heat dissipation in the power modules. The power transistors or FWDs in these power modules will fail in a few seconds if the module's heat management system deteriorates, even if the motor current is at nominal levels. A temperature sensor is used to monitor the baseplate or substrate temperature, which tracks the IGBT die's junction temperature. A two-step output can give an early warning of impending heat dissipation problems. This can be implemented with two outputs that indicate an abnormally high (125°C) temperature condition and an impending 150°C failure condition. The motor control logic or computer could also track the time from a 125-to-150°C event to determine the severity of heat rise per time of the temperature problem. A third analog output allows the computer to calibrate the temperature sensor and then track the temperature variations over time, which can predict how long the module can continue to operate until dangerous junction temperatures will occur.

Summary

This chapter has focused on the problem areas that can affect MCUs and other electronic devices in motor controls. *It is important to remember that excessive voltage levels are exceptionally dangerous to semiconductors.* Exceeding the semiconductor's breakdown voltage threshold can burn a hole into the transistor's die in a few microseconds, depending on the voltage spike's energy level. The exact failure effect correlates to the voltage fields and current densities in the semiconductor's structure. Various methods to limit voltage spikes have been shown that will greatly increase the MCU's reliability in power-laden designs such as motor controls. Depending on the application and EMI concerns, simple *RC* filters do not exhibit an abrupt voltage response, as is found with diode or zener clamping devices. Setting the frequency response of each MCU line by filter networks minimizes unwanted noise spikes from corrupting valid signals.

By incorporating the appropriate protection methods reviewed in this chapter the reliability of the MCU and the motor control application can be improved. MCUs have been successfully used by the tens of millions in many severe applications, such as automobile engine controllers, and will become more common in tough applications such as motor speed controls. Key points about this chapter are listed below.

- Intermittent connections, reversed wire hookups, power stage failures, software bugs, ESD damage, motor bearing failure and motor lockup, are a few of the more common problems that afflict motor drives.
- A voltage transient that is within the MCU's maximum electrical ratings can still suffer a momentary logic glitch. This is because these maximum ratings usually mean only that the MCU will not be destroyed if operated within the rating's limits.
- Some ways to test a motor control design for EMI tolerance include operating a cellular phone or other radio transmitter in close proximity to the control module and using a spark generator or static discharge device on the module's external lines.
- The root cause of most voltage transients in a electronic motor drive is from the high-frequency current transitions in the motor's power stage. These generate voltage transients due to the inductance in the motor and hookup wires.
- External events that are tracked by an MCU may cause program problems because these external events are not synchronized to the MCU's program, and, therefore, the MCU must interrupt its normal program and service the external event. Since this may happen at any time, there is more chance that the MCU's original program flow will be corrupted.
- One test to ensure that the MCU can operate reliably during power supply fluctuations is to slowly adjust the MCU's supply voltage from its nominal rating downward until it fails and then upward back to nominal. Most MCUs will always operate correctly when the supply voltage goes from its nominal value all the way to zero. Problems may occur when the supply voltage sags but doesn't drop to zero and then goes back up to its normal value.
- High-frequency electrical noise generated in a motor power stage can be reduced by the use of *RC* snubber networks across the power stage's transistors.
- Some cost-effective methods to protect an MCU in a motor control design include using a series rectifier in the MCU's power supply line, adding zener clamps on the power supply and any MCU line that is exposed to high voltage protects against excessive voltage, and placing *RC* filters on the MCU's input lines.
- An intermittent connection in the motor's power stage can produce failures in both the power stage and the MCU because the intermittent connection usually generates excessive voltage

spikes that will exceed the power stage's voltage ratings—this shorts out the power stage and may allow excessive voltage to be applied to the MCU.

- The top three parameters that directly affect a motor drive's reliability are excessive current levels, elevated semiconductor junction temperatures, and voltage spikes.

Acknowledgment

This chapter was reprinted and updated from the article by Richard Valentine, "Protecting MCUs in Power Circuits," *EDN Magazine* (Oct. 10, 1996), copyright 1997 Cahners Publishing Co., a division of Reed Elsevier Inc.

Further Reading

Catherwood, M., *Designing for Electromagnetic Compatibility (EMC) with HCMOS Microcontrollers*, Motorola Application Note AN1050, 1989.

Cherniak, S., *A Review of Transients and Their Means of Suppression*, Motorola Application Note AN843, rev. 1, 1991.

Reducing A/D Errors in Microcontroller Applications, Motorola Application Note AN1058, 1990.

Valentine, R., "Three Methods Protect μP I/O Lines," *EDN*, Sept. 28, 1989.

Part

4

Motor Control Electronic Devices

Chapter

13

Power Semiconductors for Motor Drives

Gary Dashney and Scott Deuty
Motorola Semiconductor Products

Introduction

Thus far, we have focused mainly on the use of microcontrollers and analog bipolar IC devices in motor controls. In this chapter our main concern will be with the discrete semiconductor devices (IGBTs, MOSFETs, rectifiers, etc.) that are commonly employed in motor drives. An understanding of the semiconductor data sheet will also be presented with an emphasis on how this information relates to a motor control design. Other issues, such as packaging, semiconductor materials, and ESD, will be discussed as well.

13.1 Semiconductor Overview

Electronic motor drives have been made possible by the development of new solid-state devices, as seen in Fig. 13.1. Large MOSFET and IGBT power devices are readily available that can switch significant

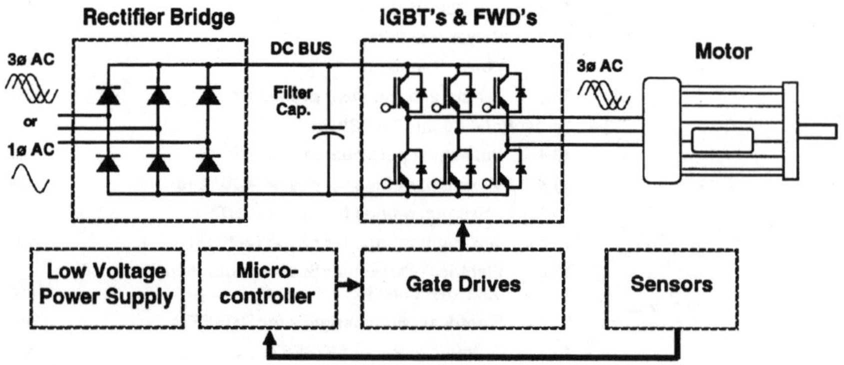

Figure 13.1 A motor drive uses several semiconductor devices.

amounts of power with the ability to block line voltages of 480 V or greater. In this section we will look at semiconductors that perform vital functions in motor drives, especially the high-power devices, as well as some low-power signal devices. These include diodes, transistors, and thyristors. Examples pertain, for the most part, to three-phase induction motors but are applicable to other motor drive technologies, including DC and brushless DC motor drives. An upcoming motor drive technology is switched reluctance, and, although this topology does not require the same degree of short-circuit protection, much of the semiconductor information provided here will be helpful.

Semiconductors perform several basic functions in motor control applications: decision making, level shifting, power management, sensing, and protection. For decision making, microcontrollers are used to handle the large amounts of information that are required to spin a motor. Some local decision making is made with comparators or transistors. Microcontrollers are low-power devices with very high integration. The power ratings of the microcontroller's output lines are kept low to minimize the MCU's power dissipation, which allows increased circuit density. Therefore, interim-level shift components are made to boost drive levels from the MCU or other logic components to the power devices.

Level shifting actually has two definitions in motor control semiconductor devices. One definition refers to a level shift in current to drive the major power devices. This level shift is managed by small signal discrete and integrated-circuit (IC) devices. The second level-shift definition refers to a voltage shift in which the power device reference point (e.g., the source on a MOSFET) floats from ground to high voltage, as in a totem-pole or bridge configuration. This type of voltage-level shift occurs frequently in gate drive circuits and is handled mainly by optocouplers, although some high-voltage bipolar ICs are resulting in monolithic silicon solutions. Power management devices include those components that are switched, such as transistors and thyristors, and those components that self-commutate, such as rectifier diodes and free-wheeling diodes. Free-wheeling diodes allow motor currents to recirculate and essentially maintain current flow in motors that appear as inductive loads.

13.1.1 Small-signal discretes

Small-signal transistors and diodes act to support the main power components (IGBTs, triacs, MOSFETs, etc.). Diodes are utilized in many motor control applications. These devices are used to block high voltages and to clamp or protect the MCU or other logic ICs from voltage spikes. Small-signal transistors are used to drive large power switching

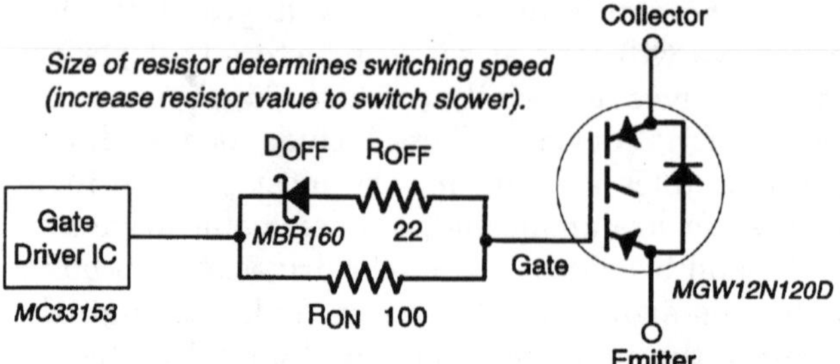

Figure 13.2 A small diode is used for faster IGBT turnoff.

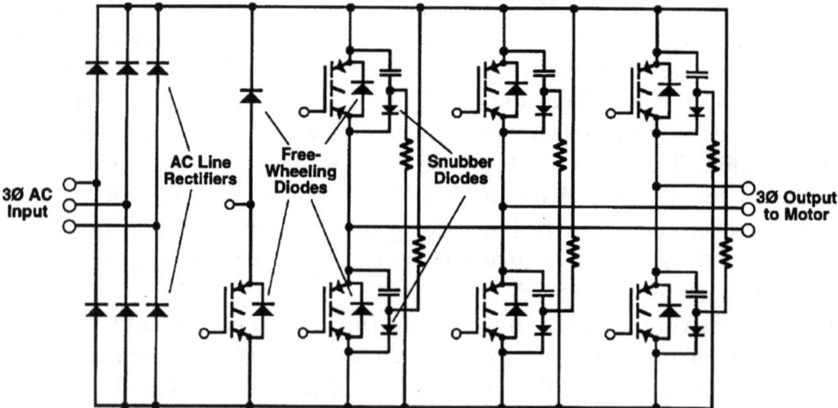

Figure 13.3 A three-phase motor control uses many power rectifiers.

devices or to shut them down during fault detection. Figure 13.2 shows one example of how a small-signal diode is used to adjust the turnoff switching time in an IGBT power stage. The diode provides a lower impedance for faster turnoff of the IGBT. This assumes that the driving device can sink the higher level of current during the turnoff period.

13.1.2 Power rectifiers

Power rectifiers, as seen in Fig. 13.3, perform rectification of line voltage and circulation of motor currents and are used in snubber networks. Power rectifiers are designed with characteristics that make them effective for these roles. In order to provide adjustable speed control to a motor, the input waveform must first be rectified to DC. The motor drive unit then recreates the sine wave by "chopping" it via pulse-width modulation (PWM). (See Chap. 8.) Line rectifier diodes for motor controls must provide low forward voltage drop to increase system efficiency. In addition, they must block high voltages (line and line transient) and be able to handle large surge currents that occur during fault conditions and when charging capacitors during initial startup.

Some motors act as inductive loads—the current will continue to flow after the main switch is turned off. This current "free wheels" through a diode. Free-wheeling diodes (FWDs) must have low forward drop and fast recovery. Typical motor control speeds vary from 4 to 40 kHz. At these speeds, free-wheeling diodes that exhibit "soft" recovery, as illustrated in Fig. 13.4, are preferred over "snappy" diodes that recover too quickly. As you can see in Fig. 13.4, there are also differences in the forward ON voltage of soft and standard ultrafast recovery diodes.

Additional tasks of rectifiers include snubbing of voltage spikes or clamping of energy. Snubber diodes must switch quickly. Voltage suppressors are used to clamp high-power voltage transients. These devices have large die that can absorb the energy of the spike, protecting the critical switching devices. This large die allows the semiconductor to survive an energy dump that would cause catastrophic failure in a smaller device. It is more economical to specify a voltage suppressor than to oversize the switching power transistor die.

13.1.3 Power transistors

Three types of power transistors are used in motor drive systems: bipolar junction transistors (BJTs), MOSFETs, and insulated gate bipolar junction transistors (IGBTs). These devices are switched with a PWM signal and are more sophisticated than SCRs or triacs, which cannot be gated off. The selection of the best power transistor type varies with the motor's power and the PWM frequency. Figure 13.5 illustrates where these different power device types generally work best.

Figure 13.6 shows the main attributes of the three power device types that are usually considered for motor drives. Bipolar junction transistors (BJTs) are the least expensive of the three transistors. The current density of a BJT is higher than that of a MOSFET yet lower than that of an IGBT. This results in effective use of silicon area. The die size of the BJT will be less than that of a MOSFET, and the price (which is based on die area) is lower. BJT devices are minority-carrier devices that require charge storage within the device in order to function. This charge storage creates a problem when switching the device fully on; the charge must be evacuated or recombined, slowing the switching speed of the device. Therefore, BJTs are not

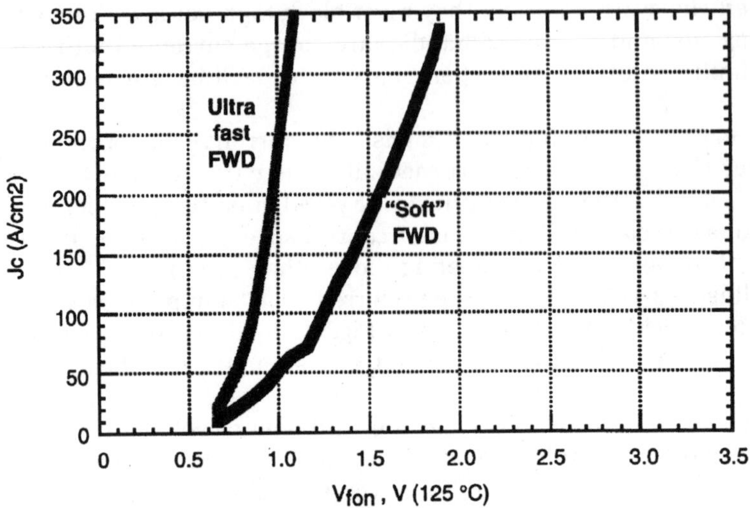

Forward ON voltage vs. die current density.

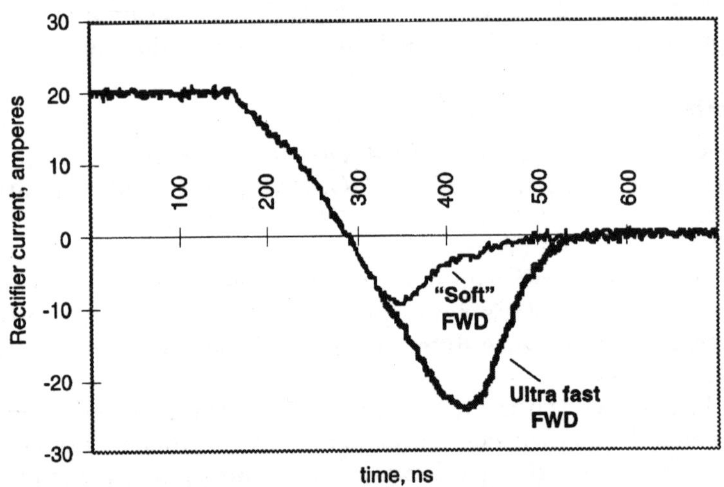

Reverse recovery waveforms of "soft" and ultra fast FWDs.

Figure 13.4 A comparison of ultrafast and soft reverse recovery rectifiers.

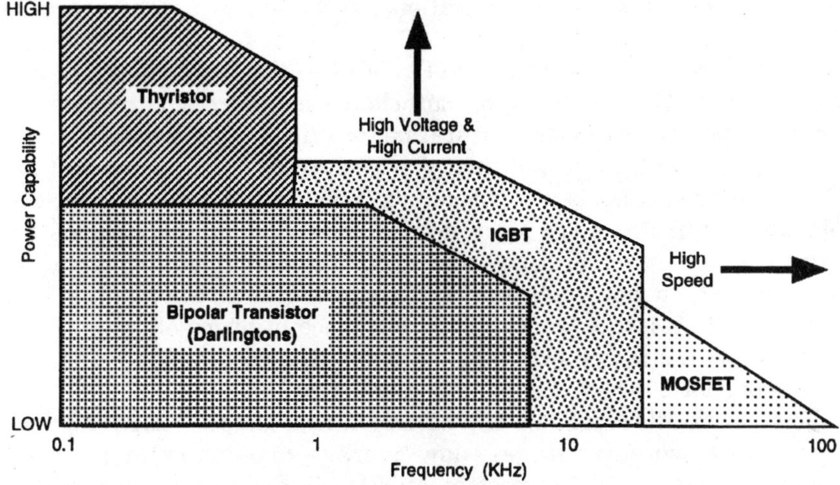

Figure 13.5 Power device selection versus speed and wattage.

Application Requirement	Power Device Technology	Attributes
LOW VOLTAGE (< 200 V)	*MOSFET*	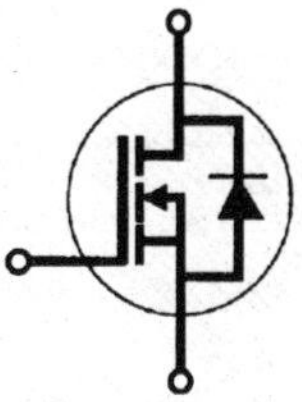 • Very low RDS_{ON} • Very High SPEED (up to 500 kHz) • Gate voltage controlled = Easy to drive
HIGH VOLTAGE LOW CURRENT	*BIPOLAR*	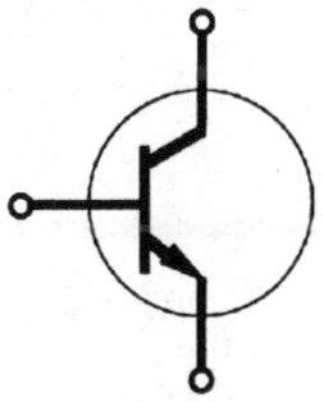 • Low ON losses • High speed (up to 100 kHz) possible • Requires complex and expensive drive
HIGH VOLTAGE HIGH CURRENT	*IGBT*	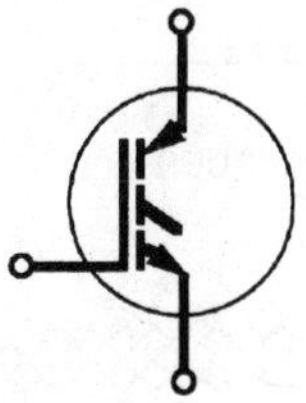 • 5x to 10x lower ON losses than MOSFET • Average speed (up to 20 kHz) • Gate voltage controlled = Easy to drive

Figure 13.6 Motor control power device attributes.

widely used in switching applications, especially at speeds above 20 kHz.

Another requirement of the BJT relates to the fact that a BJT is a current gain device. Therefore, sufficient base current is required in order to put the device into saturation (or fully on) for it to have a low forward drop and act as a switch. A base current equal to one-tenth of the collector current is not uncommon, resulting in a design that dissipates significant power in the base drive circuit. BJT transistors offer cost-effective solutions and are the alternative for moderate switching frequency applications where power dissipation in the base drive is not an issue. For better performance in terms of drive circuit dissipation and faster switching frequency, the power MOSFET is the device of choice for lower-voltage motor drives.

The power MOSFET is a transconductance, majority carrier device that offers two advantages—low average drive current and fast switching. Figure 13.7 shows the MOSFET's die cross section with a schematic overlay. Note the parasitic body diode and BJT with a shorted base. One interesting aspect of a MOSFET is that the device

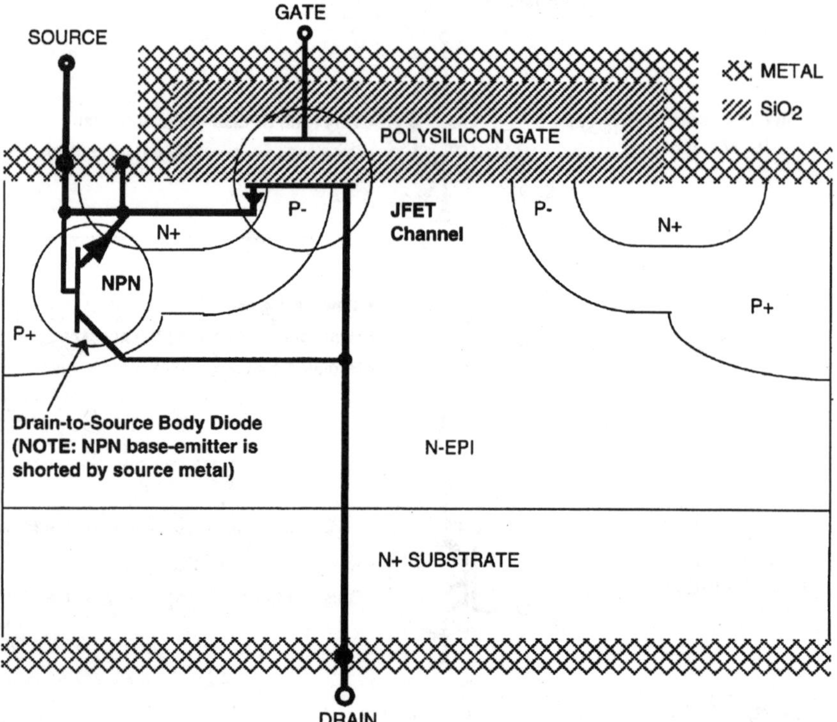

Figure 13.7 MOSFET cross section and schematic overlay.

will remain on after the gate voltage has been applied unless a means is provided to discharge the gate. Because it is a majority-carrier device, the power MOSFET switches very quickly since there is no stored charge in its drain-to-source. These characteristics make the MOSFET the device of choice for high-speed, moderate-voltage (500-V) applications. Power MOSFETs are available in a wide range of voltages (20 to 1000 V) and currents (1 to 1000 A). Typically, MOSFETs that are used as power switches in motor drives are rated at less than 200 V; otherwise, the device must be quite large to overcome the increased $R_{DS,on}$ of a high-voltage device.

The IGBT has performance similar to that of the BJT and is driven like a power MOSFET. The device construction, as seen in Fig. 13.8, is also similar to that of a power MOSFET. The IGBT has a p⁺ buffer layer so that conductivity modulation will take place. The end result is an increase in the overall current density of the IGBT to a level above that of a BJT. One limitation of the IGBT structure is the fact that stored charge becomes "trapped" in the device since there is no current path to remove it. This forces recombination within the device

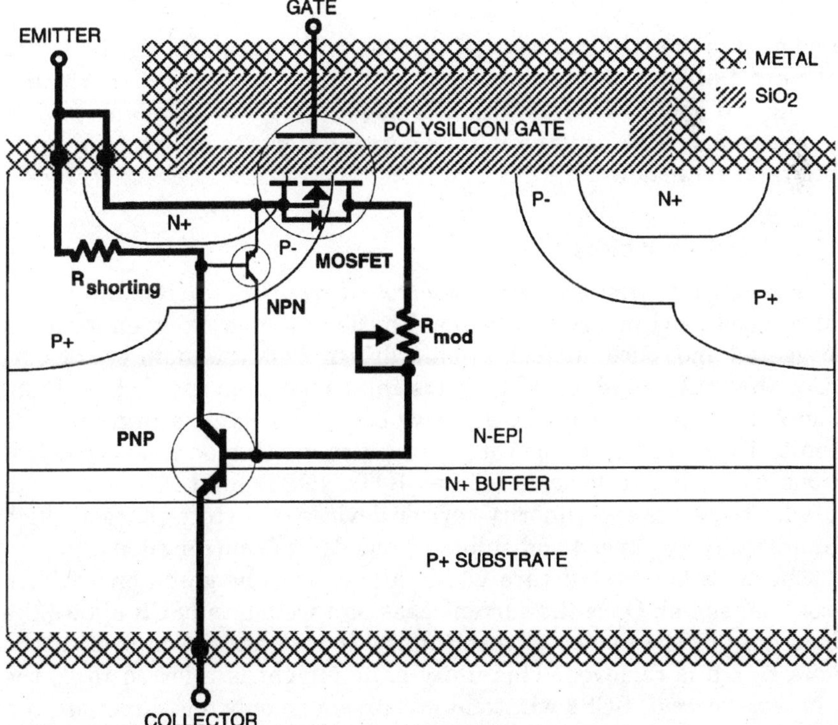

Figure 13.8 IGBT cross section and schematic overlay.

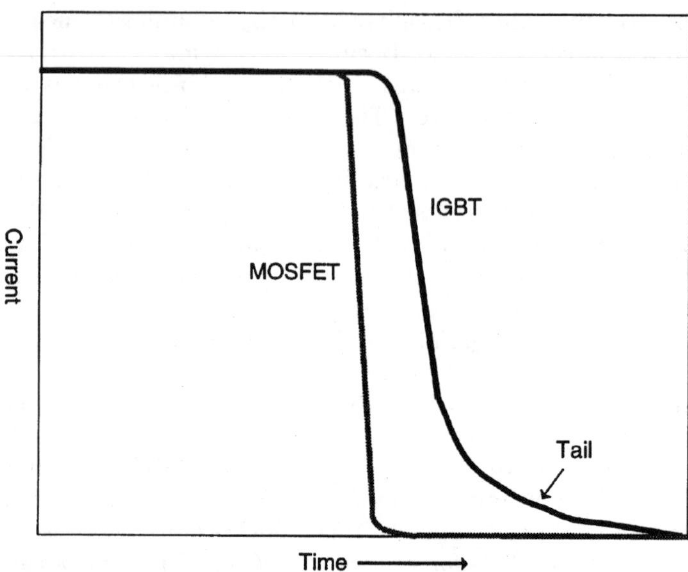

Figure 13.9 IGBT turnoff (tail) current.

and causes the IGBT to have a "current tail" as seen in Fig. 13.9. The current tail that occurs during turnoff limits the device's PWM frequency to about 40 kHz, which is adequate for most motor control applications. Note how the IGBT's current starts to switch off quickly but then slows down.

13.1.4 SCRs and triacs

Two kinds of thyristors, *silicon controlled rectifiers* (SCRs) and *triacs,* have been used in motor controls. Thyristors latch on when gated or triggered and, once latched, remain driven from the main power current that they conduct. The SCR's internal schematic in Fig. 13.10 shows how the two transistors drive each other into saturation. This limits thyristors from operating in pulse-width-modulated applications where gated devices, such as BJTs, IGBTs, and MOSFETs, are used. Thyristors are minority-carrier devices that store charge, which limits switching speed and enhances the device's current density.

The SCR has a structure which allows it to be gated on with an external signal. Once the current passing through the SCR allows the loop gain of the device to reach unity, it will remain on even if the gate signal is removed. This unity gain current is referred to as the *latching current.* SCRs will conduct current in only one direction. If a voltage larger than the breakover voltage of the device is applied, the

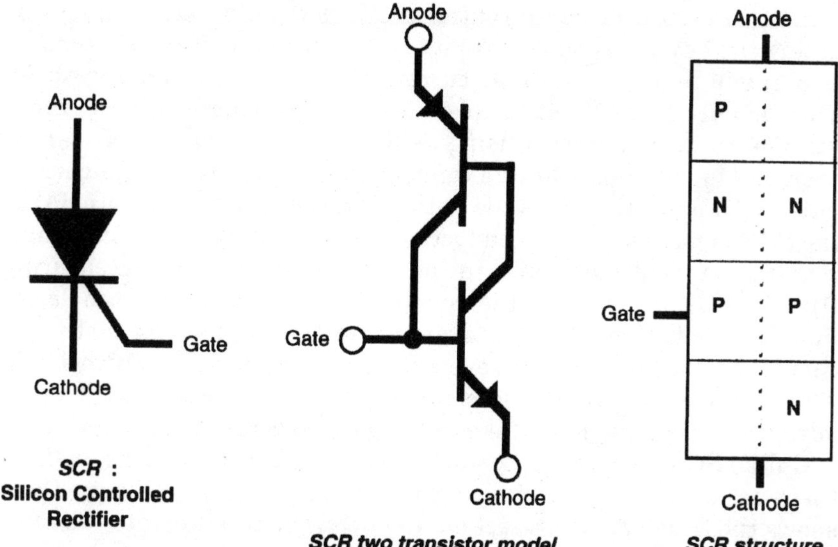

Figure 13.10 SCR schematic and structure.

SCR can be biased on without a gate current. Figure 13.11 shows this breakover voltage effect. It is worth noting that a voltage spike which exceeds the SCR's breakdown voltage can trigger the SCR into conduction. This can be a serious issue in a motor control, one that is prevented by the use of snubbers or voltage transient suppressors.

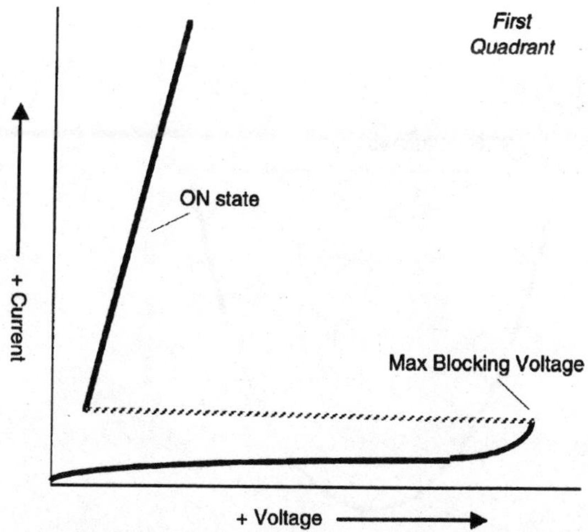

Figure 13.11 SCR voltage breakover characteristic.

SCRs are used in phase control of AC power line sine voltages by gating the device, as shown in Fig. 13.12. Once the SCR is in conduction, it will remain on until its current falls below a certain threshold. This usually happens at or near the zero crossing of the AC input voltage. In an application using SCRs, the gate drive circuit can be designed to gate the SCR into conduction with a narrow high-current pulse, or a pulse that lasts the entire duration of the time period that the SCR is meant to be in conduction. The advantages of a single narrow high-current gate pulse include savings in power and ensuring that the SCR's gate current is more than enough to trigger the device into full conduction. The disadvantage of the single-pulse method is that a momentary voltage reversal across the SCR will turn it off. This is a case in which the continuous-gate-pulse method has an advantage, even though it takes more gate power.

Unlike SCRs, triacs can conduct current in both directions, making them perfect for replacing relay contacts in AC motors. Figure 13.13 shows the triac's die cross section. In effect, the triac operates like two SCRs in parallel with one inverted. Figure 13.14 shows that the

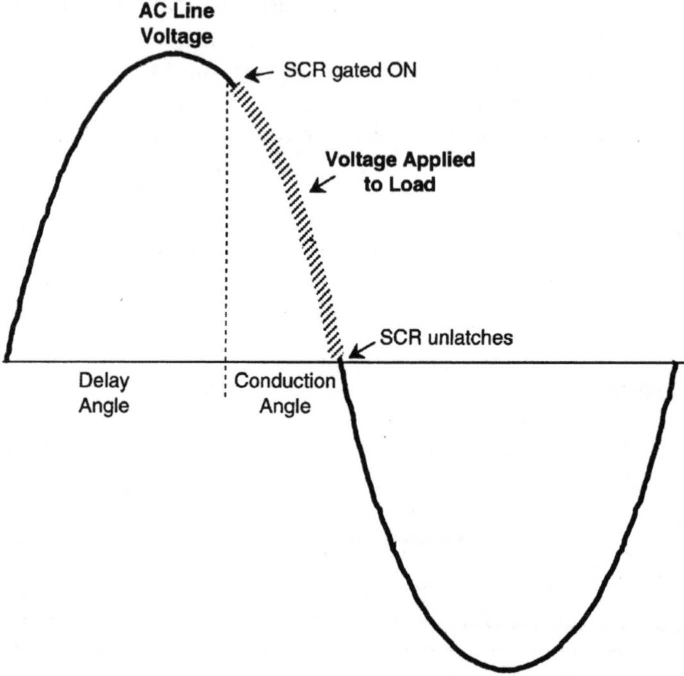

Figure 13.12 SCR conduction waveform.

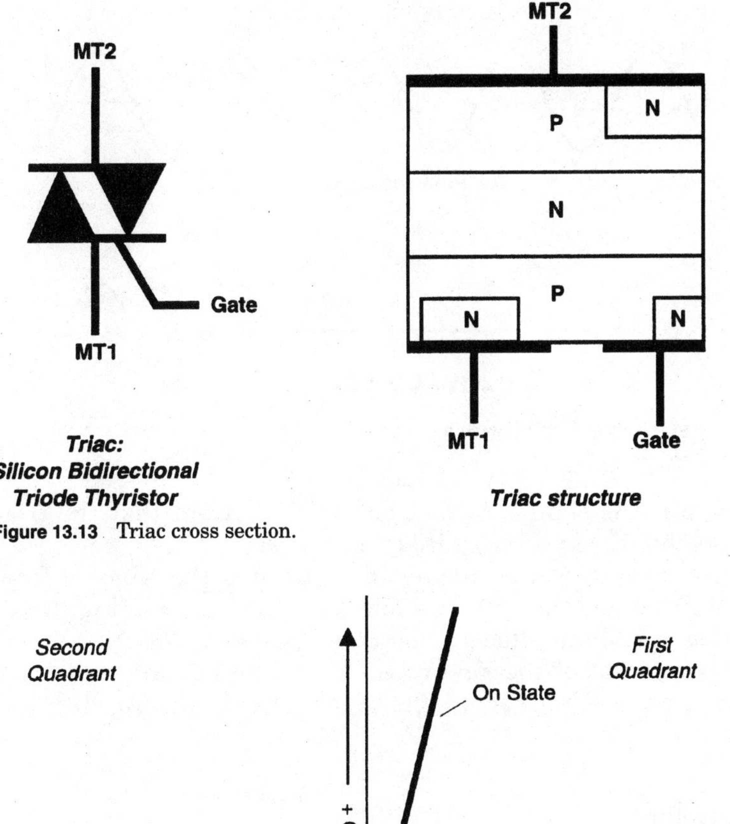

Figure 13.13 Triac cross section.

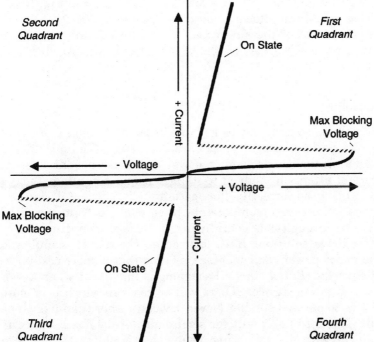

Figure 13.14 Triac voltage breakover characteristic.

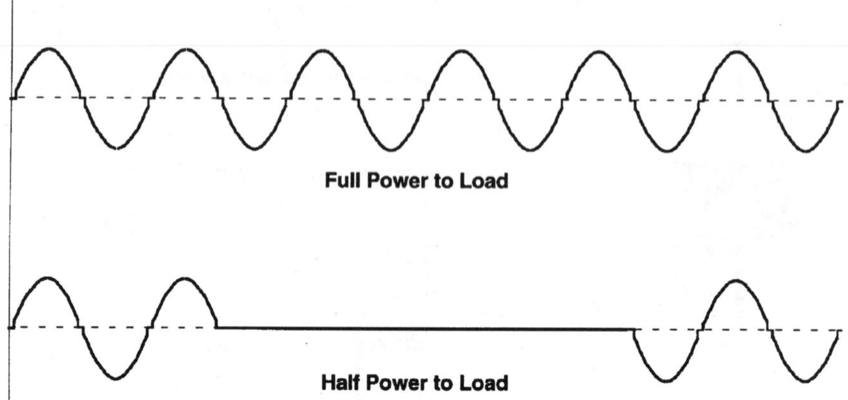

Figure 13.15 Zero-point switching with a triac.

triac's operation is similar to that of an SCR, except that the triac conducts in both the first and third quadrants.

Power to the load can be regulated by turning the triac on for a number of cycles and then off for a few cycles, as shown in Fig. 13.15. This process results in efficient operation by gating the triac at the zero-crossover point of the sine wave. There are minimal switching losses because the voltage across the triac is at or near zero when it is switched on.

13.2 Packaging

Semiconductor packaging plays a key role in the performance of power devices. Poor packaging introduces parasitic inductance and reduces heat transfer. Optimized packaging, however, does not benefit a design unless careful attention is given to its layout and mounting. Today's packages, when used correctly, offer space-friendly, efficient designs.

Packages for discrete components have evolved from mounted metal cans to through-hole components to surface mount packages. New motor drive solutions have focused on the use of modules to house the major power components, a method that is more cost-effective and efficient. (*Editor's note:* Packaging is important when developing low-cost power devices. There will be cases in which it is more cost-effective to parallel smaller power devices, rather than using one large unit. This also holds true for semiconductor die sizes. Die costs are not linear with die size. This means that doubling the die size may triple the die cost, as well as requiring a larger and more costly package. There will be certain power devices that attain an optimum balance between die size, power capability, and package type.)

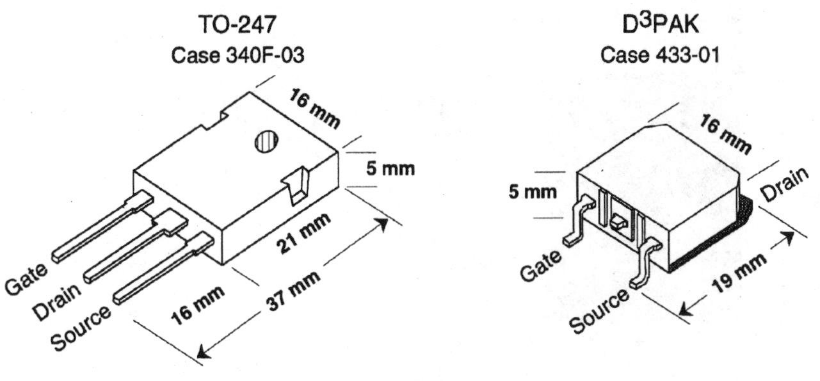

Figure 13.16 Same die in standard and surface mount package.

13.2.1 Through-hole packages

Through-hole components have traditionally been the most cost-effective way to assemble parts onto a printed-circuit board (PCB). A typical application requires the through-hole device to be inserted onto the PCB, its leads bent and trimmed, and then soldered. If necessary, the power device is mounted to a heatsink with a screw or clip fastener. Plentiful tooling and automated assembly equipment, as well as advanced industry knowledge, have made through-hole devices easy to incorporate into manufacturing. However, surface mount packages are keeping up with through-hole packages in size and performance, as can be seen in Fig. 13.16.

13.2.2 Surface mount (SMT) packages

Surface mount packages are becoming a viable alternative for discrete components, since they offer the advantages of easy assembly, reduced parasitic inductance, and higher reliability. Surface mount packages enable parts to be positioned vertically on the board instead of horizontally, reducing board profile, increasing power density, and allowing more compact designs.

Assembly of surface mount boards can be performed using tape and reel methods. Sometimes the total number of parts can be reduced in low-power applications because the surface mount device can use the PCB as a heatsink. For higher-power surface mount applications on standard FR4 circuit boards, special heatsink materials and methods have been developed. Figure 13.17 shows a typical PCB footprint for a D3PAK transistor package. Note that the device's heat dissipation is

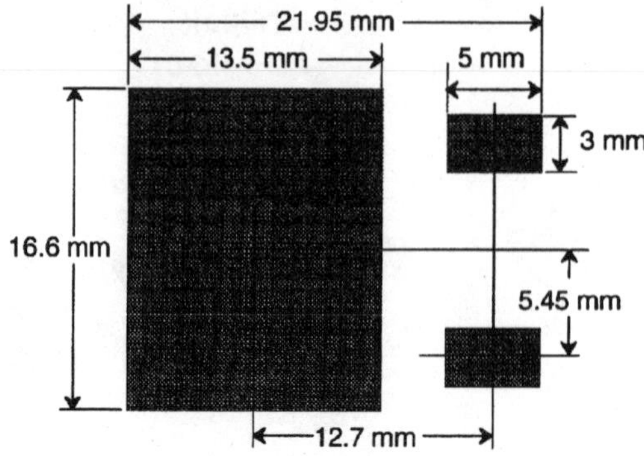

Figure 13.17 SMT power device footprint (D3PAK).

TABLE 13.1 Thermal Resistance of Materials for SMT

Substrate type	Power,* W	$R_{\theta ja}$ (°C/W)
Insulated metal substrate, aluminum	7.33	10.23
Insulated metal substrate, aluminum with heatsink	25.7	4.86
Insulated metal substrate, copper	7.56	9.92
Insulated metal substrate, copper with heatsink	27	4.60
PC board, FR4 (1.6 mm)	3.8	32.60
PC board, FR4 (1.6 mm) with heatsink	8.3	15.15
PC board, FR4 (0.15 mm) with 0.025 mm adhesive to heatsink	22.08	5.66

*Device mounted on minimum recommended footprint.

affected by the PCB materials and footprint sizes. Table 13.1 lists the thermal resistance of several materials that might be used as a heatsink with a D3PAK transistor.

Parasitic inductance and capacitance are reduced by surface mount devices. Leads are shorter, and power paths can be routed from the metal tab of the device. This allows wider electrical traces, minimizing parasitic inductance, as well as electrical and thermal impedance. Because parts are surface mounted, the parasitic coupling of the tab can be to the return rail instead of to the chassis of the assembly. This changes the electrical coupled noise from common mode to differential mode, which is much easier to compensate for and should reduce EMI problems.

Surface mount applications increase reliability in two ways. First, the reduction in parasitic inductance can minimize voltage spikes. As a result of smaller voltage spikes, less energy must be absorbed, and device temperatures are reduced. Snubber circuits can be made smaller or eliminated, increasing the overall mean time between failures (MTBF). Second, by surface mounting, the device's leads are attached to the same surface as its body. This is not the case with through-hole components that are mounted to a PCB hole. Hole mounting stresses leads and renders them susceptible to vibration damage, especially if the hole and board resonate at different frequencies. In summary, through-hole mounting that necessitates lead bending can induce die cracking or fatigue the lead's metal/solder joint.

13.2.3 Power modules

Power modules or hybrid power modules are becoming very popular in motor control. Figure 13.18 shows an example of a power module

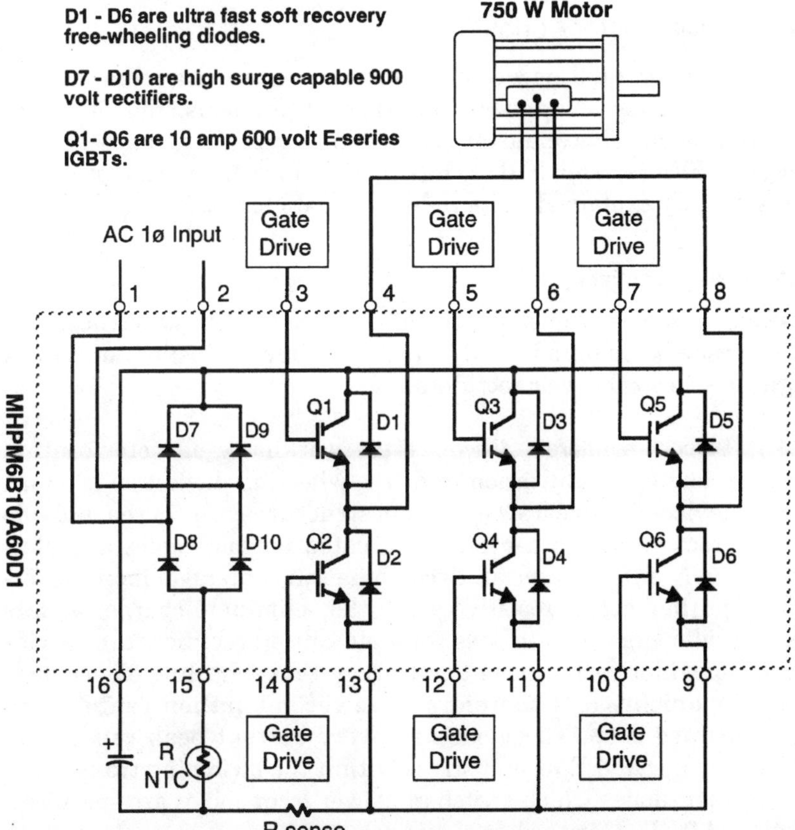

Figure 13.18 This module contains all the power devices for a motor drive.

that incorporates a number of semiconductor devices into one package. For instance, a three-phase induction motor drive may have the following die inside a module:

- Six input rectifiers for three-phase line input (four rectifiers for single phase)
- Six switching transistors (two for each phase)
- Six free-wheeling diodes across the transistors

With this many die inserted in one package, the thermal dissipation and high-energy signals are concentrated in one distinct area that reduces parasitic inductance and thermal resistance. This is an efficient alternative to 18 discrete packages, especially when considering layout and assembly. (Refer to Chap. 14 for more detailed information on power modules.)

13.3 Semiconductor Attributes

When you understand how semiconductors behave in a motor control circuit, you will be able to solve unexpected problems. You should be aware that each semiconductor type, such as rectifiers, zeners, SCRs, triacs, MOSFETs, and IGBTs, has unique properties that may not always be fully explained in its respective data sheet.

13.3.1 Power rectifiers

Rectifiers or diodes are usually considered the most basic semiconductors, almost always found in any electronic circuit. Most motor control designs use several power rectifiers.

13.3.1.1 Silicon rectifiers. Silicon is the workhorse of motor control, especially for line rectification and free-wheeling diode applications. Typical devices utilize P-N junction structures. To increase diode recovery, recombination centers are created for both holes and electrons in the N and P material. Recombination centers do increase forward drop; they act as resistive areas that eliminate charge carriers instead of allowing them to pass through. Silicon rectifiers are specifically designed for each type of application. Free-wheeling diodes need to recover quickly and therefore have recombination centers and higher forward drop. The diode's recovery current peak causes two problems: generated EMI and stress within the switching transistors. Line rectifier diodes which switch at slower frequencies are specifically designed for low forward drop and robustness.

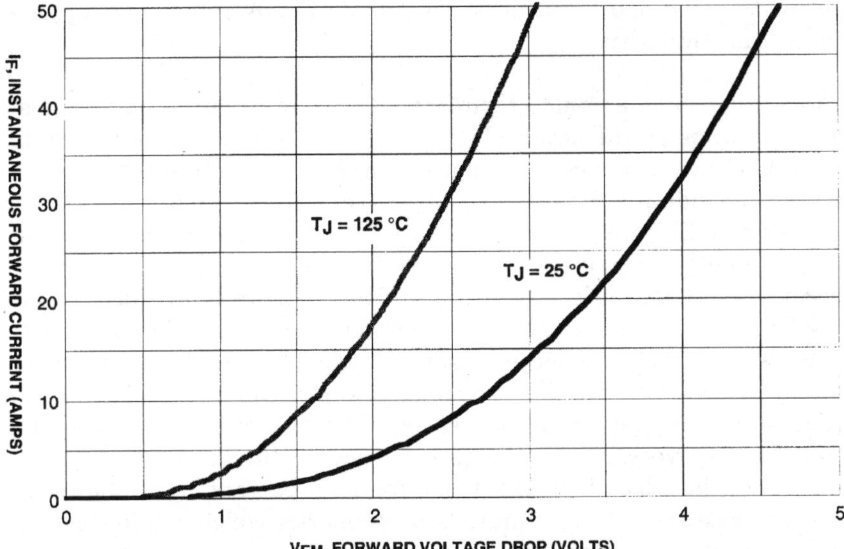

Figure 13.19 Diode voltage versus current over temperature.

Temperature effects on silicon are related, for the most part, to carrier mobility. As temperature increases, the carriers require less energy to become mobile. This results in decreased forward voltage drop as temperature increases, and leakage current increasing with temperature, as seen in Fig. 13.19. Recombination slows as temperature rises; the silicon exhibits an increased recovery current as a result. Because silicon rectifiers have a negative temperature coefficient for their ON voltages, the effect of high-temperature operation is less significant than in a MOSFET. However, at extreme temperatures, their leakage current rises. Also, if the silicon rectifier is used in some type of voltage reference circuit, its accuracy will vary significantly when wide temperature ranges occur.

13.3.1.2 Schottky rectifiers. Silicon Schottky rectifiers offer fast switching due to their design. Instead of a P-N junction, the Schottky rectifier features an abrupt junction. Although stored charge is reduced, making the Schottky a faster device than a P-N junction device, junction capacitance is greater and can limit switching speed. Schottky rectifiers also have higher leakage current that limits their voltage blocking rating. Therefore, most silicon Schottky rectifiers are rated below 200 V. Schottky rectifiers are typically used in low-voltage motor controls. It is possible to use several Schottky rectifiers in series to get a 600-V rating for a 230-V-AC line input, three-phase induction motor drive.

13.3.1.3 Gallium arsenide rectifiers. Gallium arsenide (GaAs) has recently come on the power electronics scene in the form of fast rectifiers. GaAs as a semiconductor material has several advantages over silicon, including a 6 times greater current density, 6.5 times higher carrier mobility, a higher bandgap, and better temperature characteristics. The popularity of GaAs semiconductors has been limited because the base material costs significantly more than silicon, making GaAs parts more expensive. Therefore, GaAs rectifiers are being used where performance is essential. Most three-phase induction motor drives running off 230-V lines and switching at speeds of less than 40 kHz do not require the speed of GaAs rectifiers. However, for soft, fast recovery in power factor correction converters, GaAs rectifiers take the edge off the switched inductor current. This allows the converter to run in the hundreds of kilohertz, which minimizes size and generated EMI.

GaAs rectifiers do not have the temperature dependence of silicon rectifiers. Figure 13.20 shows that leakage current in GaAs rectifiers does not vary greatly with temperature. In addition, the GaAs material can tolerate a higher junction temperature than silicon. Existing GaAs rectifiers are rated at a maximum junction of 150°C, due mainly to restrictions on package temperature limitations and not because of the GaAs material within. As module assemblies become more popular, GaAs device temperature ratings can rise because of the package's increased heat spreading capabilities. GaAs does have a higher thermal impedance than silicon. A GaAs diode's reverse recovery cur-

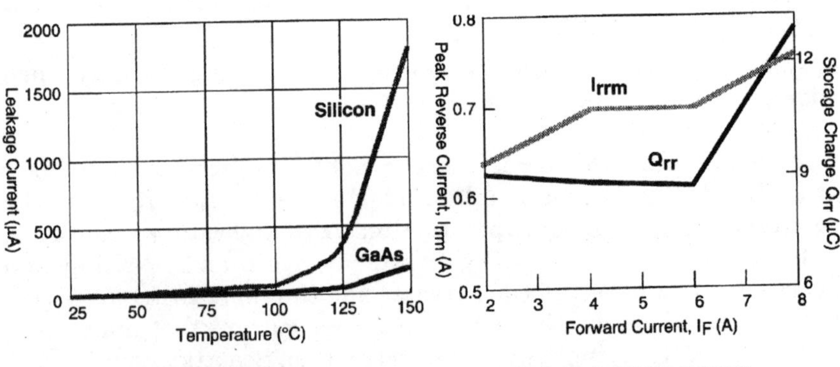

Silicon & GaAs diode leakage current over temperature.

GaAs diode reverse recovery.

Figure 13.20 Gallium-arsenide rectifier performance.

TABLE 13.2 Semiconductor Material Comparison

Property	Units	Silicon	GaAs	6H-SiC	4H-SiC
Bandgap	eV	1.11	1.43	2.9	3.2
Breakdown field	V/cm	7×10^5	7×10^5	38×10^5	35×10^5
Thermal conductivity	W/(cm °K)	1.5	0.46	4.9	4.9
Saturation velocity	cm/s	1×10^7	1×10^7	2×10^7	2×10^7
Saturation field	V/cm	8×10^3	3×10^3	52×10^3	25×10^3
Electron mobility	$cm^2/(V - s)$	1350	6000	380	800
Hole mobility	$cm^2/(V - s)$	450	330	95	120

rent is fairly stable over a wide range of forward current, as seen in Fig. 13.20.

13.3.1.4 Silicon carbide rectifiers. Silicon carbide (SiC) shows some promise as a semiconductor material for use in motor drive power devices. In future devices, SiC will allow much higher temperature capability than standard silicon. Table 13.2 compares some of the attributes of silicon, gallium arsenide, and silicon carbide semiconductor materials. Note that SiC has higher voltage capability (eV), lower ON resistance, and better thermal conductivity than does standard silicon. Obviously, SiC devices will require packages that are rated for very high temperatures.

13.3.2 Power transistors

Deciding which power transistor type to use in a motor drive depends on several factors. Each type of transistor has certain advantages and disadvantages depending on the exact motor control application. In general, MOSFETs are chosen for low-voltage motor drives, while IGBTs have moved into the forefront for high-voltage motor drives. SCRs and triacs are cost-effective replacements for mechanical contact components.

13.3.2.1 Bipolar junction transistor. The bipolar junction transistor (BJT) can be made to block high voltages and to pass high currents. Still, the bipolar transistor can go into second breakdown. (BJTs have reduced safe operating areas when compared to MOSFETs or IGBTs.) These devices exhibit low forward drop, making them favorable switching devices. However, the slow switching speed and high-power base drive limit the use of BJTs as power devices in motor drives.

13.3.2.2 SCR and triac power devices. Thyristor devices such as SCRs and triacs require less gate current as junction temperature increases, a characteristic that must be accounted for in the design. Poor design can result in false triggering at high temperatures and lack of trigger at low temperatures.

Voltage ratings of thyristors must be followed in order to avoid failure. Voltage spikes on the device that exceed the rated level will turn the device on, and the device must absorb the energy within the spike. Snubbing is recommended to provide a place for the energy to be collected. Snubbers also limit dv/dt spikes, which cause false triggering in thyristors.

Surge currents are specified as peak surge currents, meaning that the device can handle the current for a short period. Continuous or repetitive current must be considered in terms of the resulting power dissipation and junction temperature. Junction temperature is dependent on many factors, which are discussed in Chap. 15.

13.3.2.3 MOSFET power devices. MOSFETs have a parasitic body diode that can be utilized in some motor drive designs. This diode is inherent in a MOSFET structure and is rarely specifically engineered as a motor control diode. Instead, the MOSFET design parameters are optimized to operate as a switch. These parameters include $R_{DS,on}$, switching speed, and gate charge. The MOSFET body diode does not possess ideal forward drop and reverse recovery parameters for a motor control. To achieve better performance, a diode that is optimized for free wheeling is often placed in parallel with the MOSFET's body diode.

The MOSFET's body diode also determines the breakdown of the device. This is critical in circuits that have high voltage spikes such as motor drives. Since this spike forces the MOSFET's body diode into breakdown, current flows into the device and creates an energy that must be absorbed. This energy is usually a result of unclamped inductive switching (E_{as}), which is discussed later in this chapter. The design must consider this energy and specify a switching device that can handle the energy or provide a clamp diode or snubber circuit to control the energy.

MOSFET conduction losses are related to the $R_{DS,on}$ of the device. Higher-voltage MOSFETs have higher $R_{DS,on}$ as a result of the following equation:

$$R_{DS,on} / V_{DSS}^{2.7} \tag{13.1}$$

The MOSFET has a rectangular safe operating area (SOA) and does not exhibit second breakdown. When switching a power MOS-

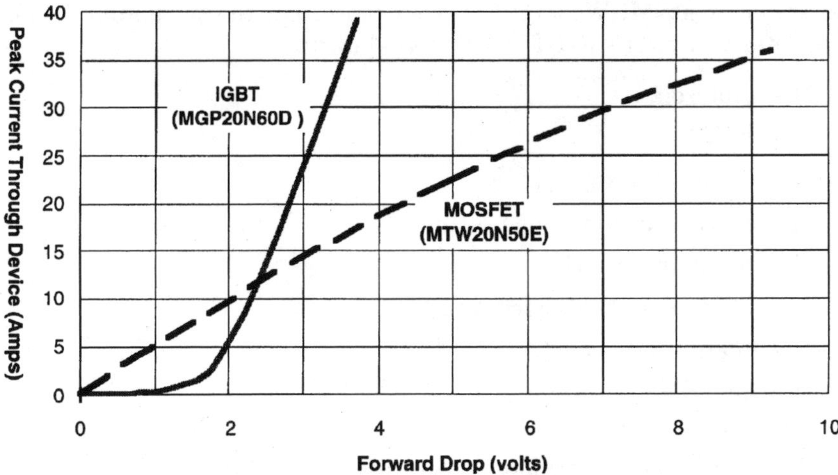

Figure 13.21 Forward ON voltage of MOSFET and IGBT.

FET, the capacitive gate must be supplied with enough initial current to charge above the device's threshold level. From there, some Miller effect charging must be considered. This means that an MCU cannot directly drive a large MOSFET, since the MCU's output port is not rated to handle the MOSFET's high gate peak currents.

13.3.2.4 IGBT power devices. IGBTs are competing with MOSFETs at voltage ratings of 250 V and above. Figure 13.21 shows a graph of MOSFET forward voltage drop compared to the forward drop of an IGBT with similar ratings. Also shown is the inherent forward "knee" of the IGBT that arises from the junction drop within the device.

IGBT design optimization is controlled by forward drop V_{CE}, switching speed/turnoff energy E_{off}, short-circuit protection time t_{sc}, and reverse blocking voltage (BV). These parameters are delicately balanced for a three-phase induction motor drive that may require BV = 600 V, E_{off} = 10 percent of losses at 4 kHz, t_{sc} = 10 μs at T_j = 125°C, and V_{CE} = 2.7 V at 20 A. All these parameters play against each other. For example, an IGBT for a switched reluctance motor might require less short-circuit protection than would a standard three-phase AC or BLM drive. This could allow the use of either faster (decreased E_{off}) or lower ON-voltage IGBTs.

The structure of an IGBT results in a lack of a parasitic body diode. In designs requiring a free-wheeling diode, this device is often placed across the IGBT in an external package. Manufacturers are including the free-wheeling diode inside the package where the letter "D" in the part number sometimes indicates a dual die (IGBT and diode). One example is an MGW21N60ED from Motorola; the part number for this manufacturer's IGBT is coded as shown:

M = Motorola

G = IGBT

W = TO-247 package size

21 = continuous amperage rating

N = N-channel IGBT

60 = voltage rating divided by 10

E = energy series part

D = dual die

13.3.3 Zeners and transient protectors

Zeners and transient protectors are diodes that are reverse-biased to the point of breakdown. The voltage is clamped at this level as current flows through the device. Zener diodes are designed to provide a regulated voltage such as that needed for a reference voltage. Zeners operate at low currents (≤10 mA) and provide a fairly stable reference overtemperature. Some zeners have a zero-temperature coefficient, but most may require a temperature compensation device if high accuracy over wide operating temperatures is needed. Transient-voltage suppressors are large die devices that absorb energy by clamping at a level below that of the device they are to protect. They may also clamp lines to absorb the inductive kick that results from switching an inductive load.

13.4 Understanding the Data Sheet

The correct interpretation of a semiconductor data sheet can mean the difference between a marginal and a robust motor drive design. Some data sheets, such as MCUs, are often hundreds of pages in length, but the critical specifications for electrical ratings are contained in just one or two pages. Power semiconductor data sheets are usually less than 10 pages, and their critical electrical ratings are usually, like those of the MCUs, only one or two pages in length. It is important to understand how to relate these ratings to a motor circuit's operation.

When comparing one manufacturer's device ratings to another manufacturer's, it is a good idea to know how the ratings were made. For instance, are the ratings based on the same exact test circuit?

More importantly, does the semiconductor's test circuit bear any resemblance or pertain to a motor drive circuit?

In this section we will discuss the various ratings found in MOSFET, IGBT, and rectifier data sheets. Other devices, such as triacs, SCRs, zeners, and bipolar transistors, will use the same data sheet format. Most semiconductor manufacturers offer application notes or technical articles that give detailed information on certain aspects of their discrete devices that may not be covered in their respective data sheets. It is wise to gather all this information on a particular device, especially if that device is one that performs a crucial function in your design.

13.4.1 Data sheet parametrics

The data sheet is a critical aspect of motor drive design. Manufacturer data sheets are meant to furnish the user with enough information to properly use the device. Parameters on the data sheet provide the maximum device ratings and information on how the part will perform. *Maximum ratings are shown to keep the user from designing a circuit that is beyond the device's capability.* Output curves are shown to illustrate typical device behavior over a range of test conditions.

Although each manufacturer may use a different format and style for their data sheets, the information provided commonly includes a headline and a verbal introduction to the part and sections specifying maximum ratings, electrical characteristics, and typical output curves. Each section will be tailored specifically to the technology and/or application for which it is intended. For components that manage the high power, current, and voltage associated with motor drives, this information is critical to the design of efficient and reliable circuitry. (*Editor's note:* Keep in mind that the circuits used to test power semiconductors are powered with pure DC, pulsed DC, or sine waves and may not duplicate some of the abnormal waveforms encountered in a motor control circuit.)

13.4.1.1 Data sheet headline information. This section of the data sheet contains the part number and general device information. Items such as package style, device technology, and a general description of the device's advantages are usually provided in this section. This section often describes the advantages of using the device in a particular application.

13.4.1.2 Absolute maximum ratings. Absolute maximum ratings represent the extreme capabilities of the device. They can best be described as device characterization boundaries and are given to facilitate worst-case design. These parameters will be different for each device technology. The only exception to this is thermal performance. Most thermal parameters are common among different technologies. (*Editor's note:* It has been the experience of the editor that using the absolute maximum ratings as normal design values will lead to an unreliable design. It is suggested that the motor control design operate only within 40 to 60 percent of the device's maximum capability. Most motors are rated to draw 6 to 8 times their normal running current when operated in a locked-rotor condition. Selecting a power device that is rated at only 2 or 3 times the motor's normal running current will not be adequate for large motor drives.)

13.4.1.3 Thermal performance maximum ratings. The thermal ratings for power semiconductors should be carefully followed. These parameters directly affect the reliability of the device. Bolting a power device onto a massive heatsink doesn't always mean that the device won't be stressed from excessive junction temperatures.

Maximum power dissipation (PD). The maximum power dissipation rating specifies the power dissipation limit which takes the junction temperature to its maximum rating while the reference temperature is being held at 25°C. It is calculated by the following equation:

$$\text{PD} = \frac{T_{\text{j,max}} - T_{\text{r}}}{R_{\text{thjr}}} \tag{13.2}$$

Maximum junction temperature $T_{j,max}$. This value represents the maximum allowable junction temperature of the device in degrees Celsius, derived from and based on reliability data. Exceeding this value will reduce the device's long-term operating life. (*Editor's note:* Some studies suggest that the device's life is doubled for each 10°C reduction of junction temperature.)

Thermal resistance R_{thjc}, R_{thja}. The quantity that resists or impedes the flow of heat energy in a device is called *thermal resistance.* Thermal resistance helps designers determine the amount of heatsinking required to maintain a given junction temperature for the amount of power that the device is dissipating. Units for thermal resistance are degrees Celsius per watt (°C/W).

(*Editor's note:* Thermal cycling is another area that is seldom addressed in a data sheet. Most manufacturers do test their power devices for thermal cycling capability. This can be a test where the device is operated in a cyclic mode at its rated power dissipation—the device heats up to its maximum T_j, usually 150°C, and is then turned off, allowing its junction temperature to drop by about 25°C. This is repeated for thousands of cycles or until the device fails. Some devices with poor die attachment or improper wire bonds will fail in less than 100 cycles. Most power devices should last for 10,000 thermal cycles. The manufacturer should be consulted if the motor design can stress the power device in this manner.)

13.4.2 Maximum ratings for power MOSFETs

With the exception of their gate structure, MOSFETs are fairly robust power devices. They can sustain some excessive drain-to-source voltage spikes. It is a good practice to ensure that the maximum voltage, current, and energy ratings are not exceeded in an application. If possible, the device should be derated by 30 to 50 percent to ensure long-term reliability.

13.4.2.1 MOSFET maximum drain-to-source voltage. This represents the absolute limit of the device's blocking voltage capability from drain to source when the gate is shorted to the source (V_{DSS}), or when a 1-MΩ gate-to-source resistor is present (V_{DGR}). It is measured at a specific leakage current and has a positive temperature coefficient. To prevent breakdown of the drain-to-source junction, the applied voltage across the power MOSFET should never exceed this rating, unless the device is rated for avalanche operation.

13.4.2.2 MOSFET maximum gate-to-source voltage. The maximum allowable gate-to-source voltage rating is either a continuous condition (V_{GS}), or a single-pulse nonrepetitive condition (V_{GSM}). Exceeding this limit will almost always result in permanent device degradation. Note that, unlike the avalanche effect that occurs when exceeding the V_{DSS} rating, an oxide punch-through is likely to occur when exceeding the maximum gate-to-source rating. Some MOSFETs have a built-in gate protection circuit to minimize this event.

13.4.2.3 MOSFET maximum continuous drain current. This rating is the maximum continuous drain current I_D level that will raise the device's junction temperature to its rated maximum while its reference temperature is held at 25°C. This can be calculated by the following equation:

$$I_D = SQRT\left(\frac{\text{PD}}{R_{\text{DS,on}} \text{ at } T_{\text{j,max}}}\right) \tag{13.3}$$

13.4.2.4 MOSFET maximum pulsed drain current. This rating gives the maximum allowable peak drain current (I_{DM}) that the device can safely handle under a 10-μs pulsed condition. It takes into consideration the device's thermal limitation, as well as $R_{DS,on}$, wire bond, and source metal limitations. The duty cycle for this rating is very narrow (i.e., 10 μs on and 1 s off). If the device is to survive a short-circuited motor condition, some type of current sensing and shutdown design that responds in less than 10 μs is necessary.

13.4.2.5 MOSFET maximum drain-to-source avalanche energy. The E_{as} specification defines the maximum allowable energy that the device can safely handle in avalanche as a result of an inductive current spike. It is tested at the I_D of the device as a single-pulse nonrepetitive condition. This value has a negative temperature coefficient. For repetitive avalanche conditions, it should be derated using the "thermal response" figure shown in the data sheet. (*Editor's note:* For maximum long-term reliability the motor control circuit should not operate the MOSFET in a continuous avalanche mode—every PWM cycle should not cause the MOSFET to avalanche. Avalanche should occur only during a random voltage spike event.)

13.4.3 Maximum ratings for power IGBTs

Like MOSFETs, IGBTs are fairly robust power devices, except for their gate structure. Unlike MOSFETs, IGBTs are not designed to operate in an avalanche mode—it is important to protect the IGBT from excessive collector-to-emitter voltages. As stated before, the IGBT should be derated by 40 to 60 percent to ensure long-term reliability.

13.4.3.1 IGBT maximum collector-to-emitter voltage. This rating represents the maximum limit of the device's blocking voltage capability from collector-to-emitter when the gate is shorted to the emitter (V_{CES}) or when a 1-MΩ gate-to-emitter resistor is present (V_{CGR}). It is measured at a specific leakage current and has a positive temperature coefficient. To prevent breakdown of the collector-to-emitter junction, the voltage across the IGBT should never exceed this rating.

13.4.3.2 IGBT maximum continuous collector current. The maximum continuous collector current I_C is the DC current level that will raise the device's junction temperature to its rated maximum while its case temperature is held at either 25°C (I_{C25}) or 90°C (I_{C90}). This can be calculated by the following equation:

$$I_{CX} = \frac{T_{j,max} - T_X}{V_{CE,on} \text{ at } (I_{CX}, T_{j,max}) * R_{thjc}} \tag{13.4}$$

13.4.3.3 IGBT maximum pulsed collector current. This rating gives the maximum allowable peak collector current I_{CM} that the device can safely handle under a 10-μs pulsed condition. It takes into consideration the device's thermal limitation, as well as $V_{CE,on}$, wire bonds, and emitter metal limitations.

13.4.3.4 IGBT maximum gate-to-emitter voltage. The maximum allowable gate-to-emitter voltage V_{GE} should never be surpassed. Exceeding this rating will usually result in permanent device degradation or failure. Like MOSFETs, some IGBTs have a built-in gate-to-emitter protective element.

13.4.3.5 IGBT maximum short-circuit withstand time. The short-circuit withstand time t_{sc} is a time rating for given values of collector-to-emitter voltage, gate voltage, and starting temperature that the device can safely handle while in a shorted load condition. (*Editor's note:* This test is important in a half H-bridge motor power stage. The t_{sc} indicates how long the IGBT devices will survive if both the upper and lower parts of the half H-bridge are inadvertently switched on simultaneously. It should be noted that t_{sc} is affected by the device's design—faster IGBTs usually have less short-circuit ability.)

13.4.4 Maximum ratings for power rectifiers

Many engineers take rectifier operation for granted. A two-terminal power rectifier device is more than just a device for converting AC into DC. Rectifiers do have several crucial attributes. *Selecting a standard rectifier for use in a high-frequency power circuit, like an electronic motor drive, can lead to disaster.* Rectifiers have switching characteristics, avalanche or nonavalanche voltage breakdown, leakage currents, and power dissipation limits. Not all rectifier data sheets use the same ratings—the more recent rectifier data sheets use ratings that relate to today's power control designs.

13.4.4.1 Rectifier maximum peak repetitive reverse voltage. The rated peak repetitive reverse voltage V_{RRM} is the maximum allowable instantaneous value of the reverse voltage, including all repetitive transient voltages, but excluding all nonrepetitive transient voltages that occur across a rectifier diode. The rating is generally based on a pulse of 100 ms at a 60-Hz rate.

13.4.4.2 Rectifier maximum peak working reverse voltage. The rated peak working reverse voltage V_{RWM} is the maximum allowable instantaneous value of the reverse voltage that can occur across the rectifier diode, excluding all repetitive and nonrepetitive transient voltages. The input voltage to the circuit must be such that the peak inverse voltage applied to the rectifier does not exceed the rated V_{RWM}. The rating is generally based on operation in a half-wave 60-Hz rectifier circuit with resistive load and applies over the entire operating range. Some rectifiers are rated for avalanche operation (W_{AVAL}) and can be useful in circuits that may have an occasional voltage spike.

13.4.4.3 Rectifier maximum DC reverse blocking voltage. The rated maximum DC reverse blocking voltage V_R is the maximum allowable DC reverse voltage, excluding all repetitive and nonrepetitive transient voltages, across a rectifier diode. The rating applies to the entire operating temperature range specified. Values for V_R are usually the same as for V_{RWM}, but V_R may be restricted to operation at a lower temperature than V_{RWM}.

13.4.4.4 Rectifier maximum peak nonrepetitive reverse voltage. The rated peak nonrepetitive reverse voltage V_{RSM} is the maximum allowable instantaneous value of reverse voltage across the rectifier, including all nonrepetitive transient voltages, but excluding all repetitive transient voltages. The rating applies to the entire operating temperature range. The nonrepetitive peak reverse voltage occurs as a random circuit transient that may originate within the equipment or may be externally generated. The V_{RSM} rating is commonly found in standard and fast-recovery rectifier data sheets, but is usually omitted for ultrafast rectifiers.

13.4.4.5 Rectifier maximum average forward current. This rating gives the maximum value of average rectified forward current I_O delivered into a resistive load at a specified case temperature in a 60-Hz half-wave circuit, unless another frequency is specified. The I_O rating is commonly found in standard and fast-recovery rectifier data sheets, but is usually omitted in recent ultrafast rectifier data sheets.

13.4.4.6 Rectifier maximum peak repetitive surge current. The maximum peak repetitive surge current I_{FRM} is the peak current that the device can safely handle. It is usually tested by applying a 20-kHz square wave at the rated V_R and $T_C = 125°C$.

13.4.4.7 Rectifier maximum nonrepetitive peak surge current. The maximum nonrepetitive peak surge current I_{FSM} is the current that the device can safely handle for a minimum of 100 times in its lifetime. It is a nonrepetitive rating in the sense that the surge may not be repeated until thermal equilibrium conditions have been restored. It is commonly tested by applying a 60-Hz sine wave with one half-cycle of surge current (I_{FSM}), followed by one half-cycle of nonrepetitive rated voltage, V_{RSM}.

13.4.5 Electrical characteristics and output curves data sheet section

These portions of a data sheet provide detailed device characterization to enable the designer to predict with a high degree of accuracy the behavior of the device in a specific application. The device parameters, test conditions, and specification limits are customarily shown; typical values measured in a test circuit in its intended application are sometimes included as well. Again, the exact parameters contained in these sections of a data sheet depend on device technology.

The last section of a power device data sheet contains actual output curves of a device's electrical and/or thermal performance. The data presented is usually based on information from a "typical" device that best represents the total distribution of the product. The advantage of these curves over single-condition electrical characteristics is that curves show device performance over a wide range of test conditions. Which curves are shown—and in what order—is highly dependent on several factors such as technology, target application, manufacturer, and the actual engineer or technician generating the data sheet. (*Editor's note:* Semiconductor manufactures are continually striving to tighten their process controls to ensure that the device's typical parameters stay within a tight boundary. In some cases, the minimum and maximum ratings in the electrical characteristics section may apply only to a very small portion of the product—the manufacturer should be consulted if a tighter guardband is required.)

13.4.6 Electrical characteristics for MOSFETs

Generally, the MOSFET's typical operation can be determined from the values shown in the electrical characterization section. It is a good idea to pay close attention to the test circuits that the manufacturer uses to make these measurements. For example, if the $R_{DS,on}$ rating were taken with a gate voltage of 10 V, and the actual motor drive stage only supplies 15 V, a lower $R_{DS,on}$ would occur.

13.4.6.1 MOSFET minimum drain-to-source breakdown voltage. The drain-to-source breakdown voltage rating $V_{BR,DSS}$ represents the lower limit of the device's blocking voltage capability from drain to source with the gate shorted to the source. It is measured at a specific leakage current and has a positive temperature coefficient.

13.4.6.2 MOSFET maximum zero-gate-voltage drain current. The zero-gate-voltage drain current rating I_{DSS} is the direct current into the drain terminal of the device when the gate-to-source voltage is zero and the drain terminal is reverse-biased with regard to the source terminal. This parameter generally increases with temperature per the temperature coefficient usually shown with the I_{DSS} rating.

13.4.6.3 MOSFET maximum gate-body leakage current. The gate-body leakage current I_{GSS} is the direct current into the gate terminal of the device. It is tested with the gate terminal biased by either a positive or negative voltage with respect to the source terminal and the drain terminal is short-circuited to the source terminal. Gate-body leakage current is useful for testing the MOSFET's gate integrity, but it is not ordinarily a consideration in power control designs.

13.4.6.4 MOSFET gate threshold voltage. This parameter is important when designing the gate driver circuit. The forward gate-to-source threshold voltage $V_{GS,th}$ is measured at a magnitude of drain current which increases to some low threshold value, usually specified as 250 μa or 1 mA. This parameter has a negative temperature coefficient that can lead to problems at extreme junction temperatures. Minimum, typical, and maximum values are commonly shown for the gate-to-source threshold voltage.

13.4.6.5 MOSFET typical and maximum drain-to-source ON resistance. The typical and maximum drain-to-source ON resistance $R_{DS,on}$ is the DC resistance between the drain-to-source terminals with a specified gate-to-source voltage applied to bias the device into the ON state. This parameter has a positive temperature coefficient. The maximum $R_{DS,on}$ value should be used when calculating conduction power losses. The typical $R_{DS,on}$ value will facilitate calculation of short-circuit current levels. Drain-to-source ON resistance is directly affected by gate-to-source voltage and junction temperature.

13.4.6.6 MOSFET typical and maximum drain-to-source ON voltage. The typical and maximum drain-to-source ON voltage $V_{DS,on}$ is the DC voltage between the drain-to-source terminals with a specified gate-to-source voltage applied to bias the device into the ON state. This parameter has a positive temperature coefficient.

13.4.6.7 MOSFET forward transconductance. The forward transconductance g_{FS} is the ratio of the change in drain current due to a change in gate-to-source voltage (i.e., $\Delta I_D/\Delta V_{GS}$). (*Editor's note:* MOSFETs with high transconductance tend to require a clean and stable gate drive—any small variation in the gate voltage can cause significant changes in the output current.)

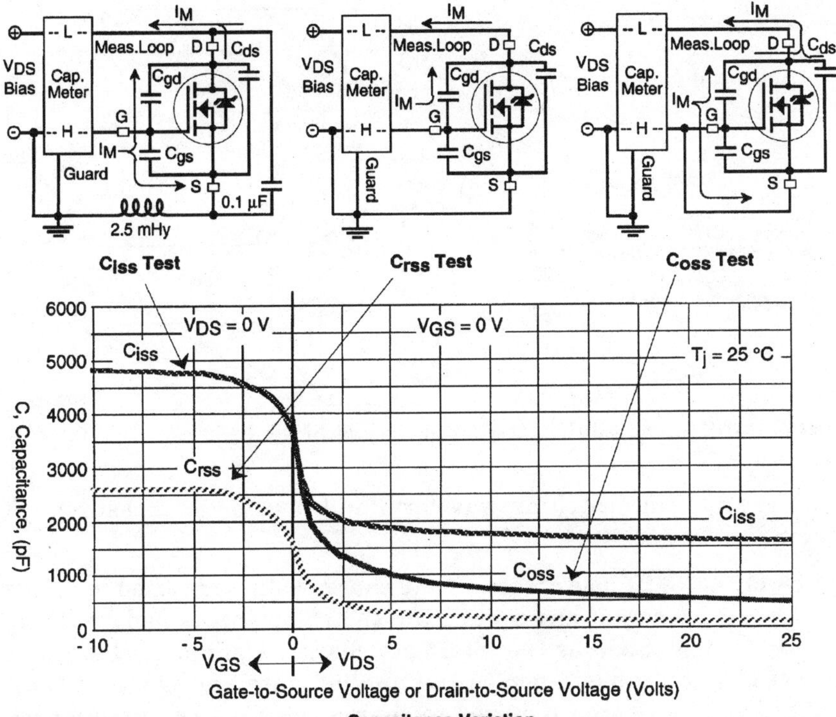

Figure 13.22 MOSFET capacitance test configurations and curves.

13.4.6.8 MOSFET device capacitance. Power MOSFET devices have internal parasitic capacitance from terminal to terminal (this capacitance is voltage-dependent): C_{ISS} is the capacitance between the gate-to-source terminals with the drain terminal short-circuited to the source terminal for alternating current, C_{OSS} is the capacitance between the drain-to-source terminals with the gate short-circuited to the source terminal for alternating current, and C_{RSS} is the capacitance between the drain-to-gate terminals with the source terminal connected to the guard terminal of a three-terminal bridge. Figure 13.22 shows test circuits used for power MOSFET capacitance measurements and also a typical MOSFET's capacitance curves.

13.4.6.9 MOSFET resistive switching. MOSFET switching speeds are faster than comparably sized bipolar transistors. It is possible to obtain switching times of less than 10 ns with power MOSFET devices if a high-frequency printed-circuit-board (PCB) layout is used. Figure 13.23 shows a typical MOSFET resistive switching test circuit;

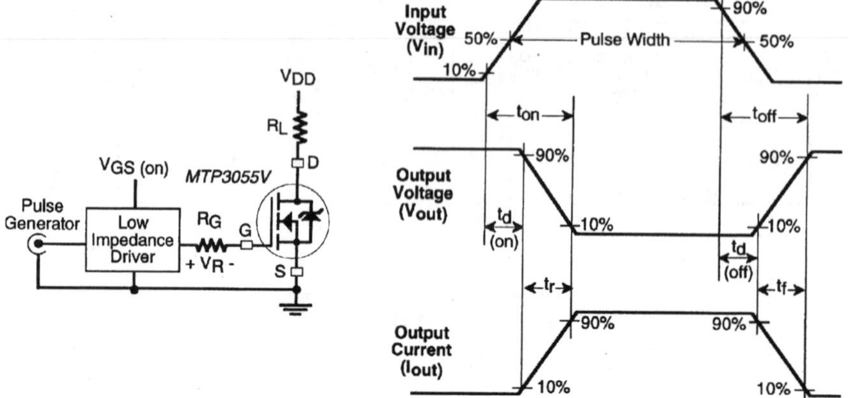

Figure 13.23 Resistive switching test circuit and switching waveforms.

note the typical switching waveform and parameter measurement points.

13.4.6.10 MOSFET gate charge. Gate charge values are used to design the gate drive circuit and to estimate switching speeds and switching losses: Q_T is defined as the total gate charge required to charge the device's input capacitance to the applied gate voltage, and Q_1 is defined as the charge required to charge the device's input capacitance to the $V_{GS,on}$ required to maintain the test current I_D. The time required to deliver this charge is called the *turnon delay time*; Q_2 is defined as the charge time required for the drain-to-source voltage to drop to $V_{DS,on}$.

13.4.6.11 MOSFET diode forward ON voltage. The MOSFET diode's forward ON voltage V_{SD} is the DC voltage between the source-to-drain terminals when the power MOSFET's intrinsic body diode is forward-biased. This is an important parameter if the diode is used as a freewheeling diode in a motor control circuit.

13.4.6.12 MOSFET diode reverse recovery time. The intrinsic body diode of a power MOSFET is a minority-carrier device and thus has a finite reverse recovery time. (*Editor's note:* If the MOSFET's internal body diode is used as a free-wheeling diode, it may have snappy reverse recovery characteristics. This can generate excessive EMI and cause other problems as well.) Here, T_a is defined as the time between the dropping I_S current's zero crossing point to the peak I_{RM}, and T_b is defined as the time between the peak I_{RM} to a projected I_{RM} zero-current crossing point through a 25 percent I_{RM} projection, as shown in Fig. 13.24. Total reverse recovery time t_{rr} is defined as the sum of t_a

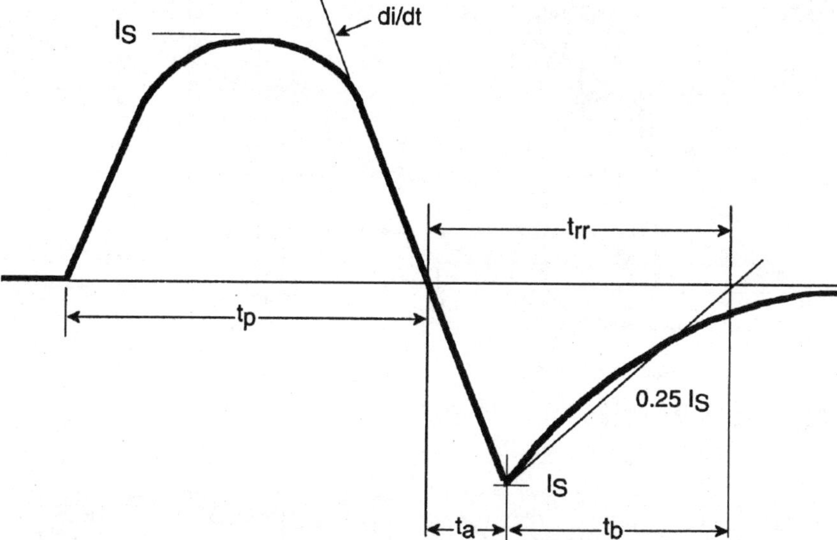

Figure 13.24 Diode reverse recovery waveform.

and t_b. Charge Q_{RR} is defined as the integral of the area made up by the I_{RM} waveform and V_R, the reapplied blocking voltage which forces reverse recovery.

13.4.7 Output curves for MOSFETs

Power MOSFET data sheets generally contain three types of output curves. These are the typical electrical characteristics, switching, and safe operating areas (SOA).

13.4.7.1 MOSFET electrical characteristic curves. These are usually *I-V* or *I-R* type curves; that is, they show device electrical output in the form of either current versus voltage or current versus ON resistance. These curves are usually self-explanatory, are easy to understand, and cover a wide range of conditions. Examples of two very common MOSFET electrical characteristic curves are shown in Fig. 13.25. The ON-region curves demonstrate the effect of gate voltage variations on the device's drain-to-source voltage over the device's rated current. Note how the current level remains constant when the gate voltage is somewhat reduced—the device is operating in a constant-current mode and will dissipate significant power. The drain-to-source resistance versus drain current and temperature is especially useful for observing the effects of varying junction temperatures. Note that the ON resistance starts to rise rapidly at high junction temperatures but

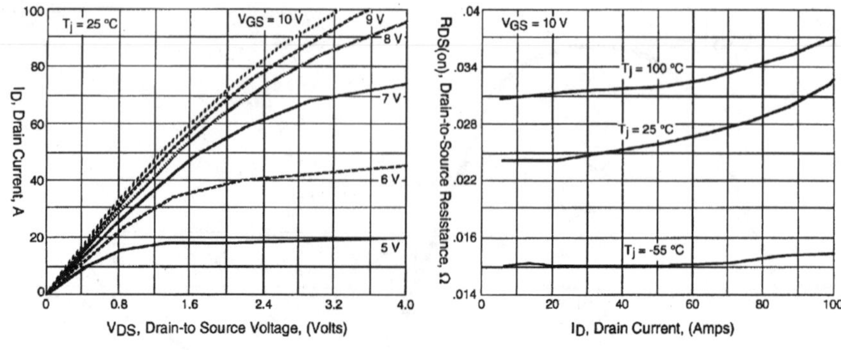

Figure 13.25 ON region and $R_{DS,on}$ versus I_c and temperature.

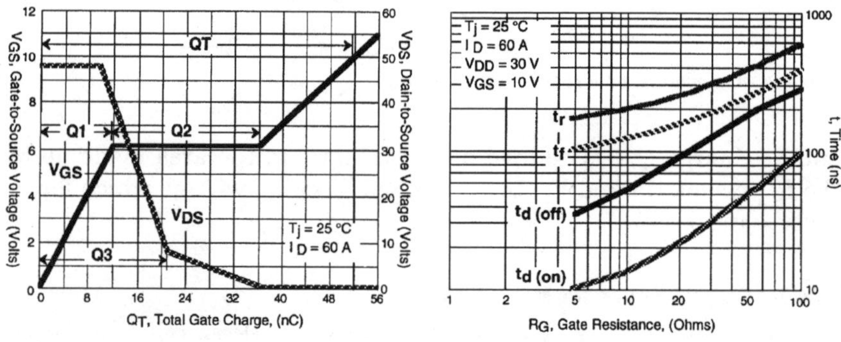

Figure 13.26 Gate charge and resistive switching.

remains fairly straight at very cold temperatures. This means that a MOSFET will conduct more current when cold, which can be significant in a motor system, as the motor's winding resistance will also decrease in value when cold.

13.4.7.2 MOSFET switching curves. Data is presented that applies to the switching capability of the device. The most common curves shown for MOSFETs are gate charge and resistive switching versus gate resistance. Examples of these curves are shown in Fig. 13.26.

13.4.7.3 MOSFET safe operating area (SOA) curves. These curves define the maximum simultaneous drain-to-source voltage and drain current that the device can safely handle. These curves are based on

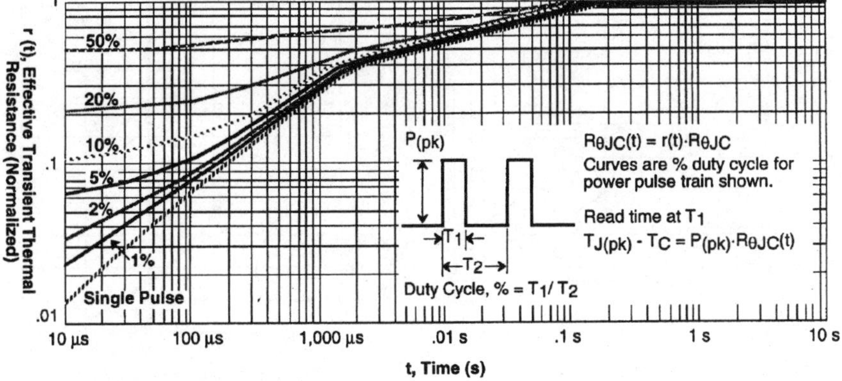

Figure 13.27 Thermal response.

the maximum peak junction temperature and a case temperature T_c of 25°C. Peak repetitive pulsed power limits are determined by using the thermal response data, as represented in Fig. 13.27. (*Editor's note:* Transient thermal response can be important for short-circuit design—the power device must be shut off abruptly in order to survive. Transient thermal response is less important in a motor control application, because a stalled motor event usually lasts for more than 100 ms. Most power devices have a transient thermal response that diminishes beyond 100 ms. Therefore, the power device will heat up and fail unless an adequate heatsink is used.)

Using the thermal response curve, the forward-biased safe operating area (FBSOA) can be calculated and curved. Since MOSFETs do not have second breakdown effects, it is a relatively simple process to calculate the curves for FBSOA. A typical FBSOA curve is shown in Fig. 13.28. There are two areas of importance in a motor application that relate to FBSOA: (1) the FBSOA should never be exceeded—there is not that much guardband at higher operating voltages in a FBSOA curve; and (2) the package limit (wirebonds, lead materials, etc.) sets the device's maximum current level—MOSFETs and IGBTs have high transconductance, which may allow significantly higher current levels to flow than their published ratings show.

Power MOSFETs are designed to be used safely in unclamped inductive switching circuits. For reliable operation, the stored energy from the circuit inductance must be less than the rated E_{as} limit and may need to be adjusted for operating conditions differing from those

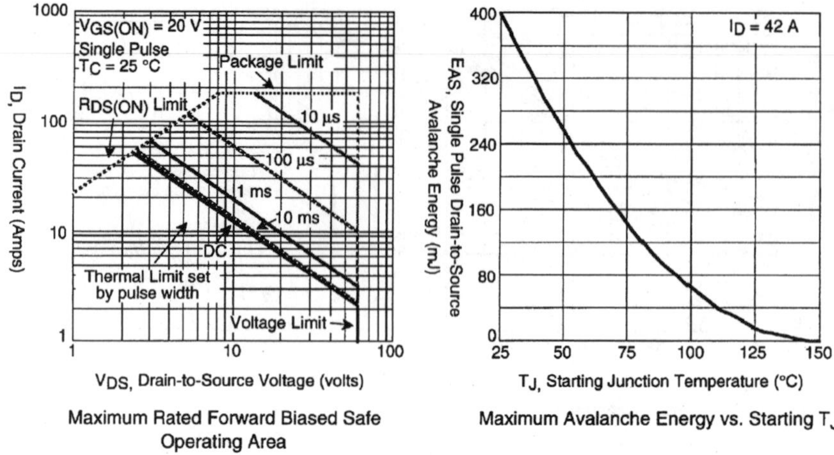

Figure 13.28 Forward-biased safe operating area and avalanche energy.

specified. The E_{as} energy rating must be derated for temperature, as shown in Fig. 13.28.

13.4.8 Electrical characteristics for IGBTs

Generally, the IGBT's typical operation can be determined from the values shown in the electrical characterization section. Large IGBTs are usually characterized for motor control circuits, while smaller IGBTs may be characterized for switching applications or automotive ignitions. Unlike MOSFETs, large IGBTs are typically tested with a 15-V gate bias. It is common for this section of the data sheet to use minimum, typical, or maximum values, depending on the test and individual manufacturer.

13.4.8.1 IGBT minimum collector-to-emitter breakdown voltage. The collector-to-emitter breakdown voltage rating (BV_{CES}) represents the lower limit or minimum of the device's blocking voltage capability from collector to emitter with the gate shorted to the emitter. It is measured at a specific leakage current and has a positive temperature coefficient ($\Delta V_{BR,CES}/\Delta T_j$), which is usually shown below the BV_{CES} rating. The positive temperature coefficient also means that, at extremely cold temperatures, the IGBT's BV_{CES} rating will decrease. For example, a 600-V breakdown IGBT at 25°C may decrease to 550 V at −55°C.

13.4.8.2 IGBT maximum zero-gate-voltage collector current. The zero-gate-voltage collector current I_{CES} rating is made by measuring the direct current into the collector terminal of the device when the gate-to-emitter voltage is zero and the collector terminal is reverse-biased with respect to the emitter terminal. This parameter, which is expressed in microamperes, generally increases with temperature. It is intended to indicate the upper limit of the leakage range at the IGBT's rated voltage and is usually shown for 25 and 125°C junction temperature.

13.4.8.3 IGBT maximum gate-body leakage current. The gate-body leakage current I_{GES} is measured in nanoamperes. It is the direct current into the gate terminal of the device when the gate terminal is biased with either a positive or negative voltage—usually 20 V—with respect to the emitter terminal and the collector terminal is short-circuited to the emitter terminal. Normally, this parameter is not used for power control design, but it can be very important for testing the integrity of the IGBT's gate structure. If the gate has suffered ESD or overvoltage damage, its leakage current will more than likely increase well beyond its I_{GES} rating.

13.4.8.4 IGBT gate threshold voltage. It is important to know this parameter when designing the gate driver circuit. The forward gate-to-emitter voltage $V_{GE,th}$ is measured at a magnitude of collector current which increases to some low threshold value, usually specified as 250 μa or 1 mA. This parameter has a negative temperature coefficient, which can lead to problems at extreme junction temperatures. Minimum, typical, and maximum values are usually shown for the gate-to-source threshold voltage.

13.4.8.5 IGBT typical and maximum collector-to-emitter ON voltage. The typical and maximum collector-to-emitter ON voltage $V_{CE,on}$ is the DC voltage between the collector-to-emitter terminals with a specified gate-to-emitter voltage applied to bias the device into the ON state. This parameter can exhibit a positive or negative temperature coefficient, depending on the collector current. The maximum $V_{CE,on}$ is used when calculating conduction loss.

13.4.8.6 IGBT forward transconductance. The forward transconductance g_{FS} is the ratio of the change in collector current due to a change in gate-to-emitter voltage (i.e., $\Delta I_E/\Delta V_{GE}$). (*Editor's note:* IGBTs with high transconductance tend to require a clean and stable gate drive—any small variation in the gate voltage can cause significant changes in the output current.)

13.4.8.7 IGBT device capacitance. Power IGBT devices have internal parasitic capacitance from terminal to terminal (this capacitance is voltage-dependent): C_{iee} is the capacitance between the gate-to-emitter terminals with the collector terminal short-circuited to the emitter

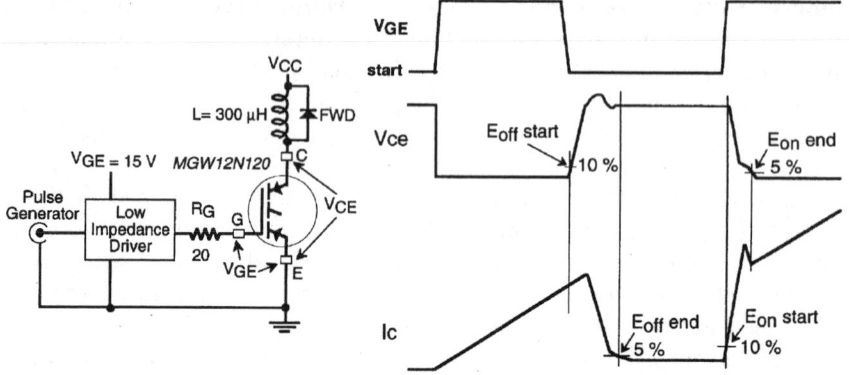

Figure 13.29 Clamped inductive switching test circuit and waveforms.

terminal for alternating current, C_{oee} is the capacitance between the collector-to-emitter terminals with the gate short-circuited to the emitter terminal for alternating current, and C_{ree} is the capacitance between the collector-to-gate terminals with the emitter terminal connected to the guard terminal of a three-terminal bridge. The test circuits used for IGBT capacitance measurements are the same as those used for MOSFETs and are shown in Fig. 13.22.

13.4.8.8 IGBT inductive switching. IGBT clamped inductive switching is tested and measured with the inductive switching test circuit, as shown in Fig. 13.29. Typical switching waveforms and the E_{on} and E_{off} parameter measurement points are also illustrated. Note that the test consists of several PWM pulses which are required to ensure that the free-wheeling diode's reverse recovery time is taken into account. The E_{on} and E_{off} values are of special interest since they reflect the power losses during the switching transitions when driving an inductive load such as a motor. Note that the E_{on} loss is related to the free-wheeling diode's reverse recovery time. The E_{off} value encompasses the total energy dissipated during turnoff, including the IGBT's tail current. These E_{on} and E_{off} values are listed for both 25 and 125°C junction temperatures on the data sheet. Some IGBT data manufacturers also show the total switching loss E_{ts}. This value is useful when the IGBT's current level is about the same for both turnon and turnoff.

Unclamped inductive switching E_{as} can be very beneficial to designers of motor drives. Some IGBTs have a built-in collector-to-gate overvoltage circuit (MGP20N40CL, for example) and are tested for E_{as}. The inductive kick of the motor creates voltage and current spikes that can destroy a power stage. By providing this parameter on a data sheet, manufacturers assist designers in their quest to ensure that the device will be robust. [*Editor's note:* If the IGBT does not contain some type of collector-to-gate overvoltage protection circuit, it will most likely fail when subjected to excessive collective voltage spikes. MOSFET power devices are usually rated to avalanche (like a zener) when their drain-to-source voltage is exceeded.]

13.4.8.9 IGBT gate charge. Gate charge values are used to design the gate drive circuit: Q_T is defined as the total gate charge required to charge the device's input capacitance to the applied gate voltage, Q_1 is defined as the charge required to charge the device's input capacitance C_{GE} to the $V_{GS,on}$ value necessary to maintain the test current I_C, and Q_2 is defined as the charge time needed to charge C_{GD} to the same value as C_{GE}. Gate charge data is used to design a drive circuit that will provide enough current to perform switching in the desired amount of time.

13.4.8.10 IGBT diode forward voltage drop. Copackage IGBT devices contain an integral free-wheeling diode between the collector-to-emitter terminals. The DC voltage between the emitter-to-collector terminals when the power diode is forward biased is called the diode forward voltage drop V_{FEC}. This parameter determines diode conduction loss.

13.4.8.11 IGBT diode reverse recovery time. The free-wheeling power diode is a minority-carrier device and thus has a finite reverse recovery time. Parameter definitions are the same as those for the MOSFET body diode, although the diode used with a motor control IGBT will usually be designed with a fast recovery and a soft knee, as shown in Fig. 13.4.

13.4.9 Output curves for IGBTs

Power IGBT data sheets generally contain the same three types of output curves as those of MOSFET data sheets. They are the typical electrical characteristics, switching, and SOA curves. Switching is a main parameter for an IGBT, so IGBT data sheets tend to show more switching information than MOSFET data sheets.

13.4.9.1 IGBT electrical characteristic curves. These are usually *I-V* or *T-V*-type curves; that is, they show device electrical output in the form of either current versus voltage or temperature versus voltage. Those curves that are similar to MOSFETs will not be repeated here.

13.4.9.2 IGBT switching curves. Data are presented which apply to

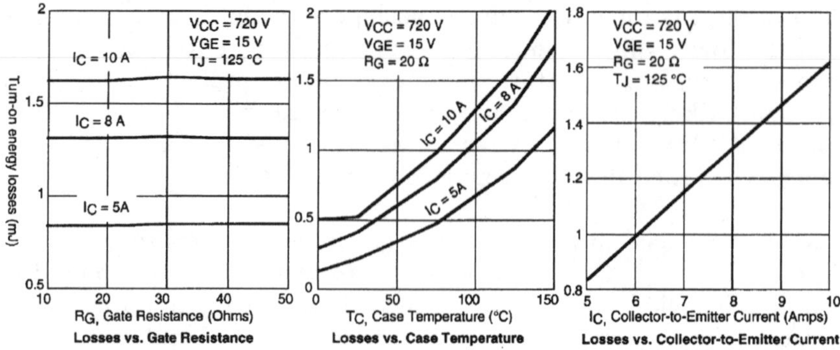

Figure 13.30 Turnoff losses.

the switching capability of the device. The most common curves shown for IGBTs are similar to those shown for MOSFETs. IGBTs, however, are used mostly in inductive load applications; therefore, most of the switching output curves reflect data obtained from such test circuits. Examples of IGBT switching curves are shown in Fig. 13.30. (*Editor's note:* These curves show turnoff switching losses versus gate resistance, turnoff losses versus case temperature, and turnoff switching losses versus collector current. Note that varying the gate resistor value does not greatly affect the turnoff loss. This is because the IGBT turnoff time is mostly controlled by the IGBT's internal design. The gate resistance does play a large role in setting the IGBT's turnon time.)

13.4.9.3 IGBT safe operating area (SOA) curves. The forward- and reverse-bias SOA curves define the maximum simultaneous collector-to-emitter voltage and drain current that the device can safely handle. These curves are based on the maximum peak junction temperature and a case temperature T_C of 25°C. Peak repetitive pulsed power limits used for a forward-bias SOA are determined by the IGBT's thermal response data. These limits are calculated and displayed similarly to those in the MOSFET data sheet (see Fig. 13.28). (*Editor's note:* Sometimes it can be confusing when determining which SOA, FBSOA or RBSOA, to use in an IGBT control circuit. Usually, FBSOA should be used for short-circuit operation, forward conduction, and turnon, or whenever the IGBT gate is forward-biased. RBSOA is used during turnoff with an inductive load. When operating an IGBT in a linear mode, both FBSOA and RBSOA come into play.)

Since IGBTs are primarily utilized in clamped inductive load circuits, IGBT data sheets usually include a reverse-biased SOA

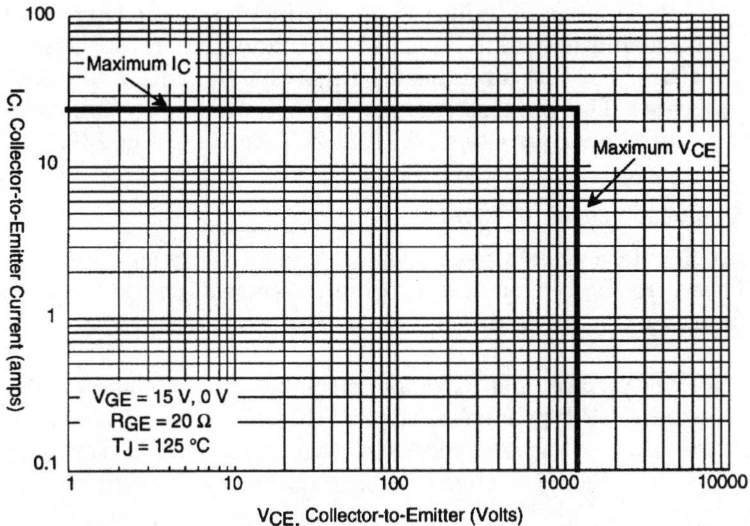

Figure 13.31 Reverse-biased safe operating area.

(RBSOA) curve. This curve is meant to show the IGBT's energy-handling capability during turnoff when driving an inductive load. RBSOA curves are usually bounded by the device's voltage and current ratings, as shown in Fig. 13.31.

13.4.10 Electrical characteristics for rectifiers

The electrical characteristics section of the rectifier data sheet usually lists typical and maximum values for instantaneous forward voltage and reverse current. Sometimes other ratings such as capacitance, reverse recovery, forward recovery, and avalanche energy are shown. (Note that the rectifier's reverse recovery characteristics have been covered in the MOSFET and IGBT sections.)

13.4.10.1 Rectifier instantaneous forward voltage. The instantaneous forward voltage rating V_F is the maximum forward voltage across the rectifier for a given test current and junction temperature. This value indicates the rectifier's voltage drop at its rated current and is usually given at 25°C and 125 or 150°C junction temperature. The instantaneous forward-voltage value is lower at the higher junction temperatures.

13.4.10.2 Rectifier maximum instantaneous reverse current. The maximum instantaneous reverse current I_R is the device's maximum reverse current. This value indicates the rectifier's reverse current at its rated voltage and is usually given at 25°C and 125 or 150°C junction temperature. The test for I_R uses a pulse (300 μs) with a low duty cycle (≤2 percent). This reverse current may increase more than thirty times at high temperature (e.g., 5 μA at 25°C to 150 μA at 150°C).

13.4.11 Output curves for rectifiers

Power rectifier data sheets generally contain three types of output curves. These are the typical and maximum electrical characteristics, current derating, and SOA curves.

13.4.11.1 Rectifier typical and maximum electrical characteristic curves. These are usually *I-V*-type curves, that is, they show device electrical output in the form of current versus voltage. Examples of two very common rectifier electrical characteristic curves are shown in Fig. 13.32.

13.4.11.2 Rectifier current derating curves. These curves show average forward current capability versus lead or case temperature for DC, sine, and square half-wave circuits. The data is based on a junction temperature equal to $T_{j,max}$. Figure 13.33 shows an example of exactly such a curve.

13.4.11.3 Rectifier safe operating area (SOA) curves. These curves define the maximum simultaneous cathode to anode voltage and drain current that the device can handle safely. These curves are based on the maximum peak junction temperature and a T_C of 25°C.

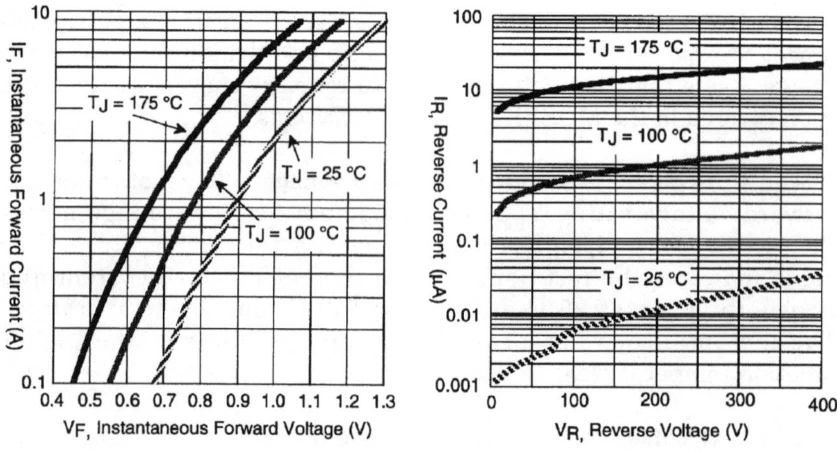

Figure 13.32 Rectifier forward and reverse voltage curves.

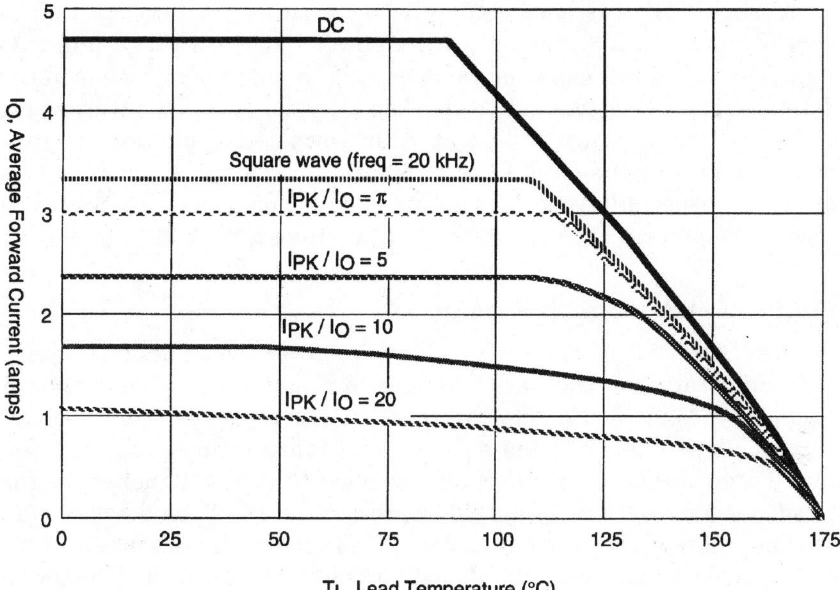

Figure 13.33 Rectifier current derating.

Peak repetitive pulsed power limits are determined by using the rectifier's thermal response data.

13.5 Device Failures

Most engineers who work with high-energy power circuits will occasionally have a power stage self-destruct for no apparent reason. In large motor drives this event can be dramatic and, in some instances, requires the use of safety measures. Large power modules can explode and fling bits of plastic, metal, and encapsulation compound over a wide distance. (It is always a good idea to wear safety glasses and work behind a protective shield when testing large motor power stages. See Chap. 17 for more information on motor control testing and instrumentation.)

13.5.1 How to check SOA

Testing for safe operating area (SOA) violations should always be done with any new motor power stage design. Computer simulation can help identify gross design problems, but the simulation is still not the same as a live bench test. The computer simulation doesn't explode and send out flames when a power device is operated beyond its SOA.

Measuring SOA is best accomplished with an oscilloscope that is hooked up in an *X-Y* fashion to the power device's voltage and current. The waveform can then be displayed in a similar manner to the data sheet's SOA curve and be checked for violations. This works best when the power stage is operated in a repetitive fashion. Digital scopes with a high sampling rate, 500 million samples per second or more, are especially good for capturing one-shot events. Some digital scopes can process the waveforms and generate a SOA-like curve.

13.5.2 Junction temperature testing

Besides SOA testing, the verification of a power semiconductor's junction temperature is the key to long-term reliability in a motor power stage. It is seldom possible to probe the power device's die without special lab instruments, but a noncontact temperature probe can suffice. Direct-contact thermocouple probes that are attached to the power devices leads or case seldom work because of the device's high voltage and switching frequencies. A noncontact temperature probe can be used to measure the device's case temperature and heatsink temperature. When the device's case temperature is known, its junction temperature can be calculated using the device's power dissipation and junction-to-case thermal resistance rating. [*Caution:* Junction temperature will rise very rapidly under certain conditions (a short circuit, for example) and may not be detected in time by the temperature probe before the device fails.]

13.5.3 ESD sensitivity and handling precautions

Electrostatic discharge (ESD) can be a silent killer of MOS power devices. It normally occurs during the handling or testing of the devices but can occur in the actual application. There are three different classifications of ESD performance according to MIL-Std-883, as shown in Table 13.3.

As can be seen from the data, most standard non-ESD-protected power MOSFET devices are considered "sensitive" in terms of ESD

TABLE 13.3 ESD Classifications. (MIL STD 883)

Classification		ESD susceptibility, V
Class 1	Sensitive	0–2000
Class 2	Nonsensitive	2000–4000
Class 3	Nonsensitive	>4000

TABLE 13.4 ESD Sensitivity by Device Type

Device type	ESD susceptibility, V
Power MOSFETs	100–2000
Power Darlingtons	20,000–40,000
JFETs	140–10,000
Zener diodes	40,000
Schottky diodes	300–2500
Bipolar transistors	380–7000
CMOS (ICs)	250–2000
ECL (ICs)	500
TTL (ICs)	300–2500

susceptibility. Realizing this, power MOSFET devices should be handled with proper caution. The same precautions that are used to handle sensitive CMOS logic devices should be taken when handling power MOSFETs. Table 13.4 identifies the sensitivity range of various other technologies.

Sensitive devices require special care to avoid ESD damage. A complete static-safe workstation should include a properly ESD-grounded conductive table top, grounded floor mats, grounded operators (wrist straps), conductive containers, and an ionized air blower to remove static from nonconductors. If these precautions are followed and the proper equipment is used, ESD-sensitive devices can be handled without being damaged.

Key factors to remember when handling ESD-sensitive devices:

- Handle all static-sensitive components in a static safeguarded work area.
- Transport all static-sensitive components in static shielded containers or packages.
- Educate all personnel in proper handling of static-sensitive components.

13.6 Advanced Motor Control Devices

Most engineers who design electronic motor controls would like to simplify their circuits, while still maintaining reliability and competitive performance. One method to accomplish this is to use higher integration in the power devices. Unfortunately, the more functionality that is put into a power device, the more limited it is in its application. Each motor drive design will be unique and may require a special feature that is not in a smart power device or an operational value different from the one that the smart device offers. Some features, such as overvoltage protection and internal temperature sensing, are necessary in most motor drives and will be integrated into future power devices.

13.6.1 Smart FETs

The die structure of a power MOSFET incorporates many cells that pass current through the device. One could generate a low-current signal for feedback to a shutdown circuitry by sensing the current through a small percentage of the cells. This technology does not tend to work unless sufficient current (over 10 percent of maximum steady-state-rated current) is flowing to allow for the current to spread across the die to the sense cells. Other forms of smart MOS technologies include device protection and gate enhancement technologies.

13.6.2 Smart IGBTs

Smart IGBTs could be developed by also sensing a percentage of the cells. However, the structure of the device allows short-circuit ratings to be incorporated by adding internal resistance to the device and keeping it from self-destruction. This type of device, of interest in induction motor drives, is referred to as a *short-circuit-rated device.*

Summary

While discrete semiconductors may not be as adventurous to design as MCUs, there would be little use for an MCU without the discrete power devices to control the motor's speed and torque. Discrete devices form the brawn of a motor drive and can cause serious problems if they fail because of a poor circuit design. In large motor drive designs the power device costs are usually several times the cost of the logic IC or MCU. Key points about this chapter are listed below.

- The two types of level shifting designs used in electronic motor control designs are current and voltage. Current shifting or amplification is required to drive power transistors from MCUs or logic ICs. Voltage shifting is necessary to drive a power stage that is floating above common.

- Small-signal transistors and diodes are frequently employed to drive large power devices and to protect logic level lines from voltage spikes.
- A total of 20 power rectifiers can be found in large three-phase motor designs. They are used for AC line rectification, free-wheeling diodes, and snubber networks.
- Bipolar junction transistors require significant drive current, whereas IGBT can usually be driven with less than 1 W of energy. IGBTs also exhibit better SOA than do BJTs.
- IGBTs exhibit a current tail during their turnoff because the IGBT's design does not allow a way to quickly discharge its internal PNP structure. It "self-discharges" or recombines and causes a current tail.
- Triacs can make excellent replacements for electromechanical contacts since they can be gated to switch only at zero crossing. Because the triac remains latched on when gated, it cannot be PWM-controlled .
- Very large power transistors sometimes cost more to manufacture because, as the semiconductor die size increases linearly, the manufacturing defects go up exponentially. Sometimes it is more cost-effective to parallel die in the same package.
- Surface mount packages allow smaller footprints, allow less lead inductance, and eliminate lead trimming.
- The main parts of a discrete semiconductor's data sheet are its maximum ratings section, which gives the absolute ratings of the device; the electrical characteristics section, which gives data operational values; and the data curves or graphs.
- The data sheet information that can be considered the most important for a motor control design is the safe operating area (SOA) and thermal characteristics. These should always be observed. Exceeding the SOA can cause instant destruction of the power device. Long-term reliability will be directly affected by the device's operating temperature.

Acknowledgments

The authors would like to thank the many contributors to this chapter. Material and technical support were provided by Ken Berringer, Mohit Bhatnager, Rodrigo Borras, Jonathon Kimball, Jeff Marvin, C.

S. Mitter, Steve Robb, Bob Rutter, Ali Salih, Warren Schultz, and George Templeton, and a host of others.

Further Reading

Borras, R., Aloisi, P., and Shumate, D., *Avalanche Capability of Today's Semiconductors,* Motorola article reprint AR598, 1993.

Clemente, S., Dubhashi, A., and Pelly, B., *IGBT Characteristics,* International Rectifier Application Note AN-983A, 1994.

Dashney, G. E., "Power MOSFETs Need ESD Protection Too," *PCIM,* Aug. 1997.

International Rectifier, *IGBT Designer's Manual,* IGBT-3, 1994.

Mitter, C. S., *Application Considerations Using Insulated Gate Bipolar Transistors,* Motorola Application Note AN1540, 1995.

Motorola Data Manual, "Theory of Thyristor Operation," *Motorola Thyristor Device Data,* DL137/D, rev. 5, 1993.

Motorola, Inc., *Motorola TMOS Power MOSFET Transistor Device Data,* DL135/D, rev. 6, 1996.

Robb, S., and Hollander, D., "Electrostatic Discharge Protection for Power MOS Devices," *Electronic Design Applications Issue,* June 24, 1996.

Salih, A., "GaAs Rectifiers for Power Supply and Motor Control Applications," *Proceedings of the Tenth Annual Applied Power Electronics (APEC) Conference,* Dallas, TX, March 5–9, 1995, pp. 743–745.

Takesuye, J., and Deuty, S., *Introduction to Insulated Gate Bipolar Transistors,* Motorola Application Note AN1541/D, 1995.

U.S. Government, MIL-STD-883C, Notice 6, Method 3015.5, *Electrostatic Discharge Sensitivity Classification,* Aug. 1987.

Chapter

14

Power Modules for Motor Control Applications

Ken Berringer
Motorola Semiconductor Products

Introduction

Large three-phase motor controls usually require six power stages with each power stage consisting of one or more power transistors and a free-wheeling diode. In this chapter we will discuss power modules that simplify the design of a motor control by combining some or all of the power stages into one package. As you will see, the topology of the power module is determined by the motor type, application, and power source.

14.1 Circuit Configurations

Power modules are available in a wide variety of configurations that accommodate most of the common motor drive topologies. As new motor controls evolve, including those for switched reluctance motors, power modules will also be designed to fulfill their specific power stage design requirements. The more common types of motor control power modules will be discussed here.

14.1.1 Six-packs

A common circuit configuration that is used for three-phase motors, both AC induction and brushless DC motors, is a six-pack. In its simplest form, a six-pack is a three-phase bridge arrangement, as shown in Fig. 14.1. The bridge consists of six transistors. The collectors of the upper transistors are all connected together to a common positive bus terminal; the emitters of the lower transistors are connected to a common negative bus terminal. The positive and negative terminals are normally connected to a DC source. The three output terminals are then connected to a three-phase motor.

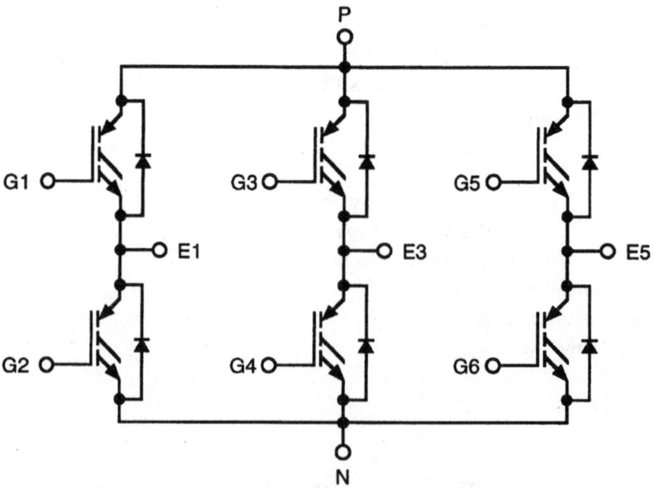

Figure 14.1 Six-pack module in its basic form.

A six-pack module could consist of IGBTs, MOSFETs, bipolar transistors, or any other type of power switching transistor. MOSFETs are commonly used for low-voltage motors. Large high-voltage motors often use an IGBT six-pack. An IGBT six-pack consists of 12 silicon die, six IGBTs, and six free-wheeling diodes. A MOSFET six-pack consists of only the six MOSFET die, as the power MOSFET has a built-in free-wheeling diode. In PWM motor drive designs IGBTs are best suited for applications that require transistors rated at 400 V or higher, while MOSFETs are most effective for applications that require <100-V transistors. In this chapter we will use the terminal names for IGBTs—gate, emitter, and collector. However, many of the circuit concepts apply equally well to other transistor types.

The six-pack configuration is nearly universal in its suitability for different three-phase motor drive systems. The six-pack configuration is applied in a wide range of market segments. Computer disk drives and tape drives utilize very small MOSFET six-packs. Appliance and consumer applications use low-cost six-packs for <750-W motors (150 to 600 W). Industrial drives employ six-packs up to 7.5-kW motors or possibly higher. Electric vehicles could utilize a very high-power six-pack (30 to 150 kW).

High-power six-packs often have a separate emitter terminal for the gate drive connections, as shown in Fig. 14.2. These terminals are often called an *emitter-Kelvin* or *Kelvin terminals*. The effect of emit-

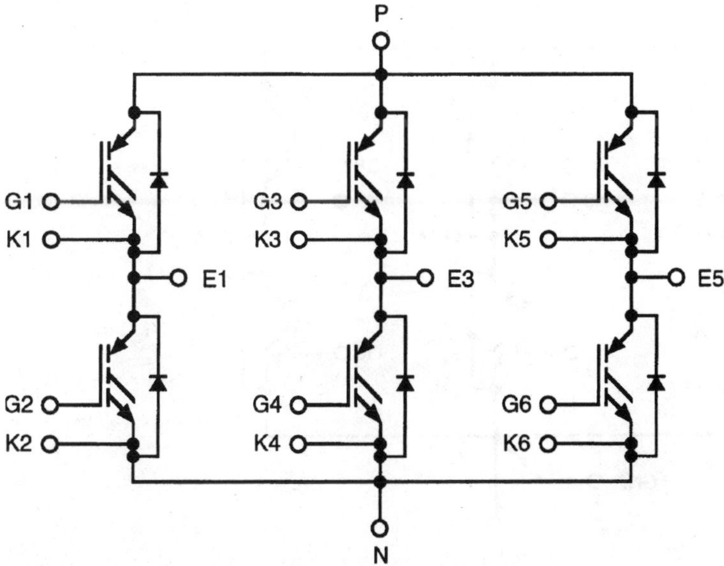

Figure 14.2 Six-pack module with emitter Kelvins.

ter inductance can be minimized by making use of this terminal. The larger currents switched by the power devices can create large voltage transients in the stray emitter inductance. The emitter-Kelvin terminals are connected as close as possible to the actual die. A true Kelvin terminal will have a separate wire connected to the emitter metalization. Grounding the gate drive at the emitter-Kelvin terminals will minimize noise and ringing on the gate drive. Optimum performance requires six floating gate drives. *The Kelvin terminals are not intended to be used for power connections; their internal wire bonds and conductors are not designed to handle high currents.*

Another configuration that is useful for slightly lower power ranges is shown in Fig. 14.3. Systems that do not use an isolated gate drive cannot benefit from separate low-side Kelvins. The circuit in Fig. 14.3 incorporates a star ground concept. The three emitters are connected together internally, and two emitter connections are brought out of the package—one for the high-current power connection and a separate connection for the gate drive ground. This will eliminate some of the inductance from the gate drive loop, specifically, the inductance in the package leads. However, it does not eliminate the wire bond inductance from the gate drive loop. Again, the gate drive emitter connection is not intended for use with high currents.

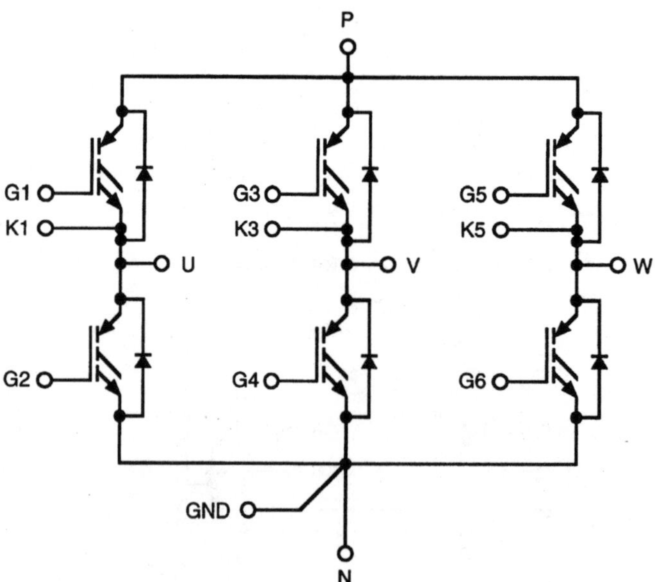

Figure 14.3 Six-pack module with star ground.

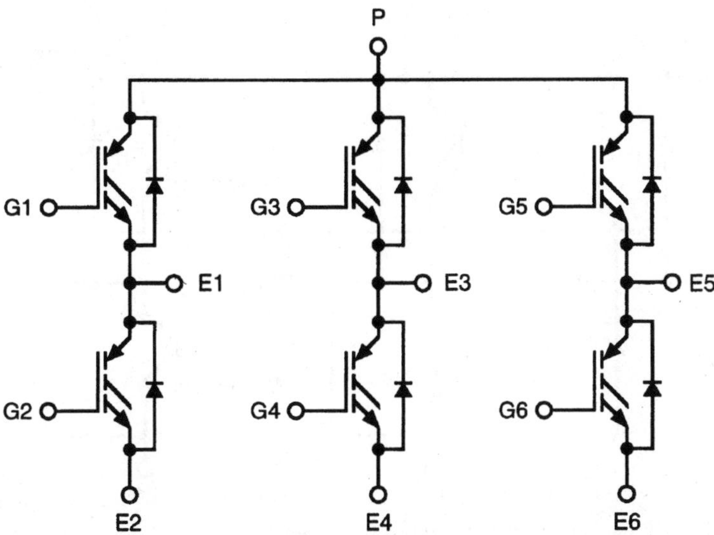

Figure 14.4 Six-pack module with separate emitters.

Another available six-pack configuration is shown in Fig. 14.4. This configuration has separate emitters for lower three transistors, enabling the current to be sensed in each of the lower transistors. A conventional six-pack with the emitters connected together allows only the DC bus to be sensed. This module with separate emitters allows some flexibility in the connection of the gate drive ground. A star ground point can be used by connecting the three emitters together on the printed-circuit board. The gate drive ground connections may also be placed in different locations to accommodate different current sensing schemes. The gate drive grounding is left to the PCB designer and is not fixed in the package.

A switched reluctance motor requires a slightly different type of six-pack. A three-phase bridge for a switched reluctance motor is shown in Fig. 14.5. The phase windings of a switched reluctance motor are normally connected between the upper and lower transistors. The SR motor windings are energized by turning on both the upper and lower transistors at the same time. The diodes are then connected to allow the motor voltage to reverse and clamp the voltage to the positive and negative bus. The anode of the upper diode is connected to the collector of the lower transistor, and the cathode is connected to the positive bus. Thus, the upper diode protects the lower transistor from excessive voltage. Correspondingly, the lower diode protects the upper transistor. (*Editor's note:* The SR motor's power stage does not suffer from the short-circuit failure mode of a half H-

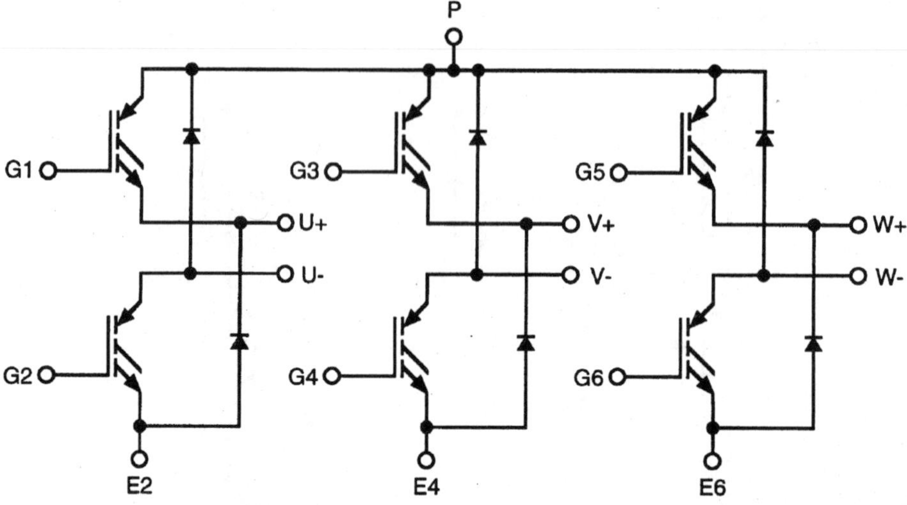

Figure 14.5 Six-pack module for switched reluctance motors.

bridge, as used in AC or BLM motors. The half H-bridge will fail when its upper and lower transistors are inadvertently turned on. The SR motor's windings are always in series with the power transistor. One possible worst-case event occurs when a short-to-ground takes place in the SR motor's winding.)

On careful examination, it is possible to use an SR-type six-pack for an AC induction or brushless DC motor (BLDC) motor. By connecting the upper emitter to the lower collector, one can obtain a simple six-pack. It has been suggested by proponents of SR motors that this configuration should be used for all six-packs. However, there are several reasons why this configuration is not optimum for AC and BLDC motors. First, the SR bridge requires three additional pins; pins add expense and size to the package. Second, the internal configuration for an SR bridge is more complex; the lower diode requires a separate isolated island and connection because the diode's cathode is not connected to the IGBT's collector. Third, the stray inductance is not optimized for AC and BLDC drives. Finally, the IGBTs are not as well protected from reverse voltages, due to the additional inductance in series with the free-wheeling diode; the loop inductance will be much higher since the current path must go through two additional pins. For these reasons, the SR bridge configuration remains a somewhat unusual configuration. The requirement for a special kind of bridge creates some resistance to the proliferation of SR drives.

Because the inductance is changing as the motor turns, switched reluctance motors often require current mode control. The separate

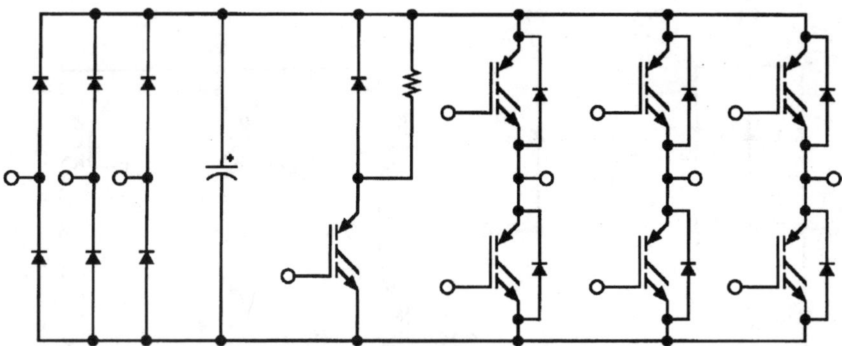

Figure 14.6 Three-phase motor power electronics.

emitter configuration permits current mode operation of each phase independently.

14.1.2 Integrated power stages

Figure 14.6 shows a basic circuit that is often used for industrial AC drive systems. This circuit illustrates all the power electronic components. There are a total of 20 power semiconductor devices in this AC drive system. Using single-transistor discrete packages is very cumbersome. A rectifier module and IGBT six-pack are often used with a discrete transistor for the brake. This simplifies matters a little, reducing the number of components that must be mounted to a heatsink and soldered.

An *integrated power stage* module is a comprehensive solution for small motor drives. A schematic of an integrated power stage is shown in Fig. 14.7. These modules contain all the power electronics for an AC drive. By combining all the power devices into a single package, the assembly of the power stage is greatly simplified. A single module, which has fewer parts than two modules, should reduce manufacturing cost and improve reliability. Small motor drives and integrated motors (motors with a drive integrated within the motor housing) benefit the most from a single power module that contains all the power devices.

Integrated power stages vary somewhat in their particular connections. The circuit in Fig. 14.7 is very flexible because it allows additional passive components to be placed between the rectifier, brake, and inverter. An integrated power stage might also incorporate separate emitters, emitter Kelvins, or a star ground for the inverter stage.

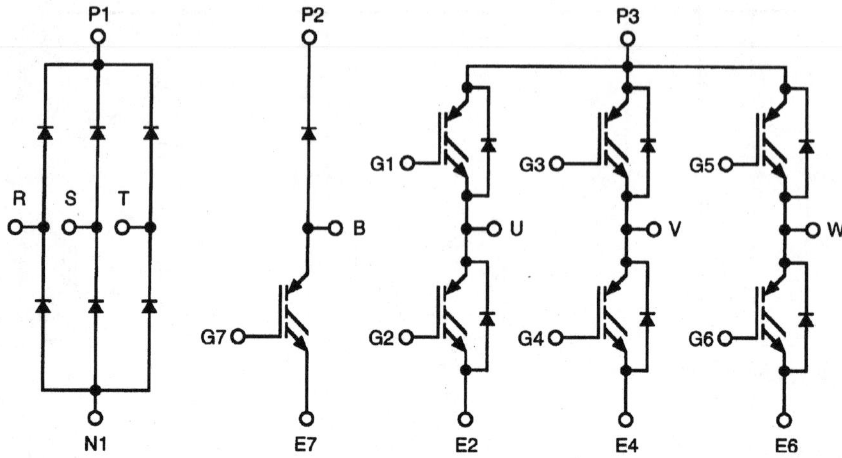

Figure 14.7 Integrated power stage for three-phase motor.

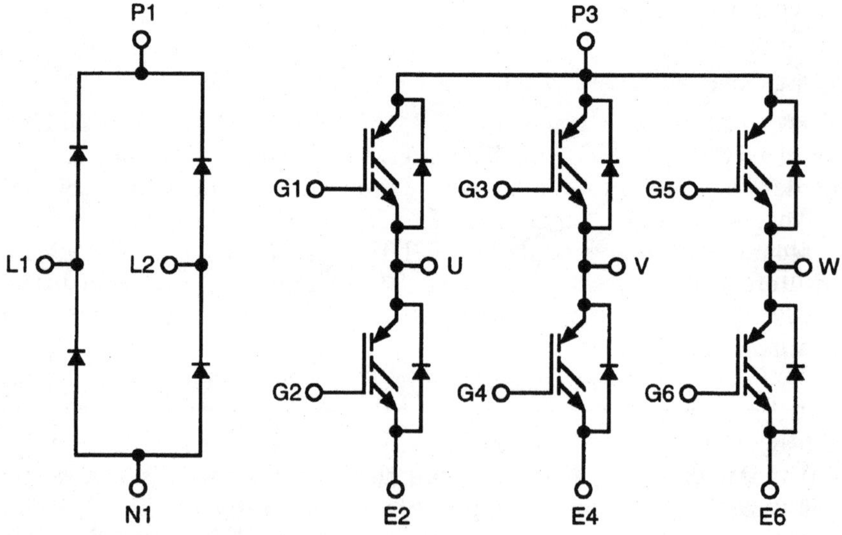

Figure 14.8 Integrated power stage for single-phase motors.

Integrated power stage modules might also include a sense resistor and a temperature sensor. These components can be used to provide status and control functions, as well as protection features.

Appliance applications and low end industrial applications often use a single-phase input bridge and do not require the brake transistor. A low-cost integrated power stage module for single-phase applications is shown in Fig. 14.8.

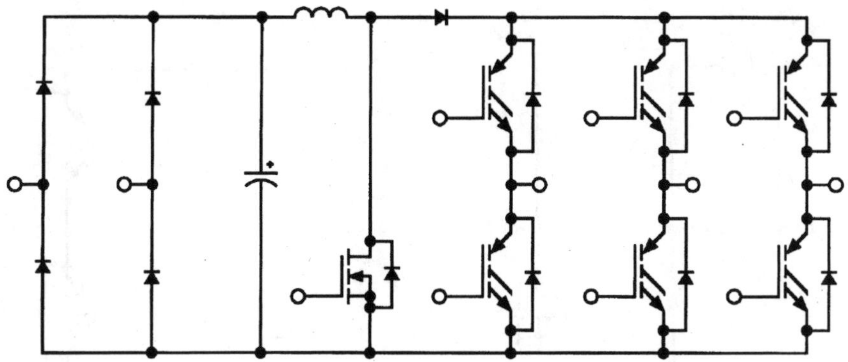

Figure 14.9 Rectifier, boost, and inverter topology for single-phase motors.

Another power topology that is useful for AC drives with single-phase input voltage is the rectifier, boost, inverter topology, shown in Fig. 14.9. This topology provides power factor correction (PFC), in addition to the rectification and inverter functions. The conventional single-phase input rectifier with a large DC capacitor results in short half-sine-wave input current pulses. Current flows only when the input voltage is higher than the DC capacitor voltage. The resulting input current has a waveform that is nonsinusoidal, a high RMS (root-mean-square) value, and a high level of harmonic content. Because the peak current is very high, the RMS value might be two or three times the motor output current. The resulting volt-ampere requirements are much higher. This requires heavier wiring and connections. The level of harmonics is also very high. Regulations that will limit the level of harmonics for electronic equipment, including motor drives, are, or soon will be, in effect. The term *power factor* is more appropriate when used to describe sinusoidal currents, such as the three-phase motor currents.

A conventional AC motor drive with a single-phase input rectifier might have good efficiency when measuring the real input power. However, the input volt-amperes and power quality are very poor. An AC motor drive with an active power factor correction circuit will have good efficiency, good power quality, and much lower volt-ampere requirements. The boost topology also provides the benefit of increased motor output voltage, improving the motor performance at higher speeds and better utilizing the energy storage capacitor and voltage capability of the power devices.

An integrated solution for the rectifier, boost, and inverter topology is shown in Fig. 14.10. The separate DC bus connections provide flexibility for current sense elements in the boost stage. The power

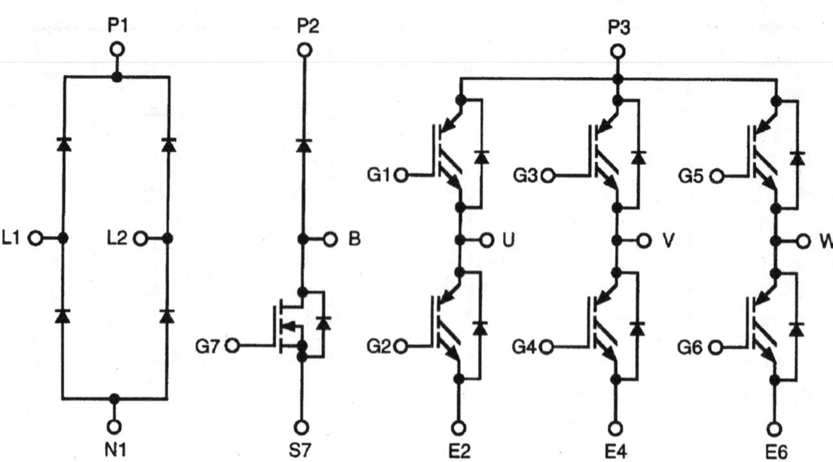

Figure 14.10 Integrated power module solution for PFC motor drive.

switch for the boost converter could be a large MOSFET of an ultrafast IGBT. The switching frequency required for the boost switch is much higher than that of the inverter switches. Higher operating frequency allows the use of a smaller inductor. The boost switch also must handle much higher peak currents. Continuous mode operation has the lowest peak current for high-power applications. Still, the current rating of the boost converter should be at least 2 times the rating of the inverter transistors. A power MOSFET provides the best high-frequency performance, but a very large MOSFET is needed. High-voltage power MOSFETs are practical for power levels up to about 300 W. A 750-W motor would require two 20-A high-voltage MOSFETs in parallel. Improvement in ultrafast IGBT technology will permit the IGBT to become a more cost-effective solution for 50- to 100-kHz operation.

The boost diode also must be a high-speed ultrafast diode. Continuous mode operation requires very low recovery losses in the diode. A diode with a reverse recovery time under 20 ns is suitable for operation up to 100 kHz. A slower diode could be used if the current were discontinuous, but this would require a larger boost transistor. (Note that the brake transistor of the integrated power stage shown in Fig. 14.7 is not generally suitable for a boost converter. It is usually too small and too slow.)

14.1.3 Dual-switch power module

Higher power motor drives often utilize dual modules. Dual modules contain two transistors and two free-wheeling diodes. Three dual modules are required to form a three-phase bridge for AC and BLDC motors. DC motors often use two dual modules to form a full H-bridge. Dual modules are also often called *half H-bridge modules,* since they are one half of a full H-bridge.

Dual modules can be optimized for inductance and will generally provide better switching performance than will two single-transistor modules in series. The flexibility and performance of dual modules have made them very popular over a wide power range. However, six-pack modules are being pushed to higher power levels and now often replace half H-bridge modules. Nevertheless, dual modules will continue to be used in the power range between six-packs and single-switch modules.

The schematic of a high-power dual module is shown in Fig. 14.11. Most dual modules have Kelvin contacts for the both upper and lower transistors. A system using dual modules should use six floating gate drives. Each of the six separate gate drive circuits should be grounded to its respective Kelvin terminal. If three dual modules are bolted together, there will be too much inductance in the emitter connections to obtain a clean gate drive signal. Dual modules also often include a collector sense for the top transistor, which can be used for voltage monitoring and desaturation protection. The emitter-Kelvin pin of the upper device can connect to a desaturation protection circuit for the lower device.

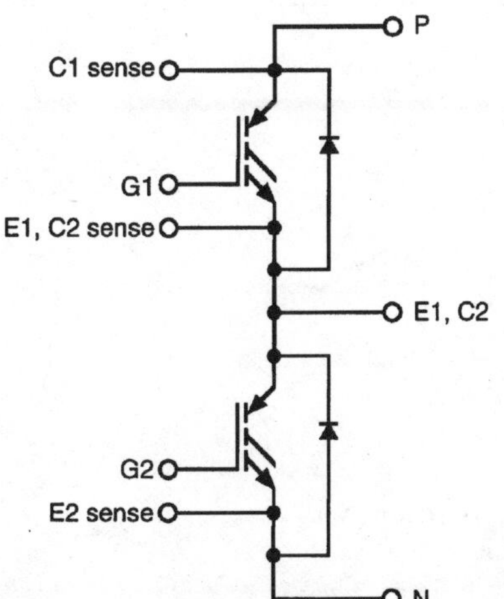

Figure 14.11 Dual-power-stage module (half H-bridge).

14.1.4 Single-switch power module

A single-switch power module is the most simple configuration possible. It is usually composed of a transistor and a free-wheeling diode in a common package. A conventional IGBT does not have a built-in free-wheeling diode. Thus, a single-switch power module is usually composed of two or more die—the power transistor and the free-wheeling diode.

At the upper end of the power spectrum, single-transistor modules are used for very high currents. Several factors limit the size of power modules. The defect density of silicon limits the size of IGBT die that can be produced with acceptable yields. The size of available ceramics is also limited. Furthermore, the stresses that occur as a result of the mismatch in thermal expansion of materials (die, solder layers, ceramic substrate, etc.) increases with the area.

When paralleling power modules, it is advisable to use low-inductance single-transistor modules. Conceivably, duals or even six-packs may also be paralleled. However, a low-inductance single-transistor module will provide the best performance and reliability. When paralleling IGBTs, it is crucial to minimize the parasitic inductance between the die that are paralleled, particularly the source inductance. IGBTs are voltage-controlled devices. As such, they are very sensitive to the voltages generated by stray inductance. When switching high currents, a few nanohenries of stray inductance can generate large voltages. For example, switching 1000 A in 200 ns results in 5 V for every nanohenry of stray inductance. Ten nanohenries can generate enough voltage (50 V) to damage the IGBT gate.

The schematic for a high-power single-transistor IGBT module is shown in Fig. 14.12. High-power singles have large screw terminals

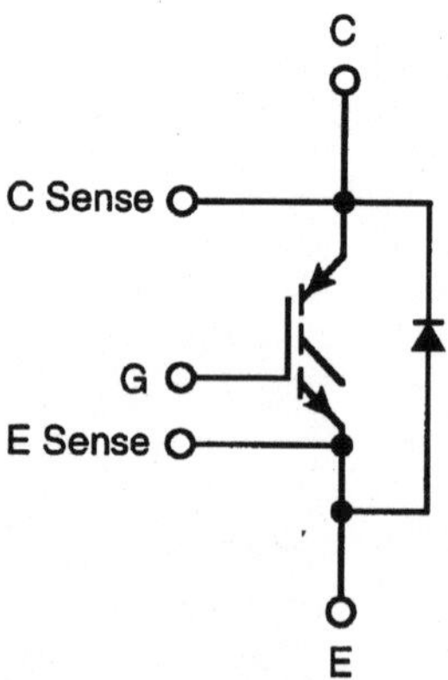

Figure 14.12 Single-IGBT and FWD power module.

for the collector and emitter. These power modules have an emitter-Kelvin or emitter return for the gate drive. Some singles also have a collector sense output, which is a small terminal located near the gate and Kelvin terminals that can be used for voltage monitoring and desaturation protection.

High-current transistors are normally composed of more than one die. Certain precautions are necessary when paralleling multiple die to ensure proper switching. The IGBT gate capacitance and the emitter inductor form a resonant *LC* circuit. The gain of the IGBT amplifies any resonant voltages and generates feedback, which can result in underdamped oscillations during switching. Conventionally, gate ballast resistors are used in series with each gate, helping to prevent oscillations and possible damage to the gate. The other important consideration is deciding how to connect the gate drive to the IGBT emitters. The conventional approach is to connect the gate drive to a point in the ceramic layout where all the IGBT emitters converge. This is difficult when the paralleled transistors are on separate ceramics. One method, in addition to the gate resistor, is to use a small resistor in series with an emitter Kelvin for each IGBT. A detailed internal circuit schematic of a single module is shown in Fig. 14.13. These emitter ballast resistors provide damping to the gate circuit and help minimize oscillations. They effectively remove the large emitter current from the path of the gate drive and also eliminate the

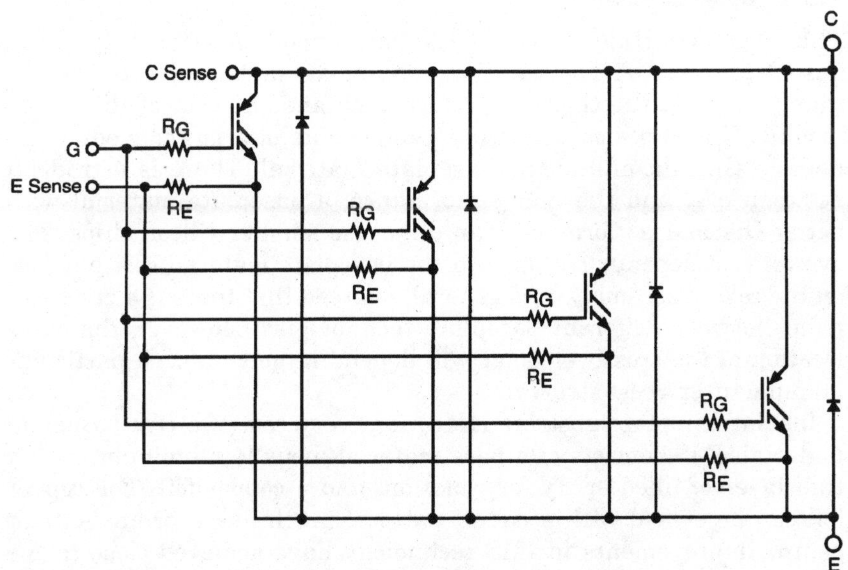

Figure 14.13 Circuit for high-power single-switch multiple-die module.

problem of deciding where to place the emitter contact. This approach allows multiple ceramics to be easily paralleled. The emitter resistors also permit the single modules to be easily paralleled. The gate drive gate and emitter terminals for two or more modules can be simply bused together. The internal gate and emitter ballast resistors provide sufficient damping for any number of power modules in parallel. (*Note:* A patent on paralleling multiple die in this fashion in a power module is held by Motorola, U.S. patent 5,373,201.)

14.2 Mechanical Construction

Power modules can be categorized by the power level and package technology. The package technology includes the baseplate material, assembly process, housing, and lead construction. There are a wide variety of package technologies available. Each package technology is most effective over a particular power range. Power modules are available over several orders of magnitude, from a few watts of output power to over a megawatt of output power. No single technology is suitable for the whole range of power levels. Industrial drive manufactures often offer a series of drives from less than 0.75 kW (1 hp) to 400 kW (500 hp). This range of power normally requires two or more package technologies.

14.2.1 Baseplate materials

The most important aspect of the power module technology is the baseplate material. The selection of baseplate material has a predominant effect on the thermal performance and the cost of the power module. The two most expensive components of a power module are usually the silicon and the baseplate material. There is a tradeoff between cost and thermal performance; a baseplate material with better thermal performance can often use smaller silicon chips. The overall cost depends not only on the baseplate material cost but also on the cost of assembly. It is generally agreed that there is a crossover point between different baseplate technologies. However, the exact location of the crossover point will depend largely on each particular manufacturer's cost structure.

Insulated metal substrate (IMS) is a very cost-effective baseplate material. IMS consists of a base material, usually aluminum, a very thin layer of filled epoxy for isolation, and a copper foil. The copper foil can be etched with a circuit pattern, much like a printed-circuit board. Improvements in IMS technology have achieved close to the same level of performance as ceramic direct-bonded copper (DBC).

The epoxy insulation is actually filled with ceramic to improve its thermal properties.

There are some significant differences between processing IMS and conventional copper-clad FR4. Conventional PCBs are processed in large panels with multiple circuits that are singulated using a router. IMS must be stamped to the final shape with a stamping tool that entails a significant capital expense. The stamping operation can affect the baseplate bow and edge quality and might possibly damage the insulation integrity. The etching process must be optimized for IMS. Special care must be taken to etch the copper without damage to the insulation or aluminum baseplate.

IMS is available with a copper thickness of 35 to 107 μm. It is also available with different baseplate thicknesses in aluminum or copper. Copper provides slightly better thermal conductivity and heat spreading. However, the judicious use of heat spreaders under the die is more effective for improving the overall thermal performance.

Molybdenum or copper heat spreaders are often used with IGBT devices. Since the copper foil is very thin, without heat spreaders, the heat flows mostly vertically. A thick copper heat spreader with a thickness of about 1 mm or greater provides the best thermal performance. However, the mismatch in thermal expansion between copper and silicon will have a deleterious effect on reliability. Molybdenum (Mo) is much closer to the coefficient of thermal expansion of silicon. However, the thermal conductivity of Mo is less than that of copper. A thin molybdenum spreader will provide good reliability and a moderate improvement in thermal performance. Small die benefit the most from heat spreaders. Large die are best used with Mo heat spreaders to reduce mechanical stress.

Higher-power modules often use direct-bonded copper technology. Thick copper foil (300 μm) is bonded directly to the ceramic in a furnace at very high temperature. The silicon die can be soldered directly to the copper. The DBC ceramic is then soldered to a thick copper plate (3 mm). The die attach process and the baseplate attach process might be done in either order, depending on the manufacturing process flow. Two different solders may be used with different melting temperatures. It is better to attach the die first with a higher-temperature solder, if possible.

Some modules are constructed using DBC without a copper baseplate. The copper on the backside of the DBC is exposed and makes direct contact with the heatsink. The plastic housing is used to press the DBC against the heatsink. The thermal performance and mounting of this kind of module are generally inferior to those of the DBC/copper modules. The copper baseplate serves two functions. It furnishes some heat spreading between the ceramic and the contact interface, and it provides a controlled baseplate bow to ensure intimate contact with the heatsink.

Ideally, the baseplate should have a positive convex bow of 5 to 10 μm/cm (0.5 to 1.0 mils/in). This will ensure that the module is in good contact with the heatsink over most of the area. A module with the proper bow should spin freely on a perfectly flat surface. If a power module is mounted using silicon grease and then subsequently removed, the contact area will normally be visible on the power module surface. A perfectly flat surface is not desirable, because it will become concave under mounting pressure—the mounting screws create a bending moment that will tend to bend the baseplate into a concave shape. A module with a slight convex bow will flatten out when properly mounted.

The baseplate bow is determined by the initial shape of the baseplate and the coefficient of thermal expansion (CTE) of the various materials when heated and cooled. The CTE is also very important in the reliability of the power module. A gross CTE mismatch in the power module's various materials results in shearing stress at the interface between these materials. This can cause voids in the solder, delamination of the DBC, or actual cracking in the die or the ceramic. The silicon die and ceramic are relatively fragile materials, similar to glass, which are bonded to heavy copper metal. The CTE and other properties of different materials used in power modules are shown in Table 14.1.

Two ceramics that are often utilized in semiconductor modules are alumina (Al_2O_3) and aluminum nitride (AlN). Aluminum nitride has a higher thermal conductivity. However, the CTE is very different from

TABLE 14.1 Properties of Power Module Baseplate Materials

Material	Density, g/cm^3	CTE, 10^{-6} K^{-1}	Thermal Conductivity, W/(m · K)	Young's modulus, GPa
SiC/Al	3	6.7	180	260
Al	2.7	27	250	62
Cu	9	17	400	140
Mo	10	5.1	160	320
AlN	3.3	4.5	180	330
Al_2O_3	3.6	6.7	17	380
Si	2.3	4.2	150	
GaAs	5.3	6.5	54	

the CTE of copper, which makes the reliability of AlN DBC much lower than that of Al_2O_3.

Silicon carbide metal matrix composites are a new materials technology with applications for power semiconductor modules. Silicon carbide (SiC) is a porous ceramic material that can be infiltrated by a metal-like aluminum. The resulting composite has a CTE that can vary depending on the relative ratio of the materials. The CTE can be adjusted to closely match the CTE of silicon or ceramics. SiC/Al can be used with AlN ceramic without some of the difficulties of copper baseplate modules.

Metal matrix composites can also be molded into complex shapes. This can make it more malleable in manufacturing, allowing new innovative design concepts. Metal matrix composites can be employed in electric vehicle modules to provide higher reliability and to allow integrated liquid cooling.

14.2.2 Examples of module types

A versatile low-power module for motors up to 7.5 kW (10 hp) is shown in Fig. 14.14. The package technology uses an insulated metal substrate (IMS) baseplate with molded in leads and a single reflow process. The package has 16 pins with uniform pin spacing and can be used with a variety of internal configurations. The standard configuration for this low-power module is the six-pack with separate emitters, as shown in Fig. 14.4. However, any of the six-pack configurations shown in Fig. 14.1 through Fig. 14.5 could be integrated in this package. In addition, the single-phase power module shown in Fig. 14.8 could also be integrated in this package. The circuit configuration is limited only by the number of pins and the internal area of the IMS substrate. A larger package with 24 pins is also available, as shown in Fig. 14.14. This package can also be configured with the more complex integrated power stage circuits previously shown in Fig. 14.7 and Fig. 14.10.

These power modules use a housing with molded-in leads. A cross section of one of these modules is shown in Fig. 14.15. A leadframe is inserted into an injection molding tool before the plastic is injected. The leadframe is then trimmed with a progressive die trim tool. The final housing provides mechanical support for the leads. Aluminum wirebonds connect the leads directly to the die or to the copper traces on the IMS. The wirebonding must be done after the housing is attached. The die and wirebonds are covered with silicone gel and sealed with a plastic cover.

A medium-power module for motor drive applications is shown in Fig. 14.16. This power module is specifically designed for the six-pack

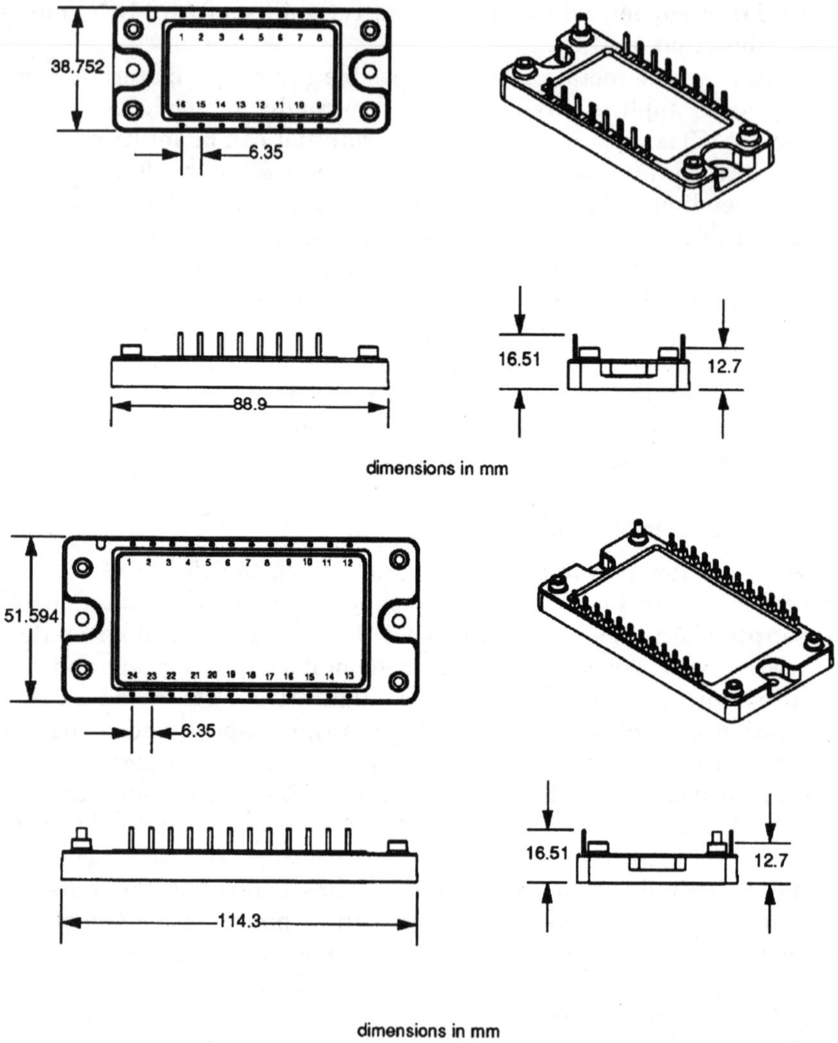

dimensions in mm

dimensions in mm

Figure 14.14 Sixteen- and twenty-four pin versatile low-power modules.

configuration with emitter Kelvins, previously shown in Fig. 14.2. This module is becoming a popular standard in Europe. It also uses molded-in leads for the pin connections. The baseplate is either IMS or DBC/copper. The IMS module uses a single-reflow process. The DBC/copper baseplate uses a two-reflow process. The wirebonding is done after the housing is attached in both cases.

A very large 1200-A 1200-V high-power single module is shown in Fig. 14.17. This high-power module uses DBC ceramics soldered to a

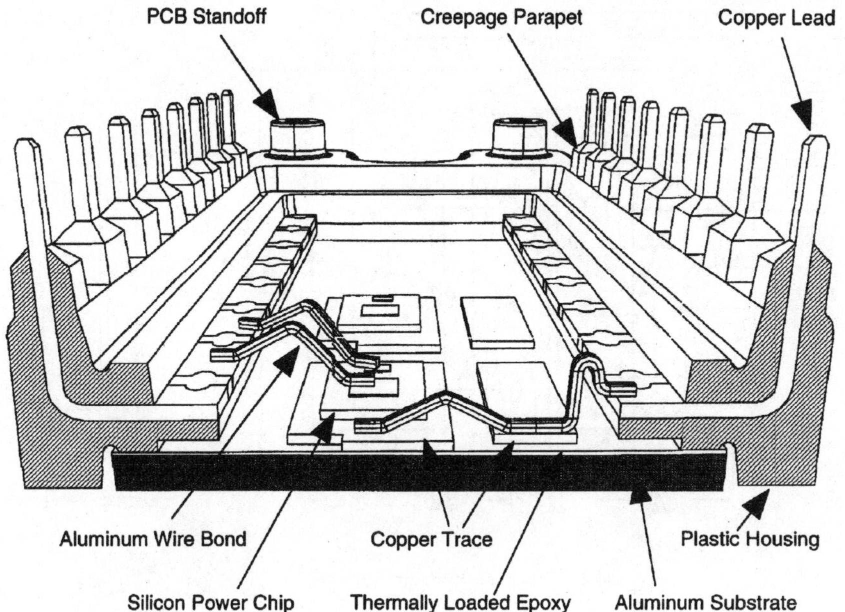

Figure 14.15 Molded-in lead module housing.

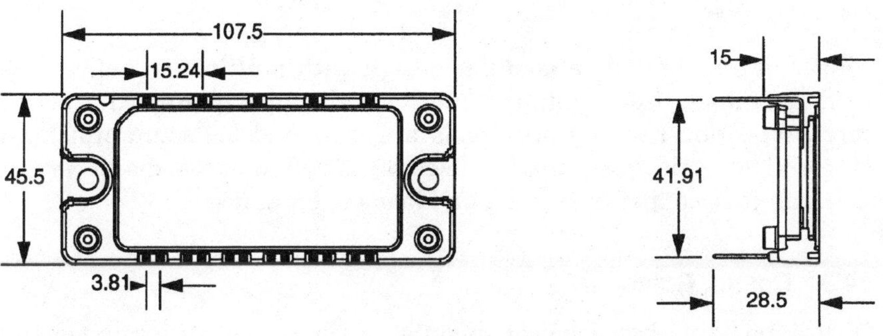

Figure 14.16 Medium-power six-pack module.

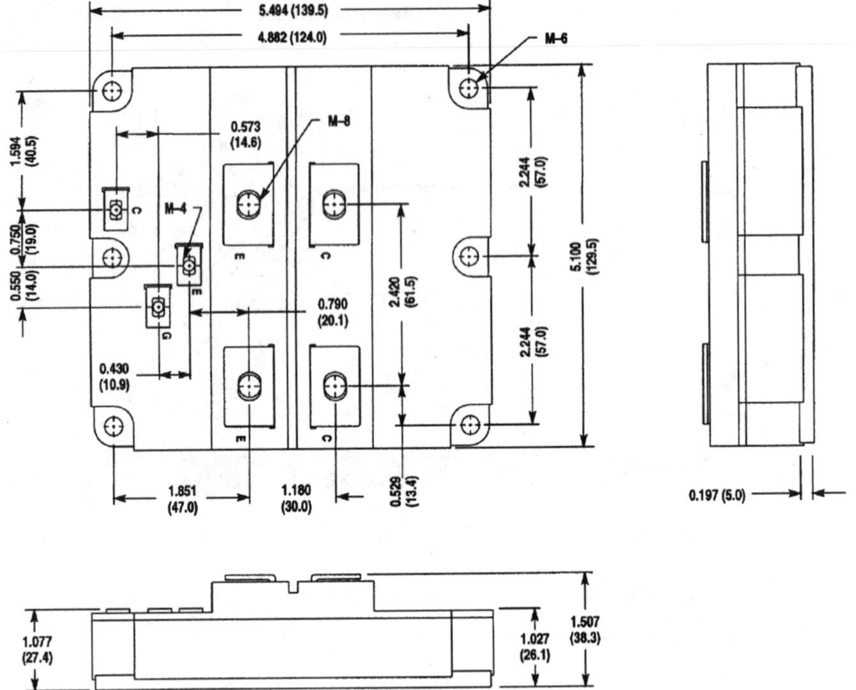

Figure 14.17 High-power module.

copper baseplate. It is a modular design with multiple ceramics. The leads are made from stamped and formed copper plates with screw terminals. The housing and leads are designed for semiautomated single-axis vertical assembly. It uses three reflow steps: die to ceramic, ceramic to copper baseplate, and leads to ceramic.

14.3 Design Rules

It is important that a power module and its associated components are designed not only to meet the appropriate standards but also to allow for manufacturing tolerances, field repairs, and other events not covered in the standards.

14.3.1 High-voltage design rules

Power modules are designed for particular voltage levels. The voltage levels affect several areas of power module design. Coordination of isolation insulation and spacing is important in order to ensure reliability and safety. There are two different types of isolation required:

functional and safety. *Functional isolation* refers to the voltages that occur between the active device terminals. Functional isolation ensures that the circuit will function reliably over the life of the product. *Safety isolation* refers to isolation between active high-voltage terminals and the end-product chassis or the low voltage user interface circuitry.

14.3.2 Safety isolation

Generally, safety isolation applies to the area between the power terminals and the baseplate of the power module. This is assuming that the baseplate is connected to the end-product chassis or an exposed heatsink. If the heatsink is completely enclosed where it cannot be touched by human hands or small pointed objects, the safety isolation requirements may not apply.

The requirements for safe isolation between the power terminals and the baseplate are the most stringent. There is a specific standard that applies to power modules from the Underwriters Laboratory (UL), which is UL1550. Most modules are tested for 2500-V or 3750-V-AC RMS. All the control terminals are connected together and grounded. The high-tension voltage is usually applied to the baseplate. This kind of testing is often called "high-pot" testing for high potential. The 2500-V-AC rating is usually specified for one minute. Actually, the modules are usually tested at 120 percent of the rated voltage for one second with controlled rampup and rampdown times. It would not be practical to test the modules for one minute on a high-volume manufacturing line.

The UL standard for power modules specifies only the test voltages. The required spacing between terminals usually depends on the use of the final product. Industrial equipment manufacturers have long used the UL508 standard. This standard has recently been updated to use the international standardized spacings listed in UL840 and the German Institute of Electrical Engineers (VDE) VDE0160 standard.

14.3.3 Functional isolation

Functional isolation is more concerned with the reliability of the product. The required spacing is smaller when using a printed-circuit board. The spacing depends on the environment in which the device will be used. The standards use what is called a *pollution degree* to describe the environment. Pollution degree has very specific requirements for potting or conformal coating.

TABLE 14.2 Creapage and Clearance Distances for Electrical Isolation

Voltage, V	Creapage, mm	Clearance, mm
10	1	—
50	1.9	0.8
63	2	—
80	2.1	—
100	2.2	0.8
125	2.4	—
150	—	1.5
160	2.5	—
200	3.2	—
250	4.0	—
300	—	3
320	5.0	—
400	6.3	—
500	8	—
600	—	5.5
630	10	—
800	12.5	—
1000	16	8.0
1500	—	14.0

SOURCES: IEC664-1 Tables 1 and 2, Overvoltage Category III, Pollution Degree 3; Table 4, Other Equipment, Pollution Degree 3, Material Group IIIa; UL840 Table 5.1, Over Voltage Category III, Pollution Degree 3; Table 6.1, Pollution Degree 3, Material Group IIIa.

Clearance is the shortest distance through air between two objects. *Creapage* is the shortest distance along a surface between two objects such as a power device's terminals. Both clearance and creapage distances are affected by the package's terminal design. Table 14.2 lists the creapage and clearance distances required for functional isolation. These tables apply only for a specific pollution degree, overvoltage category, and end product. They may or may not apply to a specific application. (*Editor's note:* For complete and up-to-date information on electrical standards, consult the appropriate standards organization.)

For calculation purposes, the voltage value selected should be the voltage that is rated for the end equipment. The higher value of the AC RMS or DC voltage should be used; the peak or absolute maximum voltage value need not be used. The spacing between terminals is based on the effects of long-term aging on insulation. The spacing can normally support voltages much higher than those specified in the tables. The typical voltage ratings are 240 V and 480 V AC in the United States. In Europe the voltage ratings are typically 230 V and 400 V AC. These voltage ratings use 600- and 1200-V IGBTs, respectively. There are some higher voltage applications up to 600 V AC that require 1600-V IGBTs. Linear interpolation is allowed.

The UL standard UL508 has specific requirements for the field wiring terminals. The field wiring terminal clearance and creapage distances should be at least 12.7 mm (0.5 in), including recommended mounting hardware and the largest recommended solid wire.

The required spacing for PCBs is smaller than the spacing required for screw terminals. Ideally, the power module should be designed so that the circuit board may be easily configured to meet the required spacing standards. But, unfortunately, the standards for high-voltage spacing can be very confusing. A conservative approach would be to make the pin spacing very large. That, however, would make the module large and costly. A better approach is to develop consistent design rules and publish the methodology. Table 14.3 lists the required distances for functional isolation for PCBs.

14.3.4 Grouping and spacing

The actual spacing used depends on the voltage of the end product. Linear interpolation is allowed from the appropriate clearance or creapage tables. The highest AC or DC voltage should be used. A normal motor control system has several high voltage nodes. The different voltages between nodes form a matrix of voltages. This matrix can be simplified by grouping together nodes that are within 50 V of one another. The gate and emitter of an IGBT are essentially at the same voltage. The entire high-side drive circuit can be considered as the same voltage as the output phase. These high-voltage groups are sometimes called voltage "class" by computer engineering tools (such as the Cooper and Chyan router). The spacing between groups is then called the *class-to-class spacing*. The basic voltage classes for an AC inverter circuit are illustrated in Fig. 14.18. The voltages and spacing requirements between these classes are also shown for a 480-V-AC RMS drive. These matrixes are suitable for automated clearance checking, using class-to-class clearance rule definitions. This tech-

TABLE 14.3 PCB Clearance and Creapage Distances

Voltage, V	Pollution degree 1, mm	Pollution degree 2, mm
10	0.025	0.04
50	0.025	0.04
63	0.04	0.063
80	0.063	0.1
100	0.1	0.16
125	0.16	0.25
160	0.25	0.4
200	0.40	0.63
250	0.56	1
320	0.75	1.6
400	1.00	2
500	1.3	2.5
630	1.8	3.2
800	2.4	4
1000	3.2	5

SOURCES: IEC664-1 Table 4, PCBs, Pollution Degree 1, 2; UL840 Table 6.2, Pollution Degree 1, 2.

nique may be employed for any high-voltage circuit. Note that the spacing requirements between voltage classes are calculated using linear interpolation from Table 14.3.

These basic rules can be further simplified for manual checking by grouping together similar class-to-class entries. There are really only a few types of voltages—input to input, output to output, DC bus, and input or output to DC bus. These groups and spacings are summarized in Table 14.4.

14.3.5 Current requirements

The use of high-current modules also requires careful design. High-current modules (>50 A) usually use screw terminals and bus bars to accommodate the very high currents. Lower-current modules (<50 A) often use PCB mounting.

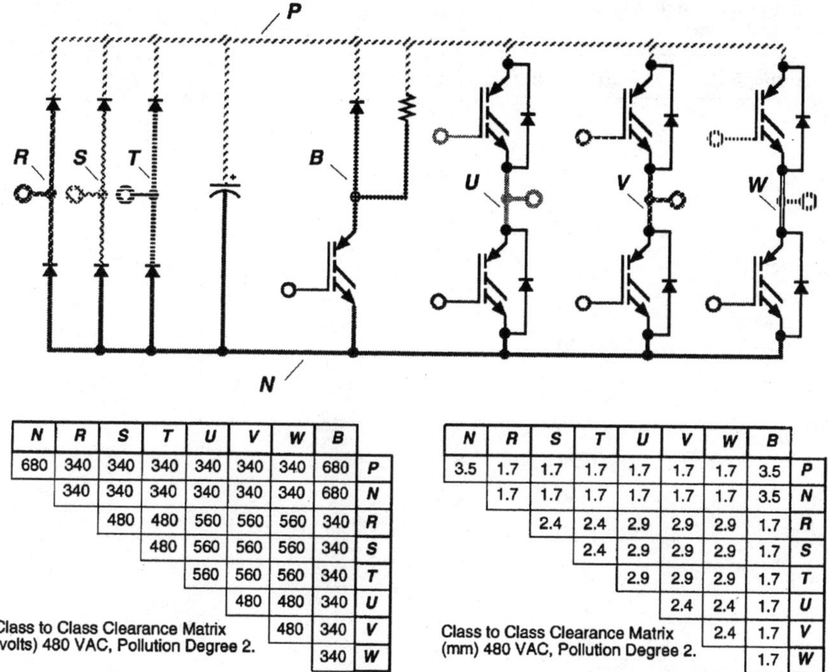

Class to Class Clearance Matrix (volts) 480 VAC, Pollution Degree 2.

N	R	S	T	U	V	W	B	
680	340	340	340	340	340	340	680	P
	340	340	340	340	340	340	680	N
		480	480	560	560	560	340	R
			480	560	560	560	340	S
				560	560	560	340	T
					480	480	340	U
						480	340	V
							340	W

Class to Class Clearance Matrix (mm) 480 VAC, Pollution Degree 2.

N	R	S	T	U	V	W	B	
3.5	1.7	1.7	1.7	1.7	1.7	1.7	3.5	P
	1.7	1.7	1.7	1.7	1.7	1.7	3.5	N
		2.4	2.4	2.9	2.9	2.9	1.7	R
			2.4	2.9	2.9	2.9	1.7	S
				2.9	2.9	2.9	1.7	T
					2.4	2.4	1.7	U
						2.4	1.7	V
							1.7	W

Figure 14.18 Inverter voltage classes and spacing requirements.

TABLE 14.4 Simplified Functional Spacing

Functional group	Voltage	Pollution degree 1, mm	Pollution degree 2, mm
1. DC bus	680 V DC	2.0	3.5
2. Input to output	560 V AC	1.6	2.9
3. Phase to phase	480 V AC	1.3	2.4
4. Phase to bus	340 V DC	0.9	1.7

Normal PCB materials can be utilized for high current levels if special design rules are followed for the track widths. The standard thickness of one ounce copper with one ounce plating (70 μm or 2.8 mils) can be used for up to 5 or 10 A. If possible, heavy copper plating is generally preferred. Two-sided boards using 2 oz starting material with 1 oz plating are fairly easy to manufacture. The total thickness

TABLE 14.5 Track Width versus Current

Current, A	Width,* mm
1	0.25
2	0.50
3	0.75
5	1.25
8	2.00
10	2.50
12	3.00
15	3.75
20	5.00
25	6.25
30	7.50
40	10.00
50	12.50
75	18.75
100	25.00

*70 μm (2 oz) copper.

is 107 μm (4.2 mils). When thick copper is used, the minimum PCB trace width, spacing, and pad ring width should all be increased by about 50 μm (2 mils). High-pin-count quad flat-pack (QFP) packages cannot be mounted on thick copper boards.

The track thickness depends on the acceptable temperature rise of the PCB. Some heat is conducted to the PCB by the leads of the package. However, this is not an effective thermal path to cool the semiconductors. The design rules summarized in Table 14.5 are based on a temperature rise of 10°C. Normal FR4 is rated at 85°C. High-temperature FR4 with a 105°C rating is better suited for high-power circuitry.

It is, of course, better to use printed-circuit tracks that are as wide as possible to conduct the current while maintaining the required high-voltage spacing. When possible, all available copper should be employed for the high-current tracks. Most PCB layout tools do not allow this level of freedom. A good method is to first route the board manually with the minimum track width and clearances and then increase the width of tracks where possible.

High-power circuit layout should be done manually. However, it is very useful to have automated design rule checking. Most PCB design tools are optimized for high-density logic board design. Only the very high-end workstation tools offer the advanced design rule checking required to automate high-voltage board design. Specifically, the class-to-class spacing rules and to/from track width rules can be used to check high-power boards. These advanced rule sets are moving down into the lower-cost PCB design tools.

Not everybody can afford the high-end workstation tools. One trick for PCB designers on a budget is to check for a particular clearance by printing out an oversized plot of the copper traces. Print out the solder mask with the oversize set to one-half the desired clearance, and then look for white space between the tracks of interest. Repeat for the input voltage, DC bus, and input to DC bus.

Summary

Standard power modules, the next level above discrete power devices, simplify the assembly of the various power stages used in a motor control. Integrated power stage modules combine different power stages into one module, greatly simplifying the manufacturing of a motor control. The utilization of a versatile power module package allows many different combinations of power devices for complex motor designs. Noteworthy points about this chapter are listed below.

- Power modules are available for most common motor drive designs: six-pack modules for three-phase BLM and induction motors, single-switch modules for DC motors, and, more recently, versatile modules for switched reluctance motors.
- Power modules simplify the design and assembly of the motor's power electronics. They offer higher reliability and less stray inductance, and are designed specifically for motor drive applications.
- The main disadvantages of power modules over discrete power devices are that the power module's initial price may be higher than a similarly rated discrete power device. If one part of the power module fails, the entire unit must be replaced, rather than an individual discrete unit. Also, the power module's footprint, while usually compact, may not adapt well to every specific application.

- The main problem to watch for when paralleling power modules is stray lead inductance. Single-switch modules normally have lower internal inductance than do dual or six-pack modules.
- The integrated power stage module contains some or all of the power devices required for a motor drive—line rectifiers, braking transistor, and inverter stage power transistors, as well as the free-wheeling diodes. A standard power module contains power devices that are configured for only one function.
- The choice of materials for a power module varies according to its power rating and application. The module's substrate should have a temperature coefficient similar to that for the silicon die in order to minimize thermal stresses. The baseplate and substrate thermal resistance affect the power dissipation capability of the module—a high thermal resistance value means higher junction temperatures.
- A perfectly flat baseplate may not make good thermal contact near its center—the area that is hottest. Ideally, the baseplate should have a positive convex bow of 5 to 10 μm/cm (0.5 to 1.0 mils/in).
- An easy method to test a power module's baseplate-to-heatsink interface is to apply a thin, even coat of white silicon grease to the module, bolt it down per the recommended torque, and then carefully remove the module. The area where the grease is thinnest (squeezed the most) is in closest contact with the heatsink. Ideally this should be in the center area of the power module.
- *Creapage* refers to the shortest surface distance between two electrical points. *Clearance* refers to the shortest distance through air between two electrical points.
- Some general rules to observe when designing PCBs that directly attach to power modules:

  A motor control power circuit board should use the largest width traces possible.

  Be sure the track-to-track spacing is adequate per the most recent standards for the country where the motor drive will be employed.

  Use high-temperature-rated PCB material (105°C or higher) and the thickest copper plating available.

Acknowledgments

The author would like to thank James M. Fusaro, Jeff Marvin, Memo Romero, and Pablo Rodriguez for their assistance with this chapter.

Further Reading

Motorola Inc., MHPM6B10A60D, 6-pack/16-pin hybrid power module; MHPM7A30E60DC3, 24-pin/3-phase input rectifier bridge and 3-phase IGBT + FWD inverter and brake and temperature sense, advance product data information in BR1480/D, rev. 2, *Silicon Solutions for Off Line Motor Drives,* 1997.

Toshiba Corporation, Semiconductor Group, *TOSHIBA Semiconductor Data Book—* GTR Module, 1992.

Chapter

15

Motor Electronics Thermal Management

Thomas Huettl and Pablo Rodriguez
Motorola Semiconductor Products

Introduction

In the previous chapters we have mentioned that the junction temperatures of the power device should never exceed their data sheet ratings. In this chapter we will focus on how to calculate the appropriate size for the heatsink in order to ensure that the heat from the power device is adequately removed. Thermal management of motor control inverter designs will be discussed as well. By reviewing different packaging and cooling techniques for semiconductor devices, a designer will have the confidence to select the appropriate thermal management system to meet an electronic motor control system's requirements.

To establish an appropriate compromise between performance, cost, and reliability, proper thermal management is essential in motor control system design. In particular, every designer should be aware that reliability can be affected when semiconductor devices are continuously operated at high junction temperatures. Service life and performance of a semiconductor device is a direct function of its junction temperature—reliability reports suggest that for every 10°C junction temperature increment above 100°C under continuous operation, the operating life of the device is halved. To ensure safe motor control operation, power devices must have their heat dissipated as quickly as possible. In the first section of this chapter, we will deal with the failure mechanisms that occur in high-power control circuits such as motor controls. In many cases, these failures are related to poor thermal management.

15.1 Failure Modes For Semiconductor Power Devices

There are three primary events or modes that can invoke a power semiconductor device to fail: (1) an electrical overstress during its normal operation, (2) high-peak-power cycling caused by low-duty-cycle PWM operation, and (3) extreme thermal cycling created by a repetition of wide temperature swings. While these modes interact

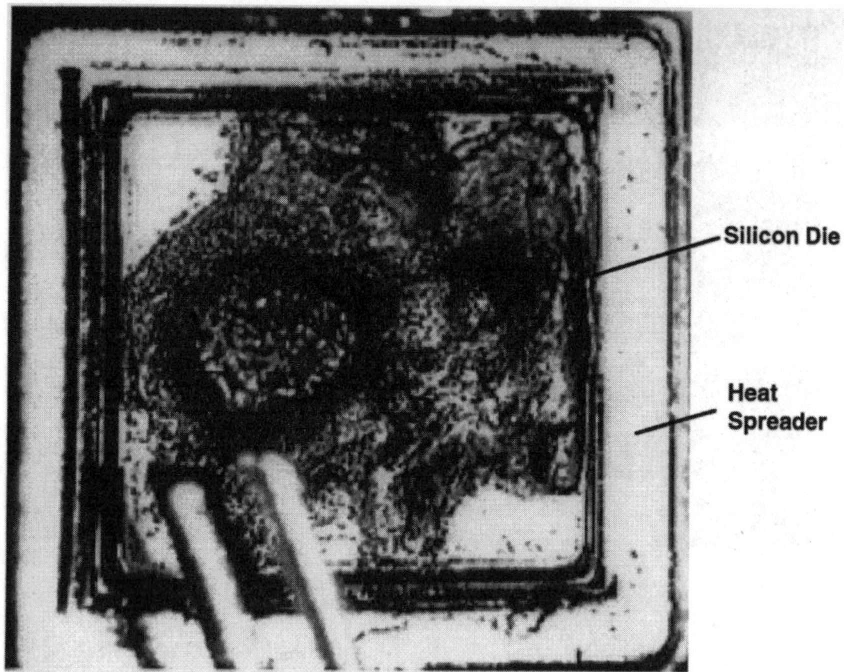

Figure 15.1 A serious power device failure.

extensively with one another, each exhibits a distinct (and usually observable) failure mechanism with the power device.

15.1.1 Electrical overstress failures

The characteristic mechanism that leads to ultimate destruction of a device lies with the electrical stress that it experiences. If the power device experiences a safe operating area (SOA) violation much beyond its ratings, failure will result, sometimes in a spectacular fashion. SOA violations include surge currents, voltage spikes, and other phenomena that may not have been expected in the normal operating environment. Figure 15.1 illustrates an extreme case in which a short circuit in the load literally obliterates the semiconductor device. Heat generated during such an event cannot be accommodated by the device or its external thermal management system—the sudden rise in junction temperature literally melts a hole in the die. (*Editor's note:* This is why current sensing and shutdown circuits are necessary to protect against such a failure. Normally, the power device can sur-

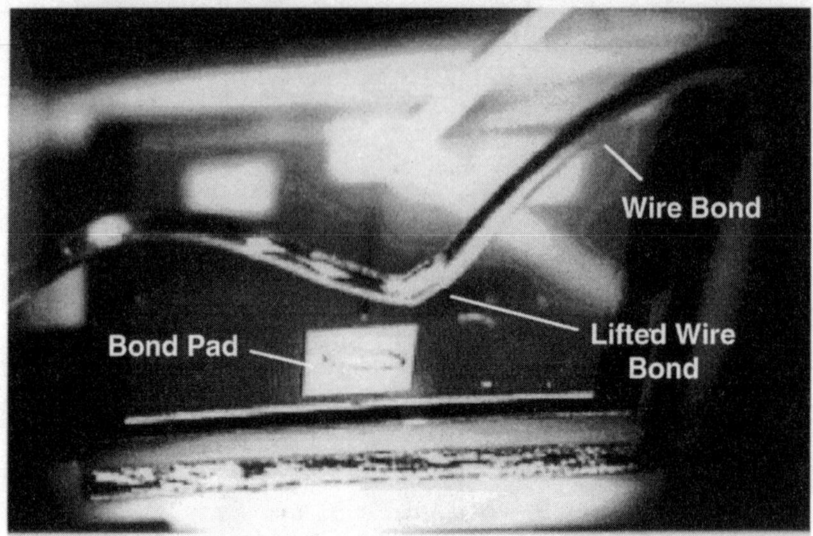

Figure 15.2 Wirebond lifting due to power cycling.

vive for up to 10 μs in a short circuit and will survive if latched off before 10 μs.)

15.1.2 Power cycling failures

The distinction between power cycling and thermal cycling failure modes is somewhat more subjective. At one extreme, power cycling primarily involves the mechanical and thermal stresses experienced by the power device in applications with small PWM duty cycle. Localized heating at the device junction leads to cyclic stresses that can lift the wirebond or crack the semiconductor device, as illustrated in Fig. 15.2. At about 190 to 230°C solder begins to reflow, which causes the bond wires to separate from the silicon die. When the bond wire lifts from the silicon die, arcing occurs, and the semiconductor surface is destroyed. It should be noted that, again, in the extreme case of small duty cycles, such a failure mode may not necessarily be accompanied by elevated case temperatures. Nevertheless, power cycling can be considered a thermal system failure because of the inability to remove the localized heat in the semiconductor die.

15.1.3 Thermal cycling failures

Thermal cycling failure is primarily invoked by large temperature variations of the semiconductor device. The variations may be caused by the device itself, from heat generated during normal operation, or by the environment. (Some applications, such as automotive motor controls, are sometimes subjected to −40°C when a vehicle is parked outside and then heat up to over 40°C when the vehicle is driven.) As the semiconductor package transitions from one temperature to another, the various materials used in the power device tend to expand (going from low to high temperatures) or contract (going from high to low temperatures). Problems arise from the different coefficient of thermal expansion (CTE) rates of the different materials. For a given temperature variation, a material will expand (or contract) an amount proportional to the temperature delta. The proportionality constant or CTE is typically unique to each material. As the power device experiences a temperature difference, the various materials forming the package will tend to grow (or contract) in different amounts, leading to stress concentration at the interfaces between the different materials. This mode of failure can be observed in cracked die or solder delamination, as illustrated in Fig. 15.3. The x-ray images show the same solder layer before thermal cycling and after. Note how the solder delaminated or separated from the edges of its ceramic substrate after only 16 thermal cycles, −55 to 125°C. This type of failure will raise the junction temperature of the device, because there is now less surface area to conduct heat from the die through the substrate and out to the heatsink.

The failure and stress modes can be summarized as follows. Typical power control applications will create various types of stresses on

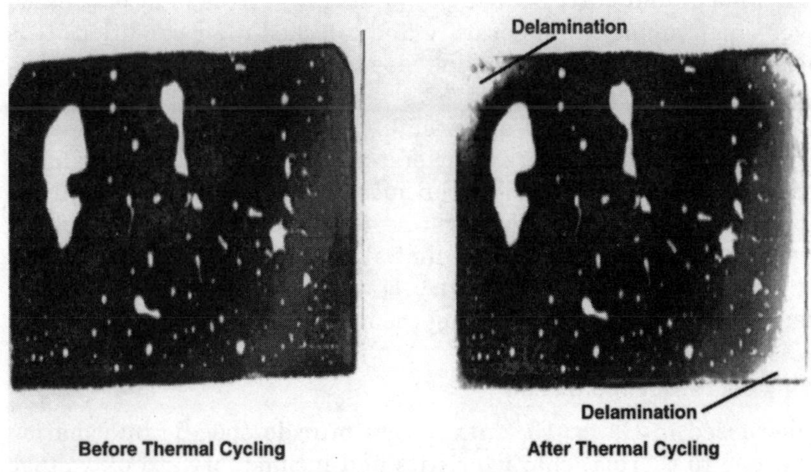

Figure 15.3 Solder bond failure due to thermal cycling.

power devices. Electrical stress (not necessarily at a destructive level), power cycling, and temperature cycling can cause a device to experience sudden or slow failure from any of the modes mentioned previously. The mode which ultimately causes the failure of the device depends on the application, SOA of the internal die, wirebonds, die attach materials, and package design. For a given application, the motor control designer may need to take precautions to ensure that these failure mechanisms are contained.

15.2 Thermal Requirements of a System

The thermal requirements of the motor's power device are usually considered after the motor's power stages have been initially defined. The design is set by the needs of the final application—they determine the power rating and the type of motor and electronics that are required. Every type of motor, whether it is AC induction, brushless DC, switched reluctance, or brush-type DC, requires a specific control strategy. The motor control strategy will affect the heat dissipation in the power stages and must be taken into account when contemplating thermal management. [*Editor's note:* Some AC motor control modulation methods can reduce power loss in the power stage by reducing the number of switching transitions. (See Chap. 8.) The type of power device used in a motor's power stage (IGBT, MOSFET, or BJT) will have the same basic thermal management requirements.]

For an electronic motor design, the thermal requirements are determined by several factors—locked-rotor condition, high power line voltage, and ambient operating temperatures, which affect the motor's maximum current. The power stage inverter is also stressed by very high ambient temperature and a locked-rotor condition. It is crucial to pick the correct heatsink size to ensure that the motor electronics will be reliable.

To determine the size of a heatsink, it is important to understand all the thermal parameters for worst-case conditions. The primary parameters which you need to consider are the power losses due to switching the semiconductor power device, conduction losses, and the maximum junction temperature limitations. Device data sheets provide valuable thermal and device characteristics, which must be taken into account when designing the motor's power stage.

15.2.1 Data sheet information

As discussed in Chap. 13, data sheets provide specific information pertaining to thermal characteristics and include forward ON voltage drops or forward ON resistances. It is important to use the data sheet ratings ($V_{DS,on}$, $R_{DS,on}$, etc.) that best represent the operating conditions in your motor control design. Some data sheets show ratings and characteristics for two or three different junction temperatures. Remember that junction temperature and gate bias voltage can directly affect the power dissipation of a power device. A MOSFET's forward ON resistance is especially sensitive to junction temperature. Both MOSFETs and IGBTs become less efficient when operated with less-than-normal gate bias voltages.

15.3 Thermal Models

An accurate thermal model of semiconductor device characteristics is helpful for designing a reliable system with a minimum amount of silicon. An understanding of how thermal models work is essential when designing an efficient heat removal system.

The cost of the power electronics may be reduced by using highly efficient thermal systems, which minimizes the size of the semiconductor devices. The cost of the thermal management system is also a factor. (*Editor's note:* A cost issue becomes evident when selecting a large heatsink and smaller power transistors or larger transistors and a smaller heatsink. With some applications, the heatsink size is limited by the motor drive's physical size, making this issue easier to resolve. In other cases, it may be useful to consider two approaches—select the largest power transistors available with a minimal heatsink, or select the smallest power transistors that will suffice for the motor's power ratings and use a large heatsink.)

15.3.1 Heat transference

Before we jump into the thermal calculations, it is useful to review some basics about heat removal. There are three methods, or modes, that allow the transfer of heat in a system: conduction, convection, and radiation. *Conduction* occurs by diffusion through solid material or in stationary liquids or gases. *Convection* involves the movement of a liquid or gas between two points. *Radiation* occurs through electromagnetic waves. *The interior of a semiconductor uses conduction, while a heatsink will use both radiated and convection methods.*

15.3.1.1 Steady-state condition. One method for modeling the thermal characteristics of a power device is accomplished during its steady-state condition. When power is applied to the semiconductor, the junction temperature will rise to a value based on the power dissipated at the junction and the ability of the device and its heatsink to conduct this heat away. A steady-state—or average—condition is

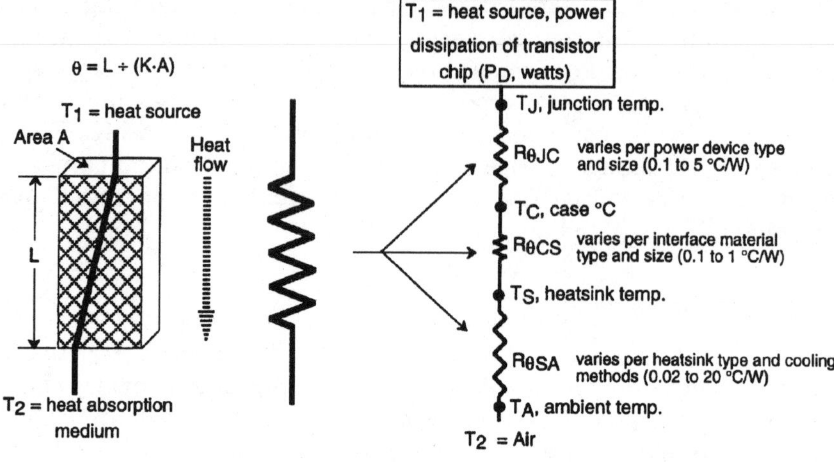

Figure 15.4 Basic heat conduction model and power stage model.

reached when the heat generated by the semiconductor power device equals the heat conducted away.

Internal thermal resistance values for semiconductor devices are a measurement of the ability of the semiconductor device, and its package, to remove heat away from the junction. These thermal resistance values can be expressed in an electrical analogy. Just as an electrical resistance is associated with the conduction of electricity, thermal resistance is associated with the conduction of heat. Figure 15.4 shows the heat conduction through a material slab, which, for explanatory purposes here, is the basic model used for thermal resistance encountered in a motor's power stage. The temperature distribution through the slab is linear and can be modeled as a resistance value. [*Editor's note:* The basic model's equation (shown in Fig. 15.4) indicates that the length or distance of the heat path can dominate the thermal resistance value, if it is larger than the area and the material's thermal conductivity constant *K*. To decrease thermal resistance, it is wise to minimize the length of the thermal path and increase its area; you want the thermal path to be as short as possible and its width as wide as possible. Normally, the motor power stage designer has little control over which materials semiconductor manufacturers have used in their power devices and modules. The designer does, however, have some choices regarding the heatsink and interface materials.]

Figure 15.5 illustrates the main components in the power stage thermal model. [*Editor's note:* There are several variables at work in

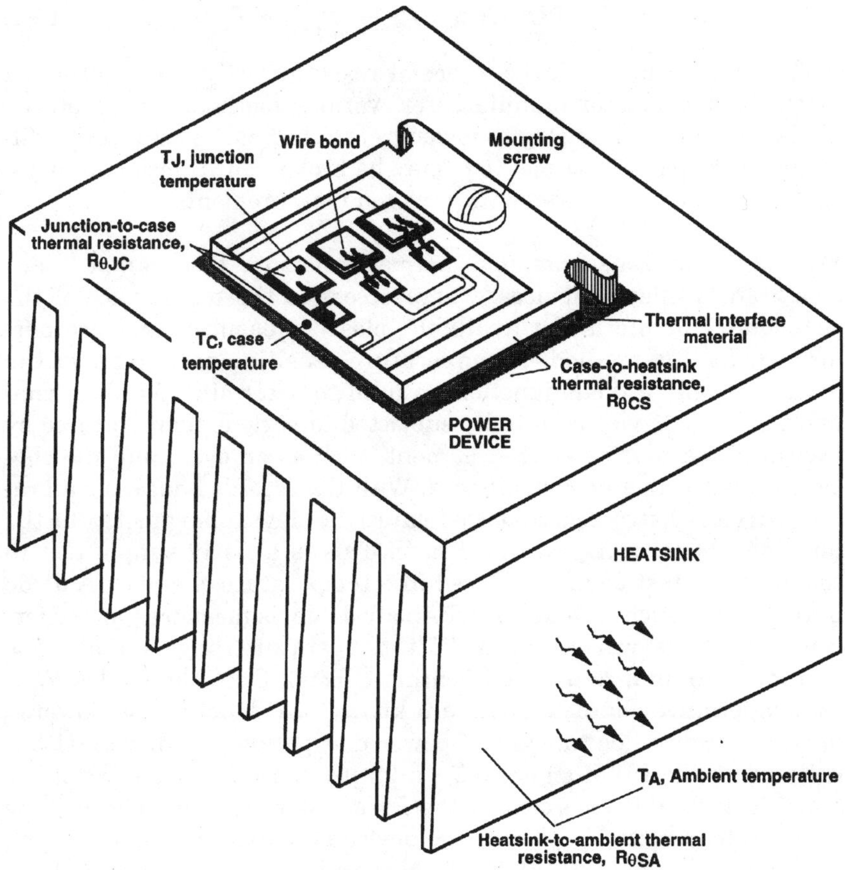

Figure 15.5 Main components of the power stage model.

this model. The power device's mounting pressure can vary according to its fastener's torque. This will change the thermal resistance between the power device's case and the heatsink. Some interface materials can vary their thermal resistance over time. The heatsink's thermal resistance (surface-to-air or liquid) will change per the air (or liquid) flow's velocity and turbulence. Finally, foremost among all these variables is the power dissipation of the power device; it will vary with the motor control's load and PWM modulation method.]

An approximation of the heatsink's thermal ratings can be made, as shown in Eq. (15.1), once the fundamental thermal resistance values of the power stage model and its maximum power dissipation are known.

$$T_j = \text{PD} \cdot (R_{\phi jc} + R_{\phi cs} + R_{\phi sa}) + T_a \tag{15.1}$$

The semiconductor device thermal resistance ($R_{\phi jc}$) is determined by the semiconductor manufacturer. Various methods are employed by the electronics industry to measure the thermal resistances of different packages. These methods may be broken down into two major categories: open measurements or closed measurements.

15.3.1.2 Closed-measurement thermal resistance test method. For small, encapsulated devices, where access to the top surface of the semiconductor die is not available, closed measurements are performed. In this method, a temperature-sensitive parameter of the device (usually a diode junction) is used to determine the die's temperature. The device is initially subjected to various temperatures by exposure to a particular environment, such as an oven, and the chosen device parameter is monitored. With the recorded data, a calibration curve is extracted that correlates the device parameter to the junction temperature. [*Editor's note:* This is a fairly simple test to conduct. The test device is placed in a temperature chamber. A diode in the device, such as a MOSFET's body diode, is measured for its forward ON voltage over a −55 to 175°C temperature range and at a low current (to minimize any self-heating effects). Once the diodes $V_{F,on}$-vs.-temperature characteristics are known, the MOSFET is powered up with a controlled amount of power dissipation and driven with a high-duty-cycle PWM signal (e.g., a 99 percent ON and 1 percent OFF cycle). During the 1 percent of the PWM OFF time, the body diode is again tested for its $V_{F,on}$. Since the device's die had little time to cool, the diode's $V_{F,on}$ measurement can be correlated to the previous $V_{F,on}$-vs.-temperature data.]

When determining thermal resistance values, you need to remember two important points during the calibration cycle: (1) heat generation due to the device itself should be minimized to prevent the introduction of error, and (2) the temperature of the entire package should be stabilized.

Thermal resistance of the package is performed by applying a given amount of power to the device and then observing the device parameter used in the calibration. The correlation between a given temperature and power is subsequently reduced. It should be noted that the closed-measurement technique will provide only bulk junction temperature data; localized temperature variations across the surface of the semiconductor die will not be accessible. Additionally, test conditions may have a substantial impact on derived values. When different published data from various sources are compared, a consistent test method should be applied to achieve a valid analysis.

15.3.1.3 Open-measurement thermal resistance test method. Open measurements are performed on larger devices in which heat conduction through a single direction in the package is the primary means of heat removal. In such a case, an open device is mounted on a test stand and allowed to generate a fixed amount of heat. Direct measurements on the surface of the die are performed by thermocouple or infrared camera measurements. While this test procedure is typically less sensitive to the test conditions, consistent test methods should still be preferred for making valid comparisons.

With both open and closed measurements, test conditions should include an environment that is as close as possible to the actual application. Substantial variations from published and measured data may be observed because of test conditions. Since the final device application will ultimately provide the operating environment of the device, final measurements of critical or marginal components are strongly encouraged.

15.3.1.4 Power device internal structures. Another factor in selecting a power semiconductor device is its internal makeup. Different power device structures are used to achieve improvements in thermal characteristics and to accommodate various design options for the physical layout of power devices. Figure 15.6 shows three of these structures currently in use for power devices. The internal junction-to-case thermal resistance value, however, is still calculated in the same way for each of the structures.

15.3.1.5 Case-to-heatsink thermal resistance calculation. Since junction-to-case thermal resistance is controlled by the power device manufacturer, its value is simply obtained from the data sheet ratings. The case-to-heatsink thermal resistance is, however, somewhat under control of the power stage designer. Case-to-heatsink thermal resistance ($R_{\phi cs}$) needs to be calculated according to the material chosen for the interface between the device and the heatsink. Most manufacturers of interface materials list the thermal attributes of their prod-

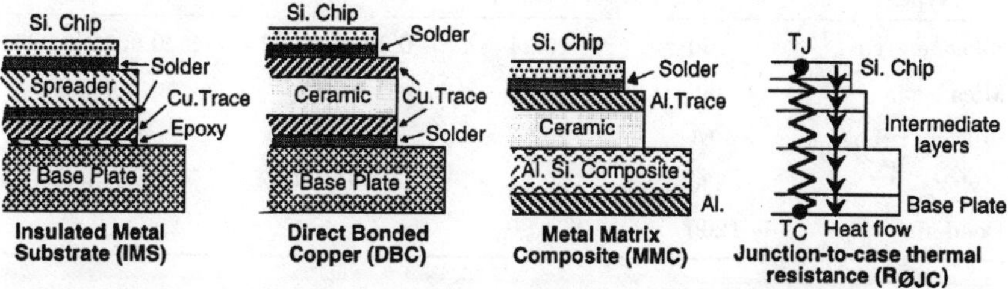

Figure 15.6 Cross sections of various power devices.

ucts. Some of those materials, such as silicone grease, are less exact, with the thermal resistance varying according to its application method.

The case-to-heatsink thermal resistance is dependent on the thermal resistance of the interface material ρ, average material thickness of the interface material *t*, and surface area *A* of the power device to be mounted. $R_{\phi cs}$ is fairly easy to calculate:

$$R_{\phi cs} = \rho \frac{t}{A} \tag{15.2}$$

where ρ = thermal resistance of the interface material, °C · in/W(°C · cm/W)
t = average material thickness
A = area of power device to be mounted on heatsink

An example of how to calculate $R_{\phi cs}$ using a typical motor control IGBT (a MGY40N60D, 40-A, 600-V IGBT from Motorola) is presented:

First, we must obtain some facts about this IGBT. It has a thermal resistance $R_{\phi jc}$ of 0.48°C/W, which we will use later for calculating $R_{\phi ja}$, and its TO-264 package's baseplate measures 28 × 19 mm (1.1 × 0.76 in), which gives a surface area of 540 mm^2 (0.84 in^2).

Now we need to know the thermal resistance ρ of the interface material. The interface material may also be required to electrically isolate the IGBT package's baseplate from the heatsink. This thermal resistance ρ value depends on the type of material chosen for the interface material. In this case, we will allow the heatsink to be electrically "hot" and just use silicone grease. Table 15.1 shows some of the values and approximate thickness of different materials used for power device-to-heatsink interfacing. From the table we can see that

TABLE 15.1 Interface Material Thermal Resistances

Material type	Resistance ρ, °C · in/W	Resistance ρ, °C · cm/W	Typical thickness	
			in	cm
Silicone grease	60	15.24	0.0005–0.002	0.0013–0.0051
Mica	66	16.76	0.002	0.00508
Silicone rubber	81	20.57	0.05–0.25	0.127–0.635
Mylar	236	59.94	0.0005–0.002	0.0013–0.0051
Dead air	1200	304.80		

silicone grease has a thermal resistance ρ equal to 15.24°C · cm/W (60°C · in/W). The silicone grease thickness is determined by its application and the amount of torque used to tighten the device's fastener to the heatsink. In this case, we will assume that the thickness t can vary from 0.0013 to 0.0051 cm (0.0005 to 0.002 in). (*Note: Always use the mounting torque parameters recommended by the device manufacturer to ensure good thermal conduction.*)

Therefore, using the formula in Eq. (15.2), we find that the $R_{\phi cs}$ will vary from 0.036 to 0.143°C/W, depending on the thickness of the silicone grease. [*Editor's note:* These 0.036 to 0.143°C/W values mean that for 100 W of IGBT power dissipation, there will be a junction-to-case temperature rise of 48°C (100 W times $R_{\phi jc}$ 0.48) and 3.6 to 14.3°C case-to-heatsink temperature rise, adding up to a junction temperature rise of 52 to 62°C—before we include the heatsink's thermal resistance. This indicates that, if the interface resistance becomes too large, it can significantly raise the IGBT's junction temperature. Silicone grease can be difficult to apply and may eventually dry out. There are some thermal conductive interface pads, commercially available, that offer similar performance and are less difficult to use in manufacturing. (Power Devices Inc., Laguna Hills, CA; Aavid, Laconia, NH; and Berquist, Edina, MN are only three of several companies that offer thermal pad products.)]

15.4 Software for Thermal Modeling

The total thermal management of the motor control's power stage can be calculated by using a spreadsheet on a personal desktop computer or with high-performance software running on a computer engineering workstation. To use either a spreadsheet or the high-performance thermal modeling program, you need sufficient parametric data about the motor's power stage components and some understanding of how the data interacts with the motor's operation.

15.5 Motor Power Stage Thermal Resistance

When you are designing a heatsink for a DC or AC electronic motor control, it is important to remember that your goal is to protect the power device or module from reaching a temperature that will damage it. One way to gain insight on how to accomplish this is by example, as discussed below.

15.5.1 Motor power stage thermal calculations

Let's say we need to determine the thermal requirements of a heatsink for a DC motor controller. (*Editor's note:* This example could apply to any motor, since the basic conditions are similar for most motor drives.) This 750-W (1-hp) DC motor is powered from 150 V DC and normally draws 5 A. Under locked-rotor conditions it can draw 40 A. The motor will be operated in an environment that may reach 40°C. The motor uses PWM for speed control, with only three speed settings, slow- (25 percent PWM), medium- (66 percent PWM), and full-speed (100 percent PWM).

The DC motor's power stage uses one MGY40N60D, 40-A, 600-V IGBT for the PWM power control and another MGY40N60D with its gate-to-emitter shorted out as the free-wheeling diode. Both devices are mounted on a heatsink whose thermal resistance needs to be calculated. The IGBT's maximum case temperature will be designed to 90°C.

To make this more interesting, we will use both the 5-A normal full-speed run current and the 40-A locked-rotor current values for calculating the appropriate heatsink thermal resistance value. We need to know one other IGBT parameter—its $V_{CE,on}$ at 5 and 40 A. The characteristics section in the data sheet lists a maximum $V_{DS,on}$ of 3.25 V at 40 A with a 15-V gate bias. The IGBT's $V_{DS,on}$ is about 1.5 V at 5 A according to the data sheet's output characteristics curves.

The first calculation that we need to determine is the IGBT's power dissipation. This involves basic Ohm's law, as shown in Eq. (15.3). Note that we are assuming that the highest-power IGBT dissipation for normal operation occurs at full speed with the PWM at 100 percent—no losses from PWM switching edges will be considered at this time. The worst-case IGBT power dissipation will be during a locked-rotor condition at 100 percent PWM.

$$\text{PD} = IE \tag{15.3}$$

$$\text{IGBT PD (normal motor operation) } 7.5 = 5 \cdot 1.5 \tag{15.4}$$

$$\text{IGBT PD (locked motor operation) } 130 = 40 \cdot 3.25 \tag{15.5}$$

where PD = IGBT power dissipation, W
I = IGBT current, A
E = IGBT forward ON voltage, $V_{CE,on}$, V

Using Eq. (15.4), we see that, during normal motor full-speed operation, the IGBT's PD is only 7.5 W, but during a locked-rotor condition it increases 14 times, to 130 W per Eq. (15.5). As we will soon see, 130 W of power dissipation will be a serious cost issue in this motor design.

The IGBT's junction-to-case temperature rise can be calculated once its PD is known. Using a PD of 7.5 or 130 W, we can calculate the IGBT's junction-to-case temperature rise per Eqs. (15.6) to (15.8). We call this a *rise,* since it rises above the case, heatsink, and ambient air temperatures.

$$T_j \text{ to } T_c = R_{\phi jc} \cdot \text{PD} \tag{15.6}$$

$$T_j \text{ to } T_c \text{ (normal operation)} \quad 3.6 = 0.48 \cdot 7.5 \tag{15.7}$$

$$T_j \text{ to } T_c \text{ (locked rotor)} \quad 62.4 = 0.48 \cdot 130 \tag{15.8}$$

where $R_{\phi jc}$ = IGBT junction-to-case thermal resistance, °C/W
T_j = IGBT junction temperature, °C
T_c = IGBT case temperature, °C
PD = IGBT power dissipation, W

We now know that the IGBT junction-to-case temperature will increase by 3.6°C during normal motor and by 62.4°C during a locked-rotor condition. We can calculate the interface material's temperature rise using the 0.143°C/W value obtained from Eq. (15.2) (since it is the same IGBT power device, and we are using the same silicone grease as an interface material). The interface material's temperature increase can be calculated according to the following equations:

$$T_c \text{ to } T_{hs} = R_{\phi cs} \cdot \text{PD} \tag{15.9}$$

$$T_c \text{ to } T_{hs} \text{ (normal motor operation)} \quad 1.07 = 0.143 \cdot 7.5 \tag{15.10}$$

$$T_c \text{ to } T_{hs} \text{ (locked-rotor operation)} \quad 18.6 = 0.143 \cdot 130 \tag{15.11}$$

where $R_{\phi cs}$ = case-to-heatsink thermal resistance, °C/W
T_c = IGBT case temperature, °C
T_{hs} = IGBT heatsink temperature, °C
PD = IGBT power dissipation, W

As we have seen from the previous formulas, there will be a heat rise across the IGBT's junction-to-case and the case-to-heatsink interfaces. The thermal resistance requirements of the heatsink can be calculated by incorporating these factors into Eq. (15.12). This equation varies from the normal textbook equations in that it uses a maximum IGBT case temperature of 90°C, rather than the traditional

150°C maximum junction temperature. Equation (15.13) indicates that a 6.5°C/W heatsink will suffice for normal motor operation. This can be a standard extrusion-type aluminum heatsink. During locked-rotor operation, a large bonded-fin 0.24°C/W heatsink will be required, as per Eq. (15.14). The 0.24°C/W heatsink will need about 22 times more surface area than the 5.4°C/W size. Its cost may also rise by a similar 22-fold factor.

$$R_{\phi sa} = \frac{T_{c,max} - (T_a + (R_{\phi cs} \cdot PD))}{PD} \tag{15.12}$$

$$R_{\phi sa} \text{ (normal operation) } 6.52 = \frac{90-(40 + (0.143 \cdot 7.5))}{7.5} \tag{15.13}$$

$$R_{\phi sa} \text{ (locked rotor) } 0.24 = \frac{90 - (40 + (0.143 \cdot 130))}{130} \tag{15.14}$$

where $R_{\phi sa}$ = heatsink-to-air thermal resistance, °C/W
$T_{c,max}$ = IGBT maximum case temperature per user, °C
T_a = ambient-temperature maximum, °C
PD = IGBT power dissipation, W

[*Editor's note:* It is always a good idea to reverse the math equations to double-check your calculations. Using the 130-W IGBT PD locked-rotor example, a 0.24°C/W heatsink will have a temperature rise of 31.2°C. Adding the interface heat rise (18.6°C) to this equals about 50°C. Adding the maximum 40°C ambient brings the total to 90°C, which is the IGBT's case temperature design goal. The IGBT's junction temperature adds up to 152.4°C (90 + ($R_{\phi jc}$ · PD)). This means that we are operating the IGBT at its maximum rated junction temperature (150°C)—and slightly beyond—during a locked-rotor condition.]

15.5.1.1 Motor power stage thermal example summary. In the previous example, several factors should have became obvious. The IGBT power dissipation has a direct relationship to the heatsink requirements. During a locked-rotor condition the IGBT's power dissipation rises significantly, creating the need for a larger heatsink and possibly a larger IGBT than was initially selected. It may also be worthwhile to add current limiting to the power stage, to 15 A, for example, which would allow a less expensive heatsink to be used.

(*Editor's note:* Adding a temperature sensor to the power device's case can help prevent a motor control failure. If the heatsink's airflow is obstructed, or something has occurred that allows the power device to reach critical temperatures, then the power stage or entire motor control can be shut down. A temperature sensor normally, however, cannot respond fast enough to protect against a short circuit.)

15.5.2 Device power dissipation (thermal generation)

Power losses in semiconductor devices are produced by conduction and switching losses. Conduction losses are a function of the current in each particular device and that device's DC electrical characteristics. The thermal generation in a semiconductor device is very straightforward in a DC or steady-state condition. The power-loss calculations are based on the forward voltage or resistance drop at a constant current. Switching power losses are a function of the switching frequency, the current in each device, and the device's dynamic characteristics. Every time a PWM cycle occurs, there will be a switching loss in the IGBT during its turnon and turnoff.

The most common devices used in today's three-phase inverters are IGBTs with internal free-wheeling diodes (FWDs). The switching transistor and FWD current waveforms in a three-phase AC induction power stage are complex. The IGBT and FWD devices may switch from a frequency range of 2 to 20 kHz. However, the motor's average current level varies over the much longer period of the motor's fundamental operating frequency, which is typically 1 to 120 Hz for AC induction motors. Power losses may be modeled by empirical analysis of the current waveforms, computer modeling, and qualitative measurements. Computer modeling is somewhat difficult because of the complexity of the current waveforms and accuracy of semiconductor models. Equations for the average and sinusoidal IGBT and FWD currents can be derived from a quantitative analysis of these current waveforms. Conduction losses in the semiconductor devices are estimated using piecewise linear approximation based on the device's ON voltage characteristics.

15.5.2.1 Conduction losses. For an example of conduction losses, the forward DC electrical characteristic of a 40-A 600-V IGBT is shown in Fig. 15.7. The IGBT ON voltage increases at a mostly linear rate at high currents. As seen in Fig. 15.7, the gate voltage plays a major role in the IGBT's and MOSFET's ON voltage.

15.5.2.2 Switching losses. Measurements of switching energy versus current can be utilized to obtain a curve of losses versus current. Integration of these losses over a full-sine-wave period is used to obtain a closed form expression for switching losses as a function of sine-wave current. The transistor and diode currents and, thus, the power losses, depend a great deal on the modulation techniques

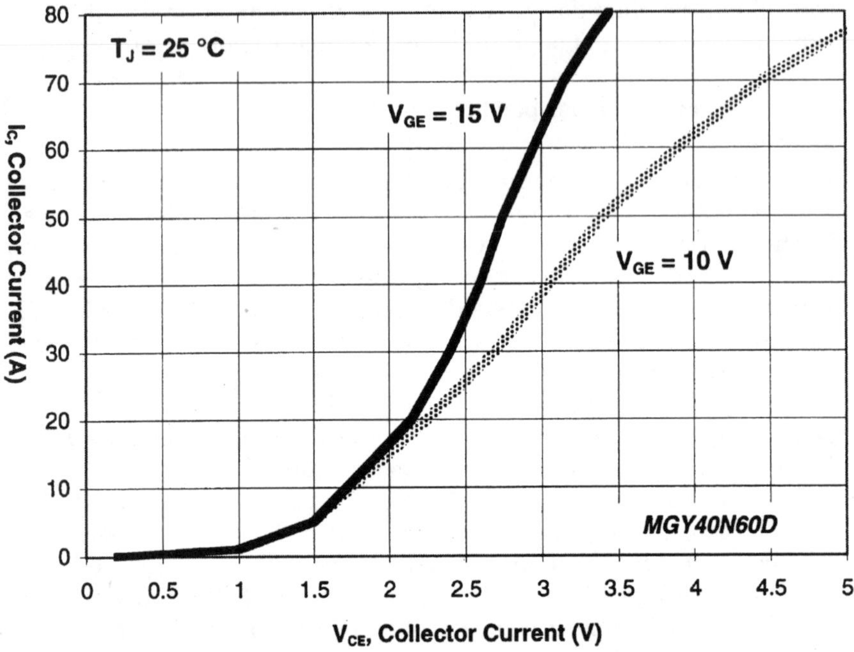

Figure 15.7 IGBT forward voltage characteristics.

employed for the inverter. Most three-phase inverters use sine-wave PWM, as discussed in Chap. 8. With this method the six inverter switches are treated as three independent phases and each phase is pulse-width-modulated by a sine-wave function. A high carrier frequency in the range of 2 to 20 kHz is used. The duty cycle of the pulse waveform varies over the longer period of the sine wave. This effectively produces sine wave voltages on all three phases that may be varied in both amplitude and frequency. The mathematical calculation of switching losses in a three-phase power stage is fairly complex. It involves interaction among various factors—the IGBT and the free-wheeling diode losses and the motor's fundamental operating frequency, as well as the PWM duty cycle and carrier frequency. (*Editor's note:* Since this is an applications-type handbook, we will not go into the theory of switching loss calculations—this information is available in many motor control engineering textbooks.)

15.5.3 Model validation

It is often difficult to isolate and accurately measure inverter power losses in a motor drive system by using mathematical calculations. Losses in the IGBTs and free-wheeling diodes must be isolated from losses in input rectifiers. Even if accurate thermal resistance ($R_{\phi jc}$) is available for each die, the designer seldom has access to the power silicon devices for measurement of junction temperatures during system operation.

The following is one example of a method that was used to measure losses. A motor drive was configured for "normal" operation. This included a motor drive system consisting of a 750-W (1-hp)-rated integrated smart power module, 750-W (1-hp) three-phase AC induction motor, microcontroller logic module, and DC supply to the inverter. The load was adjusted to give a phase current of exactly 4.5 A RMS, which is approximately 750 W (1 hp). The power device's (module) baseplate temperature was monitored with a thermocouple. The 750-W (1-hp) load was maintained until the baseplate temperature stabilized at a constant reading. This temperature was recorded at four operating frequencies: 2, 4, 8, and 16 kHz.

The next step is to accurately transform temperature rise into watts. Without knowledge of what losses are in the IGBTs or the diodes and without knowledge of the accurate thermal resistance for the heatsink, power in watts cannot be calculated from temperature rise. Alternatively, a known and controlled power can be applied to the module until the same rise in baseplate temperature is achieved. This was done by biasing all the IGBTs ON and running current through the inverter. With a current-limiting supply, the resulting $V_{CE,on}$ voltages of the top and bottom devices are measured across the inverter half-bridge. The applied current times measured voltage is the power in watts dissipated in the power devices. When thermal equilibrium with this applied DC power is equal to the temperature driving a load at 750 W (1 hp), the AC power losses have been determined. The method of applying DC power has exchanged constant IGBT conduction losses for sinusoidal waveforms and switching losses in the IGBTs and diodes in a motor drive. This simplification is valid because the thermal resistance to ambient has not changed and equivalent power dissipation will cause the same temperature rise.

The measured data is summarized in Fig. 15.8. Note that the power losses are not precisely a linear function of switching frequency, as might be predicted. This is due to the effect of increasing junction temperature on conduction and switching losses.

15.6 Choosing a Power Device Package

Unique requirements of motor control require continual development of new products, such as power linear bipolar ICs, smart power discrete products, integrated sensors, microcontrollers, and hybrid power modules, which all require specialized heat dissipation technologies.

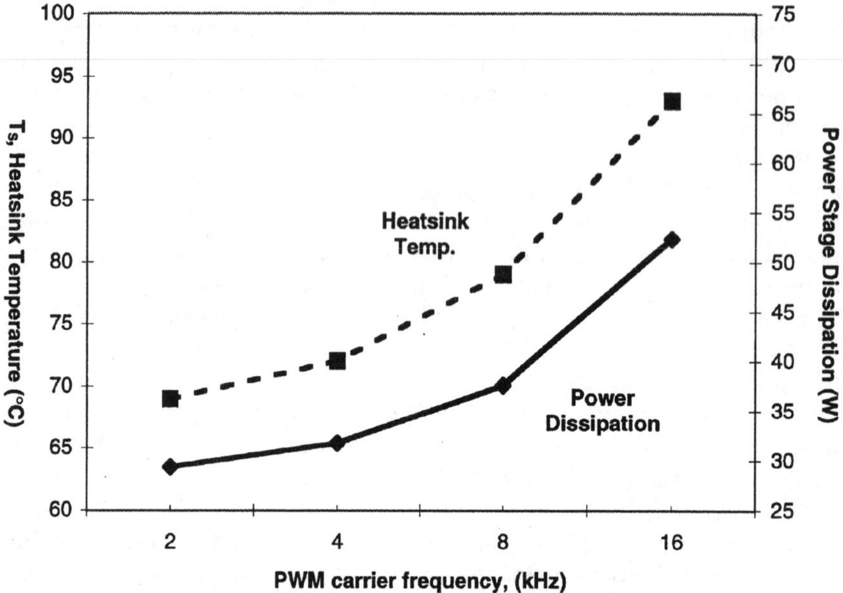

Figure 15.8 PWM frequency effect on temperature and power loss.

Some interesting and innovative work in packaging is currently under development.

15.6.1 Options for packaging

The semiconductor industry offers many package styles and materials. This can make it difficult to keep up with new innovations in package designs and materials. Names such as TO-92, TO-220, TO-247, TO-264, SOT, DPAC, DPAK, SSOP, SMD, HPM, and Case-Number are only a few of the acronyms and trade names used to describe device packaging types. After modeling and/or calculating the amount of power that will be dissipated in a motor control power device, the device's package may still be an issue. If the power dissipated is equal to or greater than the maximum value of the device that you have selected, you may need to look for a different package style, custom or hybrid. Some companies have chosen to develop their own packages combining chip-scale, flip-chip, and bump technology to place die on a substrate material or have developed system integration onto a single chip.

Chip-scale packaging is targeted for high-volume, low-cost reel-to-reel manufacturing and is set up for copper polymide flex circuit redistributed IC pads to a 0.5-mm array. Chip scale is compatible with wirebondable die and existing circuit-board assembly methods. The chip-scale packaging technology is a migration of ball grid array technology with pitches of 1.25 mm down to 0.5 mm. This technology was developed for those designers who choose not to (or cannot) take advantage of the ultrafine pitch of direct chip attachments.

15.6.2 Power module thermal considerations

Hybrid power modules offer a way to integrate system functions into a custom-designed package that provides not only a total system solution but also, in most cases, a cost-effective solution for designers, as has been discussed in Chap. 14.

One method of ensuring thermal management on all modules is to perform a heat-transfer analysis. Power module construction begins with the layout of the desired power electronics. Computer-aided design (CAD) and modeling tools are used to determine the optimum placement of the power dies in order to meet thermal and parasitic parameters.

To comprehend electronic packaging, you need to understand the impact that process-related tasks impose on product performance. In the development of power hybrids, extreme care is taken to define component prefabrication requirements prior to assembly. Using finite element methods and proven solder deformation models, numerical simulations are performed to ensure correct flatness of the final product, thus ensuring accurate thermal management. Thermal balancing of silicon devices in electronic packages is critical in minimizing the switching characteristics of paralleled die.

In electronic packages, solder is still the primary attachment scheme. Understanding the deformation kinetics of solder plays an important role in providing accurate reliability assessments of products. Solder deformation models allow the flexibility to chose the most fatigue-resistant solder alloys, while simultaneously providing a means to optimize and improve existing processes.

15.6.3 Surface mount packaging

Historically, the electronics industry has moved from using through-hole mounted components on 2.54-mm (0.1-in) centers to surface mount components using center-to-center spacing at 0.5 mm (0.0197 in). Surface mount technology continues to meet the needs of reducing the device footprint. The advantage of surface mount devices is a reduced board space (very high-density boards), flexibility in soldering, and system cost reductions. (See Chap. 13.) With new techniques such as flip-chip and chip-scale packaging, future systems will be even smaller.

In the future, wafer-level packaging for discrete transistors will provide size reduction with comparable performance and dramatically

TABLE 15.2 Heatsink Thermal Resistances

Type	Lower limit, °C/W*	Cost range*	Notes
Liquid-cooled	0.007	Highest	Pump, radiator, fan
Bonded-fin (with forced air)	0.028	High	Fan
Bonded-fin (no forced air)	0.220	Medium	Thin fins
Custom forced-air extrusion	0.092	Medium	Special-order
Flat extrusion	1.400	Lowest	Cut to length
Standard aluminum heatsinks	4.200	Low	Catalog stock
Extrusion aluminum heatsinks	0.600	Low	Catalog stock

*These values are approximations and will vary per manufacturer and application.

improved performance for the same-size die. Even in discrete semiconductors, pin count is increasing as added functionality is provided.

15.7 Types of Heatsinks

Thermal management is critical for the safe operation of semiconductors. As a designer, you must make choices in order to utilize the best method of transferring heat. Natural-air convection, forced-air convection, bonded-fin extrusions, customs, liquid cooling, and other methods offer advantages as well as disadvantages. Table 15.2 lists some of the more popular types of heatsink materials.

15.7.1 Natural-air-convection heatsinks

Natural convection is a process of allowing the surrounding air to remove heat without externally forced air movement. As the air molecules gain heat, they create a natural thermal draft, and, as cooler molecules move in, they create their own natural-air velocity. This application is usually sufficient for low-power applications.

15.7.2 Forced-air-convection heatsinks

Forced convection is employed when natural convection does not provide enough cooling by an external force. A fan or blower moves air across the heatsink, facilitating the natural process of removing heated air molecules. However, the boundary layer surrounding the heatsink is affected as well. Using too much airflow means that the maximum transfer of heat to the air molecules will not occur. The linear velocity of the air as it passes the heatsink reaches a point of diminishing return at approximately 500 m per minute.

15.7.3 Heatsink with bonded-fin extrusions

Extrusions create more surface area for additional heat exchange. Thin extrusions will add to the surface area, thus decreasing the thermal resistance. Extrusions can allow up to 4 times more cooling surface area than conventional heatsinks.

15.7.4 Heat pipe

A *heat pipe* is a closed evaporator-condenser system consisting of a sealed hollow tube whose inside walls are lined with a capillary structure or wick. Developed for aerospace applications, heat pipes transmit thermal energy at rates hundreds of times greater than the most solid conductor heatsinks. The medium inside the heat pipe that transfers thermal energy may be air, water, alcohol, acetone, or other materials.

15.7.5 Cold plate

A *cold plate* is based on conductive and convection heat transfer. Heat is conducted through the semiconductor device to the heatsink interface; it is transferred by conduction from the inner wall to the cooling passages, which is then transferred to the moving fluid. Some possible fluids include water mixtures, ethylene, or propylene glycol and more exotic fluorocarbon-based inert heat-transfer liquids. Thermal resistance can reach 0.005 to 0.05°C/W. Special attention must be made to avoid overdesigning the cooling system.

15.7.6 Immersion

Immersion is a more radical concept—it allows the semiconductor chip to be bathed or immersed. Semiconductors immersed in liquid at the die level can greatly increase their thermal handling capabilities in high-power dense electronic packages. The obvious concerns are contamination and electrical isolation.

15.7.7 Liquid spray

Another method to cool a heatsink surface is to spray liquid coolant uniformly onto the heated surface. This thin-film evaporation technique requires a spray mechanism, directional nozzles, a heat-exchanging system, pressure control, and a pump. Liquid spray systems are able to remove heat rapidly.

15.7.8 Pool boiling

Boiling heat transfer has traditionally been used to dissipate energy in high-heat-flux heat exchangers, refrigeration, air conditioning, and some electronics cooling. Without using pumps, *pool boiling* transfers heat through bubbles in a liquid. As the liquid molecules acquire heat, they will boil away from the surface. The heat-transfer coefficient is dependent on the surface coating.

15.7.9 Heat removal limitations

There are limitations to the amount of heat dissipated when using a forced-air heatsink, including the physical size of the heatsink and the amount of airflow that is used. Liquid has a heat-transfer ability that is 10 times that of air. However, most liquid-cooling methods need plumbing, a reservoir, a heat exchanger, and a pump, adding to the cost and complexity of the system. Heat pipes, spray, or boiling techniques have their advantages but may not be as cost-effective for motor controls as standard heatsinks.

Another solution is a custom mixture of heatsink technologies. One method to minimize cost and still ensure excellent heat transfer involves a combination of extrusion and liquid cooling; specifically, the power devices are mounted on a plate with the extrusion immersed in circulating liquid that removes the heat. Another method is to mount the power device onto an open chamber where spray can be directly applied on the bottom of the device.

Summary

Thermal management is a critical part of motor control design since it protects power devices from reaching temperatures that will damage them. Cost is usually a major issue when designing the thermal management system. Decisions need to be made on the basis of the system's power requirements, the composition of the heatsink, the amount of heat that is generated, and the rate at which the energy needs to be removed.

(*Editor's note:* A good prototyping method when designing a motor control power stage is to select a heatsink that is 2 to 5 times larger than you might expect. Selecting power devices that are rated to easily sustain the current level of a motor's locked-rotor condition will allow faster circuit testing and fewer blown parts. Once the circuit design is stable under all operating conditions, the power stage can be carefully evaluated with instrumentation to determine exactly how much heat is being dissipated. With these data in hand, a smaller heatsink and power devices can then be safely evaluated.)

Some of the significant points presented in this chapter are listed below.

- The heatsink protects the power stage's devices from reaching temperatures that will induce failures.
- Power semiconductor devices fail from electrical overstress, high peak power cycling, and extreme thermal cycling.
- Power device bond wires will be stressed and may separate from the semiconductor substrate if operated much beyond the device's maximum junction temperature rating.
- A *steady-state condition* is an equilibrium state or average condition and is reached when the heat generated by the semiconductor power device equals the heat conducted away.
- Conduction losses for the IGBT and diode may be found using a piecewise linear estimate of the device's DC electrical characteristics.
- The worst-case thermal requirements of an electronic motor drive are affected by locked-rotor condition, high power line voltage, and ambient temperature.
- The primary parameters warranting special consideration are the power losses due to switching of the semiconductor power device, conduction losses, and the maximum junction temperature limitation.
- The maximum junction-to-case thermal resistance is a crucial data sheet rating that is used when calculating heatsink size. The device's conduction and switching characteristics are also important when calculating its power dissipation.
- A temperature sensor on the power device's case can alert the motor's control logic when its temperature has reached a critical level. The power device can then be shut down.
- A heatsink's thermal conductivity can be improved by either increasing its surface area or forcing more air to flow across it.

Acknowledgments

The authors would like to thank James M. Fusaro, Jeff Marvin, and Ken Berringer for their assistance with this chapter.

Further Reading

Berringer, K., and Romero, G., "High Current Power Modules for Electric Vehicles," IEEE Workshop on Power Electronics in Transportation, Oct. 1994.

Jurgen, R., *Automotive Electronics Handbook,* McGraw-Hill, 1995.

Kaufman, M., and Seidman, A., *Handbook of Electronics Calculations,* 2d ed., McGraw-Hill, 1988.

Powerex Inc., *IGBT and Intelligent Power Modules, Applications and Technical Data Book,* secs. 3.4.1 and 3.4.2, 1994.

Pshaenich, A., *Basic Thermal Management of Power Semiconductors,* Motorola Application Note AN1083/D, 1990.

Pshaenich, A., and Hollander, D., *Managing Heat Dissipation in DPAK Surface Mount Power Packages,* Motorola article reprint AR323/D, 1988.

Soule, C. A., "Cooling High Density Electronics with Liquid-Cooled Cold Plates," *Powertechnics Magazine,* Aug. 1988.

Chapter

16

Motor Control Sensors

Randy Frank
Motorola Semiconductor Products

Introduction

Sensors play a critical role in modern motor controls. Typically, they are used to sense shaft speed or phase currents. New control schemes, however, are providing new applications for sensors by applying them in different ways.

In the past decade sensing technology based on semiconductor science has advanced considerably. Micromechanical structures made by a batch process called *micromachining* are among the semiconductor-based sensors. Some micromachined sensors have applications in motor controls, providing feedback or diagnostic information for a motor or a process. Solid-state sensors are proving to be more economical and more reliable than their mechanical counterparts, making them an attractive feature in a motor control system.

Software, low-cost memory, and alternatives to sensing, however, are motivating the development of "sensorless" motors. Electronic circuits measure or "sense" motor voltage and current and feed these signals to an MCU. The MCU then processes these signals into data values that are used to estimate torque and velocity error without adding sensors to the motor.

In this chapter we will discuss developments in sensing that are having an impact on motor controls, common sensing applications, evolving technologies, and new applications for sensors.

16.1 Trends in Sensors

Semiconductor technology has produced planar sensors for over two decades. IC temperature sensors, optoelectronic, and Hall-effect sensors have been used in motor controls. If a sensor can be implemented with semiconductor technology that meets the requirements of the application, improvements that can occur include ease of interfacing to other electronic circuitry, improved reliability, lower cost, and a smaller package. A key technology in producing new sensors from silicon is micromachining. Figure 16.1 illustrates a motor system that employs many types of sensors.

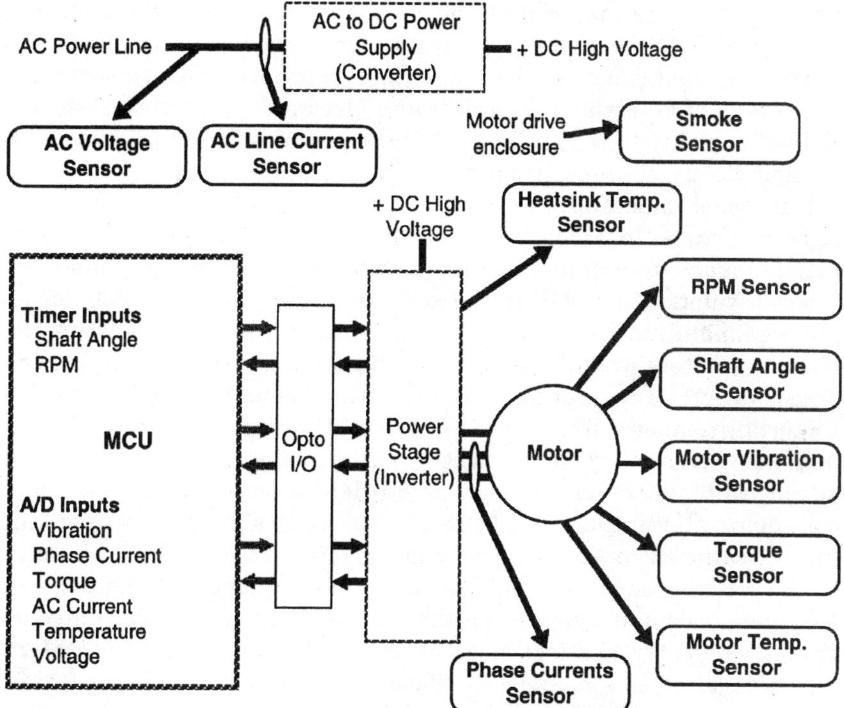

Figure 16.1 Many sensors are possible in an electronic motor control.

16.1.1 Micromachining

A *micromachine* is a three-dimensional mechanical structure or microstructure made by a chemical etching process called *micromachining*. Micromachining is compatible with semiconductor processing techniques. The manufacturing processes used to make these microstructures include photolithography, thin-film deposition, and chemical and plasma etching. Two types of micromachining are already used for several commercial products: bulk and surface micromachining. The *bulk* micromachined diaphragm in a pressure sensor is approximately 25 μm thick. In contrast, the accelerometer *structure* is only 2 μm thick and is separated by only 1 or 2 μm from the supporting substrate and the top layer.

Silicon has many mechanical properties that make it ideal for mechanical structures. It has a modulus of elasticity (Young's modulus) comparable to that of steel and a higher yield strength than in steel or aluminum. Silicon has essentially perfect elasticity, resulting in minimal mechanical hysteresis. Also, silicon's electrical properties have made it the material of choice in most integrated circuits, providing established manufacturing techniques for many aspects of micromachined sensors. Micromachined semiconductor sensors take advantage of both the mechanical and electrical properties of silicon. However, products that fully exploit the combination of the mechanical and electrical properties are still in their infancy.

Bulk micromachining is a process for making three-dimensional microstructures in which a masked silicon wafer is etched in an orientation-dependent etching solution. Using micromachining technology, several wafers can be fabricated simultaneously, and lot-to-lot consistency is maintained by controlling a minimal number of parameters. Key parameters in bulk micromachining include crystallographic orientation, etchant, etchant concentration, semiconductor starting material, temperature, and time. Photolithography techniques common in IC technology define precise patterns for etching both sides of silicon wafers. The crystallographic orientation, etchant, and semiconductor starting material are chosen by design, leaving etchant concentration, temperature, and time as lot-to-lot control parameters.

The selective etching of multiple layers of deposited thin films, or surface micromachining, allows movable microstructures to be fabricated on silicon wafers. With surface micromachining, layers of structural material, typically polysilicon, and a sacrificial material, such as silicon dioxide, are deposited and patterned. The sacrificial material acts as an intermediate spacer layer and is etched away to produce a freestanding (polysilicon) structure. Surface micromachining technology allows smaller and more complex structures with multiple layers to be fabricated on a substrate. The same batch processing controls used for bulk micromachining apply for surface micromachining. However, the processing steps for surface micromachining are more compatible with the CMOS processes used to manufacture integrated circuits. As a result, surface micromachining holds great promise for revolutionary new products in the future.

16.1.2 Integration

The relative ease of accomplishing both bulk and surface micromachining, as well as other new micromachining approaches, has led many researchers to investigate a variety of different structures and applications for these structures. The development activity for future micromachined products includes several that have potential for use in motor control systems. The extent of this activity has created a new system area based on the use of micromachining technology. The micro-level design and fabrication of mechanical structures is called *microelectromechanical systems* (MEMS). MEMS devices include sensors but extend the micromachining to the actuator portion of the system. Actuators micromachined from silicon and other semiconductor materials use electrostatic, electrothermal, thermopneumatic, electromagnetic, electroosmotic, electrohydrodynamic, and other means to provide motion. These actuators range from existing products that provide unmatched performance compared to their macroscale counterparts to intriguing lab curiosities that require significant development to become practical. Electrostatic micromotors are among the more intriguing examples of a complex MEMS device. On the basis of the size of a MEMS device, those that can be manufactured in volume will add functionality and/or reduce the cost of a motor control system.

The ultimate promise from silicon processing techniques is the monolithic combination of both the electronics and the mechanical micromachined structures. The term "smart sensor" is used to describe a sensor that has additional logic, regardless of whether or not it is apparent to the user, and even if it is not manufactured using micromachining techniques.

The smart sensor models developed by several sources have as many as five distinct elements for analog output sensors. In addition to the sensing element and its associated amplification and signal conditioning, an analog-to-digital converter, memory of some type, and logic (control) capability are included in the smart sensor. Once the signal is in digital format, it can be communicated by a digital communication protocol. (See Chap. 11 for information on motor control networking.)

Reducing the number of discrete elements in a smart sensor, or any system, is desirable in order to reduce the number of components, form factor, interconnections, assembly cost, and, frequently, the component cost as well. The choices for how this integration occurs are often a function of the original expertise of the integrator. Processing technology is a key factor. However, suppliers not only must be willing to integrate additional system components, but must also achieve a cost-effective solution. As a result, both hybrid (package-level) and monolithic integration are being pursued. Different design philosophies, and the necessity to partition the sensor/system at different points, can determine whether a smart sensor is purchased as a single component or, alternatively, designed using a sensor signal processor or other components to meet the desired performance of the control system. Those components that have the combined electronics and micromachined structures will provide the system designer more flexibility and functionality.

16.1.3 Multiple sensors

More than one sensor is frequently required to provide sufficient information for a motor control system. R&D efforts are progressing in several areas to integrate multiple sensors ultimately on the same silicon. The sensing arrays can include a number of sensors for different measurements, including pressure, flow, temperature, and vibration. Multiple sensors of a given type can be used to increase the range, provide redundancy, or capture information at different spatial points. Also, multiple sensors of a given type, with minor modifications, can measure different categories in chemical sensing applications.

In addition to the sensors, signal conditioning to amplify the signal, to calibrate offset and full-scale output, and to compensate for temperature effects could be included on these devices. Since it will probably use CMOS technology for the semiconductor process, an onboard analog-to-digital converter (ADC) could also be integrated. Other system capabilities could include fuzzy logic and neural network interpretation of the input signals. This is especially true if an array of chemical sensors is used to indicate a wide variety of chemical species and overcome many of the problems of available chemical sensing products.

16.2 Position and Velocity Sensing

Sensors provide feedback in the motor control system to implement today's more advanced control strategies. As we observed in Fig. 16.1, many sensors are key elements in a motor control system. In this section we will look at the techniques used to sense position and velocity, including traditional and more recent developments.

16.2.1 Resolvers and rotary encoders

Resolvers and rotary encoders are normally used to provide feedback to determine the rotor speed and position. A *resolver* provides one absolute signal per revolution and tolerates vibration and higher temperatures. At lower speeds, the single signal per revolution limits performance. Standard resolver-to-digital converters with 12-bit resolution and a 500 μs sampling time can achieve a minimum speed of about 30 rpm.

Rotary encoders can directly use the incremental signal. As a result, the speed control range possible with rotary encoders is 1:600,000, compared to resolvers with 1:200. The rotary encoder values apply for a maximum speed of 6000 rpm and a sampling time of 500 μs, which are common for modern machine tools.

16.2.2 Linear and rotational variable differential transformers

The linear variable differential transformer (LVDT) and rotational variable differential transformer (RVDT) are used to provide motion. An LVDT pressure sensor consists of a primary winding and two secondary windings positioned on a movable cylindrical core. The core is attached to a force collector, which provides differential coupling from the primary to the secondaries, resulting in a position output that is proportional to pressure.

An alternating current is used to energize the primary winding, inducing a voltage in each of the secondary windings. The windings are connected in a series-opposing fashion, so that the equal but opposite output of each winding tends to be canceled (except for a small residual voltage called the *null voltage*). An applied force causes the core to be displaced from its null position, and the coupling between the primary and each of the secondaries is no longer equal. The resulting output varies linearly within the design range of the transducer and has a phase change of 180° from one side of the null position to the other. Since the core and coil structures are not in physical contact, essentially frictionless movement occurs.

The output of an LVDT requires signal conditioning electronics. This electronic stage consists of an oscillator for the supply voltage, circuitry to transform the constant voltage to a constant current, amplifier with high input impedance for the output, synchronous demodulator, and a filter with characteristics designed for quasi-static or dynamic measurements.

16.2.3 Hall-effect sensors

A frequently used principle to sense position in motors is the Hall effect. The presence of a magnetic field at a right angle to an epitaxial layer with current flowing through it causes a Hall voltage to be generated. The Hall voltage is sensed by taps in the layer that are at right angles to the current flow.

The output of the Hall-effect is converted to a linear or digital output. The linear output Hall sensor adds a DC amplifier and voltage regulator to the basic Hall element to provide an output that is linear, proportional to the applied magnetic flux density, and ratiometric (proportional) to the supply voltage. A Schmitt trigger threshold detector with built-in hysteresis added to the linear Hall sensor provides digital output. The addition of an open collector NPN transistor allows the Hall switch to be compatible with digital logic.

In a brushless DC motor, three radially positioned Hall-effect sensors are mounted on the stator close to the rotor magnets. Electronic circuitry decodes the output of the Hall device and controls the direction of the currents applied to the three motor phases. Power transistors perform the actual switching.

In a closed-loop speed control it is desirable to maintain a set speed regardless of load variation. Hall-effect sensors have been used not only to commutate the motor but also as a reference frequency for speed regulation. The Hall-effect devices provide a lower frequency output than do shaft encoders or resolvers. This limits lower speeds to about 500 rpm. The Hall output can be used to regulate speed to about $\pm$ 1 to 5 percent of the speed in systems with varying loads or to about $\pm$ 0.5 to 1 percent in systems with stable loads.

16.2.4 Optical encoders

An optical encoder contains a light source, and a photodetector provides a digital means of measuring displacement or velocity when an alternating opaque and translucent grid is passed between them. Uniformly spaced apertures on a wheel (i.e., a flat disk mounted on motor shaft) allow logic circuitry to count the number of pulses in a given timeframe to determine shaft velocity or angular displacement. Linear measurements of displacement and velocity are also possible by adding two or more rows of apertures to the disk. This quadrature approach does, however, require the use of two or more slotted optocouplers. An application example of an optical sensor is shown in Fig. 6.19 (Chap. 6).

Basic presence sensing can be accomplished through the use of a single optical channel (or emitter-detector pair). Speed and incremental position sensing must use two channels. The most commonly used approach is *quadrature sensing,* in which the relative positions of the output signals from two optical channels are compared by a logic device or MCU. This then provides direction information, in addition to either of the individual channel's transition signal being used to derive either count or speed information. A typical application may utilize the direction output as an up/down input control flag, and either channel, A or B, as the count input trigger for a common up/down counter. The counter in an incremental system will increment or decrement as required to maintain a relative position or count output.

Quadrature direction sensing requires that the two optical channels create electrical transition signals which are out of phase with each other by 90° nominally. When the code wheel is spinning, the two electrical output signals, A and B, will be 90° out of phase because the wheel windows are 90° out of phase with the two sensor channels. The interpretation of whether signal A leading B is clockwise or counterclockwise is a matter of choice. The two waveforms are in four equal quadrants, exactly 90° out of phase with each other. The detector is ON when light is present and turns OFF when the web blocks the beam.

An absolute optical encoder provides instant recognition of position even if power has been interrupted. Absolute encoders are available in single- and multiturn configurations. The single-turn version utilizing optical resolver techniques can provide resolution of 2^{24} lines in one revolution. Multiturn units act as a multiplier to extend the total range through multiple shaft rotations.

16.2.5 Magnetoresistive elements

Linear position, rotational, and speed-sensing magnetoresistive sensors are being developed to overcome the limitations of inductive-type sensors and Hall-effect sensors. Inductive-type sensors require complex signal conditioning for the signal output and external temperature compensation networks. Hall-effect devices are limited because of temperature drift, temperature range, repeatability, nonlinearity, and offset adjustment. A comparison of four types of magnetic sensing technologies is shown in Table 16.1.

Magnetoresistive elements (MREs) utilize four strips of permalloy: a ferromagnetic alloy composed of 20% iron and 80% nickel, arranged

TABLE 16.1 Comparison of Magnetic Field Sensors

Type	DC operation	Package size	Signal amplitude	Sensitivity	Relative cost	Temperature stability
Giant magnetoresistive (GMR)	Yes	Small	Large	Very high	Low–medium	Good
Anisotropic magnetoresistance (AMR)	Yes	Large	Medium	High	High	Good
Hall effect	Yes	Small	Small	Low	Low	Fair
Variable reluctance (inductive)	No	Very large	Varies with frequency	Varies with size	Medium–high	Good

in a Wheatstone bridge configuration. The output signal, based on a change in resistance, is proportional to the input signal magnitude modulated by the magnitude of the magnetic field. Also known as *anisotropic magnetoresistance* (AMR) sensors, these MRE sensors rely on the change in resistance of a sense layer ($\Delta R/R$) in response to magnetic fields. These magnetic fields are applied parallel and transverse to the current flowing in the MRE sensor. To measure the speed of a rotating shaft, a gear wheel made of a ferromagnetic material is mounted on the shaft. MREs have a preferred direction of rotation so that the output signal can also be used to indicate direction of rotation. Linear characteristics up to 175°C have been demonstrated by MRE sensors.

Giant magnetoresistive (GMR) structures have alternate thin layers (in the range of 30 to 60 Å) of magnetic and nonmagnetic materials. The antiferromagnetic coupling is overcome by a magnetic field (>20 kOe) applied in the plane of the layers, which forces the layers to align in parallel with each other. Spin-dependent behavior of conduction electrons produces a GMR-like effect in many ferromagnetic/conductor superlattice combinations. The largest MR ratios are measured when adjacent magnetic layers have opposite spins (as in coupled Fe/Cr systems) because the change in spin-dependent behavior is the largest. A spin valve is produced by adding an antiferromagnetic layer to stabilize the orientation of one of the magnetic layers, leaving the other free to respond to external fields. The saturation field in the spin valve is in the 10- to 20-Oe range, making it attractive for magnetic recording heads and disk drives.

16.3 Current Sensing

The classic approach to current sensing in a motor's electronic power stage is to use a special current-sensing resistor. Because of the need for electrical isolation, other current sensors are also employed. These include Hall current sensors, current transformers, and current transducers.

16.3.1 Resistor current sensing

Current sensing for low-power motor controls, especially those that are operated at low voltages (<50 V), can use a resistor for current sensing. (Figure 10.9 in Chap. 10 shows a circuit that uses a 0.01-Ω current-sensing resistor and an op amp.) The resistor should be designed specifically for this type of application. Standard wirewound or thick carbon restores are seldom suitable for use as a current-sensing element. The wirewound resistor exhibits too much internal inductance and will generate ringing in a PWM-type motor control circuit. Standard carbon resistors are rarely available in low resistance values (0.01 to 0.1) and may not be rated to handle high current peaks. Several manufacturers offer noninductive low-resistance current sense resistors. (Caddock Electronics Inc., Roseburg, CA; Isotek Corp., Swansea, MA; and Vishay Resistors, Malvern, PA are three of many companies who offer current-sensing resistors.)

There are some circuit design methods that can help improve the resistor-type current-sensing circuit. Because a high-gain op amp or comparator is required to amplify the low-level voltages (1 to 100 mV) across the current-sensing resistor, it is important that the input lines to the op amp or comparator be located close to the resistor. The input lines should be right at the resistor's contacts, and routed away from the resistor's high-current traces. To reduce EMI problems, high-frequency filters can be added to the op amp's input, and the output of the op amp or comparator can be gated. The gate timing is set such that it occurs just after the power stage is switched on and just before it is switched off. This minimizes any electrical noise that occurs during the power stage's switching transitions. One downside of this gating method is that an overcurrent condition will not be detected when the power stage is switched off. A more elaborate gate scheme would be to gate off the current sensing element only during the switching transition times.

The main drawback to resistor current sensing is its lack of voltage isolation. It is possible to use an optocoupler on the output of a comparator that is connected to the current sense resistor. This method works well for detecting overcurrent levels. (Figure 10.13 in Chap. 10 shows a current-sensing circuit that uses an optocoupler.) For motor current measurements, the optocoupler must operate in a linear mode, which usually requires a more expensive optocoupler device.

Because of the need for high-voltage isolation and current measurement linearity, other noncontact current sensors are often employed.

16.3.2 Hall-effect current sensing

Hall-effect devices can also be used to measure the current flowing through a conductor. They provide voltage isolation and current sensing over a wide frequency, DC to over 10 kHz. Their main limitation is lack of sensitivity and stability over temperature. The Hall effect device's temperature coefficient varies from 0.02 percent/°C to −0.06 percent/°C. Hall-effect devices require an external power supply for their bias current and to power their interface circuit.

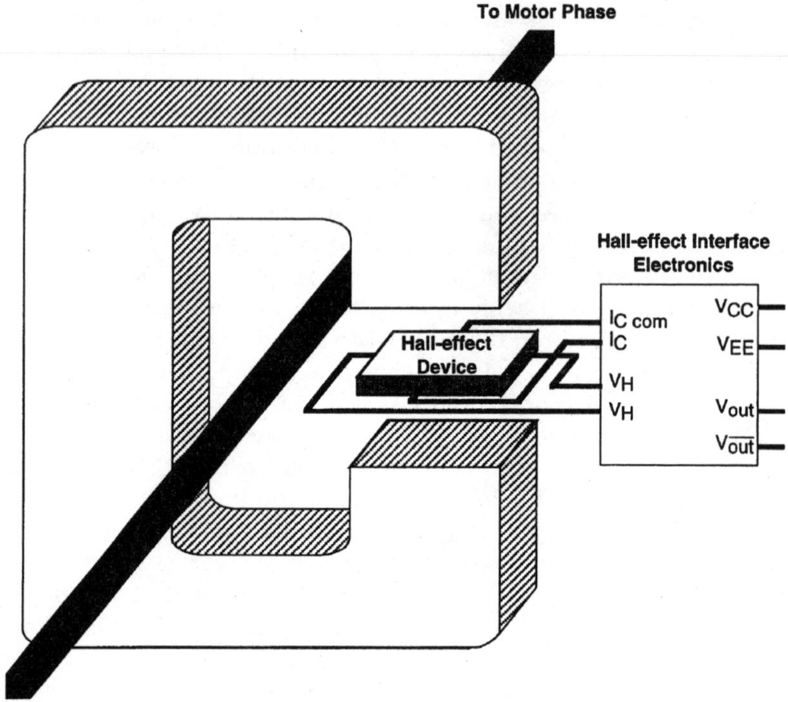

Figure 16.2 Hall effect current sensor.

To increase the sensitivity of the Hall-effect device, an external magnetic core is used, as shown in Fig. 16.2. This type of Hall-effect current sensor product is available in a wide range of currents (5 to 10,000 A) and different packages. (F. W. Bell, Orlando, FL and Jenesco, Inc., Amherst, NH are two of many companies who offer Hall-effect current sensor products.)

16.3.3 Transformer current sensing

Phase current sensing can be accomplished with current-sensing transformers. A current transformer offers simplicity, but its output signal will vary somewhat with frequency. Larger cores are required to obtain lower-frequency response. Generally, transformer-type current sensors are limited to 5 Hz. Their upper frequency range can reach in the megahertz region, making it a good sensor to detect fast high-current pulses. When dealing with bidirectional transients, it is important to avoid exceeding the internal dissipation ratings of the transformer. A peak current time rating is usually set by the manufacturer. (Amecon Inc., Anaheim, CA; Pearson Electronics Inc., Palo Alto, CA; and Smith Research & Technology, Inc. Colorado Springs, CO are three of many companies that offer transformer current sensor products.)

16.3.4 Transducer current sensors

Some manufacturers list their current sensor products as *current transducers*. This usually refers to a current sensor that has a built-in electronic interface and provides either a digital or analog output signal to the motor's control logic. The designer simply supplies power to the current transducer and does not have to worry about op amps or other interface circuits to obtain a usable current sense signal. (LEM U.S.A., Inc., Milwaukee, WI is one of many companies that offer current transducers.)

16.4 Temperature Sensing

Temperature has a direct impact on the operating life of a motor. The motor windings and bearing lubrication provide the longest service when their temperature ratings are not exceeded. The life of the components in the control electronics is also improved when temperatures are kept below their maximum ratings, as discussed in Chap. 15. Separate temperature sensors could be placed at many locations in the motor control system to indicate when temperature limits are exceeded. However, the motor control electronics, specifically the power semiconductors, experience the most significant damage with the shortest exposure to extreme temperatures. Since the power semiconductors usually constitute a significant portion of the motor control cost, it is worthwhile to provide them with protection and diagnostics.

Temperature can be sensed by temperature sensors, such as a thermistor or thermocouple, or by semiconductor techniques. Most semiconductor junction temperature sensors use a diode-connected bipolar transistor. A constant current passed through the junction produces a junction voltage between the base and emitter V_{BE} that is proportional to temperature. The overall forward voltage drop has a temperature coefficient of approximately 2 mV/°C. (A circuit design is shown in Fig. 6.3, Chap. 6, that uses a transistor's V_{BE} as the temperature sensor.)

When compared to a thermocouple or a resistive temperature device (RTD), the temperature coefficient of a semiconductor sensor is larger, but still quite small. Also, the semiconductor sensor's forward voltage has an offset that varies significantly from unit to unit. However, the linearity of the semiconductor junction voltage versus temperature is much lower than that for the thermocouple or RTD. In addition to the temperature-sensing element, circuitry is easily integrated to produce a monolithic temperature sensor with an output that can be interfaced to a microcontroller. An embedded temperature sensor circuitry with protection features can also be added to integrated circuits (ICs). (Analog Devices, Inc., Wilmington, MA; Dallas Semiconductor, Dallas, TX; Linear Technology Corp., Milpitas, CA; National Semiconductor Corp., Santa Clara, CA; Telecom Semiconductor, Inc., Mountain View, CA; Ketema Rodan Division, Anaheim, CA; Pearson Electronics Inc., Palo Alto, CA; and Unitrode Corp. Merrimack, NH are among the many companies that offer IC temperature sensor products.)

16.4.1 Built-in temperature sensing for power devices

Power MOSFETs are used extensively in lower-voltage motor controls. Power MOSFETs can be produced with integral polysilicon diodes (and resistors) that are isolated from the power MOSFET. The diodes can be used as temperature-sensing elements to determine the junction temperature of the device in an actual application. The thermal sensing that is performed by the polysilicon elements is a significant improvement over power device temperature sensing that is performed by an external temperature-sensing element. Figure 16.3 shows a temperature-sensing element integrated in a power MOSFET. By sensing with polysilicon diodes, the sensor can be located

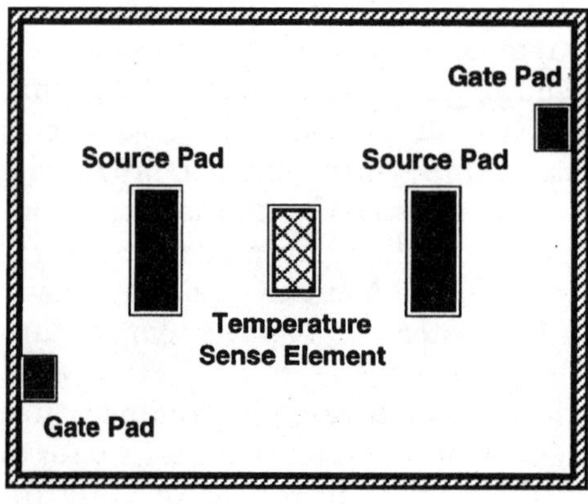

Figure 16.3 Power device layout with integral temperature sensor.

close to the center of the power device near the source bond pads, where the current density is the highest and, consequently, where the highest die temperature occurs. The thermal conductivity of the oxide that separates the polysilicon diodes from the power device is two orders of magnitude less than that of silicon. However, because the layer is thin, the polysilicon element offers an accurate indication of the actual peak junction temperature.

Sensing for fault conditions, such as a short circuit, is an integral part of many smart power (or power) ICs. A smart power IC can have multiple power drivers integrated onto a single monolithic piece of silicon. Each of these drivers can have a temperature sensor integrated to determine the proper operating status and shut off only a specific driver if a fault occurs.

16.4.2 Motor temperature compensation

Temperature compensation is required in some of the complex adaptive control algorithms in adjustable speed motors. Temperature measurements compensate for rotor resistance variations in the actual operation of the motor under various load conditions. If not taken into account, these temperature-caused variations can produce serious errors from the model.

Temperature can also be a process control element that dictates a change in motor operation. This use of sensing—separate from the motor and part of a larger system—is becoming more common because of its ability to handle a number of control variables in industrial applications, including heating and air conditioning. In the next section we will expand on the usage of sensors that are finding applications in motor controls.

16.5 Other Sensors for Feedback or Diagnostics

In a motor-controlled system, sensors provide feedback for other system parameters, including pressure, flow, and weight. Sensors also provide diagnostic warning for vibration, presence of harmful chemicals, or excessive humidity. One interesting application of sensor technology is in a vacuum cleaner. A piezoresistive material in a 90° element of the vacuum cleaner's dirt path picks up the sound of dirt going toward the bag. Circuitry in the vacuum cleaner illuminates a red LED to indicate that dirt is still being accumulated by the unit. A green LED indicates that dirt is no longer detected. In this motor-controlled application the sensor can possibly provide better job performance, as well as savings in time and energy, since it enables the user to operate the cleaner only when it is actually cleaning.

16.5.1 Pressure, flow, and weight

Pressure sensors are the most developed microelectromechanical systems (MEMS) devices, with over a decade of use in commercial and industrial applications. Since their initial introduction, the bulk micromachined devices have improved range (both higher and lower) to address more measurements, have enhanced accuracy, and have added integration for easier interface and increased functionality. Gauge, absolute and differential pressure measurements, allow process control. In addition, gauge vacuum measurements for vacuum forming, flow measurements calculated from ΔP, and weight can be determined by pressure sensors. The differential pressure ΔP is also measured across a filter element and provides an indication of the need for maintenance. When motor-driven pumps provide the flow (either liquid or air), the ΔP measurement can be used to regulate the flow. Figure 16.4 shows a motor application that incorporates a surface mount pressure sensor with integral signal conditioning.

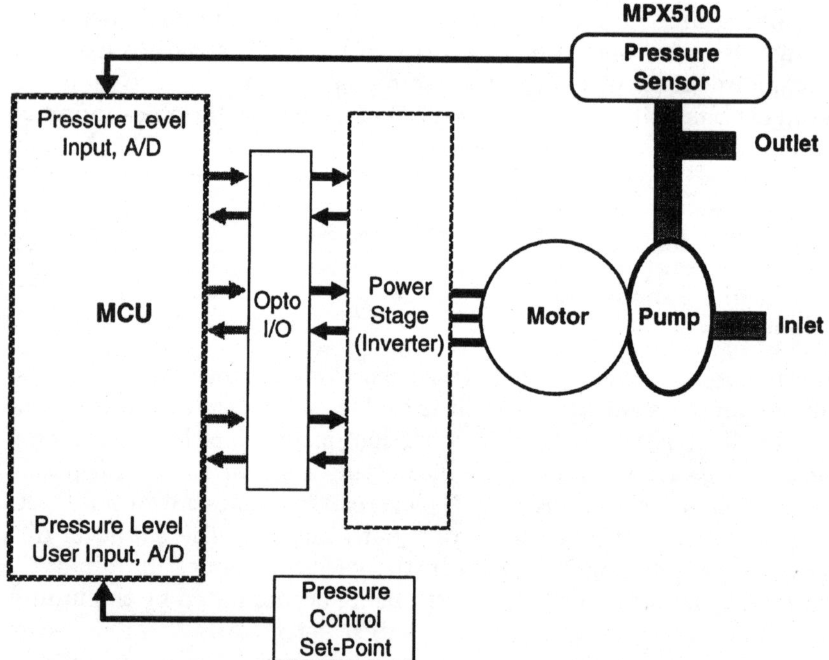

Figure 16.4 Motor control using a pressure sensor.

16.5.2 Vibration

Accelerometers are employed to detect excessive vibration in motor-driven systems, including out-of balance conditions, pending bearing failures, and shock. Oscillating vibration can be sensed and used to modify the motor's operation to avoid resonant conditions. The availability of low-cost accelerometers enables a number of desired parameters to be measured that would have been prohibitively expensive with earlier technologies.

16.5.3 Chemical and humidity sensors

One of the newest micromachined sensors is the chemical sensor. Thin-film metal oxide technology is combined with an embedded microheater on a thin micromachined silicon diaphragm. The small sample area of a micromachined sensor can be raised more readily to the higher operating temperature (about 450°C) that is required for detecting the presence and actual value of specific chemicals. One example of how this sensor type may be employed is in oil-burning furnaces. The sensor would be designed to detect excessive carbon monoxide levels and shut down the fuel supply pump motor and valve controls.

The current chemical sensors require about 45 mA of average current to operate. This is necessary to maintain a high internal temperature (450°C). The current to the sensor is usually pulsed full on for 5 s in order to burn off any surface contaminants, such as water, then it is pulsed partially on for 10 s with a reading taken at 9.5 s.

Chemical sensors can be used either as a simple diagnostic or possibly in the control loop and have the potential to significantly enhance the sensed inputs in future systems. Since sensors are the means by which a control system can "see," "feel," "hear," and "touch" its environment, the chemical sensor adds the capability for the control systems to "smell" as well. Chemical sensors have been developed for consumer and industrial applications to sense CO and methane. In the factory or office of the future, chemical sensors could detect harmful or annoying levels of these or other chemicals and ultimately prove beneficial as feedback elements in more complex control systems. Dampers in the HVAC system could be automatically opened or closed, changing the air exchange and circulation to minimize the effects of harmful chemicals or offensive odors. The combination of chemical sensors with efficient motor controls can be utilized to avoid the sick building syndrome—a problem that occurs in tightly sealed buildings.

As the number of chemicals that can be detected is expanded, it is possible that chemical sensing could be used as a process monitoring and quality control tool. Humidity sensors are among the possible sensors that could be produced by micromachining to detect the process environment and provide comfort information for HVAC motor control systems. Humidity could also be combined with multiple chemical sensors for more complex control and reduced cost per sensed parameter in future systems.

16.6 Sensorless Alternatives

Considerable development effort has been directed at alternatives to sensors in motor controls. Initial efforts were directed at replacing Hall-effect sensors. A back EMF sampler circuit obtains the position information directly from the motor windings. This information is fed into an MCU, which computes the correct phase angles for commutation using phase-locked loop techniques, and a proprietary algorithm calculates position from signals on unenergized windings. The alignment is automatic, eliminating one of the difficulties of Hall-effect sensors. Instead of the usual five wires to the motor that is required for Hall effect sensors, no extra sensor wires are needed. There are some limitations to back EMF sensing. The most obvious problem with a sensorless design is that no back EMF occurs at zero speed, and very little may occur at low rpm. This makes motor control difficult at very low shaft speeds. In some applications this is not an issue, and the sensorless method works well. The sensorless function may need to be fine-tuned for a specific motor application.

More recently, eliminating speed sensors in induction and switched reluctance motors has been the focus of many R&D labs. Several observer-based and adaptive schemes have been proposed that estimate rotor slip from the back EMF in induction machines. These approaches are inherently parameter-dependent and fail under low-speed operation. An alternate approach was developed using a fast Fourier transform (FFT), digital signal processor, and the current signal from Hall-effect sensors that accurately determined speed down to 1 Hz.

Sensorless flux vector control offers significant improvements in torque recognition without the need for external position sensor in standard AC motors. A software model of the AC motor is stored in the variable frequency drive's memory. The model contains parameters for the motor, including primary resistance, secondary resistance, leakage inductance, mutual inductance, and load inertia. During operation the microcontroller performs high-speed calculations, using the measured motor current and the motor parameters from the software model to obtain a complete picture of the actual motor and loading conditions.

In switched reluctance motors, researchers are investigating ways of sensing the variation in stator winding inductance as a function of time to indicate rotor position and to eliminate the resolver and reduce cost. A state observer has been employed by one group to estimate rotor position from measurements of the stator voltage and coil currents. However, several assumptions must be made to arrive at a control algorithm, and the assumptions are not always valid. A self-tuning algorithm based on the phase inductances has been reported to be simpler and more direct. It also may be more easily applied to mass production, where variations in production tolerances would detune the drive control.

Summary

Sensors add to the functionality of the overall motor control system and provide protection and diagnostics. However, software and the increased computational power of MCUs are being used more frequently to reduce hardware components and may eliminate the need for some sensors in a motor control design.

A few of the significant points presented in this chapter are listed below.

- Motor current and shaft speed or position are the most common applications for sensors in motor controls.
- Micromachining is a method that utilizes semiconductor processing techniques to manufacture a micromechanical structure.
- The sensor interface circuits can be integrated with the sensor element to simplify its use. No special external signal conditioning circuits are required.
- Hall-effect, variable reluctance, giant magnetoresistive (GMR), and anisotropic magnetoresistance (AMR) are magnetic field-type sensors.
- Motor control sensors need to have good performance over a wide temperature range and sufficient sensitivity; in addition, they need to offer electrical isolation, if required, and they must be cost-effective.
- One method to determine a motor's shaft position is to use a disk with two rings of slots that are offset from each other. Each ring also contains one slotted optocoupler and a special indexing slot. The optocoupler's outputs are fed to an MCU, which then computes the time from the index slot, the shaft's velocity, the motor rpm, and the direction of the shaft.

- Hall-effect current sensors typically employ a magnetic core to increase their sensitivity to a wire's magnetic flux.
- A current-sensing resistor is the most cost-effective way in which to obtain current measurements. Its main disadvantage is lack of electrical isolation from the motor's power stage.
- Temperature sensing can prevent semiconductor power device failures, and warn of impending motor problems, such as poor bearing lubrication or excessive winding/rotor heating.
- *Sensorless motor control* usually refers to a motor design that uses back EMF sensing, rather than Hall-effect units or optical disks, to determine shaft position. Some work is also under way to compute the motor's current level with back EMF sensing.

Further Reading

Allen, R., "High-Performance Bulk-Micromachined Carbon-Monoxide Sensor Developed for Commercial Use at Low Cost," *Electronic Design,* Dec. 16, 1996.

Berardinis, L. A., "Keeping the Drive Alive," *Machine Design,* Oct. 12, 1995, pp. 84–95.

Derbyshire, K., and Korczynski, E., "Giant Magnetoresistance for Tomorrow's Hard Drives," *Solid State Technology,* Sept. 1996, pp. 57–66.

Ehsani, M., and Ramani, K. R., "Direct Control Strategies Based on Sensing Inductance in Switched Reluctance Motors," *IEEE Transactions on Power Electronics,* Jan. 1996, vol. 11, no. 1, pp. 74–82.

Frank, R., *Understanding Smart Sensors,* Artech House, Norwood, MA, 1995.

Gum, J. W., "DC Brushless Motors, Rotary Encoders Boost Motion Control Growth," *Product Design and Development,* July 1988, pp. 51–67.

Hurst, K. D., and Habetler, T. G., "Sensorless Speed Measurement Using Current Harmonic Spectral Estimation in Induction Machine Drives," *IEEE Transactions on Power Electronics,* Jan. 1996, vol. 11, no. 1, pp. 66–73.

Jurgen, R. K., *Automotive Electronics Handbook,* McGraw Hill, 1994.

Lopez, M., *Using Two-Wire, Brushless DC Motors,* EDN Products ed., Sept. 6, 1996, pp. 36–38.

Meshkat, S., "Meshkat's Motion—What Is an Adaptive Controller?" *Motion,* March/April 1988, pp. 20–22.

Nana Electronics Co., Ltd., *Hall Current Sensor* product catalog, 1995.

Pearson Electronics, Inc., *Application Notes* for *Current Monitors Manufactured by Pearson Electronics, Inc.,* 1991.

Polak, P., "Rotary Encoders or Resolvers?" *Design News,* Oct. 21, 1996, p. 88.

Reimlich, N. C., "Dirt Noise Communicates," *Appliance Manufacturer,* Feb. 1997, p. 60.

Ristic, L., *Sensor Technology and Devices,* Artech House, Norwood, MA, 1994.

Scimia, A., "Sensorless Vector Control," *Motion Control,* Sept./Oct. 1995, pp. 16–18.

Smith, C. H. and Brown, J. L., *Giant Magnetoresistance Materials and Integrated Magnetic Sensors,* SAE 970603, Society of Automotive Engineers, Warrendale, PA, 1997.

Smith, D. R., "Inductive Sensors," *Measurements & Control,* June 1988, Vol. 22, No. 3, Issue 129.

Thundat, T., Oden, R., and Warmack, B., "NanoSensor Array Chips," *Appliance Manufacturer,* April 1997, pp. 57–58.

Williams, D., *A Simple Pressure Regulator Using Semiconductor Pressure Transducers,* Motorola Semiconductor Application Note, AN1307, 1995.

Part

5

Diagnostics for Motor Controls

Chapter

17

Electronic Motor Control Testing and Instrumentation

Richard J. Valentine
Motorola Semiconductor Products

Introduction

Testing an electronic motor control design for the first time can be an exciting event, the culmination of much effort and time. It can be very exasperating as well, since power semiconductors will fail in less than 100 millionths of a second, allowing little time to figure out what went wrong. In this chapter we will discuss various methods to detect problems and lessen the chances for unexpected power stage device failures.

17.1 Motor Control Test Lab

The layout and equipment of a motor test lab should be thoughtfully planned and organized. It should include a controlled power source, dynamometer, oscilloscopes, temperature meters, and digital multimeters. A computer development station should also be located in close proximity to the motor's electronics.

17.1.1 Lab layout

The first consideration for a motor test lab is its location. The motor dynamometer (dyno) needs to be accessible with motor-handling equipment. A large dyno may require a water supply and drain hookup. Smaller dynos usually are fan-cooled and may require an inlet and outlet for outside air. Figure 17.1 illustrates a lab layout for testing medium-power electronic-controlled motors. Note that the benches are arranged in a U-shaped configuration, which allows plenty of work surface while keeping everything close at hand.

Room for large power supplies is another consideration. Instead of running the motor control unit directly off the AC power line, large DC power supplies are typically used for testing the motor's power stage. For large motors these power supplies may be in the 5- to 10-kW range and weigh several hundred kilos. An AC power control switch or disconnect box to each power supply will facilitate their hookup and maintenance. Power supplies that operate the motor control's power stage circuits should be operated from a dedicated AC power source. This is important, since a major problem in the motor's power stage can "trip" the power supply's main breakers. You don't

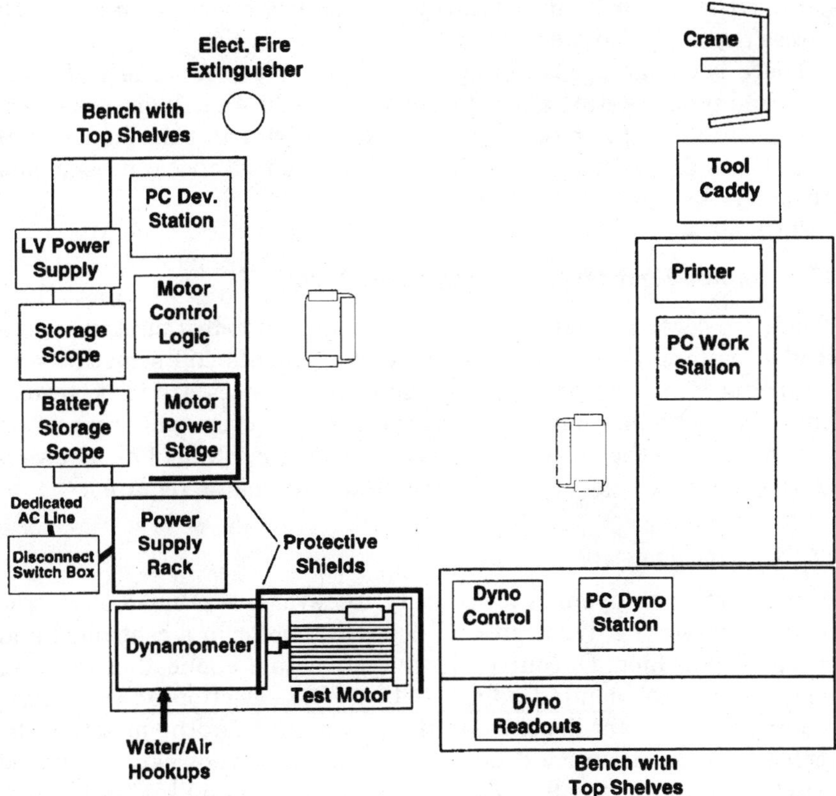

Figure 17.1 Motor test lab layout.

want to lose the power on both the computer development station and instrumentation, since they may have just detected the source of the problem in the power stage.

17.1.2 Motor test lab safety issues

In addition to meeting all building and safety regulations for the lab layout, it is also wise to add extra protection against flying shrapnel when a serious fault occurs with the motor's power stage or from the dyno. Thick plastic panels that are rated to withstand metal impacts can be placed around the motor's power stage and the dyno's rotating parts. (Some dynos already have shields on their brake unit but may not adequately shield the motor-to-dyno shaft coupler.) The power supply controls and main shutoff switches need to be located near the motor test area, preferably between the operator and the exit, and not behind the motor or in a location where the operator must reach across the motor's power stage.

There is a high probability that a motor or power stage will overheat and possibly start a fire during some tests. We will discuss methods to minimize the risk for this to happen later in this chapter, but one or two fire extinguishers that are rated for electrical fires should always be handy in any motor test lab.

17.2 Motor Electronics Test Instrumentation

There are several types of instruments required to test electronic-controlled motors. Selecting the correct instruments can sometimes be confusing, especially if a wide range of motor sizes are to be tested, but once you figure out just what type of measurements and testing are necessary, the selection process can begin. Table 17.1 shows a matrix of motor tests and the instrumentation that is required.

17.2.1 Dynamometers

A dynamometer (dyno) is a necessity to develop a reliable electronic motor control. The dyno can load down the motor in a controlled and repeatable fashion. Of course, the motor's actual application will also need to be implemented as a test bed. The selection of an electric motor dynamometer depends on the power range of the motors to be tested and the accuracy desired for the dyno. Dynos can be divided into three groups: small (<10 kW), medium (10 to 50 kW), and large (>50 kW) for the really big motors. It is important to note that a small motor cannot be tested very well on a medium or large dyno. The medium or large dyno's energy latency (the amount of energy it takes to operate the dyno) may be several hundred watts and will burn out a small motor or cause large errors. The dyno should be designed for the motor power ranges and operated within those

TABLE 17.1 Motor Control Test Equipment Requirements

Test to be performed	HV power supply	LV power supply	PC development station	Digital storage scope	Digital multi-meters	Dynam-ometer	Current probes	Temp-erature meter	Voltage probe isolation
Control logic debugging		✓	✓	✓	✓			✓	
Power stage evaluation	✓	✓	✓	✓	✓		✓	✓	✓
Motor control system test	✓	✓	✓	✓	✓	✓	✓	✓	✓

ranges. One should consider the dyno's ability to sustain a locked-rotor test and serious motor overloads. Some dynos can be overloaded for a short period of time but will burn out if operated too long in an overload condition.

17.2.1.1 Dyno protection. One simple way to minimize dyno overheating problems is to attach thermally controlled switches to areas of the dyno that will overheat first, such as the brake unit. The thermal switches can be wired to shut down the test motor's power and/or to turn on auxiliary cooling and a warning light. This simple add-on can save the expense of replacing or rebuilding an expensive dyno.

17.2.1.2 Dyno instrumentation. The dyno's instrumentation can be operated in either a manual or computer control mode. The manual control mode is beneficial for initial testing. It requires little setup time and can serve as a pretest before the computer control mode is implemented. In the manual mode test the dyno is set to load down the motor at one or two specific points. A computer test mode control normally requires a personal computer and specific software to run the dyno through a predetermined test routine. The computer records the results of the test routine and may stop the test if the results are not within a certain range. The test routine can be repeated over and over, which is important when making design changes in the motor's electronics or control strategy. The effect of the design changes can be accurately interpreted. Figure 17.2 shows a dyno test setup. The personal computer used for the dynamometer testing should be dedicated for that use only, since it is operating expensive instruments; other applications may crash the motor test software (especially the GPIB interface program) and allow the dynamometer to operate in an uncontrolled manner. An energized dynamometer should never be left unattended.

17.2.2 Motor power supplies

When testing the motor's power stage, an external DC power supply offers many advantages over a direct AC power connection. The power supply provides voltage isolation and can be current limited to minimize the chance of blowing out the motor's power stage. The voltage can be turned up slowly and can be adjusted to test the effects of under- and overvoltage conditions. The main drawback of a DC power supply for medium and large motor testing (>5 kW) is the expense.

Installing a high-voltage indictor lamp or lamps that are energized whenever the motor's high-voltage power supply is active can help increase the safety awareness of the operator and bystanders.

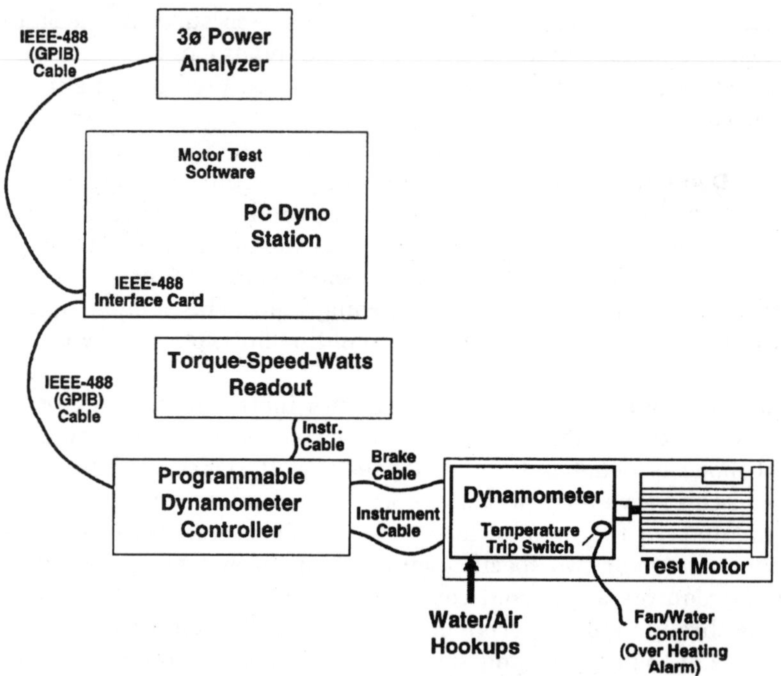

Figure 17.2 Dynamometer test setup.

17.2.2.1 AC power line controls. Another method of supplying motor power is to use a large three-phase AC variable autotransformer, which consists of round, doughnut-shaped devices with sliding arms. A large isolation transformer can provide voltage isolation when required. The variable autotransformer and isolation transformer are also useful pieces of equipment in any motor test lab. Such a setup allows the AC power line voltage to be varied but does not provide current limiting.

17.2.3 Power line test instrumentation

The AC power line will be affected by the motor's operation. Depending on the type of motor design, the AC power line will experience harmonics and voltage fluctuations that cannot exceed certain limits. There are several power line analyzers available which can measure harmonic content, power factor, peak power, true power, and voltage fluctuations. Power line analyzers can vary in their power and bandwidth ratings. It is important to select a power line analyzer that can respond to the higher frequencies associated with an IGBT-type power stage.

17.2.4 Oscilloscopes

An oscilloscope allows designers of motor controls to examine the various waveforms encountered in an electronic motor control. The oscilloscope needs to be able to capture and store high-speed single-time events. The bandwidth of a digital storage oscilloscope is less important than its digitizing or sampling rate. For example, a digital scope with a 1-GHz bandwidth and a sampling rate of 50 million per second will only provide fair performance when capturing a single-shot event with four active channels. A 200-MHz scope that has a 500-MS/s sample rate will provide good single-shot capturing ability. You should also be aware that some oscilloscopes are specified at a certain sampling rate for just one channel—the sampling rate is reduced as more channels are activated. Scopes that use separate digitizers for each channel improve the scope's single-shot capture performance.

A general-purpose digital scope can be used to debug the logic stage in motor drives that have voltage isolation between the power stages and the control logic. However, special care must be taken when examining the power stage, especially when some channels remain connected to the logic stage, since the scope's common can act as a path for noise transients. Some means of isolation is required to allow the scope to monitor both logic signals and power stage waveforms.

17.2.4.1 Oscilloscope isolation. Voltage isolation of the scope's probes can be accomplished by using special voltage probe isolation instruments. Most will handle two to four channels and will provide at least 50 MHz of bandwidth and several hundreds of volts of isolation between the scope probe's inputs and the isolation unit's outputs. (One example of such a unit is the A6907 voltage isolator manufactured by Tektronix.) Another way to obtain probe isolation is to use a handheld or battery-operated scope. Even though the logic and power stage may not be able to be probed together (the scope's common ties them together), the battery-powered scope can safely diagnose the high-voltage waveforms in the motor's power stage. The battery-powered scope does not provide a sneak path for electrical transients to flow from the scope probe's common to earth ground, as will happen with a scope that is powered from the AC power line.

One final note about scope isolation: Most serious instrument problems occur when an oscilloscope probe's commons are connected improperly. This usually destroys the scope probe, as well as the motor control electronics. It is wise to follow a basic rule: Always assume that the scope's common (which is usually connected to the scope's chassis as well) either can be tied to earth ground (if AC-powered) or, if floating, is at a high voltage that can be lethal. There are specialized handheld scopes that use isolated channels and plastic cases that help minimize the danger of a scope common that is floating at high voltages. (Tektronix THS700A series is one example.)

17.2.4.2 Voltage probes. Most scope voltage probes are designed for use in logic-level-type circuits and work well for the motor control's logic. The motor's power stage, however, may have voltages that reach several hundred volts and require special voltage probes that are rated for high voltage. The voltage rating of the probe is more important than its bandwidth rating. Normally, any probe that is rated for 50 MHz or more will suffice for power stage debugging.

A voltage probe can be used for "sniffing" EMI in a motor control circuit. This is done in two ways. The first is to add a short piece (<2 cm) of insulated wire to the probe tip and adjust the scope's sensitivity for a high setting. Moving the probe around the various components and connections will indicate which areas produce the most EMI. The second EMI test uses a voltage probe with the probe's ground clip connected to the probe's tip and, once again, the scope should be set to a high sensitivity. Remember that magnetic noise will be strongest at right angles in relation to the probe loop.

An important aspect of voltage probes is their calibration. Most voltage probes need to be calibrated for a specific scope. An improperly compensated scope probe will have a serious effect on PWM-type waveform analysis. Be sure to first check the voltage probe's compensation with a known square wave. Its voltage calibration should also be checked, using a known voltage source, before beginning circuit evaluation.

17.2.4.3 Current probes. Clamp-on current probes are useful for analyzing the motor's current waveforms. The clamp-on probe provides voltage isolation but usually requires a current probe amplifier. Current probes are available in a wide range of current levels. Care must be taken not to exceed their current ratings, or erroneous readings will occur, as well as possible probe failure. The size of the current probe should be determined by the maximum motor currents that can occur, such as during a locked-rotor event. A 15-A-rated probe to debug a 120-V 1-kW motor power stage will not work, since the motor will draw more than 40 A during locked-rotor conditions, even though it draws only about 8 A during normal operation.

Another problem with current probes is that the fast-switching edges of the motor's voltage will couple some energy into the current probe and its amplifier and may show up on the scope's screen as high-frequency noise bursts. Isolating the scope from AC power ground can help minimize this problem. These noise spikes rarely affect the scope's performance but can be a nuisance when examining low-level signals in the logic side of the motor electronics.

Current probes should be verified for accuracy with a known current level, and their high-frequency characteristics should be checked with a square-wave current generator. This is important because the motor's current waveforms may appear distorted when, in fact, a current probe problem is at fault. A current probe can incur magnetic saturation that will cause its waveform to become highly distorted.

17.2.4.4 Personal computer interface. In a lab motor test setup, it is advantageous to have a scope that can interface to a personal computer. This allows the scope's waveforms and settings to be stored in a PC format, which is very useful when the time comes to generate technical reports. Most scope manufacturers have some type of PC software available that can import and export waveforms.

17.2.5 Motor electronics handtools

It is a good idea to use insulated handtools when working with any electronic modules. More electronic circuits have probably been destroyed during testing by an accidental short from a screwdriver or set of pliers than by any other means.

17.3 Logic and Power Stage Troubleshooting

Thus far, we have discussed various instruments that can be used to test motor electronics. In this section we will give some specific tips and advice on how to locate potential problem areas with these instruments. The motor's logic elements and control circuits should be tested first—any problems in these circuits will probably damage the power stage.

17.3.1 Testing the motor's logic stage

The motor control's electronics consists of low-level logic elements, such as MCUs, and high-power devices, such as IGBT power modules. When first testing a new motor control design, the power stage should be deenergized until the logic and other low-level elements are checked.

An oscilloscope, small lab power supply, and function generator can be used to analyze the motor control's logic circuits. The function generator can supply feedback signals, such as motor speed, and the lab supply can be connected to A/D inputs to simulate current levels.

With these various signals connected to the control circuits and MCU, the PWM outputs can then be tested to see if they are correct. Essentially, all types of input conditions should be tested to verify that the PWM output signals remain stable. Remember that a glitch at this time will not blow up the power stage, which will happen when the power stage is energized.

Particular care should be taken to verify that the MCU's program does not hang up or crash during low-power-supply voltages. The PWM output signals should never simultaneously switch on the upper and lower IGBT power switches of a power stage. This might occur during MCU start-up, resets, or noisy feedback signals, for example.

Once the control logic is stable and appears to be operating as expected, the power stage can be energized, but with care.

17.3.2 Testing the motor's power stage

Testing the motor's power stage is that time of a motor control project which most experienced motor control designers dread. There is nothing more frustrating than to apply power to the motor's power electronics and have a failure occur within a few milliseconds.

Before testing of the high-voltage stage, a scope should be set up to measure the IGBT's gate signal, collector current, and collector-to-emitter voltage. When these signals are deemed satisfactory, the motor current and voltage waveforms can be analyzed. If possible, one scope should be dedicated to monitor the IGBT's voltage and current during any motor testing.

17.3.2.1 Softening the high voltage. When first testing a prototype design, one way to reduce power stage failures is to limit the high-voltage power supply current to a low level. This assumes that the motor is unloaded or not even connected. It is possible to use resistors as a load in place of the motor. These resistor values are selected to load down the motor's power stage at a low level—just enough to verify the power stage's operation.

A series resistor in the power supply's positive lead can also limit its current when using a non-current-limiting power supply. By limiting the current, any serious power stage malfunctions may be able to be detected before serious damage occurs. Besides limiting the current, the high voltage to the motor's power stage should be increased slowly. For example, a 350-V power stage should be operated at 50 V or less. (*Note:* A certain minimal voltage may be required to operate its bootstrap gate-bias circuits.) Once the power stage appears operational at reduced high voltages, the high voltage can be slowly raised—for instance, in 25-V increments—and tested at each increment.

Any abnormal waveform or unexpected audible noise should be investigated before proceeding to full voltage and current levels. Audible noises in the motor can be useful for troubleshooting.

17.3.3 Audible noise

Human hearing can often help in detecting unstable motor control operation. The human ear generally is most sensitive to frequencies in the 1- to 5-kHz range. An incorrect PWM signal can cause a chirping sound, while random ticking sounds may indicate a program error that can be difficult to track down. (See Chap. 12.) Low frequencies (<20 Hz) and high frequencies (>12 kHz) are seldom easily detected by humans. Adding a wide-range microphone or accelerometer to the motor case and connecting their output to a scope can also help detect abnormal motor operation. Some motors, such as switched reluctance types, are more prone to produce acoustic noise—the microphone's feedback can be helpful in measuring the effect of different PWM and control strategies on the motor's noise levels.

17.3.4 Temperature monitoring

Temperature probes should be attached to the power stage's heatsink and the motor. Temperatures usually change over several seconds or even minutes but should be monitored during any serious motor tests. A sudden temperature rise in the heatsink can be an indication of impending disaster. Any new motor control design should also be thermally scanned to verify that all its components are operating within their normal ranges. A noncontact IR temperature meter can be used instead of more expensive thermal scanning equipment.

Once a motor and its electronic control is placed in operation, its heatsink and motor temperature should be tested over a wide range of operating conditions to ensure that they have adequate ventilation. Another useful instrument for evaluating heatsink cooling fans is an airflow meter. An airflow meter with a small probe end can be inserted between the heatsink's fins to determine airflow patterns. A simple change in the fan's ductwork or location can often have a significant effect on the airflow and consequential heatsink temperature.

17.4 Other Motor Control Trouble Areas

There can be some interesting side effects when testing large motor controls. The magnetic fields will affect the oscilloscope's CRT (cathode-ray-tube) display if the scope is within a meter or two of the motor's leads. This is usually more of an annoyance than a serious problem.

17.4.1 EMI-related problems

Computers can also be affected by the electrical noise from the motor's power stage. It is not unusual for the MCU development station to completely lock up if any of its lines are directly interfaced to the motor's power stage. Even if isolation is used, as with optocouplers or noncontact current transducers, enough EMI can still be coupled into the development station to cause problems. The problems usually occur during high-current testing or during locked-rotor conditions, which is the worst time for the logic to fail. Operating the MCU development station from a different AC power source may help, but a battery-powered MCU development station, such as a notebook personal computer, may be required.

17.4.2 Motor insulation integrity

The motor's high-voltage fast-switching edges, normally less than 0.5 μs, stress the motor's winding insulation. This issue has recently been addressed with higher-performance winding insulation materials, but the motor's windings are still subjected to extra stress by the fast PWM transitions. Random coil windings in particular are a concern, since adjacent windings can be at full coil voltage. As we have emphasized throughout this book, there are always design tradeoffs. In this case, slowing down the IGBT's switching speed involves increased heatsink dissipation versus reduced EMI and motor winding stress.

17.4.3 Motor lead lengths

If at all feasible, the motor's power stage should be located as close as possible to the motor. Positioning the motor and the power electronics several meters apart will create problems because of the fast PWM waveforms in the motor's leads. Under certain circumstances, the motor's inverter peak voltage can double if the motor's leads or cables are too long. Another problem is created when the inverter's leads of several inverters and motors are placed in metal conduits. These leads inside the metallic conduit conduct eddy currents (from the high-frequency PWM voltages), causing the conduit's temperature to rise; in some instances, it may rise enough to cause burns, melt insulation, or produce potentially explosive situations in certain atmospheres.

17.4.4 Motor bearing degradation

The fast PWM high-voltage pulses can capacitively couple voltage to ground across the motor's bearings. The voltage is not directly conducted to ground because the bearing's grease forms a dielectric. When this dielectric is exceeded, an arc occurs that pits the bearing surfaces, eventually causing the bearings to run rough. Some methods to minimize this problem include using electrical conductive bearing grease, insulated bearings, and a shaft grounding brush mechanism.

17.4.5 Regeneration dangers

When an electronic motor control is powered from a DC power supply, the DC voltage during regeneration may greatly exceed the nominal DC power supply voltage. The DC power supply will normally act as a series rectifier and block any excess voltage, which means that the excess voltage has no return path and raises the supply voltage bus level. This will cause catastrophic results to the inverter and power supply—the filter capacitors can explode, and IGBTs can fail. An errant program can often invoke a regeneration mode during motor coastdown, and, in the case of a large dyno and motor, there will be sufficient mass to act as a flywheel. This flywheel effect provides high peak power. Figure 17.3 shows a voltage-clamping circuit that can contain excessive voltages brought on by regeneration. This clamping circuit uses lamp bulbs that can be sized according to the amount of regeneration energy. Lamp bulbs make fairly good power resistors

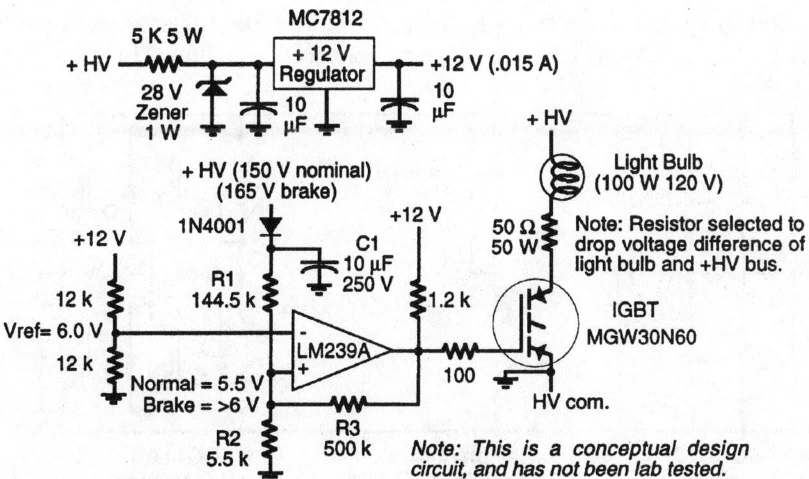

Figure 17.3 Regeneration voltage protection test circuit.

since they can dissipate over 100 W each and exhibit a positive temperature coefficient. When they are cold or have remained off for a few minutes, their internal filament resistance is about one-tenth of their nominal hot resistance. This allows the lamp to absorb a high inrush of current from a regenerating motor. A comparator with hysteresis is used to ensure that the IGBT power switch operates in a stable fashion. When the high voltage exceeds a certain value, the lamp or lamps turn on and remain on until the comparator input voltage drops below the initial trip point by about 100 mV. Resistor R3 sets this hysteresis value.

Note that a series diode and a capacitor (C1) are used as a peak detector on the high-voltage (HV) bus. The size of C1 will affect the time that the lamp bulb remains on after the HV bus has returned to its normal voltage level. This peak detector network helps to stabilize the comparator. One problem with this type of circuit is that the HV bus voltage will immediately sag when the lightbulb is energized because of the light's high inrush current. (The 50-Ω series resistor will help limit the inrush current.) When the HV bus sags, the comparator's output remains high because of the series diode and C1 capacitor. The capacitor must discharge through R1 and R2. The test circuit values are for a motor in the 500-W range with a load that acts as a small flywheel. A different motor size and its flywheel value would require different circuit values.

17.4.5.1 Motor braking. Some motor controls are deliberately operated in a regeneration mode to slow down or brake the motor. In these types of controls, a large power resistor is used to dissipate the regeneration. A special power stage switches this power resistor across the motor's high-voltage DC power when required, as shown in Fig. 17.4.

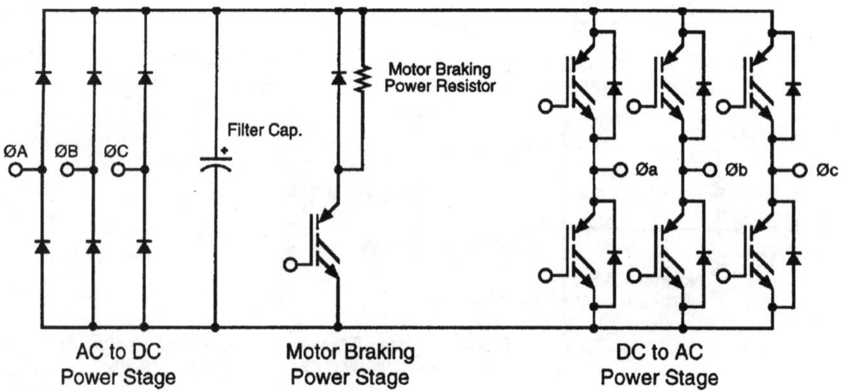

Figure 17.4 Regeneration voltage dissipation with a power resistor.

Needless to say, this can waste considerable power if the motor is continually operated in a run-brake–run-brake sequence. Using a larger-than-normal capacitor bank might be used to store more of the regeneration energy. This assumes that the capacitor's and IGBT voltage ratings are also increased.

Another way to improve the motor control system's power efficiency during braking would be to transfer the energy back into the AC power line. Unfortunately, this requires a PWM inverter power stage and AC power control logic, which will significantly increase the motor drive's cost.

A permanent magnet motor can be braked by shorting out its windings with the power stage. In a three-phase brushless DC motor design, all of the top or all of the bottom power transistors could be switched on. The downside to this method is that motor overheating may occur if it is continually operated in a run-brake–run-brake manner. The power transistor sizes may also need to be increased.

17.5 Voltage Transient Testing and Protection

Electronic circuits are inherently susceptible to sudden failure from transient overvoltages. There are two basic ways of dealing with voltage spikes. The first is to clamp or absorb the voltage spike (usually with a zener-type device). The second is to isolate the electronic circuit from the voltage spike with some type of electronic switch that opens when an overvoltage condition is detected.

Most transient-overvoltage protection methods use a clamping method rather than an isolation element. The isolation element must be rated to block the maximum voltage transient that can ever occur—over 10,000 V for typical AC power outlets. Isolation elements are used in some automotive electronics to isolate electronics on the 12-V bus from excessive voltage created from an alternator "load dump." (See Chap. 4.)

17.5.1 Voltage transients in motor controls

Voltage transients in an electronic motor control can cause damage in several areas. The AC power line can be a source of high-voltage transients that can be in excess of 10 times the normal AC line voltage. These AC power line voltage spikes can damage the motor control's converter, specifically its bridge rectifiers and filter capacitors.

Intermittent wire connections between the motor and the inverter can generate significant voltage spikes due to the inductance in the leads and motor. The IGBTs, motor windings, and free-wheeling diodes will fail if subjected to excessive voltage spikes. Because a motor produces vibration, there is a high possibility that its wire terminals will become loose over time, which will cause an intermittent connection. A loose connection can also overheat, resulting in a fire.

Other voltage transients in an electronic motor control are related to the PWM switching in the power stages. These transient voltages occur during the PWM transition times and are a different problem than the AC power line voltage spikes or intermittent connections.

17.5.2 Voltage transient suppression methods

Several devices are available to clamp transient-voltages spikes. Back-to-back zeners and MOVs (metal oxide varistors) are two popular devices that can clamp momentary voltage spikes. Back-to-back zener suppressors tend to have very accurate voltage clamping, while MOVs, although cost-effective, lack the accuracy of the zener. MOVs also exhibit a softer knee, as shown in Fig. 17.5. The voltage rating should be selected to protect the electronics. When you use either MOV or zener transient protection devices, you must take the precaution of ensuring that they have an adequate joule rating. Most voltage clamping devices are rated on a single-pulse test and have a high peak rating. Care must be taken when operating these devices in an application that continually generates voltage spikes, since their

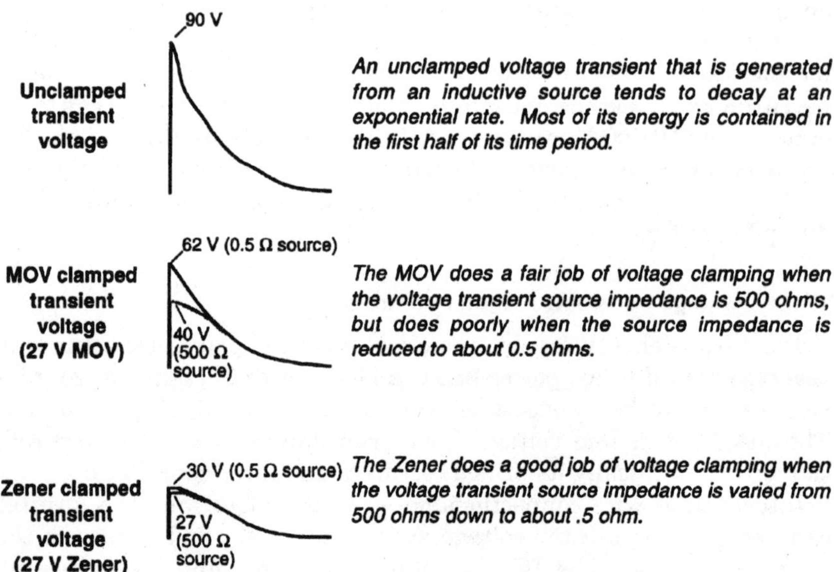

Figure 17.5 Zener versus MOVs for transient voltage protection.

average power rating will be significantly lower than their peak rating. Their energy-handling capability is determined primarily by the transient voltage's duration.

17.5.2.1 Snubbers. Snubbers are simple circuits that may consist only of a single high-frequency capacitor or may consist of a capacitor-resistor-diode network. The exact component values of a snubber depends on the application. Figure 17.6 shows four types of snubber

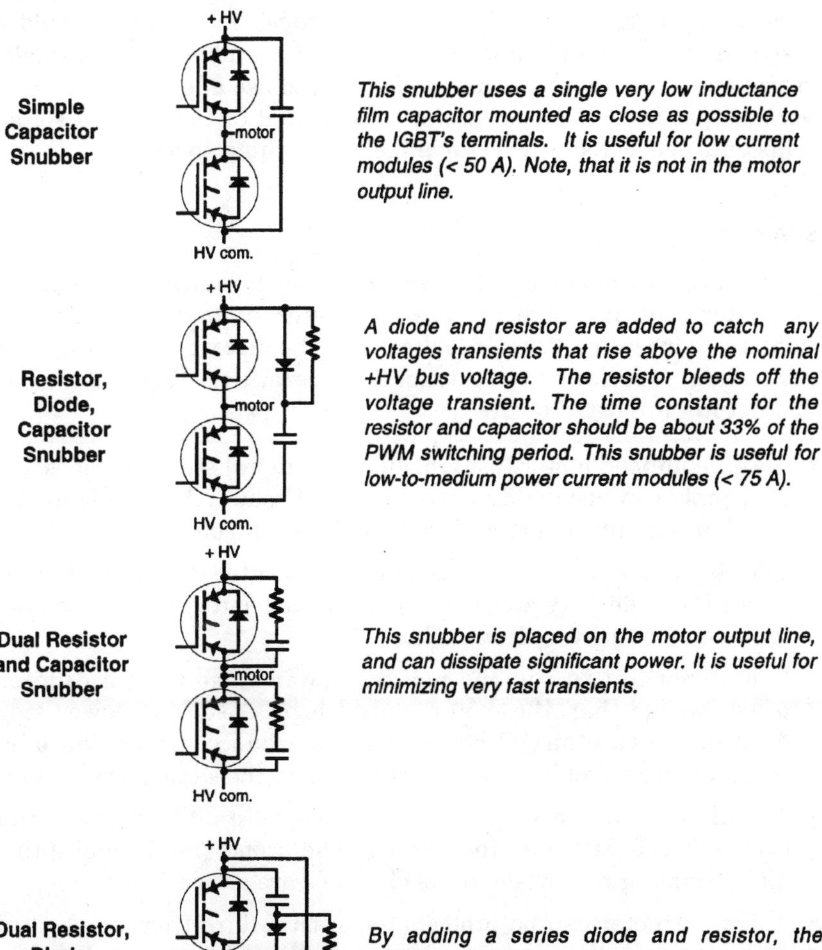

Simple Capacitor Snubber
+HV
motor
HV com.
This snubber uses a single very low inductance film capacitor mounted as close as possible to the IGBT's terminals. It is useful for low current modules (< 50 A). Note, that it is not in the motor output line.
Resistor, Diode, Capacitor Snubber
+HV
motor
HV com.
A diode and resistor are added to catch any voltages transients that rise above the nominal +HV bus voltage. The resistor bleeds off the voltage transient. The time constant for the resistor and capacitor should be about 33% of the PWM switching period. This snubber is useful for low-to-medium power current modules (< 75 A).
Dual Resistor and Capacitor Snubber
+HV
motor
HV com.
This snubber is placed on the motor output line, and can dissipate significant power. It is useful for minimizing very fast transients.
Dual Resistor, Diode, Capacitor Snubber
+HV
motor
HV com.
By adding a series diode and resistor, the snubber's power loss is reduced. This snubber is useful for high current modules.

Figure 17.6 IGBT motor snubbers.

circuits that are used in IGBT motor drives to suppress transient voltages generated by the PWM transitions. Snubbers rely on a capacitor to conduct the high-frequency components to ground or common. The placement of the snubber is therefore critical, since these high-frequency currents can again cause problems if there is too much loop inductance in the snubber network. The power dissipation of a snubber circuit can be significant if its values are incorrect. Normally, snubbers are required to “quiet down” an IGBT and freewheeling diode power if that power stage has too much internal or parasitic inductance. Ideally, the power stage's inductance should be designed to be minimal, allowing the use of a snubber that is smaller in size and less costly. It is not a wise idea to use a large snubber network to “patch over” a poor power stage layout or to filter out voltage transients generated by excessive IGBT switching speeds.

Summary

A well-equipped motor test lab with a proper layout is a vital part of the motor control design process. Computer simulations need to be validated, since unforeseen problems may be discovered during rigorous testing, especially over extended temperatures. Key points about this chapter are listed below.

- A well-equipped motor test lab will have digital storage scopes, current probes, precision dynamometers, AC power harmonic analyzers, digital multimeters, and controlled power sources.
- The dynamometer should be centrally located in the motor test area to provide easy access for motor changeouts and close hookup to the motor's power electronics.
- It is important to run the motor's power stage from a different power source than the motor control logic. A serious power stage fault may create an AC power line disturbance, which will affect the motor's control logic if it is connected to the same power source.
- The digital storage scope's ability to capture and store one-time events is critical in motor testing. The scope should maintain a high sampling rate when all its channels are active.
- A dynamometer can be protected against failure from overheating with a temperature sensor mounted on the dynamometer's hottest point. To protect against overheating, the temperature sensor can activate an alarm or trip a main power relay that shuts off the motor.

- During initial testing of the motor's power stage, the power supply should be current-limited to a minimal value, and its voltage should be slowly raised to minimize initial failures. Adding a series resistor in the high-voltage positive bus lead can also "soften" the power supply and may help prevent power stage failures until the design bugs are corrected.
- The use of battery-powered scopes and high-voltage probe isolators helps minimize EMI noise in the scope's waveforms. They also minimize grounding problems when testing the motor's high-voltage power stages.
- Several reliability concerns are associated with a motor when it is controlled by a PWM-type motor control, including motor winding insulation deterioration, bearing pitting, and metal conduit heating.
- Regeneration may occur when the motor's load has high inertia and operates as a flywheel. The motor turns into a generator and may produce excessive voltages that can destroy the power electronics.
- Voltage transient suppressors are designed to clamp or clip momentary voltage peaks, thereby protecting electronic devices from excessive voltages. Snubbers are designed to minimize high-frequency noise, usually caused by the fast PWM signals in a motor's power stage.

Further Reading

Cherniak, S., *A Review of Transients and Their Means of Suppression,* Motorola Application Note AN843, 1991.

Colby, R., Mottier, F., and Miller, T., "Vibration Modes and Acoustic Noise in a Four-Phase Switched Reluctance Motor," *IEEE Transactions on Industry Applications,* Nov./Dec. 1996, vol. 32, no. 6.

Dietz, D., and Baumgartner, H., "Analyzing Electric-Motor Acoustics," *Mechanical Engineering,* June 1997.

Jouanne, A., Enjeti, P., and Gray, W., "Application Issues for PWM Adjustable Speed AC Motor Drives," *IEEE Industry Applications Magazine,* Sept./Oct. 1996.

Petro, D., and Bell, S., "How Reflected Waves Damage AC Motors," *Power Transmission Design,* Nov. 1997.

Magtrol, Inc., *Hysteresis Absorption Dynamometers* specification sheet, March 1995.

Manz, L., "Motor Insulation System Quality for IGBT Drives," *IEEE Industry Applications Magazine,* Jan./Feb. 1997.

Motto, E., *Power Circuit Design for IGBT Modules,* Powerex Application Note, Sept. 20, 1993.

Nailen, R., "Are ASDs More Trouble Than They're Worth," *Electrical Apparatus,* Nov. 1996.

Part

6

Futuristic Motor Control

Chapter

18

Future Motor Control Electronics

Richard J. Valentine
Motorola Semiconductor Products

Introduction

Motors in the future will continue to be found in almost every kind of electrical and electronic equipment. The concept of harnessing rotating magnetic fields is still evolving, since new types of motors such as switched reluctance are just now being introduced in appliances. Electronics technology, especially microcontrollers, will allow even further advances in motor designs and their controllers. The standard AC induction motor has traditionally been designed to operate from a fixed AC line voltage and frequency. With the advent of cost-effective motor control electronics, the AC power line will no longer be the prime factor in designing motors. Since electronics can generate higher frequencies and variable power levels to match the motor's exact load requirements, smaller and lighter motors may be possible.

There are several possible motor designs that can employ high-technology components that have yet to be fully developed. One is an ultrafast motor—a motor that would operate at RF speeds. Another is a smart motor control that can adapt or be easily programmed or trained for specific applications. In this chapter we will discuss some possible futuristic motor designs and technology advances that may affect motor control designs.

18.1 Impact of Future Technology on Motor Control

It is amazing how quickly technology is advancing, especially in microelectronics. The use of finer geometries and multilayer metalization is increasing the density of microcontroller devices, as well as improving their price/performance ratio. It is inevitable that these small, but powerful, MCU devices, in addition to power IGBTs or the next generation of power devices, will be employed in many motor applications. Electronic motor controls will allow the motor to operate more efficiently and help reduce motor failures induced by operator abuse.

18.1.1 MCU advancements

Microcontrollers will be used in a wide range of motor controls. Even low-cost motors can employ an inexpensive MCU to gain power efficiency and allow simple speed and other motor control commands to be implemented. Because MCUs are processed with low-voltage materials, they will probably not be directly integrated into a high-voltage power device, but they may be added into a power module. As its complexity increases, the ease with which the MCU can be programmed and tested will be a prime issue.

Semiconductor materials and process are continually evolving, which, again, will lead to higher-performance devices. Microcontrollers will be optimized for fast throughput and have internal precision analog-to-digital (A/D) conversion units. The programmability and development tools for MCUs will also be more user-friendly. Canned software will become a viable business for embedded controllers.

Competition for the smallest memory demand and most power-efficient motor control programs will be fierce and may be licensed by the MCU manufacturers to help differentiate their MCUs from those of their competitors.

18.1.2 Power device advancements

Power IGBT technology is the current popular choice for motor power stages. A few years ago, power bipolars, Darlingtons, and thyristors were in high demand. The next generation of power devices will include more efficient IGBTs, higher-temperature materials, and improved packaging. Because of the high voltages required in motor control power transistors, their internal design is constrained by the need to block high voltages. This means that finer design rules will not have the higher-density effect that occurs in low-voltage microcontroller devices. Other advanced semiconductor materials, such as gallium arsenide and silicon carbide, may be employed in power devices.

It is possible to integrate simple circuit elements, including voltage protection or current limiting, with IGBTs. Adding a temperature sensor to large IGBTs or IGBT modules is also beginning to occur.

18.1.2.1 Power bipolar IC advancements. The MCU is limited to low voltage and current levels, making it costly to integrate high-voltage or high-current drivers into the MCU die. The IGBT power stage requires a gate driver IC and other interface devices. A power linear IC that combines the gate drives, optocouplers, and fault detection circuits is possible and would simplify a motor control design. The cost consideration when developing power analog ICs is generally set by the voltage breakdown requirements, current levels, and number of processing steps. At some point, it is usually more cost-effective to separate some of the functions and use discrete parts. The challenge for the power analog IC designer is to integrate as much as possible while maintaining a cost parity with a discrete solution. It is difficult to sell an expensive power IC when the same function can be accomplished with a low-cost IC and external SMT transistors.

18.1.3 Imaging and other advanced sensor technology

As motor drive electronics becomes more sophisticated, the sensors required for motor-based application will also increase. Digital imaging devices can "look" at a motor-controlled mechanism and instruct the motor control to take certain actions on the basis of what it "sees." For example, a digital imaging device can scan a large fan's inlet duct for large objects and try to stop the fan motor before any objects enter the rotating fan blades. Digital imaging is capable of fairly high resolution and will rely on a high-performance graphics-oriented microcomputer to decipher its scanned images. It is also possible to use infrared imaging to look for temperature-related data.

One device that is a prime candidate for cost improvement is the motor's current sensor. The present noncontact current-sensing devices are too expensive to be employed in household or small motor drives. There may be a way to integrate semiconductor designs (such as high-gain temperature compensated op amps and analog-to-digital converters) with a low-cost magnetic field sensor that could be used in 300-W to 2-kW motor drives and cost less than $1 in production volumes. There have been various methods under development to eliminate the need for motor current sensors. Most of these designs use the motor's back EMF and require sophisticated voltage processing circuits. Since the motor's back EMF can vary according to the motor's load, the sensing circuits may need to be fine-tuned for each motor system.

18.2 Advanced Motor Controls

There are several motor control advancements in progress. One of these, a switched reluctance motor drive for washing machines, has just been introduced. Other advanced motor controls, such as very high-speed motors, are used for special-purpose applications and may become widespread as their cost diminishes. Smart motor controls will allow easy programmability for individual applications. An ultrafast motor design that is truly futuristic is presented in this chapter.

18.2.1 Smart motor controls

It may soon be possible to construct a low-cost motor control with enough memory to store several built-in motor functions and control strategies. The end user would simply hook up a small personal computer to the motor drive's communication port and pick the appropriate motor control parameters required for the specific application. This would be accomplished with an easy-to-use program with graphics and would not involve programming languages. Essentially, the user would just click on the appropriate graphic boxes and enter specific values to program the motor drive. The motor drive would then retain these user-defined variables in nonvolatile memory and execute its latest program.

A motor distributor would need to stock only a minimal number of motor control systems—maybe one for small motors (<1 kW), one for medium motors (1 to 5 kW), and one for large motors (5 to 25 kW). The user could then customize the controller to each application, whether it is a temperature-controlled variable-speed fan, or a multiple-speed motor used in a factory tool.

18.2.2 Ultrafast motors

How fast can a motor be designed to rotate? Normal AC-powered motor speeds are controlled by the line frequency and the motor's poles; they usually operate at less than 10,000 rpm. Some AC and DC motors can be designed to operate at up to 100,000 rpm but are limited by mechanical constraints, mainly the brush/commutator assembly for DC motors and the bearing friction for any type of motor. Some uses for small ultrafast motors include miniature drills or cutting tools. Larger high-speed (>50,000-rpm) motors may drive turbine compressors that feed air (oxygen) to a fuel cell.

Electronic commutated motors can rotate at extremely fast speeds. Figure 18.1 shows a conceptual design for an ultrafast motor. The motor's windings may consist of just one or two turns, and its shaft will be fairly small. The bearings may need to be a well-controlled magnetic flux or may be constructed with some type of very low-friction material. Obviously, the motor's speed will be limited by the speed of electricity, which is about 30 cm (11.8 in) per nanosecond. The power stage's transistors and motor inductance will also be a factor that will limit the PWM switching speeds.

For the sake of discussion, let's assume that a power transistor that is able to switch in about 5 ns can be used. This power stage transistor would be designed as an RF device and would be optimized for switching operation. (One example of such a device and package is the MRF183 RF Power FET, a 1-GHz lateral N-channel MOSFET from Motorola Inc.) A very high-speed logic IC (>500 MHz) would create the commutation pulses. The three-phase motor's rotor material would need to be optimized for very high frequencies. The motor's winding inductance must be extremely low (<200 nH) and may consist of wide strips of copper. The motor would be operated in a vacuum, at near absolute zero temperature, with a friction bearing for initial start-up that would disengage at high speeds. A controlled magnetic flux would attempt to keep the shaft under control as it spins at extreme speeds.

Table 18.1 shows some pertinent calculations that apply to this conceptual ultraspeed motor. It is worth noting that a motor, in theory, could run at very high frequencies. What is more interesting—and may be more in the realm of science fiction—is what would happen if the motor's speed were increased to 1 billion revolutions per second. The

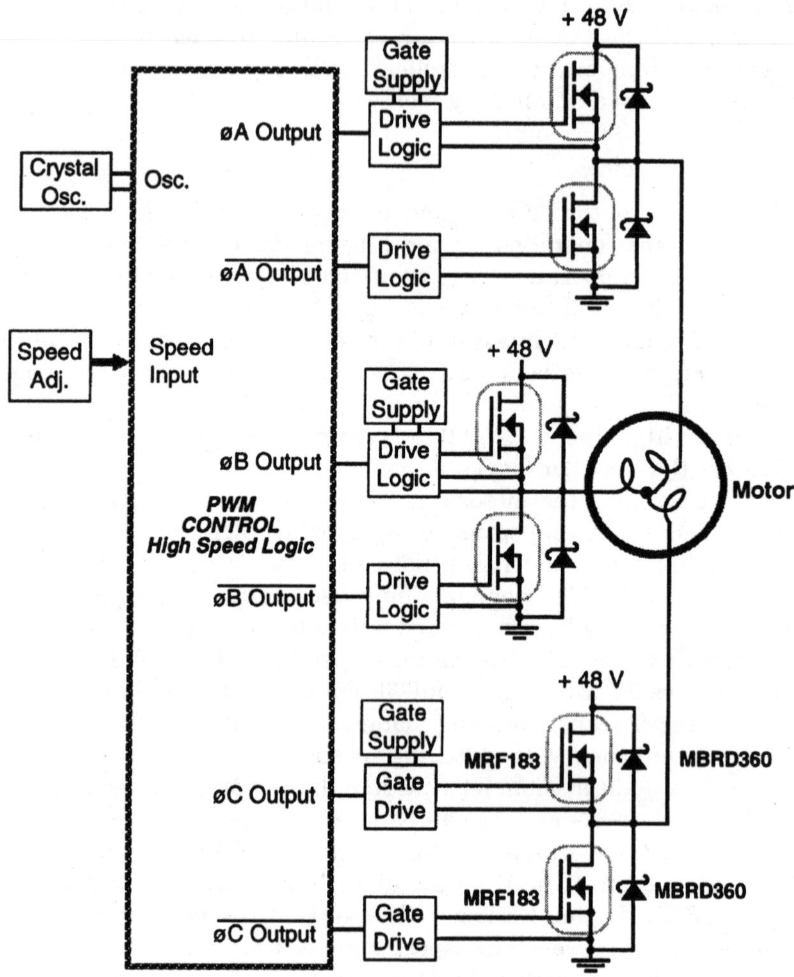

Figure 18.1 Ultrafast motor control conceptual design.

motor's speed cannot exceed the speed of light (or electricity), but there may be a way to mechanically overcome this limitation. A 10-cm (4-in)-diameter thin disk is attached to the motor's shaft. The disk's outer rim could be traveling at a rate faster (in theory, anyway) than the speed of light or electricity (31,750,000,000 cm/s or 15,500,000,000 in/s). This would be accomplished by mechanical means, and it is highly debatable whether any known materials could stand up to the stress levels caused by such extreme velocities. In any case, this supposition does show that the control of rotating magnetic fields in present-day motors

TABLE 18.1 Ultra-High-Speed Motor Design Considerations

Electronic limiting factors of motor control speed*	Parametric data
Speed of light (electricity)	299,792,458 m/s (983,571,056 ft/s) 29.98 cm/ns (11.8 in/ns)
Logic IC speed	Clock speed limited by silicon technology and design; 300–500 MHz is attainable
Power stage switching speed	Switching speed limited by materials technology and package layout; 5 ns switching speed should be attainable with RF-type MOSFETs
Motor inductance	Motor inductance would need to be <200 nH for 40 MHz and <8 nH for 1 GHz for 50 Ω impedance

*The motor's mechanical design will also be a speed limiting factor.

Motor velocity calculations	(This assumes motor control electronics operating at 1 GHz speeds)
Rotor speed	Limited by speed of electricity, logic IC, power stage, and motor design. For discussion purposes, assume 1,000,000,000 rev/s
Rotor diameter	Rotor diameter will be small [<5 mm (<0.2 in)], with a 10.11-cm (4-in)-diameter disk mounted to the rotor
Disk circumference	Circumference, cm = π*diameter; 31.75 = 3.141*10.11
Disk velocity	Motor velocity, cm/s = circumference*rev/s; 31,750,000,000 = 31.75*1,000,000,000

Motor velocity calculations	(This assumes motor control electronics operating at 200 MHz speeds)
Rotor speed	Limited by speed of electricity, logic IC, power stage, and motor design; for calculation purposes, assume 40,000,000 rev/s
Rotor diameter	Rotor diameter will be small [<5 mm (<0.2 in)], with a 10.10-cm (4-in)-diameter disk mounted to the rotor
Disk circumference	Circumference, cm = π*diameter; 31.75 = 3.141*10.11
Disk velocity	Motor velocity, cm/s = circumference*rev/s; 1,270,000,000 = 31.75*40,000,000

can be significantly enhanced by the use of RF components and design methods. Electronic circuits that operate at 1 GHz or higher are quite common in communications equipment.

18.2.3 Robotics and motors

Robots that perform routine, time-intensive tasks are under development. Indeed, some robotlike devices have been demonstrated that can even navigate stairs or investigate landscapes on distant planets. High-performance motor systems are required to make a functional robot. One basic limiting factor for robots, as well as other portable motor-powered machines, is their limited power source. Serious research is under way for cost-effective fuel cells and high-power-density batteries that may make robotics practical. The other technologies required for a robot system are already available, including digital imaging and processing, ultrapowerful microcomputers, precision servo/stepper motors, and strong, but lightweight, body materials.

18.2.4 New and improved motor control applications

Any task that involves mechanical movement is a potential application for a motor drive. As motor controls become smarter, smaller, and less costly, they will be employed whenever they can make a product easier or more convenient to use. The remote-control feature of sliding doors in passenger vans is one example in which electronics and motors have been successfully integrated.

18.2.4.1 Future automotive motor systems. Most motor controls in future vehicles will be part of a system, such as steering or braking. Electric motor-powered vehicles are not new, but their electronic controls are becoming more sophisticated, especially in hybrid systems, where both an electric motor and internal-combustion engine are used in combination. In a fully automated road system, the vehicle's direction will be managed with an electronic motor steering system. Other futuristic vehicular motor controls include electronic suspension, integral alternator-starter, and automatic tire pressure inflation. The use of multiplexing or serial data networks will allow complex motor controls to communicate with other vehicle electronics and will also provide a means for diagnostics.

Motor electronic controls in an automotive application must be especially tough but cost-effective. The automotive environment of today and the foreseeable future will continue to subject motor controls to high temperatures (130°C), vibration, fuel and oil spills, cleaning solvents, water, and overvoltage transients. Some future automotive motor control designs will operate from a higher electrical bus than 12 V, probably 42 V. This will reduce motor current levels in a steering system, for example, to levels that are more easily accommodated.

18.2.4.2 Future home appliance motor systems. Heating, ventilation, and air-conditioning are prime candidates for improved and more power-efficient motor controls. Remote-controlled duct vent motors would allow zone temperature control and would help optimize temperatures throughout a large residence. Each duct vent control would be tied in to the central HVAC control computer whose program would vary the airflow, AC or furnace temperatures, and remote duct vents. The duct vent control would also include both receiving and sending units. The sending unit would indicate the duct vent control status, duct vent position, zone ambient temperature, and duct outlet temperature.

Most household appliances are already motorized with fairly simple control elements. Future appliance motors will be more power-efficient. They will also conform to power line standards by the employment of advanced electronics and higher-performance motors. Because of gas engine noise and emission concerns, lawn equipment such as handheld hedge cutters, line trimmers, lawnmowers, and so forth, are shifting to electric motors.

Cordless shop power tools are beginning to use higher-power-density batteries. Some of these tools are prime candidates for advanced electronic controls that go beyond allowing only variable motor speed. One example would be a tool status indicator to show the amount of battery charge that is left, whether the tool is overheating, and the precise setting of the tool's torque limit (especially important for drills that drive screws or tighten fasteners). Most tools use mechanical rulers or gauges to indicate settings. It might be very possible to add an electronic digital caliper to a power tool to ensure that the settings would be easy to read, as well as highly precise.

18.2.4.3 Future industrial motor systems. The development of better motor controls in a factory usually has a positive impact on the financial balance sheet. A motor drive system that is easy to operate and allows careful optimization of a production line's speed can create a competitive edge for any company. Motors and controls may operate year after year with minimal hassle if monitored by motor electronics that provide regular maintenance messages, as well as warnings of impending failures. Reliable predictions of imminent motor failures can provide opportunities for scheduled repairs, rather than the continual threat of frantic emergencies.

Factories tend to use large motors that consume considerable electricity, usually in the hundreds of thousands of kilowatt-hours. A 5 to 10 percent gain in motor power effectiveness by the use of more efficient motors and electronic controls can result in significant cost savings.

18.2.4.4 Other future motor systems. Electronically controlled motors are used in cameras, medical equipment, office equipment, satellite and space systems, and military applications. As these products are improved, so, too, are their motor controls. The latest digital cameras use motors for their zoom and autofocus features. The most recent generation of high-speed trains employ powerful motors and electronic controls.

18.3 Motor Design Improvements

Thus far, this handbook has focused on motor control electronics. It is the author's opinion that motor fabrication will also advance to take full advantage of electronic control circuits. For example, most motors are designed for optimal performance at 50- or 60-Hz operation. An electronic motor drive can easily supply 100 to 400 Hz, or even higher frequencies, as faster MCUs and IGBTs become available. This means that the motor's inductance and materials can be optimized for higher frequencies, which will probably allow the motor size and configuration to be reduced.

Most new and improved motor designs will be based primarily on computer engineering tools, especially with the advent of inexpensive and powerful personal computers. This will probably include not only the optimization of the motor's internal magnetics but also the development of better motor materials. The motor's wire, rotor, stator metal compounds, insulation, bearings, lubrication, and frame will also be affected by computer modeling.

18.4 Ideal Motor Electronics Design

This handbook has presented many motor control designs and circuits. Most of these motor control designs require a fair amount of electronic expertise and/or patience to implement in order to create a rugged, production-worthy product. An ideal motor control design would consist of just a few components that would be designed to plug together and operate with minimal hassle. This "plug and play" concept has been evolving in the personal computer industry and has made computer users more productive. Figure 18.2 shows a conceptual motor control system that is built around a daughterboard that plugs into a submodule, which then plugs into a main power module.

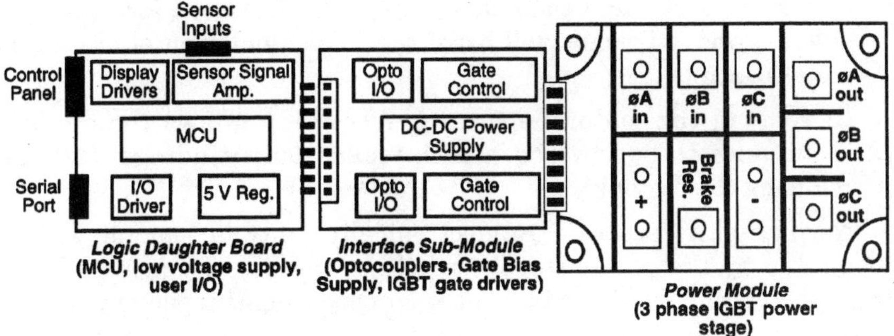

Figure 18.2 Easy-to-design motor control system.

The daughterboard contains the MCU and control logic components, while the submodule includes the interfaces, gate controls, and other high-voltage low-power motor control elements. Little circuit design effort is required—the motor user or application engineer simply picks the appropriate-size main power module and then plugs in the necessary support submodule and its logic daughterboard. The logic board would be programmed, as mentioned previously, from a personal computer.

Summary

Every factor in today's world of motor controls indicates that electronic motor controls will continue to advance, especially as better power efficiency is required by legislation. Electric motors have already become integral components in cars, homes, and industry. In some factories, electricity consumption cannot be increased, so if additional motors are required, the old motors must be replaced with highly efficient motors and electronic controls. This reduces the factory's electrical power consumption, thereby allowing more electrical loads to be added without raising the nominal power usage.

Advancements in semiconductor technology will play a valuable role in further improvements in motor efficiency, precision, and reliability. Computer devices will be utilized in all facets of motor electronics. They will be employed to design [CAE (computer-aided engineering) stations], manufacture (automated assembly and testing), market (Internet Web pages), and, finally, to control the motor in its application. Key points about this chapter are listed below.

- Advancements in electronics technology, specifically microcontrollers and power IGBTs, will have a major impact in the control of future motors and their manufacturing. The need for improved motor power efficiency will drive the implementation of electronic motor controls.

- Increasing the motor's operating frequency allows the motor's inductance to be lowered, which means less windings, thereby reducing the motor's size.
- Normally, the speed of an electronically controlled motor will be limited by the power stage's switching speeds. The motor's bearings and construction will be another speed-limiting factor.
- The integration of a microcontroller device (MCU) and the motor's power stage power transistors into one chip is seldom possible because an MCU is a low-current ($<$.01-A) and low-voltage ($\leq$5-V) device. Power transistors used in motor controls are high-current (5- to 600-A) and high-voltage (30- to 1700-V) devices.
- It is possible to combine some semiconductor devices, such as a control analog IC and two or more power transistors into one package. The challenge is to make this multichip product cost-competitive with a discrete version.
- Noncontact current sensors are a major cost factor in a motor control design.
- Switched reluctance motor systems, which offer high-torque and high-speed capabilities, have been introduced in clothes washing machines.
- A "smart" motor control could be described as one that is easily programmed to meet the exact requirements of each motor drive application. It would interface to a laptop personal computer for programming and diagnostics.
- Advanced motor controls that use microcontrollers will become part of complex systems, offering more functionality than basic motor controls.
- Motor control design can be somewhat simplified by procuring or building each of the main components of a motor control (the logic, interface, and power stage) so that they can plug together as subunits into a main module. This allows easy repair and upgrading of the individual subunits.

Acknowledgments

The author would like to thank Tom Newenhouse, Pete Pinewski, and Warren Schultz for their contributions to this chapter.

Further Reading

Buchholz, K., "Electronics Emphasis," *Automotive Engineering,* May 1997.

Emerson Motor Company, "Emerson's Inside Technology Making Your World a Better Place to Live" advertorial, Sept. 1997.

Gornick, T., "Clothes and Personal," *Appliance,* Sept. 1997.

Gottschalk, M. A., "High-Speed Motor and Drive Target Next-Generation Fuel Cells," *Design News,* Oct. 20, 1997.

Jancsurak, J., "Motors & Motor Control," *Appliance Manufacturer,* Oct. 1997.

Motorola, Inc., Motorola RF Device Data, DL110/D, rev 7.

Glossary

ABS Antilock brake system. Prevents wheel skidding during maximum braking. An ABS design may contain a motor to recirculate the brake fluid during ABS operation.

A/D converter Analog-to-digital converter. Used to convert voltage levels into a digital number. The A/D's precision is expressed in bits, and its speed is expressed as conversion time, usually in microseconds. High-performance motor controls require an A/D with 10- to 12-bit resolution and less than 2 μs conversion time.

addressing mode The method that the central processor unit (CPU) uses to determine the operand address for an instruction. Some MCUs have 16 addressing modes.

algorithm A collection of program instructions that is configured or designed to perform a task, such as solving a mathematical equation or scanning a keyboard.

application layer The top layer in the OSI reference model, consisting of the interface between the OSI environment and a user's application. It does not contain applications but provides a link from application software on one system to application software an another computer through the OSI environment.

arbitration In a data network, this is the process of gaining access to bus. Bus arbitration is usually determined by a priority method. Nondestructive bus arbitration takes place as the lower-priority senders simply wait their turn to transmit on the bus.

ASIC Application-specific integrated circuits, which are sometimes used for motor controls. ASIC devices can vary from simple digital logic gate arrays to mixed-signal devices that contain both digital and analog components.

assembly language The programming instructions that the microcontroller's CPU uses, expressed in cryptic or abbreviated words. Some programmers prefer to write their code in assembly language for maximum performance. While this increases code efficiency, it does require a high degree of familiarity with a particular MCU's instruction set and how these instructions interact with each other and the MCU's operation. Examples of assembly language instructions are BRSET or SWI. The main disadvantage of writing at the assembly language level is that the code cannot be easily adapted to another type of MCU.

asynchronous In AC motors this means that the rotor is not synchronized to the motor's applied frequency. (The difference is called "slip.") In data networking, *asynchronous* refers to data bits that are not aligned or synchronized to a clock signal. Start and stop bits are used in the serial data signal to help the receiver establish a timing reference. MCUs that use serial communication interface (SCI) are asynchronous and use extra bits for clocking.

baseplate The bottom of a power module, usually made of copper or aluminum. The proper attachment of the baseplate to the heatsink is critical to maximizing heat transfer and reliability of the power module.

baud rate The total number of bits transmitted per unit of time, usually a second. When each signal element carries one bit, the baud rate is then equal to bits per second (bps).

bipolar junction transistor (BJT) Generally, a transistor with both majority and minority carriers. BJTs offer good power density at high voltages but require complex base drive circuits for high-frequency operation.

bit A binary digit that has a value of either a logic zero or a logic one. Bits are sometimes used in MCU registers to indicate yes (logic one) or no (logic zero).

BJT See **bipolar junction transistor (BJT).**

body diode Inherent in most power MOSFETs, this is the "parasitic" drain-source diode. Because it is formed by most of the MOSFET's active silicon area, its current ratings as a diode are the same as the MOSFET's. The body diode exhibits a normal silicon diode voltage drop. IGBTs do not contain this diode and, for most motor designs, require a separate diode die to be assembled in the IGBT package.

bootstrap A gate bias method used in power control electronics. A bootstrap gate bias circuit is derived from the power stage's output and is effective only if the power stage is operating. For example, in a half H-bridge power stage, the upper transistor's gate bias is developed by applying a +15 voltage through an isolation diode that adds this voltage to a capacitor which is charged up to the HV bus. This capacitor relies on the lower half of the half H-bridge to be switched on for some minimal duty cycle. A normal bootstrap design will not work if the power stage transistors are operated in continuous ON or OFF mode.

brake-by-wire This term refers to an automotive braking system that uses electronics for control of the braking system. The brake pedal position is electronically sensed by a brake system controller that manages the braking force. The braking energy supply may be pneumatic, hydraulic, electrical, or a combination of all three.

breakpoint A specific address, data value, or combination of certain values used in a microcontroller development system to invoke a software interrupt. Utilized to debug program code. Breakpoints on certain address are employed to stop the program execution; the MCU's registers are examined and possibly modified before continuing the program execution.

brushless DC motor (BLM) The BLM is also called a *trapezoidally excited PM motor* or *switched PM motor.* This motor uses magnets on its rotor and relies on electronics for commutation of its stator windings. The stator windings are driven in sequence with PWM or rectangular pulses of current. The BLM achieves high efficiency but costs more because of its magnets and electronics. A BLM is ideal for motion control systems requiring exact speed control, since the rotor is synchronized to the stator commutation; the control electronics (MCU-based) can maintain precise speeds at higher rpm levels. (Precise control at low rpms requires additional angular shaft resolution.)

bus In digital circuits, this term refers to a set of wires that transfer logic signals; in power controls, *bus* usually refers to a power supply line, such as the high-voltage bus or the 12-V bus.

byte A byte consists of 8 bits and, as a unit, represents a number, letter, or character.

C language This is a structured, high-level programming language that is somewhat machine independent.

central processing unit (CPU) The primary functioning unit of a microcontroller. The CPU controls the execution of instructions. Generally, the CPU performs control, some arithmetic, logical operations, and input/output by executing instructions obtained from memory elements.

charge pump A circuit used for developing voltages that are normally higher than the nominal power supply voltage. Charge pumps usually require a square-wave oscillator supplying out-of-phase signals to a voltage doubler rectifier network. Small charge pumps are integrated into some ICs.

chip Refers to a single semiconductor substrate that has been fabricated into an electronic circuit. A semiconductor chip is normally an unpackaged device and is also called a die. The term *chips* has been used to describe semiconductor products.

CISC A complex instruction set computer. This is usually an MCU other than a digital signal processor (DSP) or reduced instruction set computer (RISC), although a CISC MCU may include DSP- or RISC-type elements to speed up certain tasks.

clearance In electrical equipment, this term refers to the shortest distance through open space between two electrical points, such as the terminals on a power module. There are standards for establishing minimal clearances (UL508, UL840, IEC644, and VDE0160).

clock In an MCU, a clock is a square-wave signal used to synchronize operations in logic elements. In a data signal, a clock is used to control the timing of the data signal's bits.

coefficient of thermal expansion (CTE) The degree of change in a material dimension that occurs at different temperatures. CTE matching is especially important in power module designs. Large CTE differences lead to reliability problems when the power module undergoes power cycling.

comparator A comparator measures and compares two input voltages and produces a logic output dependent on the values of the two input voltages. A comparator is usually a very high-gain operational amplifier that has minimal negative feedback.

complementary metal oxide semiconductor (CMOS) A semiconductor technology that is widely used to manufacture microcomputer or similar products.

computer operating properly (COP) A timer element that, if allowed to overflow, will invoke a reset. A COP update routine is usually written in a place in the main program that must be hit if the program flow is working properly. The COP update code normally should not be placed in an interrupt service routine.

conduction The transmission of electricity or heat by the passage of energy from particle to particle.

control loop A series of instructions that can modify an external unit. Control loops may contain several conditions for termination.

convection cooling The thermal transmission brought about by the upward movement of warm air, as occurs in a heatsink which has its fins pointed up.

converter Usually a circuit that converts AC voltage to DC voltages. In electronic motor control systems, a converter stage is used to change the AC power line into DC power that in turn is changed back by an inverter to AC by the motor's power stage.

COP See **computer operating properly (COP).**

CPU See **central processing unit (CPU).**

creapage This term refers to the shortest surface distance between two electrical points. An insulation wall is often used between electrical terminals to meet minimal creapage standards (UL508, UL840, IEC644, and VDE0160).

CTE See **coefficient of thermal expansion (CTE).**

current tail Usually refers to a power device's current during turnoff. This is the last 20 percent of the current in a power device which may take an abnormal amount of time to decay, creating a tail on the current waveform.

cycle time The period of the operating frequency: $t_{cyc} = 1 \div f_{op}$. Cycle time can also refer to the interval of time when an action is invoked and when it is completed.

cyclic redundancy check (CRC) An error-detecting code in which the code is defined to be the remainder resulting from dividing the bits to be checked in the frame by a predetermined binary number.

data-link layer The second layer of the ISO OSI model that is responsible for the transmission of information over a physical medium. After establishing the link, it ensures the error-free delivery of the information through the use of error detection, error recovery, and flow control.

DBC See **direct-bonded copper (DBC).**

debug The process of finding errors in a software program or electronic circuit.

decoder A circuit that accepts several inputs, as from a brushless motor's Hall sensors, and converts these signals into digital logic signals that are used in determining the commutation of the motor.

desaturation An operating condition that occurs when the power transistor's current conduction exceeds its gain or transconductance rating. Observed when the collector-to-emitter voltage rises beyond its normal saturated area during excessive load current conditions or reduced gate input voltage.

Di/Dt The ratio of current change to time. This term is especially important in large electronic motor drives since certain components have a definite *Di/Dt* limitation. High *Di/Dt* values also imply high magnetic fields that can interfere with nearby circuits.

direct-bonded copper (DBC) Generally, a manufacturing process used in power modules that bonds copper directly to another material such as ceramic. Semiconductor chips can be soldered directly to the copper.

distributed operating system A grouping of distributed functions that are interconnected, are resident within a well-defined boundary, are contained within a given set of physical entities, and are capable of sharing processing and information resources.

dominant bit A bit which wins arbitration when contending for the bus. For SAE J1850, a logic 0 is the dominant bit.

duty cycle The amount of time a rectangular waveform is in the high state. For example, a 75 percent duty cycle means that the waveform's high state will be on for 75 percent and off for 25 percent. A variable-duty-cycle waveform is the fundamental type of drive signal used in electronic motor controls.

Dv/Dt The ratio of voltage change and time. This term is especially important in large electronic motor drives, since certain components have a definite *Dv/Dt* limitation. Motor winding insulation is stressed by high *Dv/Dt* values, especially at high temperature. High *Dv/Dt* implies that capacitive coupling to nearby circuits can be a problem.

EEPROM Electrically erasable and programmable read-only memory. EEPROM devices provide faster prototyping than do EPROM devices and can be reprogrammed by the application. Flash is another variation of EEPROM.

EHPS See **electrohydraulic power steering (EHPS).**

electrohydraulic brake (EHB) This is a braking system that employs electronics to manage the braking system. A brake pedal position sensor detects the driver's input. The EHB uses an electric motor to drive a hydraulic pump.

electrohydraulic power steering (EHPS) Electrohydraulic power steering uses the existing power steering hydraulic components. An electric motor is added to drive the hydraulic pump. The pump and motor can be integrated into one unit and directly mounted onto the steering assembly, thereby simplifying the final assembly of the vehicle.

electromagnetic interference (EMI) Magnetic radiation and capacitive (or electrostatic) fields that interfere with nearby electronic systems. Highly accurate sensors tend to be most sensitive to EMI. Circuit-board EMI can be minimized by careful layout procedures. (See **RFI.**)

EMI Electromagnetic interference created by both magnetic and electric fields.

EPROM Electrically programmable read-only memory or erasable programmable read-only memory. EPROM devices are used for prototyping and low-volume production. Their memory contents can be erased by ultraviolet light.

EPS Electronic or electric power steering. This term has two meanings: (1) the standard hydraulic assist power steering, which uses electronics to modulate the hydraulic boost per road speed; (2) a power steering assist system that uses an electric motor and electronics to provide the power assist and contains no hydraulic components.

ESD Electrostatic discharge. ESD can cause serious damage to electronics. MOSFETs and MCUs are especially susceptible to ESD because of their high input impedance.

event timer A specialized timer useful in power control MCUs to synchronize processing events such as CPU interrupts, A/D conversion start, and PWM updates.

FBSOA See **forward-biased safe operating area (FBSOA).**

firmware Software instructions and data programmed into nonvolatile memory (ROM). New ROM is required to update firmware.

flash Memory that is similar to EEPROM but can usually be erased in large sizes and can be easily reprogrammed for the application.

forward-biased safe operating area (FBSOA) An especially important rating in power control circuits for a power transistor that gives its maximum current and voltage levels which can be safely handled when the device is forward-biased. Exceeding the FBSOA ratings will initiate the transistor to fail, especially at higher voltage levels.

forward voltage Expressed also as V_F. The nominal voltage across a semiconductor device when it is biased on. Forward voltage generally varies from 0.3 to 1 V for diodes and 0.1 to 5 V for transistors.

frame A complete message over a communications network. A frame usually consists of several bytes, including a start and stop of frame characters.

free-wheeling diode (FWD) A diode or rectifier that conducts the motor's back electromotive force voltage (CEMF). The FWD may dissipate about half as much as the power transistor's heat dissipation, depending on the PWM signal. The free-wheeling diode is used to conduct the motor's inductive current when the motor's drive power transistor is switched off.

full-step In a stepper motor, this term refers to the rotor magnetic pole position in relation to the field coils. The rotor is positioned between two energized coils, which gives high-torque operation.

fuse link A fuse link is used in automotive electrical systems to prevent catastrophic failures when a battery has been connected backward or the alternator output line is shorted-circuited. Large motor drives in automotive applications may employ a fuse link rather than a fuse. A typical fuse link is usually a short piece of wire inserted near the power source that is smaller in diameter than normal and uses flameproof insulation.

fuzzy logic A program that is based on a reasoning model rather than absolute or fixed mathematical algorithms.

gate The input line to a power MOSFET or IGBT. The MOSFET or IGBT gate behaves as a capacitive load to the gate driver circuit.

half H-bridge The half H-bridge consists of two power transistors connected in series between the power supply bus and common. The half H-bridge is used for controlling polyphase motors. Two half H-bridges make a full H-bridge.

half-step In a stepper motor, this term refers to the rotor magnetic pole position in relation to the field coils. The rotor is positioned to one energized coil which gives one-half the torque when compared to full-step operation.

Hall-effect device A frequently used device to sense position in motors. The presence of a magnetic field at a right angle to an epitaxial layer with current flowing through it causes a Hall voltage to be generated. The Hall voltage is sensed by taps in the layer, which are at right angles to the current flow.

Hall-effect sensor A specialized semiconductor device that detects the presence and polarity of magnetic fields. These sensors are commonly used in brushless motors to determine the rotor's angular position. Some Hall effect sensors include further integration of amplifiers and logic interface circuits to produce a TTL (transistor-transistor logic) output signal.

H-bridge A power stage topology that allows the motor's coil voltage to be controlled in four modes: forward, reverse, brake, and open. The H-bridge is used in PM brush motors and bipolar stepper motors. The H-bridge consists of four power transistors, configured with a pair of transistors in series between the power supply bus and common. The motor's coil is connected to the midpoint of each pair. The topology forms an "H"—thus, the name.

heatsink A device for dissipating heat generated by the motor's power stage. Heatsinks used in motor drives are usually made from an aluminum extrusion and are rated by their temperature rise per watt of power dissipation from the power transistors. Heatsinks for small (<1-kW) motor drives may range from 2 to 10°C/W, while large motor drives require less than 1°C/W rated heatsinks. Bonded-fin heatsinks are usually considered for large heatsinks and are used with forced air to further improve the heatsink's thermal performance.

hexadecimal (Hex) A base-16 numbering system that uses the digits 0 to 9 and the letters A through F. In most MCU assemblers the character $ indicates that a value is hexadecimal. (A $F would equal the decimal number 15.)

high-side switch A switch (IGBT or MOSFET) that is connected between the power supply bus and the load and is usually more difficult to gate on or off.

HVAC Heating, ventilation, and air-conditioning systems. These systems typically use three motors for a heat pump HVAC-type system consisting of an inside blower fan, an outside condenser fan, and the compressor motor. The inside blower fan may be a variable-speed unit that uses electronics to vary its speed. Variable-speed compressors are also possible with higher-power electronic speed controls.

hybrid power modules Large power modules that contain multiple semiconductors, such as IGBTs, MOSFETs, and rectifiers.

hybrid vehicles A vehicle that uses two or more forms of power generation. A typical hybrid vehicle uses an electric motor traction system and a battery pack, plus an internal combustion-engine-powered alternator. The alternator may supply sufficient energy to drive the electric motor while the battery pack supplies extra energy for acceleration.

hysteresis In the context of this handbook, *hysteresis* refers to the difference between the response of an element to an increasing and a decreasing signal. Hysteresis is used in comparator amplifier designs to minimize chatter or spurious noise, especially with signals that have slow Dv/Dt rates. The amount of hysteresis will directly affect the accuracy of a control system.

IMS See **insulated metal substrate (IMS).**

input/output (I/O) In an MCU, I/O provides logical interfaces to external constituents. The MCU's standard input lines have voltage thresholds for a 0 or 1. These thresholds vary according to the MCU specifications and operating voltages.

instruction Operations that an MCU's CPU can perform. Programmers can express instructions in assembly or higher-level languages.

insulated gate bipolar transistor (IGBT) A power semiconductor device that has the input drive characteristics of a MOSFET but the output characteristics of a MOSFET at low current and a bipolar transistor at high currents. The IGBT forward ON voltage is always offset by one diode drop or about 1 V. The IGBT offers high power density. Individual IGBT die rated at over 150 A at 600 V measure only about 12 mm square and are less than 1 mm thick. Because of the almost infinite input resistance of the IGBT's gate, static electricity or electrostatic discharge (ESD) can destroy the IGBT unless special handling and testing precautions are undertaken.

insulated metal substrate (IMS) A very cost-effective baseplate material, IMS consists of a base material, usually aluminum, a very thin layer of filled epoxy for isolation, and a copper foil. The copper foil can be etched with a circuit pattern, much like a printed-circuit board. Improvements in IMS technology have achieved close to the same level of performance as ceramic direct-bonded copper (DBC).

integrated power stage module A hybrid power module that contains additional control elements. It can be a comprehensive solution for small motor drives, as these modules contain all the necessary power electronics. When all the power devices are combined into a single package, the assembly of the power stage is greatly simplified.

interrupt (INT) A break in the normal program flow, commonly initiated by an external event that the MCU needs to service. Some MCUs can have several interrupt sources, which are prioritized if more than one interrupt occurs at the same time. An MCU's interrupt handling capability is important in motor control designs.

inverter Usually a circuit that converts DC voltage to other AC voltages. In electronic motor control system terminology, the power stage that drives the motor is called an *inverter.*

I/O See input/output (I/O).

jitter Short-term instability in an electronic circuit.

joint The area where two or more leads or contacts are joined. Since they can deteriorate over time, joints can become a reliability issue. Higher integration can usually minimize joint count, thereby increasing reliability.

jump instruction An instruction that directs the CPU to jump to another program location. Jump instructions are usually able to transverse the MCU's entire address range.

jumper A temporary wire or lead between two or more electrical lines in a circuit. To reduce costs, wire jumpers, rather than switches, are also used to configure electronic circuits.

junction The contact between different silicon layers. In reference to semiconductor thermal ratings, *junction* refers to the die.

Kelvin contact This a connection method to minimize the effects of high current levels when a low-level signal must be detected or applied to a high current line. Essentially, the Kelvin contact is made at the closest point to the power element, which minimizes voltage drops and stray inductance effects. Kelvin contacts are used in current sense resistors and in very large power transistors.

kickback voltage The voltage generated when current through a wire, motor, or any inductor is switched off. Unless contained by voltage transient suppressor devices or elements, kickback voltages can harm electronic motor control components.

latch A logic circuit that holds a certain logic state when triggered.

layout The manner in which electronic components are fixed onto a circuit board or into an assembly. Improper layout in a motor power stage can lead to instability problems and failures.

leadframe The metal frame that contains the external leads for a semiconductor device.

leakage current Usually small levels (microamperes) of currents that flow in circuits under certain conditions. Leakage currents in semiconductors change with temperature and voltage.

level shift The changing of a voltage or current amplitude by electronic devices. Interface circuits sometimes require level shifting.

limiter A circuit or MCU routine that limits a value. Motor current, torque, or speed can be limited by the motor control electronics.

linear A smooth and even change in voltage or current output that varies in direct proportion to an input signal.

load Normally, the device, such as a motor, that is driven from the power electronics or power line. *Load* can also refer to the mechanical work connected to a motor.

load dump In automotive electrical systems that employ a battery and an alternator, a load dump condition can occur when the alternator is charging the battery at a high current rate and the battery becomes disconnected because of an open cell or loose connections. When this open circuit occurs, the alternator output can reach over 45 V, and sometimes well over 85 V, depending on whether voltage clamping devices are used. It takes about 1 s for the alternator output to completely stabilize back to 12 V. The load dump event is a serious problem for the normal 12-V power electronics.

locked rotor A locked-rotor condition stresses the power stage electronics and can be a serious reliability issue. A locked rotor can be caused by a mechanical failure, such as a frozen shaft end bearing in the motor or a jammed load in the motor's application. The power stage electronics should be designed to withstand a momentary locked rotor event, and a detection circuit should remove the power to the motor to prevent a catastrophic failure. A power stage failure can mimic a locked-rotor condition. If continuous DC current is flowing through an AC induction motor's windings, as may occur with a logic fault or short-circuited power transistors, the rotor will not turn until the DC current is removed.

lookup table A collection of values in the MCU's memory that is organized for easy access by the program.

loop A state in which the MCU program repeats until certain conditions are met. Loops are sometimes used in motor control MCU programs to wait for a response, for example, from an A/D conversion.

loop execution time In MCU designs, this is the time period that a specific group of instructions requires to complete a task or respond to an external event.

low-pass filter Frequencies above a certain value are filtered out. A motor, because of its high inductance, acts like a low-pass filter to high-frequency PWM voltages.

low-side switch A switch (IGBT or MOSFET) that is connected between common and the load and is easily gated ON or OFF.

low-voltage interrupt (LIV) A circuit that forces the MCU's normal program flow to be interrupted when a low-voltage condition is occurring at the MCU's power supply bus. This gives the program time to start an orderly shutdown until the power supply voltage returns to normal.

low-voltage reset A circuit that forces the MCU to be reset until the MCU's power supply is normal.

LVDT Linear variable differential transformer. An LVDT pressure sensor consists of a primary winding and two secondary windings positioned on a movable cylindrical core. The core is attached to a force collector, which provides differential coupling from the primary to the secondaries, resulting in a position output that is proportional to pressure.

LVI See **low-voltage interrupt (LVI).**

mechatronics A term used to describe the combination of mechanical systems with electronics.

microcontroller (MCU) A MCU is usually a self-contained single-chip computer. It has built-in memory elements, general-purpose input/output ports, communication ports, analog-to-digital converters, and general-purpose timers. Its central computer core can be a complex set instruction (CISC) type, a digital signal processor (DSP), or a reduced instruction set (RISC) type. Motor control MCUs will normally have a powerful timer unit for PWM calculations and instruction sets that work well in a controller-type application.

MEMS The micro-level design and fabrication of mechanical structures is called a *microelectromechanical system* (MEMS).

metal oxide semiconductor field-effect transistor (MOSFET) A transistor that is voltage-driven. As the name indicates, the voltage field from the gate controls the transconductance of the device. MOSFETs can exhibit high transconductance, which means that a small voltage change in the gate bias can produce large voltage changes in the drain output. MOSFETs are capable of very fast switching times—10 to 100 ns is typical. Normally, power MOSFETs used in motor control drives are of an enhancement mode type, requiring a positive gate-to-source bias to switch on. Because of the bulk resistance required for higher voltages, MOSFETs are most cost-effective for lower-voltage applications, such as battery-powered or automotive systems. The MOSFET's on-resistance varies directly with its junction temperature. As a result of the almost infinite input resistance of the MOSFET's gate, static electricity or electrostatic-discharge (ESD) can destroy the MOSFET unless special handling and testing precautions are taken.

micromachining Micromachining allows a three-dimensional mechanical structure or microstructure made by a chemical etching process. The manufacturing processes used to make these microstructures include photolithography, thin-film deposition, and chemical and plasma etching.

multiplex network (MUX) A digital communications network that uses either time division or frequency division for the combination of sending several messages over the same signal path.

N-FET An N-channel field-effect transistor, as used in this handbook, is a power device whose gate input is driven from a positive voltage with respect to the FET's source. N-FET devices are useful for switching loads that are connected between their drain terminals and the positive power supply bus or as low side switches. The power density of the N-FET's silicon die is very high when processed and designed for voltages of less than 60 V. High-voltage (>200-V) N-FET devices exhibit low power density and are usually displaced by IGBT devices for high-voltage motor drives.

non-return-to-zero (NRZ) A data bit format in which the voltage or current value, which is typically voltage, determines the data bit value, specifically, one or zero.

optical encoder An optical encoder contains a light source, and a photodetector provides a digital means of measuring displacement or velocity when an alternating opaque-and-translucent grid is passed between them. Uniformly spaced apertures on a wheel (i.e., a flat disk mounted on a motor shaft) allow logic circuitry to count the number of pulses in a given timeframe to determine shaft velocity or angular displacement.

optocoupler A device that uses a light-emitting diode (LED) and a phototransistor to isolate one circuit from another. (Sometimes called *optoisolators*.)

OTPROM A one-time programmable read-only memory. Typically, an EPROM without the clear window used for UV erasing.

PFC See **power factor control (PFC).**

P-FET A P-channel field-effect transistor, as used in this handbook, is a power device whose gate-to-source voltage is driven from a negative voltage source. P-FET devices are useful for switching low-voltage (<20-V) loads from their source to common or as high-side switches. The power density of the P-FET's silicon die is about one half of a similarly rated N-FET device, which makes P-FETs less cost-effective.

power factor control (PFC) A circuit to correct the power factor (phase relationship between voltage and current) of an AC line-operated device, such as a motor.

protocol A set of rules governing the information flow within a communications infrastructure, often known as *data-link control.* Protocols control format, timing, error essential correction, and running order.

pulse-width modulation (PWM) A square wave with a variable duty cycle. The fundamental type of waveform used in electronic motor controls. A modulation method that varies the duty cycle or on time of a square wave from 0 to 100 percent.

quadrature encoding A speed and direction sensing method that has two output signals that are usually phase-shifted by 90°. This allows detection of both speed and direction, plus shaft position, if an absolute reference point is given, such as an irregular pulse caused by a missing or double tooth.

radio-frequency interference (RFI) Electrical noise that interferes with the normal reception of radio or television equipment. Electronic motor controls, by nature, are prone to generate RFI unless precautions are taken. The difficulty with minimizing RFI is that the electronic motor power levels can be in the kilowatt range, while most radio and TV receivers will detect microwatt signal levels.

RBSOA See **reverse-biased safe operating area (RBSOA).**

regeneration In an electric vehicle, regeneration is the conversion of the vehicle's kinetic energy (set per the vehicle's velocity) back into electrical energy when coasting or braking. In other words, the motor is turned into a generator by the control electronics.

resolver A resolver provides one absolute signal per revolution and tolerates vibration and higher temperatures. At lower speeds, the single signal per revolution limits performance. Standard resolver-to-digital converters with 12-bit resolution and a 500 μs sampling time can achieve a minimum speed of about 30 rpm.

reverse-biased safe operating area (RBSOA) An important rating in power control circuits for power transistors that gives maximum current and voltage levels that can be safely handled when the device is reverse-biased. Exceeding the RBSOA ratings will cause the transistor to fail.

RFI See **radio-frequency interference (RFI).**

rotary encoder Rotary encoders can directly use the incremental signal. As a result, the speed control range possible with rotary encoders is about 3000 times greater than resolvers. The rotary encoder values apply for a maximum speed of

6000 rpm and a sampling time of 500 μs, which are common for modern machine tools.

ROM Read-only memory. This memory normally contains the program code and cannot be altered.

safe operating area (SOA) A power device's voltage and current level that can be safely controlled.

sample points Time periods at which a signal is tested. In an MCU motor control design, the number of motor current sample points relate to the accuracy of the design.

SCI See **serial communications interface (SCI).**

SCR Silicon controlled rectifier; a reverse blocking triode thyristor. SCRs are designed to conduct current in one direction. They have a four-layer PNPN structure, which means that once the gate is triggered, the device has internal regenerative feedback, which latches it on until its anode-to-cathode voltage is zero. The SCR, like the triac, requires only a narrow gate pulse to latch it on.

serial communications interface (SCI) A common microcomputer interface that provides a standard mark/space (NRZ), which produces 1 start bit, 8 data bits, and 1 stop bit in a serial data format.

serial peripheral interface (SPI) A microcomputer interface that provides high-speed serial data communications with other SPI-like devices. The SPI requires a clock, data, and enable line.

servo motor A motor system that is designed to turn or operate until a certain position has been reached. The motor operates as a slave to the master control.

slip Most AC induction motors are nonsynchronous—they do not rotate in perfect sync with the line frequency. The difference between the actual shaft or rotor speed and the rotating magnetic field (which is synchronous to the AC line) speed is called "slip" and is expressed as a percentage.

slip control A motor control strategy that adjusts the slip of an induction motor.

smart FET A MOSFET that has additional functions, such as current limiting or overtemperature shutdown, integrated into its die.

smart IGBT An IGBT that has additional functions, such as current limiting or overtemperature shutdown, integrated into its die.

SMT Surface mount technology. SMT devices are soldered directly to the top or bottom of a printed-circuit board.

snubber A circuit element used to minimize high-frequency ringing in power control designs.

SOA See **safe operating area (SOA).**

source impedance The resistance of the energy source. For example, in a motor control design, the source impedance of the power supply is an important parameter to determine the maximum current level that may flow during worst-case conditions.

space vector modulation A modulation method that translates the voltage vector into inverter states. It improves bus utilization and current ripple.

SPI See **serial peripheral interface (SPI).**

SR or SRM See **switched reluctance motor (SR or SRM).**

stepper motor As the term implies, stepper motors are usually driven with voltage steps. These steps or pulses can vary from full-steps, to half-steps, to quarter-steps, all the way down to microsteps. Stepper motors blend in well with digital logic, as the number and size of the steps or pulses can easily be generated with timers. Microcontrollers are excellent for generating the steps and computing variable velocity or acceleration rates for smooth control of the stepper motor.

switched reluctance motor (SR or SRM) Switched reluctance motors need electronics for commutation. The SR power stage normally requires each winding end to be controlled by a special dual-switch power module.

synchronous rectification This normally refers to a power circuit that substitutes a MOSFET for a rectifier and is driven in synchronization with the control signals. The advantage of synchronous rectification is lower voltage loss and better control of the switching times. The disadvantage is that another power device must be controlled.

table Lookup tables are used with microcontroller programs to store data values in an organized format.

thermal cycling A temperature change that a component incurs from its own heat dissipation or from external temperature changes. Thermal cycling affects the long-term reliability of power semiconductors.

thyristor This term is associated with SCRs and triacs. Any semiconductor device whose off-to-on action depends on its internal PNPN regenerative feedback is a thyristor. Thyristor devices can be two-, three-, or four-leaded and can conduct current in either direction, depending on their design.

topology The type of design used in power control circuits. Examples are a full H-bridge, half H-bridge, or single switch.

TPU Time processor unit; used in some 16-and 32-bit microcontrollers. The TPU is programmable and, once initiated, operates independently of the MCU's internal central processor.

transconductance The gain of a MOSFET or IGBT is measured in units of transconduction, g_{FS}. The value of transconductance is defined as the ratio of drain current change corresponding to a change in gate voltage. In motor control power stages, transconductance is important when trying to determine the effects of operating the MOSFET or IGBT in a linear mode. Power devices with high transconductance values generally can conduct more current under short-circuited load conditions.

triac A bidirectional triode thyristor which can conduct AC current. The triac can be switched on by either polarity of the gate signal regardless of its main terminal's polarity. Instead of an SCR anode and cathode labels, the triac uses an MT1 label for the gate side (such as cathode for SCR) and an MT2 label for its opposite nongate side (such as anode for SCR). The triac's gate sensitivity is dependent on its signal polarity and the polarity of MT2.

ultracapacitor This refers to a very high-value capacitor (≥100,000 μF) that is used to store and then supply energy at a low rate of discharge. The type of capacitance is useful for keeping critical electronic logic powered up during power outages.

undervoltage lockout A condition that shuts off a circuit to prevent damage. Undervoltage lockout is commonly used in gate driver ICs.

update rate The frequency at which the MCU obtains new data or refreshes its data values.

variable-frequency drive (VFD) An electronic control that can change the speed of an induction motor by varying the frequency of the motor's voltage. A VFD normally consists of an AC-to-DC converter, a microcontroller or logic ICs, and a DC-to-AC inverter. Most VFDs can also adjust not only the motor's frequency but also its voltage.

vector control A motor control strategy that adjusts the motor's current vector to control torque. Speed and current sensing is required.

***V/F* control** Motor control strategy that changes the motor's voltage as its speed or frequency is varied. Sometimes called a *volts-per-hertz control*.

word A value of 2 bytes or 16 bits.

zener diode A device that, when connected to a voltage source through a series resistor, exhibits a sharp rise in current at the moment its zener voltage is reached. Zener diodes are used to regulate and clamp voltages.

Appendix

A

Switching Power Loss

Time Ref. (μs)	Vce (V)	Ic (A)	Pd (kW)
.00	300	0	*0.0*
.05	270	10	*2.7*
.10	240	20	*4.8*
.15	210	30	*6.3*
.20	180	40	*7.2*
.25	150	50	*7.5*
.30	120	60	*7.2*
.35	90	70	*6.3*
.40	60	80	*4.8*
.45	30	90	*2.7*
.50	2	100	*0.2*
	2	100	*0.2*
	2	100	*0.2*
			0.0
			0.0
	3	100	*0.3*
.00	3	100	*0.3*
.05	15	95	*1.4*
.10	30	90	*2.7*
.15	45	85	*3.8*
.20	60	80	*4.8*
.25	75	75	*5.6*
.30	90	70	*6.3*
.35	105	65	*6.8*
.40	120	60	*7.2*
.45	135	55	*7.4*
.50	150	50	*7.5*
.55	165	45	*7.4*
.60	180	40	*7.2*
.65	195	35	*6.8*
.70	210	30	*6.3*
.75	225	25	*5.6*
.80	240	20	*4.8*
.85	255	15	*3.8*
.90	270	10	*2.7*
.95	285	5	*1.4*
1.00	300	0	*0.0*

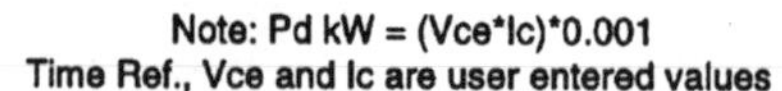
Note: Pd kW = (Vce*Ic)*0.001
Time Ref., Vce and Ic are user entered values

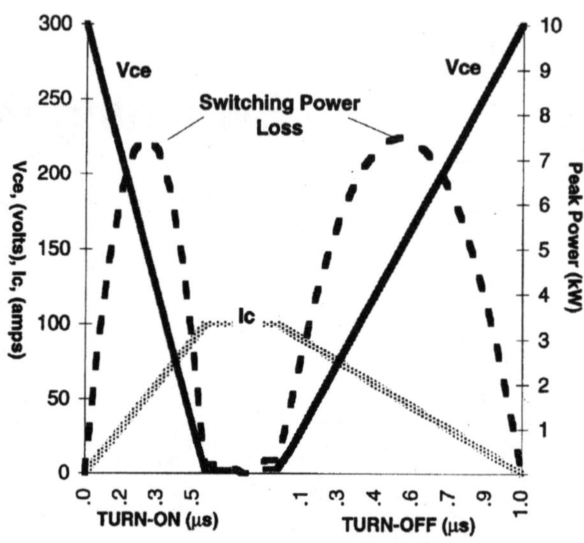

Peak Power Switching Loss Graph.

Graph reference formulas:

Vce =SERIES(,Graph!$A$9:$A$45,Graph!$B$9:$B$45,1)
Ic =SERIES(,Graph!$A$9:$A$45,Graph!$C$9:$C$45,2)
Sw pwr Loss =SERIES(,Graph!$A$9:$A$45,Graph!$D$9:$D$45,3)

(The voltage, current values, and time domain values representing the turn-on and turn-off of the power transistor are entered into this spreadsheet which allows a quick estimate of switching power loss for simple single switch motor drives. The graph shows that the transistor's collector-to-emitter voltages and collector current are triangle shaped.)

Appendix

B

Wire Inductance

AWG No.	Dia. mils	Dia. mm	Straight Wire Inductance*, μHy 25 mm (1 inch)	305 mm (1 foot)	3,005 mm (10 feet)	30,005 mm (100 feet)
1" ROD	1000.0	25.40	0.006	0.24	3.7	50.8
0000	460.0	11.68	0.011	0.28	4.2	55.4
000	409.6	10.40	0.011	0.29	4.2	56.1
00	364.8	9.27	0.012	0.30	4.3	56.8
0	324.9	8.25	0.012	0.30	4.4	57.5
1	289.3	7.35	0.013	0.31	4.4	58.2
2	258.0	6.55	0.014	0.32	4.5	58.9
4	204.3	5.19	0.015	0.33	4.7	60.3
6	162.0	4.11	0.016	0.35	4.8	61.7
8	128.5	3.26	0.017	0.36	4.9	63.1
10	101.9	2.59	0.019	0.38	5.1	64.5
12	80.8	2.05	0.020	0.39	5.2	65.9
14	64.1	1.63	0.021	0.40	5.4	67.3
16	50.8	1.29	0.022	0.42	5.5	68.7
18	40.3	1.02	0.023	0.43	5.6	70.0
20	32.0	.81	0.025	0.45	5.8	71.4
22	25.4	.64	0.026	0.46	5.9	72.8
24	20.1	.51	0.027	0.47	6.0	74.2
26	15.9	.40	0.028	0.49	6.2	75.6
28	12.6	.32	0.029	0.50	6.3	77.0
30	10.0	.25	0.030	0.52	6.5	78.4
32	8.0	.20	0.032	0.53	6.6	79.8
34	6.3	.16	0.033	0.54	6.7	81.2
36	5.0	.13	0.034	0.56	6.9	82.6
38	4.0	.10	0.035	0.57	7.0	84.0
40	3.1	.08	0.036	0.59	7.2	85.3

* Ind =(0.0002*Lmm)*(2.3026*LOG((4*Lmm/Dmm)-0.75))

Appendix

C

Motor Control Semiconductors

Supplier Name * (Internet address or telephone number)	Motor Control Semiconductor Devices					
	MCU CISC DSP RISC	Motor Control IC	Power Module	Power Discrete	Hall Sensor	Opto-coupler
Advanced Power Technology http://www.advancedpower.com/ProdInfo/ProdInfo.html			√	√		
Allegro Microsystems, Inc. http://www.allegromicro.com/control/prodline.htm		√			√	
Analog Devices Inc. http://www.analog.com/	√	√				
Atmel Corp. http://www.atmel.com/	√					
Cherry Semiconductor Corp. Http://www.cherrycorp.com		√				
Collmer Semiconductor, Inc. http://www.collmer.com/			√		√	
Eupec Inc. (Tel. 908-236-5600)			√			
Harris Semiconductor http://www.semi.harris.com/	√	√	√	√		
Hewlett-Packard Co. http://hpcc920.external.hp.com/HP-COMP/isolator/						√
Hitachi Semiconductor (America) Inc. http://www.halsp.hitachi.com/	√		√			
Intel Corporation htp://www.intel.com	√					
International Rectifier Corporation http://www.irf.com/		√		√		
IXYS Corporation htp://www.ixys.com			√	√		
Micro Linear Corporation http://www.microlinear.com/		√				
Microchip Technology Inc. http://www.microchip2.com/document.htm	√					
Mitsubishi Semiconductor Group (POWEREX) http://www.mitsubishichips.com/products/products.htm	√		√	√		
Motorola Semiconductor Products Sector http://design-net.sps.mot.com	√	√	√	√		√
National Semiconductor http://www.national.com/design/index.html	√	√		√		
NEC Electronics, Inc. http://www.ic.nec.co.jp/index_e.html	√	√		√		√
Panasonic Industrial Co. http://www.panasonic.com/pic/index-semico.html	√	√		√	√	√
Philips Semiconductors http://www.semiconductors.philips.com/	√	√		√		
SanRex Corp. Semiconductor Division (Tel: 516-352-3800)			√			
Semkikron Inc. http://www.semikronusa.com/			√	√		
SGS-Thomson Microelectronics Inc. http://www.st.com/	√	√		√		
Siemens Microelectronics Inc. http://www.sci.siemens.com/	√	√	√	√		√
Teccor Electronics, Inc. http://www.teccor.com/				√		
Temic Semiconductors http://www.temic.com/	√	√		√		√
Texas Instruments Inc. http://www.ti.com/corp/docs/prodserv.html	√	√				√
Toshiba America Electronic Components, Inc. http://www.toshiba.com/taec	√	√	√	√	√	
Unitrode Integrated Circuits http://www.unitrode.com/prods.htm		√		√		

* Partial listing of suppliers, devices may vary from what is shown.

Appendix

D

Motor Control Standards

Standards organizations that affect motor controls*	Notes
American National Standards Institute (ANSI) http://www.ansi.org/docs/home.html	ANSI administrates and coordinates a variety of standards
American Society for Testing & Materials http://www.astm.org	Provides over 10,000 technical standards
Canadian Standards Association (CSA)	standards affecting motor power efficiency
Institute of Electrical and Electronic Engineers, Inc. (IEEE) http://standards.ieee.org/catalog	Many standards pertaining to motors, IEEE 841, IEEE Std. 519-1992
International Electrotechnical Commission (IEC) http://www.iec.ch/home-e.htm	Numerous standards for motors, IEC 34-17, IEC 65, IEC 228, IEC 317, IEC 349-2, IEC 664-1, IEC 947-2 and IEC 1136-1
International Standards Organization (ISO) http://www.iso.ch/welcome.html	Promotes voluntary adaptation and implementation of standards. 112 counties are members. ANSI represents United States.
National Electrical Code	Several Articles, especially Article 430, describe general rules and applications of motors
National Electrical Manufacturers Association (NEMA)	motor standards (MG1-1993,MG3-1974)
SAE http://www.sae.org	Automotive related standards including automotive electrical and electronics
U.S. Environmental Protection Agency (EPA) **Department of Energy** (DOE)	Motor energy efficiency regulations, Energy Policy Act of 1992
Underwriters Laboratories Inc. (UL) http://www.ul.com	UL 508 and UL 840 are used with many types of electronic equipment
Verband Deutscher Elektrotechniker (VDE) e.V.	voltage spacing standards (VDE 0160)

* Partial listing only   *Reference* : Bonnett, A., "Regulatory Impact on the Application of AC Induction Motors," IEEE Industry Applications Magazine, March/April 1996.

Appendix

E

Electric Vehicle AC Induction Motor Program

MC68332 VECTOR CONTROL CODE, MAIN ASSEMBLY FILE

```
***************************************************************************
*  MC68332 Example Vector Control System for an EV traction drive        *
*  Author: Peter Pinewski, Motorola Semiconductor Products Sector        *
*  Created: 11/10/94                                                     *
*  Last updated:  2/23/96                                                *
*                                                                        *
*  Vector Control Code with Space Vector Modulation (Null = V0)          *
*   - Field weakening starts at 1920 RPM.                                *
*   - Minimal amount of regen when motor is coasting.                    *
*   - If over current fault (reduce commanded current by 18Amps ($200)   *
*   - Torque is given by IQ_cmd = sqrt(Is^2 - Id^2)                      *
***************************************************************************
       include regs.h       * MC68332 register definitions.
A_HIGH     equ      pram+$20 * ---|
B_HIGH     equ      pram+$60 *    |  HIGH TIME REGISTERS IN TPU
C_HIGH     equ      pram+$a0 * ---|
TST_14     equ      pram+$e4
SPEED      equ      pram+$f4 *     SPEED SENSING REGISTER IN TPU
SINE_VEC   equ      $4e*4    * tpu int: level 4 ,vec# $40 - $4f
STACK_PTR  equ      $3000    * point stack to ram area
*********************************************************************
*                           SOURCE CODE                             *
*********************************************************************
       org      $60000
       dc.l     STACK_PTR    *  DEFINE
       dc.l     RESET        *  THE RESET
       dc.l     GEN_SVC      *  VECTORS
       dc.l     GEN_SVC
*-------------------------------------------------------------------*
* SETTING UP THE MCU SYSTEM:                                        *
*-------------------------------------------------------------------*
RESET:
      move.w    #$2700,sr            * disable all interrupts
      move.b    #$06,SYPCR           * SW watchdog disabled;Bus monitor enabled
      ori.w     #$4000,SYNCR         * system clock = 16.67 MHz
      move.w    #$60ce,MCR
* - SET UP CHIP SELECTS - *
      move.w    #$00ff,CSPAR0        * CSBOOT,CS0,CS1,CS2 = 16-bit CS
      move.w    #$0000,CSPAR1        *  all other chip selects are I/O
      move.w    #$0003,CSBAR0        *
      move.w    #$0003,CSBAR1        * CS0,CS1,CS2=64K at address $0000
      move.w    #$0003,CSBAR2        *
      move.w    #$5030,CSOR0         * CS0: upper; W only; 0 wait
      move.w    #$3030,CSOR1         * CS1: lower; W only; 0 wait
      move.w    #$6830,CSOR2         * CS2: both; R/W; 0 wait
      move.l    #$060468b0,CSBARBT   * CSBOOT=128K at address $60000
*                                    * async; both; R only; AS; 2 wait; S/U
* - SET UP VECTOR TABLE - *
      move.w    #255,d7
      move.l    #0,a0                * vector table starts at address 0
```

```
vec_tbl:
      move.l    #GEN_SVC,(a0)+        * initialize vector table to
      dbf       d7,vec_tbl            *     generic service routine
      move.l    #SINE_SVC,SINE_VEC    * initialize tpu ch0 vector
* - SET UP I/O PORTS - *
      clr.b     PORTE                 * drive POTRE low
      clr.b     PEPAR                 * Use PORT E for I/O
      move.b    #$f8,DDRE             * make PORT E outputs
      clr.b     PORTF                 *
      clr.b     PFPAR                 *
      move.b    #$b8,DDRF             * PF7 used as PWM enable (output)
PERIOD      equ     $0200   * Use 8.14kHz switching (512*240ns = 122.88us)
INT_RATE    equ     $0200   * Interrupt every 2 periods |int_rate:xx|-->01:xx
FIFTY_PCT   equ     $0100   * 50% duty cycle
DEADx2      equ     $1000   * deadtime = $08*240ns = 1.92us --> |deadtime:xx|
SPEED_INT   equ     $0800   * speed sense interval = 2048*3.84us = 7.865ms
DELAY       equ     $00c0   *
   include  TST.INI
   include  QSM.INI
      move.w    #$0002,XMIT_0       * start byte
      move.w    #$0000,XMIT_1       *
      move.w    #$005f,XMIT_2       * O
      move.w    #$0071,XMIT_3       * F
      move.w    #$0071,XMIT_4       * F
      bset.b    #7,SPCR1
      move.w    #$30,d7             * 2s system stabilization delay
dly2  move.w    #$ffff,d6
dly   dbf       d6,dly
      dbf       d7,dly2
      move.w    #AD_CH0,XMIT_7      * Throttle pot
      move.w    #AD_CH2,XMIT_8      * Phase A
      move.w    #AD_CH3,XMIT_9      * Phase B
      move.w    #AD_CH4,XMIT_10     * extra A/D reading (extra POT)
      move.w    #$0a00,SPCR2
      bclr.b    #7,PORTF            * disable system (PWMs)
      bsr       spi_xfer
      bsr       spi_xfer
      move.w    RCV_9,d0
      move.w    d0,IA_ZERO          * get zero current for phase A
      move.w    RCV_10,d1
      move.w    d1,IB_ZERO          * get zero current for phase B
      clr.w     fm
      clr.w     theta
      clr.w     old_id              * clear critical system variables
      clr.w     old_iq
      clr.w     VQ
      clr.w     VD
      clr.w     fault
      clr.w     ID_cmd
      clr.w     IQ_cmd
      bset.b    #7,SPCR1            * start spi transfers
      move.w    #$2300,sr           * start sin wave creation
**********************************************************
*     MAIN:   OUTER LOOP                                 *
```

```
*           - establish ID_CMD  (implement flux and    *
*           - establish IQ_CMD     torque controllers) *
*********************************************************
main:
      move.w    fm,d7           * read speed
      tblu.w    MAGNET,d7       * get flux current (ID_cmd)
      move.w    d7,d5           * d7,d5 = ID_cmd
      muls.w    d5,d5           *
      lsl.l     #1,d5           * ID_cmd^2
      move.w    throttle,d6
      lsl.w     #3,d6           * Scale A/D for table look up.
      tblu.w    CRNT,d6         * throttle pos is IS^2
      bne       motor           * if commanded current > 0 then "motoring"
      tst.w     fm              * else tst motor speed
      beq       zero_I          * if speed=zero clear ID_cmd (stop mtr current)
regen:
      move.w    #$400,d6        * if speed <> zero, then regenerate
      neg.w     d6              * d6 = IQ_cmd (negative torque)
      bra       update
motor:
      cmp.l     d5,d6           * check if ID_cmd^2 > Is^2
      bcc       ok              * if not then ok
      clr.w     d6              * if so IQ_cmd = 0
      bra       update
ok:
      sub.l     d5,d6           * d6 = Is^2 - ID_cmd^2
      lsl.l     #1,d6
      swap      d6
      tblu.w    SQRT,d6         * d6 = sqrt(Is^2 - ID_cmd^2)
      lsr.w     #1,d6
      bra       update
zero_I:
      clr.w     d7              * motor is not turning so shut off field
      clr.l     VD              * clear both current controllers
      clr.l     VQ
update:
      tst.w     fault           * before update check over current condition
      beq       good            * if no over current then good.
      sub.w     #$200,d6        * else limit IQ_cmd by 18AMPS
good:
      move.w    d6,IQ_cmd       * save to IQ_cmd
      move.w    d7,ID_cmd
      bra       main
      nop
*****************************************************************
* ISR:  INNER LOOP (Vector control code executed every 245us) *
*           - Inverter error checking                          *
*           - Slip calculator                                  *
*           - Speed sensing and integrator                     *
*           - Determine SIN_x and COS_x                        *
*           - 3ph to 2ph Transformation                        *
*           - PI Controllers                                   *
*           - Rectangular to Polar conversion                  *
*           - Implement Space Vector Modulation (null=V0)      *
```

```
*******************************************************************
SINE_SVC:
      bset.b   #7,PORTE           * !!!! TIMING MARKER !!!!
      bset.b   #6,PORTE           * !!!! TIMING MARKER !!!!
      movem.l  d0-d7,-(sp)
*-------------------------------------------------------*
* ERROR CHECKING                                        *
*-------------------------------------------------------*
error_test:
      btst.b   #6,PORTF           * check for fault condition
      bne      okay               * if no fault continue (okay)
      tst.w    fault              * check for previous fault
      bne      error              * if previous fault continue with error
      btst.b   #6,PORTF           * if no previous fault retest for fault
      bne      okay               * if no fault branch to ok
      move.w   #$ffff,fault       * set fault flag
error:
      bclr.b   #7,PORTF           * disable pwms
      move.w   #$0002,XMIT_0
      move.w   #$0020,XMIT_1
      move.w   #$005f,XMIT_2
      move.w   #$0059,XMIT_3      * if fault - display -OC- message
      move.w   #$0020,XMIT_4      *           - go_on with pwms disabled
      bra      go_on
okay:
      bset.b   #7,PORTF
      move.w   fs,d7              * display commanded current
      bsr      display            * display fs
go_on:
      move.w   RCV_7,d7           *  -  read extra A/D
      move.w   RCV_8,d6           *  -  read THROTTLE pot
      move.w   d6,throttle
      move.w   d7,extra
*------------------------------------------------------------------*
* SLIP CALCULATOR                                                  *
* ws = 1/tr (Iq/Id), fs = 1/(2*PI*tr) (Iq/Id), fs = .3183*(Iq/Id) *
* K = .3183 approx  where  tr = .5s approx                         *
*------------------------------------------------------------------*
SLIP_CALC:
      move.w   ID_cmd,d6          * d6 = ID_cmd
      move.w   IQ_cmd,d7          * d7 = IQ_cmd
      ext.l    d7
      bne      calc_SC            * check if IQ_cmd = 0
*        clr.w    ID_cmd
      clr.w    d7
      bra      done_SC
calc_SC:
      asl.l    #4,d7              * shift by 4 because $10 = 1 herz
      divs.w   d6,d7              * d7 = Iq/Id
      muls.w   #$517c,d7          * fs = .3183(Iq/Id)
      swap     d7
done_SC:
      move.w   d7,fs              * store slip command
* d6 = xxxx
```

```
* d7 = fs (slip frequency)
*-----------------------------------------------------------------*
* SPEED SENSING + INTEGRATOR                                      *
* The TPU executes a timed transition count (TTC) primitive       *
* written by Peter Pinewski to determine motor speed.             *
*   - time interval for counting is every 7.8ms                   *
*   - Speed sensing element has 128 teeth                         *
*   - the TTC counts every edge (256 edges)                       *
* Value in SPEED (TRANS_CNT) correlates to motor freq as follows *
*         $0001 =    1Hz motor freq =    30 RPM (4 pole motor)    *
*         $0010 =   16Hz motor freq =   480 RPM (4 pole motor)    *
*         $0100 =  256Hz motor freq =  7680 RPM (4 pole motor)    *
*-----------------------------------------------------------------*
SPEED_SENSE:
      move.w   SPEED,d5           * d5 = fm (read motor freq)
      lsl.w    #4,d5              * scale speed reading
      move.w   d5,fm              * save motor freq
      move.w   d5,d4              * d4 = motor freq (fm)
      lsr.w    #4,d4              * scale fm for MUX controller ($1=30 RPM)
      move.w   d4,XMIT_6          * send fm to MUX controller
      add.w    d7,d5              * d5 = fe = fm + fs
      move.w   d5,fe              * save fe for integration
integrate:
      move.w   theta,d6           * d6 = theta
      add.w    d5,d6              * integrate fe  (theta = theta + fe)
      move.w   d6,theta           * save theta
* d4 = xx
* d5 = fe
* d6 = theta
*-----------------------------------------------------*
* TABLE LOOKUP and INTERPOLATE                        *
*   find SIN(theta), find COS(theta)                  *
*-----------------------------------------------------*
      move.w   d6,d7
      add.w    #$4000,d7
      tbls.w   SIN_X,d6           * d6 = SIN(x)
      tbls.w   SIN_X,d7           * d7 = COS(x)
* d4 = xx
* d5 = fe
* d6 = SIN(x)
* d7 = COS(x)
*-----------------------------------------------------*
      bclr.b   #6,PORTE           * timing marker
*-----------------------------------------------------*
* CURRENT READING and NORMALIZATION                   *
*  d0 = Ia read from A/D  \___ range =   $000 - $3ff  *
*  d1 = Ib read from A/D  /            -600A - 600A   *
*  d0 = Ia normalized     \___ range =  $8000 - $7fff *
*  d1 = Ib normalized     /            -600A - 600A   *
*-----------------------------------------------------*
      move.w   RCV_9,d0           *  -  read Ia
      move.w   RCV_10,d1          *  -  read Ib
      bsr      i_norm             * A/D normalization
* d0 = Ia normalized
```

```
* d1 = Ib normalized
* d4 = xx
* d5 = fe
* d6 = SIN(x)
* d7 = COS(x)
*-------------------------------------------------------------*
* 3ph-2ph TRANSFORMATION (ABC_DQ + VECTOR ROTATOR)            *
*  Transformation equations:                                  *
*     IQ = Ia*COS(theta) + (1.1547*Ib + .57735*Ia)*SIN(theta)  *
*     ID = Ia*SIN(theta) - (1.1547*Ib + .57735*Ia)*COS(theta)  *
* or                                                          *
*     IQ = Ia*COS(theta) + (2*.55735*Ib + .57735*Ia)*SIN(theta) *
*     ID = Ia*SIN(theta) - (2*.55735*Ib + .57735*Ia)*COS(theta) *
*  1.1547 = 2/SQRT(3)                                         *
*  .57735 = 1/SQRT(3)                                         *
* NOTE: Since CPU32 does not have extension bits, ID and IQ are *
*       scaled such that the radix point is between 14:13.    *
*       This gives two extension bits.                        *
*-------------------------------------------------------------*
xform3_2:
      move.w   #$49e6,d4        * d4 = 1.1547  (radix between 14:13)
      move.w   #$24f3,d5        * d5 = 0.57735 (radix between 14:13)
      muls.w   d0,d5            * d5.1 = .55735 * Ia
      muls.w   d1,d4            * d5.1 = 1.1547 * Ib
      add.l    d4,d5            * d5.1 = 1.1547*Ib + .55735*Ia
      asl.l    #1,d5            * shift to make fractional mpy
      swap     d5
      move.w   d5,d4            * d4,d5 = 1.1547*Ib + .55735*Ia
      move.w   d0,d1            * d0,d1 = Ia
      muls.w   d7,d0            * d0 = Ia * COS(theta) (radix between 14:13)
      muls.w   d6,d1            * d1 = Ia * SIN(theta) (radix between 14:13)
      muls.w   d7,d4            * d4 = (1.1547*Ib + .57735*Ia) * COS(theta)
      asl.l    #1,d4
      muls.w   d6,d5            * d5 = (1.1547*Ib + .57735*Ia) * SIN(theta)
      asl.l    #1,d5
      add.l    d5,d0            * d0 = Iq
      sub.l    d4,d1            * d1 = Id
      swap     d0
      swap     d1
      move.w   d0,IQ_act
      move.w   d1,ID_act
* d0 = IQ  (-2 to 2)  sync ref frame
* d1 = ID  (-2 to 2)
      bset.b   #7,SPCR1         *  -  start A/D xfers again
*---------------------------------------------------------------*
* PI CONTROLLERS                                                *
* This code does the current control.  It compares the commanded *
* current (i_cmd) with the actual current (i_act) and adjusts the *
* voltage accordingly                                           *
*---------------------------------------------------------------*
      move.w   IQ_cmd,d6
      move.w   ID_cmd,d7
P_gain   move.w   #$1200,d4
I_gain   move.w   #$0140,d3         ccl:
```

```
        sub.w     d0,d6                  * d6 = err = IQ_cmd - iq_act
        move.w    old_iq,d5              * d5 = err(z^-1)
        move.w    d6,old_iq
        move.w    d6,d2
        sub.w     d5,d6                  * d6 =  err - err(z^-1)
        muls.w    d3,d2                  * I*err
        asl.l     #2,d2
        muls.w    d4,d6                  * P*(err-err(z^-1))
        asl.l     #5,d6
        add.l     d2,d6                  * P*(err-err(z^-1)) + I*err
        neg.l     d6                     * - (P*(err-err(z^-1)) + I*err)
        move.l    VQ,d5
        add.l     d6,d5
        bvc       VL1
        bpl       vq_neg_lim
        move.l    #$7fffffff,d5
        bra       VL1
vq_neg_lim:
        move.l    #$80000000,d5
VL1:    move.l    d5,VQ
cc2:
        sub.w     d1,d7                  * d7 = err = ID_cmd - id_act
        move.w    old_id,d5              * d5 = err(z^-1)
        move.w    d7,old_id
        move.w    d7,d2
        sub.w     d5,d7                  * d5 = err - err(z^-1)
        muls.w    d3,d2                  * P gain
        asl.l     #2,d2
        muls.w    d4,d7                  * I gain
        asl.l     #5,d7
        add.l     d2,d7
        neg.l     d7
        move.l    VD,d5
        add.l     d7,d5
        bvc       VL2
        bpl       vd_neg_lim
        move.l    #$7fffffff,d5
        bra       VL2
vd_neg_lim:
        move.l    #$80000000,d5
VL2:    move.l    d5,VD
        move.l    VQ,d2                  * VD,VQ are 32-bit quantities
        move.l    VD,d1
        swap      d2                     * scale to 16-bits
        swap      d1                     * scale to 16-bits
* d1 = VD                                  * $8000 - $7fff
* d1 = VQ
*---------------------------------------------------*
* RECTANGULAR TO POLAR CONVERSION                   *
*---------------------------------------------------*
        bsr       sqrroot
        bsr       arctan
* d0 = angle
* d1 = vd
```

```
* d2 = vq
* d4 = Vmag
      move.w   theta,d7
      sub.w    d0,d7     * add vector angle to angle from
      move.w   d4,d6     * the integrator
*--------------------------------------------------------*
* SPACE VECTOR MODULATION SCHEME                         *
* Input Vmag and Theta                                   *
*--------------------------------------------------------*
chk_s6:
      cmp      #$d555,d7          * $d555 = 300 degrees
      bcc      s6
      cmp      #$aaaa,d7
      bcc      s5
      cmp      #$8000,d7
      bcc      s4
      cmp      #$5555,d7
      bcc      s3
      cmp      #$2aaa,d7
      bcc      s2
      bra      s1
s6    move.w   #$d555,d0          * sector 6
      bsr      calculate
      add.w    d1,d2
      move.w   d2,A_HIGH
      clr.w    B_HIGH
      move.w   d1,C_HIGH
      bra      past
s5    move.w   #$aaaa,d0          * sector 5
      bsr      calculate
      add.w    d2,d1
      move.w   d2,A_HIGH
      clr.w    B_HIGH
      move.w   d1,C_HIGH
      bra      past
s4    move.w   #$8000,d0          * sector 4
      bsr      calculate
      add.w    d1,d2
      clr.w    A_HIGH
      move.w   d1,B_HIGH
      move.w   d2,C_HIGH
      bra      past
s3    move.w   #$5555,d0          * sector 3
      bsr      calculate
      add.w    d2,d1
      clr.w    A_HIGH
      move.w   d1,B_HIGH
      move.w   d2,C_HIGH
      bra      past
s2    move.w   #$2aaa,d0          * sector 2
      bsr      calculate
      add.w    d1,d2
      move.w   d1,A_HIGH
      move.w   d2,B_HIGH
```

```
      clr.w    C_HIGH
      bra      past
s1    move.w   #$0000,d0          * sector 1
      bsr      calculate
      add.w    d2,d1
      move.w   d1,A_HIGH
      move.w   d2,B_HIGH
      clr.w    C_HIGH
past:
exit:
      bclr.b   #6,CISR            * clear tpu interrupt
      movem.l  (sp)+,d0-d7        * restore regs used by this routine
      bclr.b   #7,PORTE           * TIMING MARKER !!!!!!!
      rte
**********************************************
*  SUBROUTINES                                *
*   - spi_xfer, - display, - I_norm           *
**********************************************
* SPI  TRANSFER  SUBROUTINE:
spi_xfer:
      bset     #7,SPCR1           * initiate QSPI transfer
wait: btst.b   #7,SPSR            * check if done
      beq      wait
      bclr.b   #7,SPSR            * clear QSPI finished flag
      rts
*---------------------------------------------*
* LCD DISPLAY SUBROUTINE:                     *
*---------------------------------------------*
display:
      movem.l  d0-d2/a1,-(sp)
      lea.l    LCD_LUT,a1         * A1 points to LCD conversion table
      move.w   #$0002,XMIT_0      * start byte for LCD display
      move.w   d7,d0
      move.w   d7,d1
      move.w   #$000f,d2
      and.w    d2,d1
      move.w   (a1,d1.w*2),XMIT_4
      lsr.w    #4,d0
      move.w   d0,d1
      and.w    d2,d1
      move.w   (a1,d1.w*2),XMIT_3
      lsr.w    #4,d0
      move.w   d0;d1
      and.w    d2,d1
      move.w   (a1,d1.w*2),XMIT_2
      lsr.w    #4,d0
      move.w   d0,d1
      and.w    d2,d1
      move.w   (a1,d1.w*2),XMIT_1
      clr.w    XMIT_5
      movem.l  (sp)+,d0-d2/a1
      rts
*--------------------------------------------------------------------------*
* I_NORM SUBROUTINE:                                                       *
```

```
* This subroutine takes the current measurements from the          *
* A/D and normalizes them to 16 bit signed fractions.              *
*  CALLED BY: --> TPU_SVC                                          *
*  REGS USED: --> d0,d1,d7                                         *
*        IN: --> d0       Ia - read from A/D  (0 - $3ff)           *
*                d1       Ib - read from A/D  (0 - $3ff)           *
*       OUT: --> d0       Ia - 16 bit signed fraction ($8000 to $7fff) *
*                d1       Ib - 16 bit signed fraction ($8000 to $7fff) *
*------------------------------------------------------------------*
i_norm:
      sub.w    IA_ZERO,d0         * subtract mid reading of A/D
      asl.w    #6,d0              * d0 = Ia (16 bit signed fraction)
      bvc      ib_chk
      bpl      a_neg_limit
a_pos_limit:
      move.w   #$7fff,d0
      bra      ib_chk
a_neg_limit:
      move.w   #$8000,d0
ib_chk:
      sub.w    IB_ZERO,d1         * subtract mid reading of A/D
      asl.w    #6,d1              * d1 = Ib (16 bit signed fraction)
      bvc      norm_ok
      bpl      b_neg_limit
b_pos_limit:
      move.w   #$7fff,d1
      bra      norm_ok
b_neg_limit:
      move.w   #$8000,d1
norm_ok:
      rts
*--------------------------------------------------------*
* Calculate Subroutine:                                  *
* This subroutine calculates T1,T2 and T0 for the space  *
* vector modulation technique                            *
*--------------------------------------------------------*
calculate:
      move.w   d7,d2              * d2 = theta
      move.w   #$2aaa,d1          * d1 = 60 degrees
      sub.w    d0,d2              * d2 = delta_theta
      sub.w    d2,d1              * d1 = 60 - delta_theta
      tbls.w   SIN_X,d1           * d1 = SIN(60-x)
      tbls.w   SIN_X,d2           * d2 = SIN(x)
      move.w   #PERIOD,d3
      mulu.w   d3,d6
      lsl.l    #1,d6
      swap     d6                 * d6 = T*m
      muls.w   d6,d1
      lsl.l    #1,d1              * d1 = T1 = T*m*sin(60-x)
      muls.w   d6,d2
      lsl.l    #1,d2              * d2 = T2 = T*m*sin(x)
      swap     d1
      swap     d2
      sub.w    d1,d3
```

```
        sub.w    d2,d3
        rts
*-------------------------------------------------------*
* Square Root Subroutine                                *
* This routine performs the square root function through *
* a lookup table.                                       *
*-------------------------------------------------------*
sqrroot:
        move.w   d3,-(sp)
        move.w   d1,d3
        move.w   d2,d4
        muls.w   d3,d3
        muls.w   d4,d4
        add.l    d3,d4
        swap     d4
        tblu.w   SQRT,d4
        bpl      no_limit
        move.w   #$7fff,d4
no_limit:
        move.w   (sp)+,d3
        rts
*-------------------------------------------------------*
* Arctangent Subroutine                                 *
* This routine performs the arctan function it is used in *
* the rectangular to polar conversion section.          *
*-------------------------------------------------------*
arctan:
        movem.w  d1/d2,-(sp)
        move.w   d1,d0
        or.w     d2,d0
        beq      done
        tst.w    d1
        smi.b    vd_sgn
        bpl      vd_pos
        neg.w    d1
vd_pos:
        tst.w    d2
        smi.b    vq_sgn
        bpl      vq_pos
        neg.w    d2
vq_pos:
        cmp.w    d1,d2
        bcc      x0_45
x90_45:
        swap     d2
        clr.w    d2
        divu.w   d1,d2
        bvc      ok1
        move.w   #$ffff,d2
ok1:
        lsr.w    #1,d2
        tbls.w   ATAN,d2
        move.w   #$4000,d0
        sub.w    d2,d0
```

```
      bra      quad_tst
x0_45:
      swap     d1
      clr.w    d1
      divu.w   d2,d1
      bvc      ok2
      move.w   #$ffff,d1
ok2:
      lsr.w    #1,d1
      tbls.w   ATAN,d1
      move.w   d1,d0
quad_tst:
      tst.b    vd_sgn
      beq      q_1
      neg.w    d0
q_1:
      tst.b    vq_sgn
      beq      done
      sub.w    #$8000,d0
      neg.w    d0
done: movem.w  (sp)+,d1/d2
      rts
**********************************************************************
*  GEN_SVC:                                                          *
*  This service routine is used to handle all unexpected interrupts  *
*  or exceptions.  The routine will initialize the QSM in assumption *
*  that it is not initialized.  It will then display on the LCD the  *
*  vector number of the exception that occurred.                     *
**********************************************************************
GEN_SVC:
      bclr.b   #7,PORTF          * Disable inverter (major error)
      move.w   #$2700,SR         * Disable all interrupts.
      move.w   #$008d,QMCR       * Normal op; ignore freeze; restricted
      move.b   #$F3,QPDR         * SCK,PCS0 default 0; all others high
      move.b   #$FE,QDDR         * MISO = input; all others output
      move.b   #$7B,QPAR         * PCS0-PCS3,PCS1,MISO,MOSI assign to QSPI
      move.w   #$A80A,SPCR0      * MSTR; BITS=10; CPOL:CPHA=0:0;SPBR=833 KHz
      move.w   #$0f0f,CMD_0      * LCD display : LCD display
      move.w   #$0f0f,CMD_2      * LCD display : LCD display
      move.w   #$0f0f,CMD_4      * LCD display : LCD display
      clr.l    XMIT_0            * XMITRAM initially all zeros
      clr.l    XMIT_2            *      (clears LCD display for
      clr.l    XMIT_4            *       syncronization to micro)
      move.w   #$0500,SPCR2      * NO wrap;  que length 0-5 (LCD dsply);
      clr.b    SPCR3             * Clear SPIF flag
      bsr      spi_xfer
      move.w   6(sp),d0          * FORMAT: VECTOR OFFSET
      andi.w   #$0fff,d0         * mask off FORMAT code
      lsr.w    #2,d0             * divide by 4 to get vector number
      lea.l    LCD_LUT,a1
      move.w   #$0002,XMIT_0     * start byte
      move.w   #$0079,XMIT_1     *   E
      move.w   #$0020,XMIT_2     *   -
      move.w   d0,d1             * display 2-digit
```

```
       andi.w   #$00f0,d0          *     vector number (in hex)
       lsr.w    #4,d0
       move.w   (a1,d0.w*2),XMIT_3
       andi.w   #$000f,d1
       move.w   (a1,d1.w*2),XMIT_4
       clr.w    XMIT_5
       bset.b   #7,SPCR1           * display E-## (## = vector number)
self   bra      self
       rte
*!!!!!!!!!!!!!!!!!!!!!!!!!!!!!!!!!!!!!!!!!!!!!!!!!!!!!!!!!!!!!!!!!!!!!!!!!!!!*
*!!!!!!!!!!!!!!!!!!!!!!!!!!!!!!!          ROM area      !!!!!!!!!!!!!!!!!!!!!!!!!!*
*!!!!!!!!!!!!!!!!!!!!!!!!!!!!!!!!!!!!!!!!!!!!!!!!!!!!!!!!!!!!!!!!!!!!!!!!!!!!*
CRNT:
       dc.w     $0000,$0000,$0200,$0400,$0800,$0a00,$0c00,$0e00
       dc.w     $1000,$1200,$1400,$1600,$1800,$1a00,$1c00,$1e00
       dc.w     $2000,$2400,$2800,$2c00,$3000,$3200,$3280,$3300
       dc.w     $3380,$3400,$3480,$3500,$3540,$3580,$35c0,$3600,$3600
MAGNET:
       dc.w     $0500,$0500,$0500,$0500,$0500,$0400,$0355,$02db
       dc.w     $0280,$0238,$0200,$01d1,$01aa,$0189,$016d,$0155,$0140,$012d
LCD_LUT:
       dc.w     $005f,$0006,$003b,$002f,$0066,$006d,$007d,$0007,$007f,$006f
       dc.w     $0077,$007c,$0059,$003e,$0079,$0071
SIN_X:
       dc.w     $0000,$0324,$0647,$096a,$0c8b,$0fab,$12c8,$15e2
       dc.w     $18f8,$1c0b,$1f19,$2223,$2528,$2826,$2b1f,$2e11
       dc.w     $30fb,$33de,$36ba,$398c,$3c56,$3f17,$41ce,$447a
       dc.w     $471c,$49b4,$4c3f,$4ebf,$5133,$539b,$55f5,$5842
       dc.w     $5a82,$5cb4,$5ed7,$60ec,$62f2,$64e8,$66cf,$68a6
       dc.w     $6a6d,$6c24,$6dca,$6f5f,$70e2,$7255,$73b5,$7504
       dc.w     $7641,$776c,$7884,$798a,$7a7d,$7b5d,$7c29,$7ce3
       dc.w     $7d8a,$7e1d,$7e9d,$7f09,$7f62,$7fa7,$7fd8,$7ff6
       dc.w     $7fff,$7ff6,$7fd8,$7fa7,$7f62,$7f09,$7e9d,$7e1d
       dc.w     $7d8a,$7ce3,$7c29,$7b5d,$7a7d,$798a,$7884,$776c
       dc.w     $7641,$7504,$73b5,$7255,$70e2,$6f5f,$6dca,$6c24
       dc.w     $6a6d,$68a6,$66cf,$64e8,$62f2,$60ec,$5ed7,$5cb4
       dc.w     $5a82,$5842,$55f5,$539b,$5133,$4ebf,$4c3f,$49b4
       dc.w     $471c,$447a,$41ce,$3f17,$3c56,$398c,$36ba,$33de
       dc.w     $30fb,$2e11,$2b1f,$2826,$2528,$2223,$1f19,$1c0b
       dc.w     $18f8,$15e2,$12c8,$0fab,$0c8b,$096a,$0647,$0324
       dc.w     $0000,$fcdb,$f9b8,$f695,$f374,$f054,$ed37,$ea1d
       dc.w     $e707,$e3f4,$e0e6,$dddc,$dad7,$d7d9,$d4e0,$d1ee
       dc.w     $cf04,$cc21,$c945,$c673,$c3a9,$c0e8,$be31,$bb85
       dc.w     $b8e3,$b64b,$b3c0,$b140,$aecc,$ac64,$aa0a,$a7bd
       dc.w     $a57d,$a34b,$a128,$9f13,$9d0d,$9b17,$9930,$9759
       dc.w     $9592,$93db,$9235,$90a0,$8f1d,$8daa,$8c4a,$8afb
       dc.w     $89be,$8893,$877b,$8675,$8582,$84a2,$83d6,$831c
       dc.w     $8275,$81e2,$8162,$80f6,$809d,$8058,$8027,$8009
       dc.w     $8000,$8009,$8027,$8058,$809d,$80f6,$8162,$81e2
       dc.w     $8275,$831c,$83d6,$84a2,$8582,$8675,$877b,$8893
       dc.w     $89be,$8afb,$8c4a,$8daa,$8f1d,$90a0,$9235,$93db
       dc.w     $9592,$9759,$9930,$9b17,$9d0d,$9f13,$a128,$a34b
       dc.w     $a57d,$a7bd,$aa0a,$ac64,$aecc,$b140,$b3c0,$b64b
       dc.w     $b8e3,$bb85,$be31,$c0e8,$c3a9,$c673,$c945,$cc21
```

```
        dc.w    $cf04,$d1ee,$d4e0,$d7d9,$dad7,$dddc,$e0e6,$e3f4
        dc.w    $e707,$ea1d,$ed37,$f054,$f374,$f695,$f9b8,$fcdb
        dc.w    $0000
ATAN:
        dc.w    $0000,$0051,$00a3,$00f4,$0146,$0197,$01e9,$023a
        dc.w    $028b,$02dc,$032d,$037e,$03cf,$0420,$0470,$04c1
        dc.w    $0511,$0561,$05b1,$0601,$0651,$06a0,$06ef,$073e
        dc.w    $078d,$07dc,$082a,$0878,$08c6,$0914,$0961,$09ae
        dc.w    $09fb,$0a48,$0a94,$0ae0,$0b2c,$0b77,$0bc2,$0c0d
        dc.w    $0c57,$0ca1,$0ceb,$0d34,$0d7d,$0dc6,$0e0f,$0e56
        dc.w    $0e9e,$0ee5,$0f2c,$0f73,$0fb9,$0fff,$1044,$1089
        dc.w    $10ce,$1112,$1156,$1199,$11dc,$121f,$1261,$12a3
        dc.w    $12e4,$1325,$1366,$13a6,$13e6,$1425,$1464,$14a2
        dc.w    $14e0,$151e,$155b,$1598,$15d5,$1611,$164c,$1688
        dc.w    $16c2,$16fd,$1737,$1770,$17aa,$17e2,$181b,$1853
        dc.w    $188a,$18c1,$18f8,$192e,$1964,$199a,$19cf,$1a04
        dc.w    $1a38,$1a6c,$1a9f,$1ad3,$1b05,$1b38,$1b6a,$1b9c
        dc.w    $1bcd,$1bfe,$1c2e,$1c5e,$1c8e,$1cbe,$1ced,$1d1b
        dc.w    $1d4a,$1d78,$1da5,$1dd3,$1dff,$1e2c,$1e58,$1e84
        dc.w    $1eb0,$1edb,$1f06,$1f30,$1f5a,$1f84,$1fae,$1fd7
        dc.w    $2000
SQRT:
        dc.w    $0000,$1000,$16A0,$1BB0,$2000,$23C0,$2730,$2A50
        dc.w    $2D40,$3000,$3290,$3510,$3760,$39B0,$3BD0,$3DF0
        dc.w    $4000,$41F0,$43E0,$45B0,$4780,$4950,$4B00,$4CB0
        dc.w    $4E60,$5000,$5190,$5320,$54A0,$5620,$57A0,$5910
        dc.w    $5A80,$5BE0,$5D40,$5EA0,$6000,$6150,$62A0,$63E0
        dc.w    $6530,$6670,$67B0,$68E0,$6A20,$6B50,$6C80,$6DB0
        dc.w    $6ED0,$7000,$7120,$7240,$7360,$7470,$7590,$76A0
        dc.w    $77B0,$78C0,$79D0,$7AE0,$7BE0,$7CF0,$7DF0,$7EF0
        dc.w    $8000,$80F0,$81F0,$82F0,$83F0,$84E0,$85D0,$86D0
        dc.w    $87C0,$88B0,$89A0,$8A90,$8B70,$8C60,$8D40,$8E30
        dc.w    $8F10,$9000,$90E0,$91C0,$92A0,$9380,$9460,$9530
        dc.w    $9610,$96F0,$97C0,$98A0,$9970,$9A40,$9B20,$9BF0
        dc.w    $9CC0,$9D90,$9E60,$9F30,$A000,$A0C0,$A190,$A260
        dc.w    $A320,$A3F0,$A4B0,$A580,$A640,$A700,$A7C0,$A890
        dc.w    $A950,$AA10,$AAD0,$AB90,$AC50,$AD10,$ADC0,$AE80
        dc.w    $AF40,$B000,$B0B0,$B170,$B220,$B2E0,$B390,$B440
        dc.w    $B500
*!!!!!!!!!!!!!!!!!!!!!!!!!!!!!!!!!!!!!!!!!!!!!!!!!!!!!!!!!!!!!!!!!!!!!!!!!!!!*
*!!!!!!!!!!!!!!!!!!!!!!!!!!!!!      RAM area    !!!!!!!!!!!!!!!!!!!!!!!!!!!!!!*
*!!!!!!!!!!!!!!!!!!!!!!!!!!!!!!!!!!!!!!!!!!!!!!!!!!!!!!!!!!!!!!!!!!!!!!!!!!!!*
        org     $5000
throttle ds.w   1
extra   ds.w    1
theta   ds.w    1
omega   ds.w    1
SIN_PT  ds.w    1
SIN_FREQ ds.w   1
SIN_AMP ds.w    1
IA_ZERO ds.l    1
IB_ZERO ds.l    1
i_act   ds.w    1
i_cmd   ds.w    1
```

```
old_id   ds.w    1
old_iq   ds.w    1
fm       ds.w    1
fe       ds.w    1
fs       ds.w    1
fault    ds.w    1
ID_cmd   ds.w    1
IQ_cmd   ds.w    1
ID_act   ds.w    1
IQ_act   ds.w    1
VD       ds.l    1
VQ       ds.l    1
vd_sgn   ds.b    1
vq_sgn   ds.b    1
  end
```

MC68332 VECTOR CONTROL CODE
INCLUDE FILE: TPU.INI

```
*-----------------------------------------------------------------------*
* TPU INITIALIZATION CODE                                               *
* Author:  Peter Pinewski - Motorola Semiconductor Products Sector       *
* Created: 11/10/94                                                     *
* Last Modified:  2/13/96                                               *
* This code is responsible for downloading the TPU microcode into TPU   *
* ram and setting up the TPU registers for six center-aligned PWMs.     *
* ch0 through ch12 are used for PWM generation                          *
* ch13 is a spare PWM for test purposes                                 *
* ch14 is used for event timing (generates the CPU interrupt by a delay *
*  from the center-aligned PWMs (requires ch0 to be connected to ch14)  *
*  this is for event synchronization (minimizing system delays          *
* ch15 is used for speed sensing.                                       *
* external references: FIFTY_PCT                                       *
*                      PERIOD                                           *
*                      DEADx2                                           *
*                      SPEED_INT                                        *
*                      INT_RATE                                         *
*                      DELAY                                            *
*-----------------------------------------------------------------------*
*  TPU_UCODE INIT:                                                      *
*  This is UCODE data and the initialization code to load the TPU ucode *
*  into emulation ram.                                                  *
*         3 functions:                                                  *
*         ============                                                  *
*    - MCPWM   Multi-channel Centered PWM (written by MOTOROLA        )  *
*    - TTC     Timed Transition Counter   (written by Peter Pinewski )  *
*    - PWM     Pulse Width Modulation     (written by MOTOROLA        )  *
*    - INT     Interrupt delay            (written by Peter Pinewski )  *
*                                                                       *
* TPU microcode consists of 84 longwords:     (ram offset = $000)       *
*         - MCPWM = 15. (beta version)   32 long words                  *
*         - ITC   = 14. (  rev 0     )   11 long words                  *
*         - PWM   = 13. (  standard  )   24 long words                  *
*         - INT   = 12. (  rev 0     )    6 long words                  *
```

```
*                                          =============                  *
*                                          73 long words                  *
*                                                                         *
* TPU entry points consist of 24 longwords:  (ram offset = $780)          *
*        - 16 entry points INT              8 long words     ($780)       *
*        - 16 entry points PWM              8 long words     ($7a0)       *
*        - 16 entry points TTC              8 long words     ($7c0)       *
*        - 16 entry points MCPWM            8 long words     ($7e0)       *
*-------------------------------------------------------------------------*
          bra       DOWNLOAD
TPUCODE:
MCPWM     dc.l      $7FFFFEFE,$3FFFFFFE,$BE05464C,$185FF007,$09FFF007
          dc.l      $3C7FFA03,$20DFD007,$22DC0FFF,$961DFEFF,$107FF003
          dc.l      $D01FF807,$BC0D46BC,$BFFF477C,$AE1DFFC7,$1E5FF203
          dc.l      $105FF003,$101FF00F,$023FF80F,$B21DFEFF,$1C5E7203
          dc.l      $367FDFFF,$303C8203,$32EE8FFF,$8E1BFEFF,$307EBFFF
          dc.l      $32EFFFFF,$307FFFFF,$B41EFFFF,$3C7EF807,$5C583EFE
          dc.l      $3C7EF80B,$5C583EFA
TTC       dc.l      $BFFFFFF8,$09FFF807,$5C58FEFF,$3EFFF00E,$A427FFFF
          dc.l      $30FFD00F,$7FFBFFFE,$29FFF00B,$1EFFF807,$3FFFF00F
          dc.l      $5C5CFFFE
PWM       dc.l      $E1E401C7,$8E2FFEF8,$7859FEFF,$7A59FEFF,$3C7FF807
          dc.l      $D432FFFF,$525CB5FA,$163FF00B,$101DF80F,$8639FFFF
          dc.l      $36FEB013,$37FC4FFF,$843AFFFF,$D9FF1FFF,$545CF18A
          dc.l      $545CF14A,$A42FFFFF,$D030FFFF,$A440FEFF,$9C40FEFF
          dc.l      $3C7FF807,$D432FFFF,$545CF3FA,$505DF3FE
INT       dc.l      $BFFFFFF8,$7FF9FEFE,$A448FFFF,$3C5E3FFF,$5FF9FEFE,$7FF9FEFA
ENT_PTS:
int       dc.l      $ee00ee00,$ee000043,$20452045,$20452045
          dc.l      $ee01ee01,$ee01ee01,$20452045,$20452045
pwm       dc.l      $283D283B,$002BFE01,$9042FE3F,$302F302F
          dc.l      $FE00FE00,$FE00FE00,$FE00FE00,$FE00FE00
ttc       dc.l      $EE00EE00,$EE000020,$60246024,$60246024
          dc.l      $EE01EE01,$EE01EE01,$60246024,$60246024
mcpwm     dc.l      $a80ba80b,$880c8802,$a80da804,$880d8804
          dc.l      $ee00ee00,$ee00ee00,$a80da804,$880d8804
DOWNLOAD:
          move.w    #$0200,RAMBAR      * initialize standby ram to $20000
          lea.l     TPUCODE,A0
          lea.l     $020000,A1         * LOAD uCODE.
          moveq.l   #72,d0             * 73 longwords of microcode.
ld_ucode  move.l    (a0)+,(a1)+
          dbf       d0,ld_ucode
          lea.l     ENT_PTS,A0
          lea.l     $020780,A1         * LOAD ENTRY POINTS.
          moveq.l   #31,d0             * 32 longwords of entry points.
ld_entpt  move.l    (a0)+,(a1)+
          dbf       d0,ld_entpt
*---------    UCODE initialization complete    ----------------------*
*-------------------------------------------------------------------------*
*  TPU_INIT:                                                              *
*  Channels Used:                                                         *
*    mcpwm  ch0    = master channel                  tcr1                 *
*    mcpwm  ch1,2  = phase A_top (inverted)          tcr1                 *
```

```
*    mcpwm  ch3,4    = phase A_bot                   tcr1              *
*    mcpwm  ch5,6    = phase B_top (inverted)        tcr1              *
*    mcpwm  ch7,8    = phase B_bot                   tcr1              *
*    mcpwm  ch9,10   = phase C_top (inverted)        tcr1              *
*    mcpwm  ch11,12 = phase C_bot                    tcr1              *
*      pwm  ch13     =  -- spare PWM for test --     tcr1              *
*      int  ch14     = interrupt dealy               tcr1              *
*      ttc  ch15     = speed sensing                 tcr2              *
*  tcr1 = 240ns resolution (prescaler is 1)=(system_clk/4 )            *
*  tcr2 =  4us  resolution (prescaler is 8)=(system_clk/64)            *
*  TPU interrupt level = 4                                             *
*      vector numbers  = 40  (SINE_SVC)                                *
*  Upon completion of this code section the tpu should be generating   *
*  the waveforms with a duty-cycle of 50%.                             *
*---------------------------------------------------------------------*
        move.w   #$1ecf,TMCR      * tcr1 = 240ns; tcr2 = 4us; emu=1
        clr.w    CIER             * disable all tpu interrupts
        move.w   #$0440,TICR      * tpu int lvl = 4 ; vec # tpu ch0 = 40.
        move.w   #$ffff,CFSR3     * set tpu channels
        move.w   #$ffff,CFSR2     *  ch 0 = MCPWM master channel
        move.w   #$ffff,CFSR1     *  ch 1,2,3,4,5,6,7,8,9,10,11,12 = MCPWM
        move.w   #$ecdf,CFSR0     *  ch  13 = PWM
*                                 *  ch  14 = INT
*                                 *  ch  15 = TTC*
*              |--------------- Slave Channels ------------------|
*  cont        | B   A | B   A | B     A | B   A | B   A | B   A |Master
*   no         |       |       |         |       |       |       | Chan
*  links       | Ph- C | Ph-!C |  Ph- B  | Ph-!B | Ph- A | Ph-!A |
*  |-----------------------------| |-----------------------------|
*  | 01 xx : xx 10 : 01 10 : 01 10 | | 01 10 : 01 10 : 01 10 : 01  xx |
*  |_15__________________________8_| |_7___________________________0_|
*              HSQR0                           HSQR1
        move.w   #$6664,HSQR1      * ch 1,3,5,7,9,11  = slave type_A
        move.w   #$4266,HSQR0      * ch 2,4,6,8,10,12 = slave type_B
*   Ch 0   (-- master channel --)
        move.w   #PERIOD,pram+$00  * set up period
        move.w   #INT_RATE,pram+$02  * set table step rate
        move.w   #$0004,pram+$08   * rise_time_ptr to master ch.
        move.w   #$0006,pram+$0a   * fall_time_ptr to master ch.
* PHASE A_top - ram initialization * (--- INVERTED ---)
*   Ch 1   (-- slave type_a --)
        move.w   #PERIOD,pram+$10 * PWM period
        move.w   #$0021,pram+$16  * no dead; hi_time ptr to type_b (ch2)
        move.w   #$0006,pram+$18  * fall_time_ptr to master channel (ch0)
        move.w   #$0004,pram+$1a  * rise_time_ptr to master channel (ch0)
*   Ch 2   (-- slave type_b --)
        move.w   #FIFTY_PCT,A_HIGH * hi_time for __PHASE A__ (par ram: $20)
        move.w   #$0012,pram+$28  * B_fall_time_ptr to type_a channel (ch1)
        move.w   #$0014,pram+$2a  * B_rise_time_ptr to type_a channel (ch1)
* PHASE A_bot - ram initialization
*   Ch 3   (-- slave type_a --)
        move.w   #PERIOD,pram+$30 * PWM period
        move.w   #$0023,d0
        ori.w    #DEADx2,d0
```

```
        move.w   d0,pram+$36     * 2us dead; current_hi_time ptr to (ch2)
        move.w   #$0004,pram+$38  * rise_time_ptr to master channel (ch0)
        move.w   #$0006,pram+$3a  * fall_time_ptr to master channel (ch0)
*    Ch 4   (-- slave type_b --)
        move.w   #$0000,pram+$40  * hi_time  (!!!! --- NOT USED --- !!!!)
        move.w   #$0034,pram+$48  * B_fall_time_ptr to type_a channel (ch3)
        move.w   #$0032,pram+$4a  * B_rise_time_ptr to type_a channel (ch3)
* PHASE B_top  - ram initialization * (---  INVERTED  ---)
*    Ch 5   (-- slave type_a --)
        move.w   #PERIOD,pram+$50 * PWM period
        move.w   #$0061,pram+$56  * no dead; hi_time ptr to type_b (ch6)
        move.w   #$0006,pram+$58  * fall_time_ptr to master channel (ch0)
        move.w   #$0004,pram+$5a  * rise_time_ptr to master channel (ch0)
*    Ch 6   (-- slave type_b --)
        move.w   #FIFTY_PCT,B_HIGH * hi_time for __PHASE B__ (par ram: $60)
        move.w   #$0052,pram+$68  * B_fall_time_ptr to type_a channel (ch5)
        move.w   #$0054,pram+$6a  * B_rise_time_ptr to type_a channel (ch5)
* PHASE B_bot  - ram initialization
*    Ch 7   (-- slave type_a --)
        move.w   #PERIOD,pram+$70 * PWM period
        move.w   #$0063,d0
        ori.w    #DEADx2,d0
        move.w   d0,pram+$76      * 2us dead; current_hi_time ptr to (ch6)
        move.w   #$0004,pram+$78  * rise_time_ptr to master channel (ch0)
        move.w   #$0006,pram+$7a  * fall_time_ptr to master channel (ch0)
*    Ch 8   (-- slave type_b --)
        move.w   #$0000,pram+$80  * hi_time  (!!!! --- NOT USED --- !!!!)
        move.w   #$0074,pram+$88  * B_fall_time_ptr to type_a channel (ch7)
        move.w   #$0072,pram+$8a  * B_rise_time_ptr to type_a channel (ch7)
* PHASE C_top  - ram initialization * (---  INVERTED  ---)
*    Ch 9   (-- slave type_a --)
        move.w   #PERIOD,pram+$90 * PWM period
        move.w   #$00a1,pram+$96  * no dead; hi_time ptr to type_b (ch10)
        move.w   #$0006,pram+$98  * fall_time_ptr to master channel (ch0)
        move.w   #$0004,pram+$9a  * rise_time_ptr to master channel  (ch0)
*    Ch 10 (-- slave type_b --)
        move.w   #FIFTY_PCT,C_HIGH * hi_time for __PHASE C__ (par ram: $a0)
        move.w   #$0092,pram+$a8  * B_fall_time_ptr to type_a channel (ch9)
        move.w   #$0094,pram+$aa  * B_rise_time_ptr to type_a channel (ch9)
* PHASE C_TOP  - ram initialization
*    Ch 11 (-- slave type_a --)
        move.w   #PERIOD,pram+$b0 * PWM period
        move.w   #$00a3,d0
        ori.w    #DEADx2,d0
        move.w   d0,pram+$b6      * 2us dead; current_hi_time ptr to (ch10)
        move.w   #$0004,pram+$b8  * rise_time_ptr to master channel (ch0)
        move.w   #$0006,pram+$ba  * fall_time_ptr to master channel  (ch0)
*    Ch 12 (-- slave type_b --)
        move.w   #$0000,pram+$c0  * hi_time  (!!!! --- NOT USED --- !!!!)
        move.w   #$00b4,pram+$c8  * B_fall_time_ptr to type_a channel (ch11)
        move.w   #$00b2,pram+$ca  * B_rise_time_ptr to type_a channel (ch11)
*    Ch 13 (-- current monitoring (PWM) --) (--- ID current ---)
        move.w   #$0092,pram+$d0  * capture and match tcr1
        move.w   #$0400,pram+$d4
```

```
        move.w    #$0800,pram+$d6
*    Ch 14 (-- interupt delay (INT) --)
        move.w    #$0007,pram+$e0  * capture & match tcr1 detect rising edge
        move.w    #DELAY,pram+$e2
*    Ch 15 (-- speed sensing (TTC) --) (--- SPEED sensing ---)
        move.w    #$006f,pram+$f0  * detect both edges. cap and match tcr2
        move.w    #SPEED_INT,pram+$f2  * set count interval for speed detect
        move.w    #$0000,pram+$f4  * clear pulse count
        move.w    #$0000,pram+$f6  * clear pulse count temporary
* initialize as: master = 11
*                slave  = 10
*                slave  = 01 (INVERTED)
*
*   ttc int     s   s   s   s   s     s   s   s   s   s   s   s    m
*  |-------------------------------| |-------------------------------|
*  | 11 11 : 10 10 : 10 01 : 01 10 | | 10 01 : 01 10 : 10 01 : 01  11 |
*  |_15___________________________8_| |_7___________________________0_|
*              HSRR0                              HSRR1
*
*   INITIALIZE:
        move.w    #$9697,HSRR1      *  ch 0 = master
        move.w    #$fa96,HSRR0      *  ch 1,2,3,4,5,6,7,8,9,10,11,12 = slaves.

*   itc         s   s   s   s   s     s   s   s   s   s   s   s    m
*  |-------------------------------| |-------------------------------|
*  | 10 11 : 01 10 : 10 10 : 10 10 | | 10 10 : 10 10 : 10 10 : 10  11 |
*  |_15___________________________8_| |_7___________________________0_|
*             CPR0                               CPR1
        move.w    #$4000,CIER       * enable interrupts for master  (ch0)
        move.w    #$ffff,CPR1
        move.w    #$b7ff,CPR0       * enable all MCPWM channels as high.
*                                   * enable qpm channel as middle.
*                                   * enable pwm channels as low.
*-----------------  TPU initialization complete  -----------------*
```

MC68332 VECTOR CONTROL CODE
INCLUDE FILE: QSM.INI

```
*************************************************************************
* QSM INITIALIZATION CODE                                               *
* Author:  Peter Pinewski - Motorola Inc.                               *
* Created: 11/10/94                                                     *
* Last Modified: 11/10/94                                               *
* This section initializes and enables the QSPI.                        *
* The code below will initialize the SPI que with the correct command   *
* data and initialization data. This code will set up the first six     *
* entries in the que with zeros to sync. the LCD display dvr. to MCU.   *
*   PCS0:   LCD display chip select   (active high)                     *
*   PCS1:   a/d chip select           (active low)                      *
*   PCS2:   HC11V8 select             (active low)                      *
*   PCS3:   not used                                                    *
* The LCD driver requires:                                              *
*         - 6 bytes of data (start_byte, 4 digits, and end_byte)        *
*         - an active high Chip select.                                 *
* The A/D requires:                                                     *
```

```
*         - an active low chip select                                    *
*         - 10-bit transfers                                             *
*         - Chip select to SCK delay of 4us                              *
*         - Delay after transfer to allow for the conversion time        *
*         - Ch0 - Throt Pot                                              *
*         - Ch1 - Throt Pot error signal(signals short or open circuits) *
*         - Ch2 - Phase_A                                                *
*         - Ch3 - Phase_B                                                *
*         - Ch4 - extra Pot (TEST PURPOSES)                              *
* The HC11V8 requires:                                                   *
*         - an active low chip select                                    *
*         - 8 bit transfers                                              *
*         - 32us Delay after transfer (allow HC11 code to refill SPI)    *
*   SCI: int_lvl = 3;  Vector_# = $50                                    *
*  QSPI: int_lvl = 3;  Vector_# = $51                                    *
*  IARB: D                                                               *
* external references: NONE                                             *
*------------------------------------------------------------------------*
* A/D  channel  conversion  codes
AD_CH0   equ      $0000
AD_CH1   equ      $0040
AD_CH2   equ      $0080
AD_CH3   equ      $00C0
AD_CH4   equ      $0100
AD_HS    equ      $02C0
AD_ZERO  equ      $0300
AD_FS    equ      $0340
* To  be  loaded  into  SPCR1
DT88us   equ      $4a2e  * CS to SCLK dly=4.425us min, Dly after Xfer=88us
DT32us   equ      $4a11  * CS to SCLK dly=4.425us min, Dly after Xfer=32.6us
DT22us   equ      $4a0c  * CS to SCLK dly=4.425us min, Dly after Xfer=22.88us
DT17us   equ      $4a09  * CS to SCLK dly=4.425us min, Dly after Xfer=17.17us
         move.w   #$008D,QMCR   *Normal op; ignore freeze; restricted
         move.w   #$2D50,QILR   *QSPI int=5 vec#=$51; SCI int=5 vec#=$50
         move.b   #$F3,QPDR     *SCK,PCS0 default 0; all others high
         move.b   #$FE,QDDR     *MISO = input; all others output
         move.b   #$7B,QPAR     *PCS0-PCS3,PCS1,MISO,MOSI assign to QSPI
         move.w   #$A808,SPCR0  *MSTR; BITS=10; CPOL:CPHA=0:0 ;SPBR=1.05 MHz
         move.w   #DT22us,SPCR1 *A/D conv.delay=21us (17us DT + 4.425us DSCK)
*       CMDRAM (FFFFFD40)   XMITRAM (FFFFFD20)    RCVRAM (FFFFFD40)
* ==========================================================================
*     C B D D P P P P
*     o i t s C C C C
*     n t   c S S S S
*     t s   k 3 2 1 0
*   |<--- 8-BITS  --->|          |<---- 16-BITS ---->| |<---- 16-BITS ---->|
*   -------------------          --------------------- ---------------------
*  0| 0 0 0 0 1 1 1 1 |   LCD   0|  0 0    |   0 2   | |      - - - -      |
*   -------------------          --------------------- ---------------------
*  1| 0 0 0 0 1 1 1 1 |   LCD   2|       DIG1        | |      - - - -      |
*   -------------------          --------------------- ---------------------
*  2| 0 0 0 0 1 1 1 1 |   LCD   4|       DIG2        | |      - - - -      |
*   -------------------          --------------------- ---------------------
*  3| 0 0 0 0 1 1 1 1 |   LCD   6|       DIG3        | |      - - - -      |
```

```
*   ------------------        --------------------- ---------------------
*  4| 0 0 0 0 1 1 1 1 |  LCD    8|        DIG4        | |      - - - -      |
*   ------------------        --------------------- ---------------------
*  5| 0 0 0 0 1 1 1 1 |  LCD   $a|  0 0     |   0 0   | |      - - - -      |
*   ------------------        --------------------- ---------------------
*  6| 0 0 0 0 1 0 1 0 | HC11V8 $c|          |   RPM   | |         |  TEMP   |
*   ------------------        --------------------- ---------------------
*  7| 0 1 1 1 1 1 0 0 | 10-BTS  $e|      | A/D-1      | |      | result-4   |
*   ------------------        --------------------- ---------------------
*  8| 0 1 1 1 1 1 0 0 | 10-BTS $10|      | A/D-2      | |      | result-1   |
*   ------------------        --------------------- ---------------------
*  9| 0 1 1 1 1 1 0 0 | 10-BTS $12|      | A/D-3      | |      | result-2   |
*   ------------------        --------------------- ---------------------
* $a| 0 1 0 1 1 1 0 0 | 10-BTS $14|      | A/D-4      | |      | result-3   |
*   ------------------        --------------------- ---------------------
* $b| 0 0 0 0 1 1 1 0 |        $16|                    | |                   |
*   ------------------        --------------------- ---------------------
* $c| 0 0 0 0 1 1 1 0 |        $18|                    | |                   |
*   ------------------        --------------------- ---------------------
* $d| 0 0 0 0 1 1 1 0 |        $1a|                    | |                   |
*   ------------------        --------------------- ---------------------
* $e| 0 0 0 0 1 1 1 0 |        $1c|                    | |                   |
*   ------------------        --------------------- ---------------------
* $f| 0 0 0 0 1 1 1 0 |        $1e|                    | |                   |
*   ------------------        --------------------- ---------------------
          move.w   #$0f0f,CMD_0      * LCD display : LCD display
          move.w   #$0f0f,CMD_2      * LCD display : LCD display
          move.w   #$0f0f,CMD_4      * LCD display : LCD display
          move.w   #$0a7c,CMD_6      * HC11V8      : A/D conv
          move.w   #$7c7c,CMD_8      * A/D conv    : A/D conv
          move.w   #$5c0e,CMD_10     * A/D conv    : unused
          move.w   #$0e0e,CMD_12     * unused      : unused
          move.w   #$0e0e,CMD_14     * unused      : unused
          clr.l    XMIT_0            * XMITRAM initially all zeros
          clr.l    XMIT_2            *      (clears LCD display for
          clr.l    XMIT_4            *       syncronization to micro)
          move.w   #$0500,SPCR2      * NO wrap;  que length 0-5 (LCD dsply);
          clr.b    SPCR3             * Clear SPIF flag
          bset.b   #7,SPCR1          * Initiate transfers to sync LCD display
ini_wait  btst.b   #7,SPSR           * Check if transfer is complete
          beq      ini_wait
          bclr.b   #7,SPSR           * Clear  SPI finished flag
*---------------------- end QSM initialization ----------------------*
**********************************************************************
```

Appendix

F

Space Vector Modulation Routine, MC68HC08MP16

```
;*****************************************************************************
; SVM interrupt service routine (ISR)                                        *
; Author: Peter Pinewski, Motorola Inc.                                      *
;   This assembly file implements the ISR to implement the alternating-reversing *
;   SVM technique.  It takes the frequency (incval) and voltage values passed.   *
;   from main and updates the PWM high times. (The voltage range is 0-256)       *
;   256 point table lookup (90 degrees of sine)                              *
;   NO INTERPOLATION                                                         *
;   variable PWM frequency                                                   *
;   execution time = 48us (approx)                                           *
;*****************************************************************************
        xdef    _pwm
        xref.b _PCTL1
        xref.b _PCTL2
        xref.b _PORTC
        xref.b _COPCTL
        xref.b _PMOD
        xref.b _PVAL1
        xref.b _PVAL3
        xref.b _PVAL5
        xref.b _newPMOD
        xref.b _newinc
        xref.b _ldfreq
        xref.b _mod_type
        xref.b _direction
        xref.b _theta
        xref.b _volts
        xref.b _incval
        xref.b _flag
        xref    _sin
        xref    _third
_pwm:
        pshh
        bset    5,_PORTC        ; set timing marker (Green LED)
        clr     _COPCTL         ; service COP timer
        bclr    4,_PCTL1        ; clear PWM reload flag
        ais     #-6             ; open scratch area on stack
        tst     _flag           ; check if PWM freq needs updating
        beq     _isr_cont
_isr_pwm_update:
        mov     _newPMOD+1,_PMOD+1   ; update PMOD (pwm freq)
        mov     _newPMOD,_PMOD
        mov     _newinc+1,_incval+1  ; update incval
        mov     _newinc,_incval
        lda     _ldfreq
        lsla
        lsla                    ; ldfreq needs to be
        lsla                    ; shifted 5 times for
        lsla                    ; loading into PCTL2
        lsla
        sta     _PCTL2          ; update LDFREQ bits
        clr     _flag           ; clear update flag
_isr_cont:
        tst     _direction      ; check motor direction
        beq     _isr_rev
_isr_fwd:
```

```
        lda     _theta+1
        add     _incval+1       ; perform integration (forward)
        sta     _theta+1
        lda     _theta          ; theta = theta + incval
        adc     _incval
        sta     _theta
        bra     _isr_svm
_isr_rev:
        lda     _theta+1
        sub     _incval+1       ; perform integration (reverse)
        sta     _theta+1
        lda     _theta          ; theta = theta - incval
        sbc     _incval
        sta     _theta
; ---------------------------------
; svm calculations
; ---------------------------------
_isr_svm:
        ldhx    #$2aaa          ; check if in sector 1
        cphx    _theta
        bhi     _isr_sec1
        ldhx    #$5555          ; check if in sector 2
        cphx    _theta
        bhi     _isr_sec2
        ldhx    #$8000          ; check if in sector 3
        cphx    _theta
        bhi     _isr_sec3
        ldhx    #$aaaa          ; check if in sector 4
        cphx    _theta
        bhi     _isr_sec4
        ldhx    #$d555          ; check if in sector 5
        cphx    _theta
        bhi     _isr_sec5
        ldhx    #$0000          ; sector 6
        bra     _isr_sec6
; ---------------------------------
; Calculate subroutine
; Input:  hx = vect angle (V2)
; Output: T1,T2,T0 on stack area
;         sp+6 = T1  (sp+8)
;         sp+5 = T2  (sp+7)
;         sp+4 = T0  (sp+6)
;     (T1,T2,T0 are byte values)
; ---------------------------------
_calculate:
        txa                     ; a = low portion of vect angle (V2)
        sub     _theta+1
        tax                     ; x = L (V2-theta)
        pshh
        pula                    ; a = high portion of vect angle (V2)
        sbc     _theta          ; a = H (V2-theta)
        stx     3,sp            ; store angle (this angle is effectively
        sta     4,sp            ;               60-delta_theta)
        lslx                    ; scale angle for table lookup
        rola                    ; requires two left shifts for
        lslx                    ; the 256 point 90 degree sine
        rola                    ; table.
        tax                     ; x = upper byte of angle (index portion)
```

```
        clrh
        lda     _sin,x          ; a = sin(60-delta_theta)
        ldx     _volts          ; x = voltage (modulation index, m)
        mul                     ; x:a = m*sin(60-delta_theta)
        txa                     ; use most significant byte only
        sta     8,sp            ; store T1 value
        lda     #$aa            ; a = L (60 degrees)
        sub     3,sp            ; a = L (60 - (60-delta_theta))
        tax                     ; x = L (delta_theta)
        lda     #$2a            ; a = H (60 degrees)
        sbc     4,sp            ; a = H (delta_theta)
        lslx                    ; scale angle for table lookup
        rola                    ; requires two left shifts for
        lslx                    ; the 256 point 90 degree sine
        rola                    ; table.
        tax
        lda     _sin,x          ; a = sin(delta_theta)
        ldx     _volts          ; x = voltage (modulation index, m)
        mul                     ; x:a = m*sin(delta_theta)
        txa                     ; use most significant byte only
        sta     7,sp            ; store T2 value
        add     8,sp            ; a = T1+T2
        nega                    ; a = -(T1+T2)
        add     #$ff            ; a = T0 = 1-(T1+T2)
        sta     6,sp            ; store T0
        rts
; ----------------------------------
_isr_sec6:
        bsr     _calculate
        lda     4,sp
        lsra                    ; a = .5T0
        sta     2,sp            ; save b_high
        add     6,sp            ; a = T1+.5T0
        sta     3,sp            ; save c_high
        add     5,sp            ; a = T1+T2+.5T0
        sta     1,sp            ; save a_high
        bra     _isr_ht_a
_isr_sec5:
        bsr     _calculate
        lda     4,sp
        lsra                    ; a = .5T0
        sta     2,sp            ; save b_high
        add     5,sp            ; a = T2+.5T0
        sta     1,sp            ; save a_high
        add     6,sp            ; a = T1+T2+.5T0
        sta     3,sp            ; save c_high
        bra     _isr_ht_a
_isr_sec4:
        bsr     _calculate
        lda     4,sp
        lsra                    ; a = .5T0
        sta     1,sp            ; save a_high
        add     6,sp            ; a = T1+.5T0
        sta     2,sp            ; save b_high
        add     5,sp            ; a = T1+T2+.5T0
        sta     3,sp            ; save a_high
        bra     _isr_ht_a
_isr_sec3:
```

```
        bsr     _calculate
        lda     4,sp
        lsra                    ; a = .5T0
        sta     1,sp            ; save a_high
        add     5,sp            ; a = T2+.5T0
        sta     3,sp            ; save c_high
        add     6,sp            ; a = T1+T2+.5T0
        sta     2,sp            ; save b_high
        bra     _isr_ht_a
_isr_sec2:
        bsr     _calculate
        lda     4,sp
        lsra                    ; a = .5T0
        sta     3,sp            ; save c_high
        add     6,sp            ; a = T1+.5T0
        sta     1,sp            ; save a_high
        add     5,sp            ; a = T1+T2+.5T0
        sta     2,sp            ; save b_high
        bra     _isr_ht_a
_isr_sec1:
        bsr     _calculate
        lda     4,sp
        lsra                    ; a = .5T0
        sta     3,sp            ; save c_high
        add     5,sp            ; a = T2+.5T0
        sta     2,sp            ; save b_high
        add     6,sp            ; a = T1+T2+.5T0
        sta     1,sp            ; save a_high
; -------- high time calculations --------
_isr_ht_a:                      ; DETERMINE HIGHTIME - Phase A
        lda     1,sp            ; a = va
        ldx     _PMOD           ; x = PMOD_H
        mul                     ; x = HH a = HL
        stx     5,sp            ; save HH
        sta     4,sp            ; save HL
        lda     1,sp            ; a = va
        ldx     _PMOD+1         ; x = PMOD_L
        mul                     ; x = LH a = LL
        lsla                    ; set C bit for rounding (discard LL)
        txa                     ; a = LH
        adc     4,sp            ; a = HL + LH + C (LL discarded)
        sta     _PVAL1+1
        clra
        adc     5,sp            ; a = HH + 00 + C
        sta     _PVAL1
_isr_ht_b:                      ; DETERMINE HIGHTIME - Phase B
        lda     2,sp
        ldx     _PMOD           ; x = PMOD_H
        mul                     ; x = HH a = HL
        stx     5,sp            ; save HH
        sta     4,sp            ; save HL
        lda     2,sp
        ldx     _PMOD+1         ; x = PMOD_L
        mul                     ; x = LH a = LL
        lsla                    ; set C bit for rounding (discard LL)
        txa                     ; a = LH
        adc     4,sp            ; a = HL + LH + C (LL discarded)
        sta     _PVAL3+1
```

```
        clra
        adc     5,sp            ; a = HH + 00 + C
        sta     _PVAL3
_isr_ht_c:                      ; DETERMINE HIGHTIME - Phase C
        lda     3,sp
        ldx     _PMOD           ; x = PMOD_H
        mul                     ; x = HH a = HL
        stx     5,sp            ; save HH
        sta     4,sp            ; save HL
        lda     3,sp
        ldx     _PMOD+1         ; x = PMOD_L
        mul                     ; x = LH a = LL
        lsla                    ; set C bit for rounding (discard LL)
        txa                     ; a = LH
        adc     4,sp            ; a = HL + LH + C (LL discarded)
        sta     _PVAL5+1
        clra
        adc     5,sp            ; a = HH + 00 + C
        sta     _PVAL5
; ------ high time calculations done ------
        ais     #6              ; remove stack scratch area
        bset    1,_PCTL1        ; set load ok (LDOK) bit
        bclr    5,_PORTC        ; clear timing marker (Green LED)
        pulh
        rti
;*****************************************************************
        end
```

Appendix

G

Brush DC Motor Program, MC68HC05MC4

```
*****************************************************************************************
*       Note: See Disclaimer in Preface                                                 *
* ITC122 AND ITC127 board set                                                           *
* MC68HC05MC4 MCU BRUSH TYPE MOTOR CONTROLLER, ASSEMBLY SOURCE CODE                     *
* VERSION 4/28/96                                                                       *
* THE POSITIVE MOTOR WIRE IS CONNECTED TO THE ITC122 BOARD TERMINAL Aout                *
* THE NEGATIVE MOTOR WIRE IS CONNECTED TO THE ITC122 BOARD TERMINAL Cout                *
* ITC127 TO ITC122 BRUSH MOTOR I/O ASSIGNMENTS:                                         *
* LINES IN THE I/O ASSIGNMENT PRECEDED BY ** IN THE FIRST TWO COLUMNS                   *
* ARE CONNECTED BY THE RIBBON CABLES, BUT NOT UTILIZED IN THIS SOFTWARE                 *
* IMPLEMENTATION OF A BRUSH TYPE MOTOR DRIVE.                                           *
**      IRQ,PB6     PHASE A HALL SENSOR                                                 *
**      TCAP1,PB7   PHASE B HALL  SENSOR                                                *
**      TCAP2,PA0   PHASE C HALL  SENSOR                                                *
*       PWMA1       PHASE A BOTTOM DRIVE (PA1)                                          *
**      PWMA2       PHASE B BOTTOM DRIVE (PA3)                                          *
*       PWMA3       PHASE C BOTTOM DRIVE (PA5)                                          *
*       PWMB1       PHASE A TOP DRIVE (PA2)                                             *
**      PWMB2       PHASE B TOP DRIVE (PA4)                                             *
*       PWMB3       PHASE C TOP DRIVE (PA6)                                             *
**      PA7         PULLED-UP TO +5 VOLTS WITH 10K AND TERMINATED AT PROTOTYPE AREA     *
**      PC0/AD0     POWER BOARD BUFFERED B+ FEEDBACK                                    *
**      PC1/AD1     POWER BOARD CURRENT FEEDBACK                                        *
**      PC2/AD2     POWER BOARD TEMPERATURE DIODE                                       *
*       PC3/AD3     SPEED CONTROL POT                                                   *
*       PC4/AD4     DIRECTION CONTROL SWITCH                                            *
*       PC5/AD5     RUN/STOP CONTROL SWITCH                                             *
*****************************************************************************************
* MINIMUM AND MAXIMUM ALLOWABLE PWM SETPOINT VALUES
MINPWM EQU    6    ;A SETPOINT VALUE LESS THAN THIS PLACES 0 IN PWMAD
MAXPWM EQU    249  ;A SETPOINT VALUE GREATER THAN THIS PLACES $FF IN PWMAD
* I/O REGISTERS
PORTA   EQU     $00             ;DATA REGISTER FOR PORT A
PORTB   EQU     $01             ;DATA REGISTER FOR PORT B
PORTC   EQU     $02             ;DATA REGISTER FOR PORT C
PORTD   EQU     $03             ;DATA REGISTER FOR PORT D
DDRA    EQU     $04             ;DATA DIRECTION REGISTER FOR PORT A
DDRB    EQU     $05             ;DATA DIRECTION REGISTER FOR PORT B
DDRC    EQU     $06             ;DATA DIRECTION REGISTER FOR PORT C
DDRD    EQU     $07             ;DATA DIRECTION REGISTER FOR PORT D
CTCSR   EQU     $08             ;CORE TIMER CONTROL AND STATUS REGISTER
CTCR    EQU     $09             ;CORE TIMER COUNTER REGISTER
PWMAD   EQU     $10             ;PWM A DATA REGISTER
PWMAI   EQU     $11             ;PWM A INTERLOCK REGISTER
PWMBD   EQU     $12             ;PWM B DATA REGISTER
PWMBI   EQU     $13             ;PWM B INTERLOCK REGISTER
CTLA    EQU     $14             ;PWM A CONTROL REGISTER
CTLB    EQU     $15             ;PWM B CONTROL REGISTER
RATE    EQU     $16             ;PWM RATE REGISTER
UPDATE  EQU     $27             ;PWM UPDATE REGISTER
TCR     EQU     $17             ;TIMER CONTROL REGISTER
TSR     EQU     $18             ;TIMER STATUS REGISTER
ICRH2   EQU     $19             ;INPUT CAPTURE 2 REGISTER - HIGH BYTE
ICRH1   EQU     $1B             ;INPUT CAPTURE 1 REGISTER - HIGH BYTE
OCRH    EQU     $1D             ;OUPUT COMPARE REGISTER - HIGH BYTE
TMRH    EQU     $20             ;TIMER REGISTER - HIGH BYTE
```

```
ACRH     EQU     $22              ;ALTERNATE TIMER REGISTER - HIGH BYTE
ADDR     EQU     $24              ;A/D CONVERTER DATA REGISTER
ADSCR    EQU     $25              ;A/D CONVERTER STATUS AND CONTROL REGISTER
COCO     EQU     7                ;A/D CONVERSION COMPLETE BIT IN ADSCR
ISCR     EQU     $0F              ;IRQ STATUS AND CONTROL REGISTER
ABOT     EQU     8                ;PATTERN TO TURN ON PHASE A BOTTOM OUTPUT
BBOT     EQU     $10              ;PATTERN TO TURN ON PHASE B BOTTOM OUTPUT
CBOT     EQU     $20              ;PATTERN TO TURN ON PHASE C BOTTOM OUTPUT
ATOP     EQU     1                ;PATTERN TO TURN ON PHASE A BOTTOM OUTPUT
BTOP     EQU     2                ;PATTERN TO TURN ON PHASE B BOTTOM OUTPUT
CTOP     EQU     4                ;PATTERN TO TURN ON PHASE C BOTTOM OUTPUT
CTLMSK   EQU     $B8              ;PWM CONTROL REGISTER COMMON MASK
MOR      EQU     $F00             ;MASK OPTION REGISTER
SWITCH   EQU     5                ;RUN/STOP SWITCH CONNECTED TO PORT C BIT 5
DIRECT   EQU     4                ;FORWARD/REVERSE SWITCH CON. TO PORT C BIT 4
* VARIABLES
         ORG     $50
DIRSW    RMB     1                ;LAST DIRECTION INPUT. USED FOR CHANGE OF STATE
DLYTMR   RMB     1                ;COUNTER INCREMENTS EACH REAL TIME INTERRUPT
FLAGS    RMB     1                ;GENERAL PURPOSE FLAGS. NEXT EQU'S DEFINE THEM
RUNCOS   EQU     0                ;USED FOR SMOOTH STARTUP. 1 = RUN, 0 = STOP
PWRUP    EQU     1                ;FORCES SMOOTH STARTUP AFTER RESET
SETPT    RMB     1                ;PWM SETPOINT VALUE BETWEEN MINPWM AND MAXPWM
SENS     RMB     16               ;BUFFER TO HOLD A/D INPUTS
SENSEND  EQU     *-1              ;END OF BUFFER
ADTEMP   RMB     2                ;PLACE TO DO A/D AVERAGE MATH
*        ORG     MOR              ;MOR REGISTER
*        FCB     0                ;DISABLE THE WATCHDOG TIMER
* THE PROGRAM CODE STARTS HERE
         ORG     $100
START    EQU     *
         RSP
         LDX     #$50             ;ZERO MEMORY FROM $50 TO $FF
CLRMEM   CLR     0,X
         INCX                     ;BUMP THE POINTER TO THE NEXT LOCATION
         CMPX    #$0              ;DONE YET?
         BNE     CLRMEM           ;NOT YET..KEEP CLEARING RAM
         LDA     PORTC            ;GET THE FORWARD/REVERSE DIRECTION SWITCH INPUT
         AND     #$10             ;RETAIN BIT 4 MOTOR DIRECTION SWITCH
         STA     DIRSW            ;NEED THIS FOR THE FIRST TIME THROUGH MAIN
         BRCLR   DIRECT,PORTC,GOREV   ;IF THE BIT IS LOW..RUN COUNTER CLOCKWISE
         JSR     A_TO_C
         BRA     CONT
GOREV    JSR     C_TO_A           ;ELSE COUNTER CLOCKWISE
CONT     LDA     #0               ;INITIALIZE AT MINIMUM PWM DUTY CYCLE
         STA     PWMBD            ;TUCK IT AWAY IN THE PWM REGISTER   WLL was pwmad
         STA     PWMAD            ;TUCK IT AWAY IN THE PWM REGISTER   WLL was pwmad
         LDA     #$10             ;ENABLE THE REAL TIME INTERRUPT @ 5.5 MS.
         STA     CTCSR
         CLI                      ;TURN ON INTERRUPTS
*******************************************************************************
* THE MAIN LOOP WILL MONITOR PC3/AD3 TO DETERMINE THE DESIRED SPEED, PC4      *
* TO DETERMINE THE DIRECTION AND PC5 TO STOP OR ROTATE THE MOTOR. MAIN        *
* WILL BE INTERRUPTED FROM FOUR SOURCES (IRQ, TCAP1, TCAP2, TOF AND RTI)      *
*******************************************************************************
MAIN     EQU     *
         LDX     #SENS            ;START OF THE A/D BUFFER
MVLOOP   LDA     1,X              ;GET THE SECOND ENTRY
```

```
        STA     0,X                 ;MOVE UP ONE
        INCX                        ;POINT TO NEXT
        CMPX    #SENSEND            ;END OF TABLE?
        BNE     MVLOOP              ;NO
        LDA     #$23                ;ENABLE CHANNEL 3 OF A/D CONVERTER
        STA     ADSCR               ;TURN ON A/D CONVERTER AND START A CONVERSION
ADLOOP  BRCLR   COCO,ADSCR,ADLOOP   ;LOOP ON THE A/D CONVERSION COMPLETE BIT
        BRSET   SWITCH,PORTC,OFF    ;IS THE RUN/STOP SWITCH OFF?
        LDA     ADDR                ;GET THE POT VALUE
        STA     SENSEND
* AVERAGE THE LAST 16 ANALOG INPUTS
        CLR     ADTEMP              ;ZERO THE ACCUMULATOR
        CLR     ADTEMP+1
        LDX     #SENS               ;STAR OF BUFFER
        CLC                         ;CLEAR FOR THE FIRST ADD
        LDA     0,X                 ;FIRST PAIR OF INPUTS
ADDLOOP ADD     1,X                 ;ADD THE NEXT IN LINE
        BCC     NOCARRY
        INC     ADTEMP
NOCARRY STA     ADTEMP+1
        INCX                        ;POINT TO THE NEXT ENTRY
        CMPX    #SENSEND            ;DONE YET
        BNE     ADDLOOP             ;NO
* DIVIDE BY 16
        LDA     #4                  ;LOOP COUNTER
DVLOOP  LSR     ADTEMP              ;SHIFT MSB 1 BIT TO THE RIGHT INTO CARRY
        ROR     ADTEMP+1            ;SHIFT THE LSB TO THE RIGHT
        DECA                        ;BUMP THNE COUNTER DOWN BY 1
        BNE     DVLOOP              ;DONE YET?
        LDA     ADTEMP+1
        BRSET   PWRUP,FLAGS,NORAMP  ;NOT FIRST TIME THROUGH AFTER RESET
        LDA     ADDR                ;GET THE RAW POT VALUE
        BSET    PWRUP,FLAGS         ;INHIBIT THE ABOVE TWO LINES AFTER ONCE THROUGH
NORAMP  STA     SETPT               ;SAVE THE INTEGRATED POT VALUE
        CMP     #MINPWM             ;CHECK MINIMUM PWM VALUE
        BCC     CHKHI               ;NOT TOO LOW..CHECK FOR MAX PWM VALUE
        CLR     SETPT               ;FORCE ZERO PWM
CHKHI   CMP     #MAXPWM             ;CHECK MAXIMUM PWM VALUE
        BCS     RUN                 ;VALUE WITHIN LIMITS
        LDA     #$ff                ;FORCE MAXIMUM PWM
        STA     SETPT
        BRA     RUN                 ;CONTINUE TO RUN THE MOTOR
OFF     CLR     DLYTMR              ;DEBOUNCE THE RUN/STOP SWITCH
OFFDLY  LDA     DLYTMR              ;GET THE CURRENT COUNT
        CMP     #2                  ;EACH COUNT = 5.5 MS.
        BNE     OFFDLY              ;LOOP UNTIL 2 COUNTS ACCUMULATE (11 MS.)
        BCLR    RUNCOS,FLAGS        ;SET FOR THE NEXT STOP TO RUN TRANSITION
        LDA     #CTLMSK             ;DISABLE ALL TOP AND BOTTOM PWM OUTPUTS
        STA     CTLB
        STA     CTLA
        CLR     PWMBD               ;wll was pwmad
        CLR     PWMAD               ;wll was pwmad
        LDA     PORTC               ;GET THE FORWARD/REVERSE DIRECTION SWITCH INPUT
        AND     #$10                ;RETAIN BIT 4 MOTOR DIRECTION SWITCH
        CMP     DIRSW               ;CHANGE IN THE DIRECTION SWITCH?
        BEQ     MAIN                ;NO CHANGE OF STATE..CONTINUE
        BRA     COS                 ;CHANGE OF DIRECTION SWITCH
MAINEXT JMP     MAIN
```

```
RUN     BRSET   SWITCH,PORTC,MAIN ;IF THE RUN/STOP SWITCH IS OFF CONTINUE MAIN
        BRSET   RUNCOS,FLAGS,NONEW
        BRCLR   DIRECT,PORTC,GOREV1 ;IF THE BIT IS LOW..RUN COUNTER CLOCKWISE
        JSR     A_TO_C
        BRA     CONT1
GOREV1  JSR     C_TO_A              ;ELSE COUNTER CLOCKWISE
CONT1   BSR     ACCEL               ;RAMP UP TO THE REQUESTED POT SPEED
        BSET    RUNCOS,FLAGS        ;THE MOTOR IS UP TO SPEED DON'T RAMP NEXT LOOP
NONEW   LDA     SETPT               ;GET THE LATEST SPEED REQUEST
        STA     PWMBD               ;UPDATE THE PWM REGISTER  wll was pwmad
        STA     PWMAD               ;UPDATE THE PWM REGISTER  wll was pwmad
        LDA     PORTC               ;GET THE FORWARD/REVERSE DIRECTION SWITCH INPUT
        AND     #$10                ;RETAIN BIT 4 MOTOR DIRECTION SWITCH
        CMP     DIRSW               ;CHANGE IN THE DIRECTION SWITCH?
        BEQ     MAINEXT             ;NO CHANGE OF STATE..CONTINUE
* A CHANGE IN DIRECTION HAS BEEN DETECTED.
COS     STA     DIRSW               ;SAVE FOR THE NEXT CHANGE OF DIRECTION
        LDA     #CTLMSK             ;DISABLE ALL TOP AND BOTTOM PWM OUTPUTS
        STA     CTLB
        STA     CTLA
        CLR     PWMBD               ;INITIALIZE AT MINIMUM SPEED  wll was pwmad
        CLR     PWMAD               ;INITIALIZE AT MINIMUM SPEED  wll was pwmad
        CLR     DLYTMR              ;TIMER IS USED FOR CHANGE IN DIRECTION DELAY
COSDLY  LDA     DLYTMR              ;GET THE TIMER VALUE
        CMP     #150                ;HAVE WE WAITED LONG ENOUGH
        BCS     COSDLY              ;NOT DONE YET
        BRCLR   DIRECT,PORTC,REV ;GET THE DIRECTION SWITCH
        JSR     A_TO_C              ;SET THE OUTPUTS TO Aout = + AND Cout = -
REVACC  BSR     ACCEL               ;RAMP UP TO THE REQUESTED SPEED
        JMP     MAIN
REV     JSR     C_TO_A              ;SET THE OUTPUTS TO Aout = + AND Cout = -
        BRA     REVACC
************************************************************************************
*  THIS IS A LIMITED ATTEMPT AT CONTROLLED ACCELERATION. THE FOLLOWING            *
*  CODE WILL PREVENT LARGE CURRENTS WHEN SWITCHING MOTOR DIRECTION AND            *
* UPON INITIAL STARTUP                                                             *
************************************************************************************
ACCEL   BRSET   SWITCH,PORTC,EXIT ;IF THE RUN/STOP SWITCH IS OFF CONTINUE MAIN
        LDA     PWMBD               ;GET THE CURRENT PWM VALUE  wll was pwmad
        CMP     SETPT               ;IS IT AT THE SETPOINT YET?
        BEQ     EXIT                ;NO ACCELERATION OR DEACCELERATION NECESSARY
        BCS     ACDLY1              ;NO..CONTINUE TO ACCELERATE
        LDA     SETPT               ;LET THE MOTOR SLOW DOWN
        STA     PWMBD               ;UPDATE THE PWM REGISTER  wll was pwmad
        STA     PWMAD               ;UPDATE THE PWM REGISTER  wll was pwmad
        BRA     EXIT                ;RETURN TO MAIN
ACDLY1  CLR     DLYTMR              ;START AT A COUNT OF ZERO
ACDLY   LDA     DLYTMR              ;GET THE CURRENT COUNT
        CMP     #2                  ;EACH COUNT = 5.5 MS.
        BNE     ACDLY               ;LOOP UNTIL 2 COUNTS ACCUMULATE (11 MS.)
        LDA     PWMBD               ;GET THE CURRENT PWM RATE  wll was pwmad
        CMP     #MAXPWM             ;HAVE WE EXCEEDED THE MAXIMUM PWM RATE
        BCC     EXIT                ;NO
        INC     PWMBD               ;BUMP THE PWM. INITIALLY IT WILL BE MINPWM  wll was pwmad
        INC     PWMAD               ;BUMP THE PWM. INITIALLY IT WILL BE MINPWM  wll was pwmad
        BRA     ACCEL
EXIT    RTS
***************************************************************************
```

```
* COMMUTATION POSITION 0 FOR THE MOTOR. COMMUTATE                  *
* THE MOTOR TO ITS NEXT POSITION.                                  *
********************************************************************
* A_TO_C DRIVES + OUT OF ITC122 TERMINAL Aout AND - OUT OF TERMINAL Cout
* TO DRIVE THE MOTOR CLOCKWISE

A_TO_C  LDA     #CTLMSK^CBOT     ;PA5=PWM (PHASE C BOTTOM)
        LDX     #CTLMSK^ATOP     ;PA6=1 (C TOP), PA4=1 (B TOP), PA2=0 (A TOP)
POSX    STX     CTLB             ;UPDATE PA2, PA4, AND PA6
        STA     CTLA             ;UPDATE PA1, PA3, AND PA5
        RTS
* C_TO_A DRIVES + OUT OF ITC122 TERMINAL Cout AND - OUT OF TERMINAL Aout
* TO DRIVE THE MOTOR COUNTER COUNTER CLOCKWISE
C_TO_A  LDA     #CTLMSK^ABOT     ;PA5=PWM (PHASE A BOTTOM)
        LDX     #CTLMSK^CTOP     ;PA6=0 (C TOP), PA4=1 (B TOP), PA2=1 (A TOP)
        BRA     POSX
*********************************************************************
* CORE TIMER'S REAL TIME INTERRUPT SERVICE ROUTINE. USED FOR        *
* ACCELERATION.                                                     *
*********************************************************************
RTI     INC     DLYTMR           ;BUMP THE COUNTER
        BCLR    6,CTCSR          ;CLEAR THE INTERRUPT FLAG
        RTI                      ;RETURN
BADTOF  EQU     *                ;TIMER OVERFLOW IS NOT USED
BADCAP1 EQU     *                ;INPUT CAPTURE1 IS IS NOT USED
BADCAP2 EQU     *                ;INPUT CAPTURE2 IS NOT USED
BADIRQ  EQU     *                ;IRQ IS NOT USED
BADSWI  EQU     *                ;SWI IS NOT USED
BADSCI  EQU     *                ;SCI IS NOT USED
        RTI     *                ;JUST IGNORE INTERRUPT EXCEPT THE
*                                ;REAL TIME INTERRUPT
***************************************************************
* RESET/INTERRUPT VECTORS.                                    *
* NOTE: THE SCI AND SWI VECTORS ARE NOT USED IN THIS APPLICATION. *
***************************************************************
        ORG     $0FF0
        FDB     RTI              ;CORE TIMER VECTOR
        FDB     BADSCI           ;SCI VECTOR - NOT USED
        FDB     BADTOF           ;TIMER VECTOR 1 - TIMER OVERFLOW
        FDB     BADCAP1          ;TIMER VECTOR 2 - TCAP1
        FDB     BADCAP2          ;TIMER VECTOR 3 - TCAP2
        FDB     BADIRQ           ;IRQ VECTOR
        FDB     BADSWI           ;SWI VECTOR
        FDB     START            ;RESET VECTOR - NOT USED
***********************************************************************************
```

Appendix H

Brushless DC Motor Program, MC68HC05MC4

```
*******************************************************************************
*      Note: See Disclaimer in Preface                                        *
* ITC122 REV. 1.0 AND ITC127 REV. 1.1                                         *
* MC68HC05MC4 3 PHASE BRUSHLESS MOTOR CONTROLLER, ASSEMBLY SOURCE CODE        *
* INTERRUPT DRIVEN                                                            *
* VERSION 4/28/96                                                             *
* ITC127 BRUSHLESS I/O ASSIGNMENTS:                                           *
*      IRQ,PB6      PHASE A HALL SENSOR                                       *
*      TCAP1,PB7    PHASE B HALL  SENSOR                                      *
*      TCAP2,PA0    PHASE C HALL  SENSOR                                      *
*      PWMA1        PHASE A BOTTOM DRIVE (PA1)                                *
*      PWMA2        PHASE B BOTTOM DRIVE (PA3)                                *
*      PWMA3        PHASE C BOTTOM DRIVE (PA5)                                *
*      PWMB1        PHASE A TOP DRIVE (PA2)                                   *
*      PWMB2        PHASE B TOP DRIVE (PA4)                                   *
*      PWMB3        PHASE C TOP DRIVE (PA6)                                   *
*      PA7          PULLED-UP TO +5 VOLTS WITH 10K AND TERMINATED AT PROTOTYPE AREA *
*      PC0/AD0      POWER BOARD BUFFERED B+ FEEDBACK                          *
*      PC1/AD1      POWER BOARD CURRENT FEEDBACK                              *
*      PC2/AD2      POWER BOARD TEMPERATURE DIODE                             *
*      PC3/AD3      SPEED CONTROL POT                                         *
*      PC4/AD4      DIRECTION CONTROL SWITCH                                  *
*      PC5/AD5      RUN/STOP CONTROL SWITCH                                   *
*******************************************************************************
* MINIMUM AND MAXIMUM ALLOWABLE PWM SETPOINT VALUES
MINPWM  EQU     6               ;A SETPOINT VALUE LESS THAN THIS PLACES 0 IN PWMAD
MAXPWM  EQU     249             ;A SETPOINT VALUE GREATER THAN THIS PLACES $FF IN PWMAD
* I/O, TIMER, PWM, SCI, A/D REGISTERS
PORTA   EQU     $00             ;DATA REGISTER FOR PORT A
PORTB   EQU     $01             ;DATA REGISTER FOR PORT B
PORTC   EQU     $02             ;DATA REGISTER FOR PORT C
PORTD   EQU     $03             ;DATA REGISTER FOR PORT D
DDRA    EQU     $04             ;DATA DIRECTION REGISTER FOR PORT A
DDRB    EQU     $05             ;DATA DIRECTION REGISTER FOR PORT B
DDRC    EQU     $06             ;DATA DIRECTION REGISTER FOR PORT C
DDRD    EQU     $07             ;DATA DIRECTION REGISTER FOR PORT D
CTCSR   EQU     $08             ;CORE TIMER CONTROL AND STATUS REGISTER
CTCR    EQU     $09             ;CORE TIMER COUNTER REGISTER
BAUD    EQU     $A
SCCR1   EQU     $B
SCCR2   EQU     $C
SCSR    EQU     $D
SCDR    EQU     $E
PWMAD   EQU     $10             ;PWM A DATA REGISTER
PWMAI   EQU     $11             ;PWM A INTERLOCK REGISTER
PWMBD   EQU     $12             ;PWM B DATA REGISTER
PWMBI   EQU     $13             ;PWM B INTERLOCK REGISTER
CTLA    EQU     $14             ;PWM A CONTROL REGISTER
CTLB    EQU     $15             ;PWM B CONTROL REGISTER
RATE    EQU     $16             ;PWM RATE REGISTER
UPDATE  EQU     $27             ;PWM UPDATE REGISTER
TCR     EQU     $17             ;TIMER CONTROL REGISTER
TSR     EQU     $18             ;TIMER STATUS REGISTER
ICRH2   EQU     $19             ;INPUT CAPTURE 2 REGISTER - HIGH BYTE
ICRH1   EQU     $1B             ;INPUT CAPTURE 1 REGISTER - HIGH BYTE
OCRH    EQU     $1D             ;OUTPUT COMPARE REGISTER - HIGH BYTE
```

```
TMRH    EQU     $20             ;TIMER REGISTER - HIGH BYTE
ACRH    EQU     $22             ;ALTERNATE TIMER REGISTER - HIGH BYTE
ADDR    EQU     $24             ;A/D CONVERTER DATA REGISTER
ADSCR   EQU     $25             ;A/D CONVERTER STATUS AND CONTROL REGISTER
COCO    EQU     7               ;A/D CONVERSION COMPLETE BIT IN ADSCR
ISCR    EQU     $0F             ;IRQ STATUS AND CONTROL REGISTER
ABOT    EQU     8               ;PATTERN TO TURN ON PHASE A BOTTOM OUTPUT
BBOT    EQU     $10             ;PATTERN TO TURN ON PHASE B BOTTOM OUTPUT
CBOT    EQU     $20             ;PATTERN TO TURN ON PHASE C BOTTOM OUTPUT
ATOP    EQU     1               ;PATTERN TO TURN ON PHASE A BOTTOM OUTPUT
BTOP    EQU     2               ;PATTERN TO TURN ON PHASE B BOTTOM OUTPUT
CTOP    EQU     4               ;PATTERN TO TURN ON PHASE C BOTTOM OUTPUT
CTLMSK  EQU     $B8             ;PWM CONTROL REGISTER COMMON MASK
MOR     EQU     $F00            ;MASK OPTION REGISTER
HALL1   EQU     0               ;HALL 1 SENSOR CONNECTED TO PORTA BIT 0
HALL2   EQU     7               ;HALL 2 SENSOR CONNECTED TO PORT B BIT 7
HALL3   EQU     6               ;HALL 3 SENSOR CONNECTED TO PORT B BIT 6
SWITCH  EQU     5               ;RUN/STOP SWITCH CONNECTED TO PORT C BIT 5
DIRECT  EQU     4               ;FORWARD/REVERSE SWITCH CON. TO PORT C BIT 4
* VARIABLES
        ORG     $50
DIRSW   RMB     1               ;LAST DIRECTION INPUT. USED FOR CHANGE OF STATE
DLYTMR  RMB     1               ;COUNTER INCREMENTS EACH REAL TIME INTERRUPT
ERROR   RMB     1               ;HALL SENSOR INPUT ERROR COUNTER
FLAGS   RMB     1               ;GENERAL PURPOSE FLAGS. NEXT EQU'S DEFINE THEM
FORREV  EQU     0               ;ROTATION DIRECTION FLAG 1 = CW, 0 = CCW
RUNCOS  EQU     1               ;USED FOR SMOOTH STARTUP. 1 = RUN, 0 = STOP
PWRUP   EQU     2               ;FORCES SMOOTH STARTUP AFTER RESET
LAST1   RMB     1               ;PREVIOUS HALL SENSOR INPUT
SETPT   RMB     1               ;PWM SETPOINT VALUE BETWEEN MINPWM AND MAXPWM
TIMEOUT RMB     1               ;COUNTER FOR TIMER OVERFLOW TIMEOUTS
TMP     RMB     1               ;TEMP FOR JMP TABLE CALCULATION
SENS    RMB     16              ;BUFFER TO HOLD A/D INPUTS
SENSEND EQU     *-1             ;END OF BUFFER
ADTEMP  RMB     2               ;PLACE TO DO A/D AVERAGE MATH
IOBUF   RMB     50              ;I/O TEST PRINT BUFFER
CRFLAG  RMB     1               ;ENTER KEY FLAG
*       ORG     MOR             ;MOR REGISTER
*       FCB     0               ;DISABLE THE WATCHDOG TIMER
* THE PROGRAM CODE STARTS HERE
        ORG     $100            ;START OF EPROM
START   EQU     *
*******************************************************************************
* This code executes at power-up to test the SCI port on the ITC127 board.   *
* It will execute if the RUN/STOP is in the STOP position.                    *
* The baud rate is determined at processor reset time by the                  *
* FORWARD/REVERSE switch. If the switch is in the forward position, the      *
* baud rate will be 1802. If the switch is in the reverse position, the      *
* baud rate will be 112.7. These baud rates assume that that the processor   *
* is being clocked at 6.0 MHz.                                                *
*******************************************************************************
        BRCLR   7,PORTA,SCITST  ;IF THE RUN/STOP SWITCH IS IN RUN..EXIT
        JMP     SYSTEM          ;NORMAL SYSTEM OPERATION
SCITST  LDA     #$80            ;DISABLE THE EXTERNAL IRQ
        STA     ISCR
        LDA     #$10            ;ENABLE THE REAL TIME INTERRUPT @ 5.5 MS.
        STA     CTCSR
        CLI
```

```
        CLR     CRFLAG             ;ZERO THE ENTER FLAG
        LDA     PORTC              ;READ THE FOR/REV SWITCH
        AND     #$10
        BEQ     LOW                ;REVERSE
        LDA     #$33               ;1802 BAUD
        BRA     CONT               ;FORWARD
LOW     LDA     #$37               ;112.7 BAUD
CONT    STA     BAUD
        CLR     SCCR1
        LDA     #$2C               ;RIE,TE,RE
        STA     SCCR2
        CLRX
TXMSG   LDA     T_TEXT,X           ;OUTPUT THE GREETING MESSAGE
        BEQ     STAYSCI            ;AT THE END OF THE STRING YET ?
        BSR     SCITX              ;OUTPUT THE BYTE
        INCX                       ;POINT TO THE NEXT CHARACTER
        BRA     TXMSG              ;CONTINUE
STAYSCI LDA     CRFLAG             ;EXIT THE SCI TEST IF THE ENTER KEY IS PRESSED
        BEQ     STAYSCI            ;LOOP ON THE SCI TEST
TSTIO   JMP     IOTEST             ;YES...EXIT TO THE I/O TEST
SCITX   BRCLR   7,SCSR,SCITX       ;WAIT UNTIL TRANSMITTER READY
        STA     SCDR               ;OUTPUT THE CHARACTER
        RTS
SCIIRQ  EQU     *
        LDA     #$D                ;LINE FEED
        BSR     SCITX
        LDA     #$A                ;CARRIAGE RETURN
        BSR     SCITX
        LDA     SCSR               ;START TO CLEAR THE INTERRUPT
        LDA     SCDR               ;COMPLETE THE CLEARING AND GET THE CHARACTER
        BSR     SCITX              ;ECHO IT
        CLR     CRFLAG
        CMP     #$D                ;WAS THE LAST CHARACTER = ENTER
        BNE     EXIRQ              ;NO
        LDA     #1                 ;YES
        STA     CRFLAG
EXIRQ   RTI
* CONVERTS THE LOWER NIBBLE IN A TO ASCII
FMTMSB  LSRA                       ;SHIFT MSN TO LSN
        LSRA
        LSRA
        LSRA
FMTLSB  AND     #$0F               ;ISOLATE THE LOWER NIBBLE
        ADD     #$30               ;ADD '0' OFFSET
        CMP     #$39               ;A-F ?
        BLS     NOTAF              ;NO
        ADD     #$07;              ;YES, ADD 'A' OFFSET
NOTAF   RTS
* CONVERT THE ANALOG ON THE MULTIPLEXER IN A
GET_A2D STA     ADSCR              ;TURN ON A/D CONVERTER AND START A CONVERSION
A2DLOOP BRCLR   COCO,ADSCR,A2DLOOP ;LOOP ON THE A/D CONVERSION COMPLETE BIT
        LDA     ADDR               ;GET THE VALUE
        RTS
IOTEST  EQU     *
        CLRX
IOMSG   LDA     A_TEXT,X           ;OUTPUT THE HEADER
        BEQ     IO01               ;AT THE END OF THE STRING YET ?
        BSR     SCITX              ;OUTPUT THE BYTE
```

```
        INCX                    ;POINT TO THE NEXT CHARACTER
        BRA     IOMSG           ;CONTINUE
IO01    CLRX                    ;FILL THE OUTPUT BUFFER WITH ASCII BLANKS
        LDA     #$20            ;ASCII BLANK
IOFILL  STA     IOBUF,X
        INX
        CPX     #45             ;AT THE END OF THE BUFFER YET?
        BNE     IOFILL
        CLR     0,X             ;FORCE EOT INTO THE BUFFER
* GET THE ANALOGS AND PLACE IN THE PRINT BUFFER
        LDA     #$20            ;Vbus
        BSR     GET_A2D
        TAX
        BSR     FMTMSB
        STA     IOBUF+1
        TXA
        BSR     FMTLSB
        STA     IOBUF+2
        LDA     #$21            ;Isense
        BSR     GET_A2D
        TAX
        BSR     FMTMSB
        STA     IOBUF+8
        TXA
        BSR     FMTLSB
        STA     IOBUF+9
        LDA     #$22            ;TEMPERATURE
        BSR     GET_A2D
        TAX
        BSR     FMTMSB
        STA     IOBUF+15
        TXA
        BSR     FMTLSB
        STA     IOBUF+16
        LDA     #$23            ;SPEED POT
        BSR     GET_A2D
        TAX
        BSR     FMTMSB
        STA     IOBUF+23
        TXA
        BSR     FMTLSB
        STA     IOBUF+24
        LDA     #"F"
        BRSET   4,PORTC,SWFOR
        LDA     #"R"
SWFOR   STA     IOBUF+34
        LDA     #"S"
        BRSET   5,PORTC,SWRUN
        LDA     #"R"
SWRUN   STA     IOBUF+43
        CLR     IOBUF+44        ;FORCE EOT
        LDA     #$0D
        JSR     SCITX           ;OUTPUT THE BYTE
        LDA     #$0A
        JSR     SCITX           ;OUTPUT THE BYTE
        CLRX
IOTXT   LDA     IOBUF,X         ;OUTPUT THE HEADER
        BEQ     IO03            ;AT THE END OF THE STRING YET ?
```

```
        JSR     SCITX           ;OUTPUT THE BYTE
        INCX                    ;POINT TO THE NEXT CHARACTER
        BRA     IOTXT           ;CONTINUE
IO03    LDA     BAUD
        CMP     #$33            ;1800 BAUD
        BNE     IO02
        CLR     DLYTMR
IODLY   LDA     DLYTMR          ;GET THE CURRENT COUNT
        CMP     #91             ;EACH COUNT = 5.5 MS.
        BNE     IODLY           ;LOOP UNTIL 91 COUNTS ACCUMULATE (~500 MS.)
IO02    JMP     IOTEST          ;LOOP ON THE I/O TEST
A_TEXT  FDB     $0D0A
        FCC     "Vbus  Isense  Temp  SPEED POT  FOR/REV  RUN/STOP"
        FCB     0
T_TEXT  FCC     $87             ;BELL CODE
        FDB     $0D0A           ;LINE FEED/CARRIAGE RETURN
        FCC     "ITC127 SERIAL PORT TEST"
        FDB     $0D0A
        FCC     "Type characters on the keyboard and they will echo"
        FDB     $0D0A
        FCC     "to the first column on your terminal. Press"
        FDB     $0D0A
        FCC     "ENTER to exit this test and start the I/O test."
        FDB     $0D0A
        FCB     $0A
        FCB     0               ;EOT
SYSTEM  EQU     *
        LDX     #$50            ;ZERO MEMORY FROM $50 TO $FF
CLRMEM  CLR     0,X
        INCX                    ;BUMP THE POINTER TO THE NEXT LOCATION
        CMPX    #$0             ;DONE YET?
        BNE     CLRMEM          ;NOT YET..KEEP CLEARING RAM
        LDA     PORTC           ;GET THE FORWARD/REVERSE DIRECTION SWITCH INPUT
        AND     #$10            ;RETAIN BIT 4 MOTOR DIRECTION SWITCH
        STA     DIRSW           ;NEED THIS FOR THE FIRST TIME THROUGH MAIN
        BRCLR   DIRECT,PORTC,GOREV ;IF THE BIT IS LOW..RUN COUNTER CLOCKWISE
        BSET    FORREV,FLAGS    ;INITIALIZE THE FLAG TO ROTATE CLOCKWISE
GOREV   LDA     #CTLMSK         ;DISABLE ALL POWER OUTPUTS
        STA     CTLB            ;ALL PWM AND TOP OUTPUTS WILL BE HIGH (OFF)
        STA     CTLA
        CLR     PWMBD           ;INITIALIZE AT MINIMUM PWM DUTY CYCLE
        CLR     PWMAD           ;INITIALIZE AT MINIMUM PWM DUTY CYCLE
        LDA     #$E0            ;ENABLE TOF, TCAP1, AND TCAP2 INTERRUPTS
        STA     TCR
        LDA     #$10            ;ENABLE THE REAL TIME INTERRUPT @ 5.5 MS.
        STA     CTCSR
        CLI                     ;TURN ON INTERRUPTS
*******************************************************************************
* THE MAIN LOOP WILL MONITOR PC3/AD3 TO DETERMINE THE DESIRED SPEED, PC4 *
* TO DETERMINE THE DIRECTION AND PC5 TO STOP OR ROTATE THE MOTOR. MAIN   *
* WILL BE INTERRUPTED FROM FOUR SOURCES (IRQ, TCAP1, TCAP2, TOF AND RTI) *
*******************************************************************************
MAIN    EQU     *
        LDX     #SENS           ;START OF THE A/D BUFFER
MVLOOP  LDA     1,X             ;GET THE SECOND ENTRY
        STA     0,X             ;MOVE UP ONE
        INCX                    ;POINT TO NEXT
        CMPX    #SENSEND        ;END OF TABLE?
```

```
        BNE     MVLOOP          ;NO
        LDA     #$23            ;ENABLE CHANNEL 3 OF A/D CONVERTER
        JSR     GET_A2D         ;READ THE SPEED POT
        BRSET   SWITCH,PORTC,OFF ;IS THE RUN/STOP SWITCH OFF?
        STA     SENSEND
* AVERAGE THE LAST 16 ANALOG INPUTS
        CLR     ADTEMP          ;ZERO THE ACCUMULATOR
        CLR     ADTEMP+1
        LDX     #SENS           ;STAR OF BUFFER
        CLC                     ;CLEAR FOR THE FIRST ADD
        LDA     0,X             ;FIRST PAIR OF INPUTS
ADDLOOP ADD     1,X             ;ADD THE NEXT IN LINE
        BCC     NOCARRY
        INC     ADTEMP
NOCARRY STA     ADTEMP+1
        INCX                    ;POINT TO THE NEXT ENTRY
        CMPX    #SENSEND        ;DONE YET
        BNE     ADDLOOP         ;NO
* DIVIDE BY 16
        LDA     #4              ;LOOP COUNTER
DVLOOP  LSR     ADTEMP          ;SHIFT MSB 1 BIT TO THE RIGHT INTO CARRY
        ROR     ADTEMP+1        ;SHIFT THE LSB TO THE RIGHT
        DECA                    ;BUMP THE COUNTER DOWN BY 1
        BNE     DVLOOP          ;DONE YET?
        LDA     ADTEMP+1        ;GET THE INTEGRATED SETPOINT VALUE
        BRSET   PWRUP,FLAGS,NORAMP ;IF JUST POWERED-UP, RAMP UP TO POT SPEED
        LDA     ADDR            ;USE THE RAW A/D FOR THE FIRST TIME
        BSET    PWRUP,FLAGS     ;FORCE A JUMP AROUND THE POWERED-UP CODE ABOVE
NORAMP  STA     SETPT           ;SAVE THE INTEGRATED VALUE
        CMP     #MINPWM         ;CHECK MINIMUM PWM VALUE
        BCC     CHKHI           ;NOT TOO LOW..CHECK FOR MAX PWM VALUE
        CLR     SETPT           ;FORCE ZERO PWM
CHKHI   CMP     #MAXPWM         ;CHECK MAXIMUM PWM VALUE
        BCS     RUN             ;VALUE WITHIN LIMITS
        LDA     #$ff            ;FORCE MAXIMUM PWM
        STA     SETPT
        BRA     RUN             ;CONTINUE TO RUN THE MOTOR
OFF     CLR     TIMEOUT         ;START WITH 0 WHEN SWITCH SAYS RUN
        CLR     DLYTMR          ;DEBOUNCE THE RUN/STOP SWITCH
OFFDLY  LDA     DLYTMR          ;GET THE CURRENT COUNT
        CMP     #2              ;EACH COUNT = 5.5 MS.
        BNE     OFFDLY          ;LOOP UNTIL 2 COUNTS ACCUMULATE (11 MS.)
        BCLR    RUNCOS,FLAGS    ;SET FOR THE NEXT STOP TO RUN TRANSITION
        LDA     #CTLMSK         ;DISABLE ALL TOP AND BOTTOM PWM OUTPUTS
        STA     CTLB
        STA     CTLA
        CLR     PWMBD
        CLR     PWMAD
        LDA     PORTC           ;GET THE FORWARD/REVERSE DIRECTION SWITCH INPUT
        AND     #$10            ;RETAIN BIT 4 MOTOR DIRECTION SWITCH
        CMP     DIRSW           ;CHANGE IN THE DIRECTION SWITCH?
        BEQ     MAIN            ;NO CHANGE OF STATE, CONTINUE
        BRA     COS             ;CHANGE OF DIRECTION SWITCH
RUN     BRSET   SWITCH,PORTC,MAIN ;IF THE RUN/STOP SWITCH IS OFF CONTINUE MAIN
        BRSET   RUNCOS,FLAGS,NONEW
        SWI                     ;THIS CAUSES AN IMMEDIATE CALL OF POS SO THE
*                               ;MOTOR WILL START UP SMOOTHLY
        BSR     ACCEL           ;RAMP UP TO THE REQUESTED POT SPEED
```

```
        BSET    RUNCOS,FLAGS        ;THE MOTOR IS UP TO SPEED DON'T RAMP NEXT LOOP
NONEW   LDA     SETPT               ;GET THE LATEST SPEED REQUEST
        STA     PWMBD               ;UPDATE THE PWM REGISTER
        STA     PWMAD               ;UPDATE THE PWM REGISTER
        LDA     PORTC               ;GET THE FORWARD/REVERSE DIRECTION SWITCH INPUT
        AND     #$10                ;RETAIN BIT 4 MOTOR DIRECTION SWITCH
        CMP     DIRSW               ;CHANGE IN THE DIRECTION SWITCH?
        BEQ     MAIN2               ;NO CHANGE OF STATE, CONTINUE STRETCH THE BEQ
* A CHANGE IN DIRECTION HAS BEEN DETECTED.
COS     STA     DIRSW               ;SAVE FOR THE NEXT CHANGE OF DIRECTION
        LDA     #CTLMSK             ;DISABLE ALL TOP AND BOTTOM PWM OUTPUTS
        STA     CTLB
        STA     CTLA
        CLR     PWMBD               ;INITIALIZE AT MINIMUM SPEED
        CLR     PWMAD               ;INITIALIZE AT MINIMUM SPEED
        CLR     DLYTMR              ;TIMER IS USED FOR CHANGE IN DIRECTION DELAY
COSDLY  CLR     TIMEOUT             ;PREVENT TIMEOUT DURING THIS DELAY
        LDA     DLYTMR              ;GET THE TIMER VALUE
        CMP     #150                ;HAVE WE WAITED LONG ENOUGH
        BCS     COSDLY              ;NOT DONE YET
        BRCLR   4,DIRSW,REV
*       BRCLR   DIRECT,PORTC,REV ;GET THE DIRECTION SWITCH
        BSET    FORREV,FLAGS        ;SET THE CLOCKWISE FLAG BIT
REVACC  BSR     ACCEL               ;RAMP UP TO THE REQUESTED SPEED
MAIN2   JMP     MAIN
REV     BCLR    FORREV,FLAGS        ;CLEAR THE COUNTER CLOCKWISE FLAG BIT
        BRA     REVACC
*********************************************************************
* THIS IS A LIMITED ATTEMPT AT CONTROLLED ACCELERATION.  *
* THE FOLLOWING CODE WILL PREVENT LARGE CURRENTS WHEN    *
* SWITCHING MOTOR DIRECTION AND UPON INITIAL STARTUP     *
*********************************************************************
ACCEL   BRSET   SWITCH,PORTC,EXIT ;IF THE RUN/STOP SWITCH IS OFF CONTINUE MAIN
        LDA     PWMAD               ;GET THE CURRENT PWM VALUE
        CMP     SETPT               ;IS IT AT THE SETPOINT YET?
        BEQ     EXIT                ;NO ACCELERATION OR DEACCELERATION NECESSARY
        BCS     ACDLY1              ;NO, CONTINUE TO ACCELERATE
        LDA     SETPT               ;LET THE MOTOR SLOW DOWN
        STA     PWMBD               ;UPDATE THE PWM REGISTER
        STA     PWMAD               ;UPDATE THE PWM REGISTER
        BRA     EXIT                ;RETURN TO MAIN
ACDLY1  CLR     DLYTMR              ;START AT A COUNT OF ZERO
ACDLY   LDA     DLYTMR              ;GET THE CURRENT COUNT
        CMP     #2                  ;EACH COUNT = 5.5 MS.
        BNE     ACDLY               ;LOOP UNTIL 2 COUNTS ACCUMULATE (11 MS.)
        LDA     PWMBD               ;GET THE CURRENT PWM RATE
        CMP     #MAXPWM             ;HAVE WE EXCEEDED THE MAXIMUM PWM RATE
        BCC     EXIT                ;NO
        INC     PWMBD               ;BUMP THE PWM. INITIALLY IT WILL BE MINPWM
        INC     PWMAD               ;BUMP THE PWM. INITIALLY IT WILL BE MINPWM
        BRA     ACCEL
EXIT    RTS
*****************************************************************************
* IRQ INTERRUPTS WILL OCCUR WHEN THE HALL SENSOR ON THE IRQ PIN TOGGLES.*
* THIS ROUTINE WILL CLEAR THE INTERRUPT, TOGGLE THE SENSITIVITY OF THE  *
* THE NEXT TRIGGERED INTERRUPT, AND THEN BRANCH TO A ROUTINE WHICH WILL *
* SENSE THE POSITION OF THE MOTOR.                                      *
*****************************************************************************
```

```
IRQ     MUL                     ;A LITTLE DELAY FOR NOISE IF ANY
        MUL
        BRSET   5,ISCR,IRQRISE  ;IF HIGH, WE ARE LOOKING FOR A 0 -> 1 IRQ
        BRSET   6,PORTB,NOIRQ   ;IF SET...MUST HAVE BEEN A GLITCH
IRQOK   LDA     #$20            ;TOGGLE IRQ SENSITIVITY
        EOR     ISCR
        STA     ISCR
        BSET    2,ISCR          ;ACKNOWLEDGE INTERRUPT
        BRA     POS
IRQRISE BRCLR   6,PORTB,NOIRQ   ;IF CLEAR...MUST HAVE BEEN A GLITCH IGNORE IT
        BRA     IRQOK           ;MUST BE A VALID TRANSITION
NOIRQ   BSET    2,ISCR          ;ACKNOWLEDGE INTERRUPT
        RTI                     ;AND JUST RETURN
*****************************************************************************
* INPUT CAPTURE INTERRUPTS WILL OCCUR WHEN EITHER SENSOR IN TCAP1 AND       *
* TCAP2 PINS TOGGLES.  THIS ROUTINE WILL FIRST CHECK THE SOURCE OF THE      *
* INTERRUPT - TCAP1 OR TCAP2, THEN CLEAR THE INTERRUPT, TOGGLE THE          *
* SENSITIVITY OF THE NEXT TRIGGERED INTERRUPT, AND THEN BRANCH TO A         *
* ROUTINE WHICH WILL SENSE THE POSITION OF THE MOTOR.                       *
*****************************************************************************
ICISR   MUL                     ;A LITTLE DELAY FOR NOISE IF ANY
        MUL
        LDA     TSR             ;FIRST PART OF CLEARING FLAG
        BPL     TC1
*  TCAP2  HALL1
        LDA     ICRH2+1
        BRSET   1,TCR,IC2RISE   ;IF HIGH, WE ARE LOOKING FOR A 0 -> 1 TCAP2
        BRSET   0,PORTA,NOICX   ;IF SET...MUST HAVE BEEN A GLITCH
IC2OK   LDA     #$02            ;TOGGLE EDGE SENSITIVITY
        EOR     TCR
        STA     TCR
        BRA     POS
NOICX   RTI                     ;JUST RETURN
IC2RISE BRCLR   0,PORTA,NOICX   ;IF CLEAR...MUST HAVE BEEN A GLITCH IGNORE IT
        BRA     IC2OK
*  TCAP1  HALL2
IC1RISE BRCLR   7,PORTB,NOICX   ;IF CLEAR...MUST HAVE BEEN A GLITCH IGNORE IT
        BRA     IC1OK
TC1     LDA     ICRH1+1         ;AND CLEAR ANY FLAGS
        BRSET   2,TCR,IC1RISE   ;IF HIGH, WE ARE LOOKING FOR A 0 -> 1 TCAP1
        BRSET   7,PORTB,NOICX   ;IF SET...MUST HAVE BEEN A GLITCH
IC1OK   LDA     #$04            ;TOGGLE EDGE SENSITIVITY
        EOR     TCR
        STA     TCR
*****************************************************************************
* THIS ROUTINE WILL SENSE THE POSITION OF THE MOTOR SHAFT BY READING        *
* PORT B.  THE ROUTINE WILL THEN USE THE PORT B (BITS 0,7) AND PORTB        *
* (BIT 5) VALUE TO  GENERATE AN INDEX USED TO JUMP TO THE PROPER            *
* COMMUTATION SEQUENCE.                                                     *
*****************************************************************************
POS     EQU     *
        CLR     TMP
        BRCLR   HALL1,PORTA,NOT1  ;HALL 1 GRAY WIRE NOT = 1
        BSET    0,TMP
NOT1    BRCLR   HALL2,PORTB,NOT2  ;HALL 2 BLUE WIRE NOT = 1
        BSET    1,TMP
NOT2    BRCLR   HALL3,PORTB,NOT3  ;HALL 1 WHITE WIRE NOT = 1
        BSET    2,TMP
```

```
NOT3    EQU     *
        LDA     TMP             ;GET THE PATTERN WE JUST BUILT
        BRSET   FORREV,FLAGS,CW ;DIRECTION FLAG SAYS GO CLOCKWISE
        EOR     #7              ;TO REVERSE THE MOTOR, INVERT THE HALL PATTERN
CW      CMP     #0              ;CHECK FOR 0, IT IS AN ILLEGAL PATTERN
        BEQ     ERRHALL         ;GOT AN ERROR. IGNORE THIS INPUT
        CMP     #7              ;CHECK FOR 7, IT IS AN ILLEGAL PATTERN
        BEQ     ERRHALL         ;GOT AN ERROR. IGNORE THIS INPUT
        CMP     LAST1           ;IS THE INPUT THE SAME AS LAST TIME..ERROR
        BNE     INPTOK          ;INPUT IS REASONABLE
        INC     ERROR           ;KEEP TRACK OF ERRORS
        BRA     ERRHALL         ;SERVICE THE ERROR
INPTOK  STA     LAST1           ;SAVE FOR COMPARE WITH THE NEXT HALL INPUT
        DECA                    ;COMPUTE THE INPUT FOR THE JUMP TABLE
        STA     TMP             ;THE VALUE HERE WILL BE 0..5
        LSLA                    ;MULTIPLY BY 2
        ADD     TMP             ;PLUS THE ORIGINAL = MUL * 3
        TAX                     ;MOVE ACCUMULATOR TO INDEX REGISTER
        JMP     JMPTAB,X        ;USE THE CLOCKWISE ROTATION TABLE
ERRHALL TAX
        LDA     #CTLMSK         ;DISABLE ALL TOP AND BOTTOM PWM OUTPUTS
        STA     CTLB            ;ERROR DETECTED. TURN OFF PWM GENERATOR AND
        STA     CTLA            ;IGNORE THIS INPUT
        TXA
        BRA     INPTOK          ;EXIT
**************************************************
* COMMUTATION POSITION 0 FOR THE MOTOR.          *
* COMMUTATE THE MOTOR TO ITS NEXT POSITION.*
**************************************************
A_TO_B  LDA     #CTLMSK^BBOT    ;PA3=PWM (PHASE B BOTTOM);
        LDX     #CTLMSK^ATOP    ;PA6=1 (C TOP), PA4=1 (B TOP), PA2=0 (A TOP)
POSX    STX     CTLB            ;UPDATE PA2, PA4, AND PA6
        STA     CTLA            ;UPDATE PA1, PA3, AND PA5
        RTI
**************************************************
* COMMUTATION POSITION 1 FOR THE MOTOR.          *
* COMMUTATE THE MOTOR TO ITS NEXT POSITION.*
**************************************************
A_TO_C  LDA     #CTLMSK^CBOT    ;PA5=PWM (PHASE C BOTTOM)
        LDX     #CTLMSK^ATOP    ;PA6=1 (C TOP), PA4=1 (B TOP), PA2=0 (A TOP)
        BRA     POSX
**************************************************
* COMMUTATION POSITION 2 FOR THE MOTOR.          *
* COMMUTATE THE MOTOR TO ITS NEXT POSITION.*
**************************************************
B_TO_C  LDA     #CTLMSK^CBOT    ;PA5=PWM (PHASE C BOTTOM)
        LDX     #CTLMSK^BTOP    ;PA6=1 (C TOP), PA4=0 (B TOP), PA1=0 (A TOP)
        BRA     POSX
**************************************************
* COMMUTATION POSITION 3 FOR THE MOTOR.          *
* COMMUTATE THE MOTOR TO ITS NEXT POSITION.*
**************************************************
B_TO_A  LDA     #CTLMSK^ABOT    ;PA1=PWM (PHASE A BOTTOM)
        LDX     #CTLMSK^BTOP    ;PA6=1 (C TOP), PA4=0 (B TOP), PA2=1 (A TOP)
        BRA     POSX
**************************************************
* COMMUTATION POSITION 4 FOR THE MOTOR.          *
* COMMUTATE THE MOTOR TO ITS NEXT POSITION.*
```

```
*****************************************************
C_TO_A  LDA     #CTLMSK^ABOT     ;PA5=PWM (PHASE A BOTTOM)
        LDX     #CTLMSK^CTOP     ;PA6=0 (C TOP), PA4=1 (B TOP), PA2=1 (A TOP)
        BRA     POSX
*****************************************************
* COMMUTATION POSITION 5 FOR THE MOTOR.    *
* COMMUTATE THE MOTOR TO ITS NEXT POSITION.*
*****************************************************
C_TO_B  LDA     #CTLMSK^BBOT     ;PA3=PWM (PHASE B BOTTOM)
        LDX     #CTLMSK^CTOP     ;PA6=0 (C TOP), PA4=1 (B TOP), PA2=1 (A TOP)
        CLR     TIMEOUT          ;CLEAR TIMEOUT COUNTER
        JMP     POSX
* JUMP TABLE FOR ROTATION
JMPTAB  EQU     *
        JMP     A_TO_B           ;TURN ON Atop AND Bbot DRIVERS
        JMP     B_TO_C           ;TURN ON Btop AND Cbot DRIVERS
        JMP     A_TO_C           ;TURN ON Atop AND Cbot DRIVERS
        JMP     C_TO_A           ;TURN ON Ctop AND Abot DRIVERS
        JMP     C_TO_B           ;TURN ON Ctop AND Bbot DRIVERS
        JMP     B_TO_A           ;TURN ON Btop AND Abot DRIVERS
TOFISR  LDA     TSR              ;CLEAR FLAG
        LDA     TMRH+1
        LDA     TIMEOUT          ;TIMEOUT?
        CMP     #$3
        BCS     TOFX             ;NO -> EXIT
        JMP     POS              ;MOVE TO NEXT POSITION
TOFX    INC     TIMEOUT ;INCREMENT TIMEOUT COUNTER
        RTI
*****************************************************************
* CORE TIMER'S REAL TIME INTERRUPT SERVICE ROUTINE. *
* USED FOR ACCELERATION.                             *
*****************************************************************
RTI     INC     DLYTMR           ;BUMP THE COUNTER
        BCLR    6,CTCSR          ;CLEAR THE INTERRUPT FLAG
        RTI                      ;RETURN
BADSCI  RTI                      ;SCI IS NOT USED, SHOULD NEVER VECTOR HERE
******************************************************************************
* RESET/INTERRUPT VECTORS.                                                    *
* NOTE: THE SCI AND SWI VECTORS ARE NOT USED IN THIS APPLICATION. *
******************************************************************************
        ORG     $0FF0
        FDB     RTI              ;CORE TIMER VECTOR
        FDB     SCIIRQ           ;SCI VECTOR
        FDB     TOFISR           ;TIMER VECTOR 1 - TIMER OVERFLOW
        FDB     ICISR            ;TIMER VECTOR 2 - TCAP1
        FDB     ICISR            ;TIMER VECTOR 3 - TCAP2
        FDB     IRQ              ;IRQ VECTOR
        FDB     POS              ;SWI VECTOR
        FDB     START            ;RESET VECTOR - NOT USED
******************************************************************************
```

Appendix

I

Three-Phase AC Motor Program, MC68HC08MP16

```
/*****************************************************************************************
* Note: See Disclaimer in Preface                                                        *
*                3 PHASE AC INDUCTION MOTOR DEMONSTRATION CODE                           *
* ITC137 and ITC132 evaluation boards                                                    *
* MC68HC08MP16 Motor Control Microprocessor                                              *
* ROM: 14983 BYTES(2738 for motor drive, 12245 for user interfaces and ITC137 PCB test   *
* RAM:  170 BYTES                                                                        *
* Description:                                                                           *
*  11,10,9,8,or 7 bit PWM, sine, 3rd harmonic, or SVM waveforms                          *
*  input required from A/D and port A                                                    *
* Author: James Gray, Motorola Inc. Last modified : 03/26/97, Compiler: Cosmic Ver. 4.0  *
* MODULE SUMMARY:                                                                        *
*  MAIN                                                                                  *
*    -Initializes PWM and SCI(UART) units                                                *
*    -Resets communication, A/D data, and waveform data table pointers                   *
*    -Enables interrupts for PWM and Communication                                        *
*    -Enters SCAN loop                                                                   *
*  PWM                                                                                   *
*    -PWM interrupt handler                                                              *
*    -Services COP                                                                       *
*    -Passes data table pointer to QUADZ for each phase                                  *
*    -Loads PVALX registers from global ram value pwmmod                                 *
*    -Exchanges two phases if reverse direction is set                                   *
*    -Maintains waveform data table pointers                                             *
*    -sets PWM unit LDOK bit                                                             *
*  QUADZ                                                                                 *
*    -Accepts waveform data table pointer                                                *
*    -Translates full waveform pointer into quadrant pointer                             *
*    -Selects sine or 3rd harmonic injection according to settings                       *
*    -Calls SINSCALE to scale table value with global ram value vscale                   *
*    -modifies global ram value pwmmod                                                   *
*  SINSCALE                                                                              *
*    -Accepts waveform data and scaling value                                            *
*    -Scales with 24 bit accuracy                                                        *
*    -returns integer formatted for use in PVALX registers                               *
*  SVM                                                                                   *
*    -Accepts waveform data table pointer                                                *
*    -calculates SVM time segments via CALCULATE function                                *
*    -modifies PWM PVALX registers                                                       *
*  CALCULATE                                                                             *
*    -calculates t times for SVM modulation                                              *
*  SCAN                                                                                  *
*    -Scans hardware for speed, PWM, etc. settings                                       *
*    -Scans serial communication buffer for commands via GETCH                           *
*    -Parses commands, sets control flags                                                *
*    -Calls RECALC to execute setting changes                                            *
*    -Calls MENU to reflect changes back to terminal user                                *
*  RECALC                                                                                *
*    -Recalculates correct PWM modulus, load frequency, and interrupt frequency          *
*    -Recalculates correct data table pointer increment value                            *
*    -Updates PWM registers and ram variables coherently with interrupt mask             *
*  MENU                                                                                  *
*    -Transmits command menus and current settings to terminal user                      *
*  RECEPT                                                                                *
*    -SCI interrupt handler-Writes to buffer                                             *
*    -Maintains pointer                                                                  *
*  PUTCHAR                                                                               *
*    -SCI transmit                                                                       *
*  GETCH                                                                                 *
*    -Parses input string in buffer                                                      *
*****************************************************************************************/
#include <math.h>
#include <float.h>
#include <stdlib.h>
#include <mp16io.h>          /*register definitions for HC708MP16*/
#include <spursint.h>        /*default handlers for spurious interrupts*/
#include <10btsqd.h>  /*contains  table for one quadrant*/
#include <10bt3qd.h>  /*10btsqd.h for sine, 10bt3qd.h for third
                              harmonic table*/
#include <ptest.c>           /*ITC137 self test-factory use*/
```

```
#pragma space [] @tiny       /*instruct compiler to use direct page*/
/*      assembly constructs to authorize and mask interrupts
 */
#define cli()  _asm("cli\n")
#define sei()   _asm("sei\n")
#define SIZE   16     /* buffer size */
#define SCTE   0x80   /* transmit ready bit */
#define VEC1     0x0000               /*     0 degrees                        */
#define VEC2     0x2aaa               /*    60 degrees                        */
#define VEC3     0x5555               /*   120 degrees                        */
#define VEC4     0x8000               /*   180 degrees                        */
#define VEC5     0xaaaa               /*   240 degrees                        */
#define VEC6     0xd555               /*   300 degrees                        */
#define SIXTY_DEG 0x2aaa
/*      communications variables
 */
char buffer[SIZE];    /* reception buffer */
char * ptlec;         /* read pointer */
char * ptecr;         /* write pointer */
unsigned char letter;
unsigned char temp;
/*  pwm variables
 */
unsigned char tblptra;               /*array index*/
unsigned int vscale;         /* 0 to 100% modulation*/
unsigned char table;         /*select data table*/
unsigned char svm_type;
unsigned char quad;
unsigned char mbyte;
unsigned char hbyte;
unsigned int center;         /*center pwm for each modulus*/
unsigned int waveptra;               /*main period index*/
unsigned int waveptrb;
unsigned int waveptrc;
unsigned int theta;
unsigned int incval;         /*1 gives .06Hz, 1092 gives 60Hz, etc.*/
unsigned int pwmmod;
struct {
               unsigned char anlog :1;
               unsigned char ramp  :1;
               unsigned char start :1;
               unsigned char term  :1;
               unsigned char tcmd  :1;
               unsigned char spvt  :1;
               }flag;
typedef struct{
unsigned char ptru;
unsigned char ptrl;
}ptrbyte;
typedef union{
unsigned int ptrint;
ptrbyte ptbyte;
}ptrunion;
ptrunion ptrtemp;
/*variables for terminal and stand alone mode interfaces*/
float newmod;
float newpwm;
float ncenter;
```

```
float nincval;
float nvscale;
float newfreq;        /*drive frequency temporary*/
float freqcmd;
float hldfreq;
float boost;
float sum;
float sum2;
float sum3;
float sum4;
char option;
char noption;
char * adptr;         /*A to D result arrays*/
char * adptr2;
char * adptr3;
char * adptr4;
char avgadr [8];
char avgadr2 [8];
char avgadr3 [8];
char avgadr4 [8];
unsigned char rnstp;
unsigned char newdir;
unsigned char fwdrev;
unsigned char fmodpt;
unsigned char ofmodpt;
unsigned char j;
unsigned char dldsp;
unsigned int i;
unsigned int tmod;                      /*temporary modulus*/
unsigned char npctl2;         /*new pwm load/interrupt frequency*/
float ldfreq;
unsigned char svm_type;
unsigned int t0,t1,t2,h_t0;
/*PROTOTYPES*/
unsigned int sinscale (unsigned int,unsigned int);
void menu (void);
void quadz (unsigned int);
void scan (void);
void recalc (void);
void svm (void);
@interrupt    void   pwm(void)
       {
       PORTC=PORTC & 0xdf;           /*Green LED on*/
       COPCTL=0x00;
       PCTL1=PCTL1 & 0xef;   /*clear PWMF bit*/
       if (flag.spvt)
              svm();
       else
       {
/*PHASE  A*/
       quadz (waveptra);
       PVAL1=pwmmod;
/*PHASE B*/
       quadz (waveptrb);
       PVAL5=pwmmod;
/*PHASE C*/
       quadz (waveptrc);
       PVAL3=pwmmod;
```

```
	}
	if (fwdrev)
		{
		waveptra=waveptra+incval;
		waveptrb=waveptrb+incval;
		waveptrc=waveptrc+incval;
		}
	else
		{
		waveptra=waveptra-incval;
		waveptrb=waveptrb-incval;
		waveptrc=waveptrc-incval;
		}
	PCTL1=PCTL1 | 0x2;		/*set LDOK bit*/
	PORTC=PORTC | 0x20;			/*Green LED off*/
	}
void quadz (unsigned int waveptr)
			{
			if (waveptr<0x4000)
				{
				ptrtemp.ptrint=waveptr;
				quad=0;
				}
			else if (waveptr <=0x7fff)
			    {
				ptrtemp.ptrint=32767-waveptr;
				quad=1;
				}
			else if (waveptr<=0xbfff)
				{
				ptrtemp.ptrint=waveptr-0x8000;
				quad=2;
				}
			else
	/*(ptrtemp.ptrint<=0xffff)*/
				{
				ptrtemp.ptrint=0xffff-waveptr;
				quad=3;
				}
				ptrtemp.ptrint=ptrtemp.ptrint<<2;		/*divide pointer by
64*/
				tblptra=ptrtemp.ptbyte.ptru;
				if (table==0)
				pwmmod =(sinscale (sinquad[tblptra],vscale));
				else
				pwmmod =(sinscale (harmquad[tblptra],vscale));
				if(quad<=1)
				pwmmod = pwmmod+center;
				else
				pwmmod = center-pwmmod;
			}
unsigned int sinscale (unsigned int num1 ,unsigned int num2)
	{
/*	sine table to pwm scaling routine
	scales table data for modulation percentage
	and pwm modulus full scale adjustment
	- 1st operand in X:A
	- 2nd operand on stack
```

```
 */
        _asm("stx _mbyte\n"); /* save MSB1*/
        _asm("ldx 3,SP\n");          /* load MSB2*/
        _asm("pshx\n");                     /* save*/
        _asm("pulh\n");                     /* in h*/
        _asm("psha\n");                     /* save LSB1*/
        _asm("pshh\n");                     /* transfer MSB2*/
        _asm("pulx\n");                     /* to x*/
        _asm("lda _mbyte\n"); /* get MSB1*/
        _asm("mul\n");               /* H * H*/
        _asm("ldx 5,SP\n");          /* get LSB2*/
        _asm("psha\n");                     /* save low byte of H*H*/
        _asm("lda _mbyte\n"); /* load MSB1*/
        _asm("mul\n");               /* H * L*/
        _asm("sta     _mbyte\n");    /* save in place*/
        _asm("stx _hbyte\n"); /* save hbyte */
        _asm("pula\n");                     /* load H * H */
        _asm("add _hbyte\n"); /* accumulate hbyte*/
        _asm("sta _hbyte\n"); /* save in place*/
        _asm("pshh\n");                     /* transfer MSB2*/
        _asm("pulx\n");                     /* to X*/
        _asm("pula\n");                     /* load LSB1*/
        _asm("psha\n");                     /* keep it for reuse*/
        _asm("mul\n");               /* L * H*/
        _asm("add     _mbyte\n");    /* accumulate*/
        _asm("sta     _mbyte\n");    /* and store*/
        _asm("clra\n");
        _asm("adc _hbyte\n");        /* carry to hbyte*/
        _asm("sta _hbyte\n"); /* save in place*/
        _asm("txa\n");               /* transfer hi byte L * H */
        _asm("add _hbyte\n"); /* accumulate*/
        _asm("sta _hbyte\n"); /* save in place*/
        _asm("pula\n");                     /* load LSB1*/
        _asm("ldx 4,sp\n");          /* load LSB2*/
        _asm("mul\n");               /* L * L*/
        _asm("txa\n");               /* transfer hi byte L * L*/
        _asm("add _mbyte\n"); /* accumulate*/
        _asm("sta _mbyte\n"); /* save in place*/
        _asm("clra\n");
        _asm("adc _hbyte\n");        /* carry to hbyte*/
        _asm("tax\n");               /* move hbyte to x */
        _asm("lda _mbyte\n"); /* load mbyte*/
        _asm("lsrx\n");                     /*align result */
        _asm("rora\n");                     /*for use in*/
        _asm("lsrx\n");                     /*PWM PVAL */
        _asm("rora\n");                     /*registers*/
        }
void recalc (void)
            {
             newmod=((1.0/newpwm)/2.7127e-7);    /*calculate new modulus*/
             tmod=newmod;                             /*and convert to int*/
             ncenter=newmod/2;                        /*calculate new 50% point*/
             if (tmod>=0x300)                         /*set new interrupt rate*/
                  {                                            /*based on new
modulus*/
                  npctl2=0;
                  ldfreq=1;                                        /*change at
4822,9670,19290Hz*/
```

```
                }
         else if (tmod>=0x17f)
                {
                 npctl2=0x40;
                 ldfreq=2;
                }
         else if (tmod>=0xc0)
                {
                 npctl2=0x80;
                 ldfreq=4;
                 }
         else if (tmod>=0x7f)
                {
                npctl2=0xc0;
                ldfreq=8;
                }
        /*calculate table increment value for new settings*/
        nincval = (65536/newpwm)*(newfreq*ldfreq);
        /*calculate voltage scaling from modulus and drive Hz*/
        /*for full modulation at 60 or 120 HZ*/
        if (fmodpt)
        {
        if (newfreq>=60)
                nvscale=ncenter;
        else
                {
                nvscale=(ncenter*(newfreq/60)); /*no boost*/
                nvscale=((((boost/100)*(60-newfreq)/60)*ncenter)+nvscale);
                }
        }
        else
                {
                nvscale=(ncenter*(newfreq/120)); /*no boost*/
                nvscale=((((boost/100)*(120-newfreq)/120)*ncenter)+nvscale);
                }
        sei ();    /*apply new settings*/
        PMOD=tmod;
        PCTL2=npctl2;
        center=ncenter;
        incval=nincval;
        vscale=nvscale;
        cli ();
        }
void svm(void)
        {
/* SVM calculate */
        theta = waveptra - 0x4000;
        if (theta < VEC2)
                {
                calculate(VEC1);
                switch (svm_type)
                        {
                        case 2:
                        case 4:
                                {PVAL1=t1+t2;
                                 PVAL3=t2;
                                 PVAL5=0;
                                 }break;
```

```
                        case 3:
                        case 5:
                                {PVAL1=PMOD;
                                 PVAL3=t0+t2;
                                 PVAL5=t0;
                                 }break;
                        default:
                                {h_t0=t0>>1;
                                 PVAL1=t1+t2+h_t0;
                                 PVAL3=t2+h_t0;
                                 PVAL5=h_t0;
                                 }
                        }
                }

else if (theta < VEC3)
                {
                calculate(VEC2);
                switch (svm_type)
                        {
                        case 2:
                        case 5:
                                {PVAL1=t1;
                                 PVAL3=t1+t2;
                                 PVAL5=0;
                                 }break;
                        case 3:
                        case 4:
                                {PVAL1=t0+t1;
                                 PVAL3=PMOD;
                                 PVAL5=t0;
                                 }break;
                        default:
                                {h_t0=t0>>1;
                                 PVAL1=t1+h_t0;
                                 PVAL3=t1+t2+h_t0;
                                 PVAL5=h_t0;
                                 }
                        }
                }

else if (theta < VEC4)
                {
            calculate(VEC3);
                switch (svm_type)
                        {
                        case 2:
                        case 4:
                                {PVAL1=0;
                                 PVAL3=t1+t2;
                                 PVAL5=t2;
                                 }break;
                        case 3:
                        case 5:
                                {PVAL1=t0;
                                 PVAL3=PMOD;
                                 PVAL5=t0+t2;
                                 }break;
```

```
                default:
                        {h_t0=t0>>1;
                         PVAL1=h_t0;
                         PVAL3=t1+t2+h_t0;
                         PVAL5=t2+h_t0;
                         }
                }
        }

else if (theta < VEC5)
        {
        calculate(VEC4);
        switch (svm_type)
                {
                case 2:
                case 5:
                        {PVAL1=0;
                         PVAL3=t1;
                         PVAL5=t1+t2;
                         }break;
                case 3:
                case 4:
                        {PVAL1=t0;
                         PVAL3=t0+t1;
                         PVAL5=PMOD;
                         }break;
                default:
                        {h_t0=t0>>1;
                         PVAL1=h_t0;
                         PVAL3=t1+h_t0;
                         PVAL5=t1+t2+h_t0;
                         }
                }
        }

else if (theta < VEC6)
        {
        calculate(VEC5);
        switch (svm_type)
                {
                case 2:
                case 4:
                        {PVAL1=t2;
                         PVAL3=0;
                         PVAL5=t1+t2;
                         }break;
                case 3:
                case 5:
                        {PVAL1=t0+t2;
                         PVAL3=t0;
                         PVAL5=PMOD;
                         }break;
                default:
                        {h_t0=t0>>1;
                         PVAL1=t2+h_t0;
                         PVAL3=h_t0;
                         PVAL5=t1+t2+h_t0;
                         }
```

```
                        }
                }

        else
                {
                calculate(VEC6);
                switch (svm_type)
                        {
                        case 2:
                        case 5:
                                {PVAL1=t1+t2;
                                 PVAL3=0;
                                 PVAL5=t1;
                                 }break;
                        case 3:
                        case 4:
                                {PVAL1=PMOD;
                                 PVAL3=t0;
                                 PVAL5=t0+t1;
                                 }break;
                        default:
                                {h_t0=t0>>1;
                                 PVAL1=t1+t2+h_t0;
                                 PVAL3=h_t0;
                                 PVAL5=t1+h_t0;
                                 }
                        }
                }

        }
/*=================================================================
 *     calculate t1,t2 and t0 for the space vector modulation
 *     methods.
 *================================================================*/
void calculate(vector)
        {
        static unsigned int angle1,angle2;
        static unsigned char t1_tmp,t2_tmp;

        angle2 = theta - vector;        /* determine angle for table  */
        angle1 = SIXTY_DEG - angle2;  /* lookup instructions             */
          /* since we have a 90 degree sine table with 256 entries
           * and a 16-bit angle variable, an angle value of 0x4000
           * (90 degrees) must correspond with a value of 0x100.
           * Therefore, a right shift of 6 bits is needed */
        angle1 >>= 6;  /* scale angle1 variable for table lookup   */
        angle2 >>= 6;  /* scale angle2 variable for table lookup   */
        t1_tmp = ((sinquad[(char) angle1]))>>2;      /* do table lookup */
        t2_tmp = ((sinquad[(char) angle2]))>>2;      /* do table lookup */
        t1 = ((vscale>>2) * t1_tmp) >> 5;
        t2 = ((vscale>>2) * t2_tmp) >> 5;
        t0  =  PMOD - (t1 + t2);
        }
void menu (void)
        {
        printf("\n\rEnter first letter of setting to change:\n\r");
        printf("(c) control mode of ITC137 board");
        if (flag.term)
```

```
printf("  -now in TERMINAL MODE\n\r");
else
printf("  -now in STAND ALONE MODE\n\r");
printf("(f) frequency  -now %i Hz\n\r",(int)freqcmd);
printf("(v) voltage  -now %i percent\n\r",(int)((nvscale/ncenter)*100));
printf("(p) pwm rate  -now %i Hz\n\r",(int)newpwm);
printf("(d) direction");
if (newdir)
printf("  -now forward\n\r");
else
printf("  -now reverse\n\r");
printf("(r) run/stop");
if (rnstp)
printf("  -now stop\n\r");
else
printf("  -now run\n\r");
printf("(b) boost low speed voltage");
printf("  -now %2i percent boost\n\r",(int)boost);
printf("(s) set full modulation frequency");
if (fmodpt)
printf("  -now 60 Hz\n\r");
else
printf("  -now 120 Hz\n\r");
printf("(m) modulation type");
switch (table)
       {
       case 0:
       printf("  -now sine\n\r");
       break;
       case 1:
       printf("  -now 3rd harmonic\n\r");
       break;
       case 2:
       printf("  -now SVM, null = V0\n\r");
       break;
       case 3:
       printf("  -now SVM, null = V7\n\r");
       break;
       case 4:
       printf("  -now SVM, null = V0,V7\n\r");
       break;
       case 5:
       printf("  -now SVM, null = V7,V0\n\r");
       break;
       case 6:
       printf("  -now SVM, null = alternate reverse\n\r");
       break;
       default:
       printf("\n\r");
       }
printf("(a) analog readings from ITC132\n\r");
printf("(h) help- Prints this menu.\n\n\r");
if ((flag.anlog)&&(!(flag.ramp)))
 {
 printf("percent full scale-   temperature:    ");
 printf("  bus voltage:    bus current:\r\n");
 }
}
```

```
/*      character reception routine.
 *      This routine is called on interrupt
 *      It puts the received char in the buffer
 */
@interrupt void recept(void)
        {
        *ptecr = SCS1;                  /* clear interrupt */
        *ptecr++ = SCDR;                /* get the char */
        if (ptecr >= &buffer[SIZE])    /* put it in buffer */
                ptecr = buffer;
        }
/*      send a char to the SCI
 */
void putchar(char c)
        {
        while (!(SCS1 & SCTE))                  /* wait for READY */
                ;
        SCDR = c;                       /* send it */
        }
/*      character buffer reception.
 */
char getch(void)
        {
                char c;                                 /* character to be returned */
                c = *ptlec++;                   /* get the received char */
                if (c == 0x08)                  /*if backspace adjust pointers*/
                        {
                        if (ptlec>buffer+1)
                        {
                        ptlec=ptlec-2;
                        ptecr=ptecr-2;
                        }
                        }
            if (c == 0x0d)                      /* if cr, send new line, terminate*/
                  {putchar(0x0a);               /* buffer string with null and set*/
                  *(ptecr-1) = 0x00;  /* command pending flag.             */
                  flag.tcmd=1;
                  }
                if (ptlec >= &buffer[SIZE])  /* maintain pointer and return */
                        ptlec = buffer;                 /* char on stack.
*/
                return (c);
        }
        void scan (void)
        {
        unsigned int i;
        if (!(ptlec==ptecr))  /*echo received characters while*/
                putchar(getch());     /*scanning for commands.        */
        /* Decode command and call appropriate adjustment function.*/
        if (flag.tcmd)
        {
        ptlec=ptecr=buffer;
        flag.tcmd=0;
        letter = *ptlec;
        switch (letter)
                {
                case 'c' :
                case 'C' :
```

```
        {
        printf("enter 0 for stand alone mode, 1 for terminal mode \n \r");
        while (!(flag.tcmd))
              {
              if (!(ptlec==ptecr))
              putchar(getch());      /*scanning for commands.*/
              }
         if (ptlec > (buffer+1))
         {
  temp = atoi(buffer);
         if (temp)
              {
              flag.term=1;   /*set stand alone and DIP flags*/
              menu();
              }
              else
              {
              flag.term=0;           /*clear terminal mode and*/
              flag.start=1;          /*cause scan of dip switch*/
              flag.spvt=0;           /*clear space vector*/
              }
         }
         else
         menu ();
        ptlec=ptecr=buffer;
        flag.tcmd=0;
        }
        break;
        case 'f' :
        case 'F' :
        {
        if (flag.term)               /*if in terminal mode*/
        {
        printf("frequency is currently set to %i Hz \n\r",(int)freqcmd);
        printf("input frequency 0 to 120 Hz \n \r");
        while (!(flag.tcmd))
              {
              if (!(ptlec==ptecr))
              putchar(getch());      /*scanning for commands.*/
              }
        if (ptlec > (buffer+1))
        {
freqcmd = atoi(buffer);
        printf("\nramping frequency...\n\r");
       /*ramp to new setting*/
        flag.ramp=1;                              /*set ramping flag*/
        if (freqcmd>130)
        freqcmd=130;
        hldfreq=freqcmd;
        recalc();                         /*recalculate modulus and scaling*/
        }
        if(!(flag.ramp))
        menu();
        }
        else
              {
              printf("\n  NOT IN TERMINAL MODE!\n\r");
              printf("enter the letter (c) to change control mode\n\r");
```

```
        menu();
        }
 ptlec=ptecr=buffer;
 flag.tcmd=0;
 }
 break;
 case 'v' :
 case 'V' :
 {
 if (flag.term)              /*if in terminal mode*/
  {
        printf("enter percent voltage scale 0-100 \n \r");
        printf("!!! WARNING !!! This will override the correct\n\r");
        printf("voltage for the current frequency. A high voltage\n\r");
        printf("combined with a low frquency will destroy the\n\r");
        printf("power inverter stage.\n\r");
        while (!(flag.tcmd))
              {
              if (!(ptlec==ptecr))
              putchar(getch());     /*scanning for commands.*/
              }
        if (ptlec > (buffer+1))
        {
nvscale = atoi(buffer);
        nvscale=(nvscale*ncenter)/100;
        vscale=nvscale;
    }
        menu();
   }
  else
        {
        printf("\n  NOT IN TERMINAL MODE!\n\r");
        printf("enter the letter (c) to change control mode\n\r");
        menu();
        }
 ptlec=ptecr=buffer;
 flag.tcmd=0;
 }
 break;
 case 'p' :
 case 'P' :
 {
 if (flag.term)              /*if in terminal mode*/
  {
        printf("PWM rate is currently set to %i Hz \n\r",(int)newpwm);
        printf("enter PWM rate, 1800-28800.\n \r");
        while (!(flag.tcmd))
        {
        if (!(ptlec==ptecr))
        putchar(getch());     /*scanning for commands.*/
        }
        if (ptlec > (buffer+1))
        {
newpwm = atoi(buffer);
        if (newpwm<1800)      /*limit PWM settings*/
        newpwm=1800;
        if (newpwm>28800)
        newpwm=28800;
```

```
        recalc();                          /*recalculate modulus and scaling*/
        }
        menu();
   }
   else
        {
        printf("\n  NOT IN TERMINAL MODE!\n\r");
        printf("enter the letter (c) to change control mode\n\r");
        menu();
        }
   ptlec=ptecr=buffer;
   flag.tcmd=0;
  }
  break;
  case 'd' :
  case 'D' :
  {
  if (flag.term)              /*if in terminal mode*/
   {
        printf("enter direction- 1 for forward, 0 for reverse\n\r");
        while (!(flag.tcmd))
        {
        if (!(ptlec==ptecr))
        putchar(getch());     /*scanning for commands.        */
        }
        if (ptlec > (buffer+1))
newdir = atoi(buffer);
        menu();
   }
   else
        {
        printf("\n  NOT IN TERMINAL MODE!\n\r");
        printf("enter the letter (c) to change control mode\n\r");
        menu();
        }
   ptlec=ptecr=buffer;
   flag.tcmd=0;
  }
  break;
  case 'r' :
  case 'R' :
  {
  if (flag.term)              /*if in terminal mode*/
   {
        printf("enter run/stop- 1 for stop, 0 for run\n \r");
        while (!(flag.tcmd))
        {
        if (!(ptlec==ptecr))
        putchar(getch());     /*scanning for commands.        */
        }
        if (ptlec > (buffer+1))
rnstp = atoi(buffer);
        menu();
   }
   else
        {
        printf("\n  NOT IN TERMINAL MODE!\n\r");
        printf("enter the letter (c) to change control mode\n\r");
```

```
        menu();
        }
  ptlec=ptecr=buffer;
  flag.tcmd=0;
 }
 break;
 case 's' :
 case 'S' :
 {
 if (flag.term)              /*if in terminal mode*/
  {
        printf("enter full modulation point- 0 for 120Hz, 1 for 60Hz\n \r");
        printf("!!! WARNING !!! voltage will change instantly-no ramp\n\r");
        while (!(flag.tcmd))
        {
        if (!(ptlec==ptecr))
        putchar(getch());     /*scanning for commands.        */
        }
        if (ptlec > (buffer+1))
fmodpt = atoi(buffer);
        recalc();                               /*recalculate modulus and scaling*/
        menu();
  }
  else
        {
        printf("\n  NOT IN TERMINAL MODE!\n\r");
        printf("enter the letter (c) to change control mode\n\r");
        menu();
        }
  ptlec=ptecr=buffer;
  flag.tcmd=0;
 }
 break;
 case 'b' :
 case 'B' :
 {
 if (flag.term)              /*if in terminal mode*/
  {
        printf("enter voltage boost- 0 to 20 percent\n \r");
        printf("!!! WARNING !!! voltage will change instantly-no ramp\n\r");
        while (!(flag.tcmd))
        {
        if (!(ptlec==ptecr))
        putchar(getch());     /*scanning for commands. */
        }
        if (ptlec > (buffer+1))
        {
boost = atoi(buffer);
        if (boost>20)
        boost=20;
        recalc();                               /*recalculate modulus and scaling*/
        }
        menu();
  }
  else
        {
        printf("\n  NOT IN TERMINAL MODE!\n\r");
        printf("enter the letter (c) to change control mode\n\r");
```

```
        menu();
        }
  ptlec=ptecr=buffer;
  flag.tcmd;
 }
 break;
 case 'm' :
 case 'M' :
 {
 if (flag.term)              /*if in terminal mode*/
  {
        printf("enter 0 for sine\n\renter 1 for 3rd harmonic\n\r");
        printf("enter 2 for SVM, null = V0\n\r");
        printf("enter 3 for SVM, null = V7\n\r");
        printf("enter 4 for SVM, null = V0,V7\n\r");
        printf("enter 5 for SVM, null = V7,V0\n\r");
        printf("enter 6 for SVM, null = alternate reverse\n\r");
        printf("!WARNING! SVM types [V0], [V0,V7], and [V7,V0]\n\r");
        printf("require addition of at least 220 uf to the\n\r");
        printf("ITC132 bootstrap circuit for proper operation\n\r");
        while (!(flag.tcmd))
        {
        if (!(ptlec==ptecr))
        putchar(getch());     /*scanning for commands.        */
        }
        if (ptlec > (buffer+1))
{
        table = atoi(buffer);
               if (table>6)
               table=6;
               if (table > 1)
               {
        svm_type=table;
               flag.spvt=1;
               }
               else
               flag.spvt=0;
        }
        menu();
  }
  else
        {
        printf("\n  NOT IN TERMINAL MODE!\n\r");
        printf("enter the letter (c) to change control mode\n\r");
        menu();
        }
  ptlec=ptecr=buffer;
  flag.tcmd=0;
 }
 break;
 case 'a' :
 case 'A' :
 {
 if (flag.term)              /*if in terminal mode*/
  {
        printf("enter 1 to display analog readings, 0 to cancel\n \r");
        while (!(flag.tcmd))
        {
```

```
                if (!(ptlec==ptecr))
                putchar(getch());     /*scanning for commands.*/
                }
                if (ptlec > (buffer+1))
            temp = atoi(buffer);
                if (temp)
                flag.anlog=1;  /*set analog display flag*/
                else
                flag.anlog=0;  /*else clear*/
                menu();
             }
             else
                {
                printf("\n  NOT IN TERMINAL MODE!\n\r");
                printf("enter the letter (c) to change control mode\n\r");
                menu();
                }
             ptlec=ptecr=buffer;
             flag.tcmd=0;
            }
            break;
            default:
            menu ();
           }
    }
/*monitor ITC137 switches and update operation accordingly*/
    for (i=0; i<20000; i++)              /*ramp delay*/
          {;
          }
          *adptr++=ADR;                  /*get a to d speed reading to average*/
          if (adptr>=&avgadr[8])         /*maintain pointer*/
                adptr=&avgadr[0];
          sum=0;                              /*clear sum*/
          for(j=0; j < 8; j++)  /*average 8 adr readings*/
                sum=sum+avgadr[j];
          sum=sum/8;
          if (flag.anlog)                     /*if analog display is set*/
          {
          ADSCR=0x01;
          while (!(ADSCR&0x80))
          {;}
          *adptr2++=ADR;                 /*get a to d temp reading to average*/
          if (adptr2>=&avgadr2[8])       /*maintain pointer*/
                adptr2=&avgadr2[0];
          sum2=0;                             /*clear sum*/
          for(j=0; j < 8; j++)  /*average 8 adr readings*/
                sum2=sum2+avgadr2[j];
          sum2=sum2/20.4;
          ADSCR=0x02;
          while (!(ADSCR&0x80))
          {;}
          *adptr3++=ADR;                 /*get a to d bus volt reading to average*/
          if (adptr3>=&avgadr3[8])       /*maintain pointer*/
                adptr3=&avgadr3[0];
          sum3=0;                             /*clear sum*/
          for(j=0; j < 8; j++)  /*average 8 adr readings*/
                sum3=sum3+avgadr3[j];
          sum3=sum3/20.4;
```

```
ADSCR=0x03;
while (!(ADSCR&0x80))
{;}
*adptr4++=ADR;                  /*get a to d bus amp reading to average*/
if (adptr4>=&avgadr4[8])        /*maintain pointer*/
        adptr4=&avgadr4[0];
sum4=0;                                 /*clear sum*/
for(j=0; j < 8; j++)  /*average 8 adr readings*/
        sum4=sum4+avgadr4[j];
sum4=sum4/20.4;
dldsp++;
if (dldsp>16)
{dldsp=0;
if(!(flag.ramp))
printf("\t\t\t%3i\t\t  %3i\t\t  %3i\r",(int)sum2,(int)sum3,(int)sum4);
}
ADSCR=0x20;                             /*set back to cont. convert, chan 0*/
}
if (!(flag.term))                       /*if in stand alone mode*/
        {freqcmd=(sum*120)/256;                 /*convert to int Hz*/
         hldfreq=freqcmd;                       /*save frequency*/
         rnstp=PORTA & 0x80;            /*read run/stop switch*/
         newdir = (PORTA & 0x40);      /*read in direction switch*/
         if (newdir)                            /*convert to logical*/
         newdir=1;
         fmodpt=(PORTA&0x20);
         if (fmodpt)
         fmodpt=1;
         if (flag.start)
         ofmodpt=fmodpt;
         if (!(fmodpt==ofmodpt))
           {recalc();                           /*recalculate modulus and scaling*/
               ofmodpt=fmodpt;
               menu();
               }
        }
if (rnstp)
        freqcmd=0;
        else
        freqcmd=hldfreq;                        /*restore frquency*/
if (!(newdir==fwdrev))                  /*if direction is changed*/
        {                                                       /* ramp down speed
and change*/
                                                                /*phase switch*/
        freqcmd=0;
        if (newfreq == 0)
               {
               fwdrev=newdir;
               freqcmd=hldfreq;                 /*restore frquency*/
               }
        }
if (!((int)freqcmd==(int)newfreq))  /*if frequency setting has changed*/
        {if(!(flag.ramp))                               /*if not already ramping,
print*/
         printf("\nramping frequency...\n\r");
        /*ramp to new setting*/
        flag.ramp=1;                                    /*set ramping flag*/
        if (((int)freqcmd)>((int)newfreq))
```

```
				newfreq=newfreq+1;
			if (((int)freqcmd)<((int)newfreq))
				newfreq=newfreq-1;
			recalc();				/*recalculate modulus and scaling*/
			}
		else if (flag.ramp)			/*done ramping*/
			{flag.ramp=0;
			 menu();
			}
	if (!(flag.term))   /*if in stand alone mode*/
	{
	if (PORTA & 0x10)      /*test DIP2 for waveform table selection*/
	table=0;
	else
	table=1;
	noption=PORTA & 0x0e; /*get pwm setting from DIP switch*/
	if((!(noption==option)) || (flag.start))

	{					/*if startup set or pwm setting has
changed*/
	flag.start=0;			/*clear startup*/
	option=noption;				/*update pwm option*/
	switch (option)				/*and set pwm variable*/
		{
		case 0:
			{
			newpwm=2000;
			}
			break;
		case 2:
			{
			newpwm=4000;
			}
			break;
		case 4:
			{
			newpwm=8000;
			}
			break;
		case 6:
			{
			newpwm=12000;
			}
			break;
		case 8:
			{
			newpwm=16000;
			}
			break;
		case 10:
			{
			newpwm=18000;
			}
			break;
		case 12:
			{
			newpwm=20000;
			}
```

```
                        break;
                case 14:
                        {
                        newpwm=22000;
                        }
                        break;
                }
        recalc();                               /*recalculate modulus and scaling*/
        menu();
        }
    }
  }
/*      MAIN program
 *      sets up pwm unit, monitors switches or terminal commands
 *      and recalculates scaling and frequency parameters
 */
void main(void)
        {
        char k;
    PCTL = 0x30;        /*use PLL clock*/
        PBWC = 0x80;
    if(!(PORTD & 0x40)) // if prototype pad "N" is low perform production test
      ptest();  // production test code
        waveptra  = 0;
        center = 512;
        waveptrb = 0x5555;
        waveptrc = 0xaaaa;
        table = 0;
        incval = 0;                     /*initial step size through 1024(65536) table*/
        vscale = 0x00;                  /*initial voltage scaling*/
        freqcmd=0;                      /*set speed to 0*/
        newfreq=0;
        newpwm=3600;
        boost=0;
        newdir = (PORTA & 0x40);        /*read in direction switch*/
        if (newdir)
        newdir=1;
        fwdrev=newdir;
        CONFIG = 0x60;          /*set center pwm mode,negative polarity, cop on*/
        PMOD =0x3ff;                    /*set up 10 bit modulus*/
        PVAL1=center;           /*set pwm to 50%*/
        PVAL3=center;
        PVAL5=center;
        PCTL1=PCTL1 | 0x20; /*enable pwm interrupts*/
        DEADTM=15;                      /*2 microsecond deadtime*/
        DISMAP=0x00;            /*clear disable map*/
        PCTL1=PCTL1 | 2;        /*set LDOK bit*/
        PCTL2=0x00;                     /*interrupt every 1 pwm loads*/
        PCTL1=PCTL1 | 1;         /*enable pwm*/
        DDRC=0x60;                      /*setup LED drive and*/
        PORTC=PORTC | 0x60;     /* ensure LEDs are off*/
        ADCLK=0xf0;                     /*set up A to D */
        ADSCR=0x20;
        adptr=avgadr;           /*initialize result pointers*/
        adptr2=avgadr2;
        adptr3=avgadr3;
        adptr4=avgadr4;
        DDRA=0;
```

```
        flag.anlog=0;                  /*set startup, clear other software flags*/
        flag.ramp=0;
        flag.start=1;
        flag.term=0;
        flag.tcmd=0;
        flag.spvt=0;
        ptecr = ptlec = buffer;        /* initialize pointers for SCI */
        SCC1 = 0x40;                   /*enable SCI*/
        SCBR = 0x12;                   /* 9600 bauds */
        SCC2 = 0x2c;                   /* parameters for SCI interrupt */
        cli();                                /*enable interrupts-cop serviced in
                                                    pwm interrupt handler*/
        for (k=0; k<8; k++)
        *adptr++=0;                           /*clear potentiometer array*/
        adptr=avgadr;
        printf("\n\rMotorola MC68HC08MP16 Demonstration Code V 0.6v8\n\r");
        while (1)                      /*scan switches or terminal forever*/
         {
         scan();
         }
        }
#pragma space []
```

Index

ABOUT THE EDITOR AND CONTRIBUTORS

Richard Valentine is technical group leader of Motorola's GENESIS advanced vehicular electronics team. Holder of three patents and author of 50 papers on electronic controls, he has 30 years' experience with power electronics and semiconductor devices used in industrial, computer, consumer, and automotive systems. The other contributors to this book—**Ken Berringer, Gary Dashey, Scott Deuty, Randy Frank, Jim Gray, Bill Lucas, Thomas Huettl, Peter Pinewski, Chuck Powers, Pablo Rodriguez, Warren Schultz,** and **Dave Wilson**—are all distinguished members of Motorola's advanced engineering and technical staff.